Lecture Notes in Computer Science

Edited by G. Goos and J. Hartmanis

W0049889

442

M. Main A. Melton M. Mislove
D. Schmidt (Eds.)

Mathematical Foundations of Programming Semantics

5th International Conference
Tulane University, New Orleans, Louisiana, USA
March 29–April 1, 1989
Proceedings

Springer-Verlag
Berlin Heidelberg New York London
Paris Tokyo Hong Kong Barcelona

Editors

Michael G. Main
Department of Computer Science, University of Colorado
Boulder, CO 80309, USA

Austin C. Melton
David A. Schmidt
Department of Computing and Information Sciences
Kansas State University
Manhattan, KS 66506, USA

Michael W. Mislove
Department of Mathematics, Tulane University
New Orleans, LA 70118, USA

CR Subject Classification (1987): D.3.1, D.3.3, F.1.1–2, F.3.1–2, F.4.1

ISBN 3-540-97375-3 Springer-Verlag Berlin Heidelberg New York
ISBN 0-387-97375-3 Springer-Verlag New York Berlin Heidelberg

Printed in Germany

Printing and binding: Druckhaus Beltz, Hemsbach/Bergstr.
2145/3140-543210 – Printed on acid-free paper

Preface

The Fifth International Conference on the Mathematical Foundations of Programming Semantics was held on the campus of Tulane University, New Orleans, Louisiana from March 29 to April 1, 1989. The major goal of this workshop-conference series is to bring together computer scientists who work in programming semantics and mathematicians who work in areas which might impact programming semantics so that they may share ideas and discuss problems of mutual interest. By letting mathematicians see applications of their work to programming semantics and by letting computer scientists see their ideas and intuitions expressed in pure mathematics, the organizers have sought to improve communication among the researchers in these areas and to establish ties between related areas of research. With these goals in mind, the invited speakers for the conference were:

Samson Abramsky, *Imperial College*
Luca Cardelli, *DEC Research*
Peter Johnstone, *University of Cambridge*
Robin Milner, *University of Edinburgh*
Peter Freyd, *University of Pennsylvania*
John Reynolds, *Carnegie-Mellon University*

In addition, there were contributed talks by sixteen researchers, as well as a number of shorter presentations. These last were presented during the *Organizers' Sessions*, which were a new and innovative feature of the conference. They were designed to add flexibility to the program to accommodate interesting new developments which were not available at the time of the *Call for Papers*. Two papers from that session, those by John Gray and by A. J. Power, are included in this volume. In addition, the paper by C. A. R. Hoare and He Jifeng has evolved from a series of invited lectures which the authors presented at the Fourth MFPS workshop, which was held in Boulder, Colorado in 1988.

An informal preconference meeting took place on March 27 and 28, and it was at this gathering that Samson Abramsky presented a talk on the Kahn Principle, which is the topic of his contribution to these Proceedings. Also, a session organized by Carl Gunter (University of Pennsylvania) on the semantics of inheritance took place during the fifth conference and presented new results in this emerging area.

The Program Committee was chaired by Austin Melton and David Schmidt. In addition to the editors of this volume, the Committee consisted of Boumediene Belkhouche, Steve Brookes, Carl Gunter, Jimmie Lawson, Frank Oles, George Revesz, Teodor Rus, Robert Tennent and Eric Wagner. The editors wish to express their thanks to the other members of the Committee for their efforts in reviewing the papers submitted for presentation at the Conference. Additional thanks are due to Boumediene Belkhouche, who also served so capably as the Local Arrangements Chairman for the Conference.

The Conference was supported by funds from the Office of Naval Research and from the National Science Foundation. We thank these organizations for their generous support of the Conference.

Thanks are due to the many people who helped make the conference run so smoothly. These include Michael Huth, Peggy Jordan, John Kozma, Marguerite Saacks and Han Zhang. Finally, we all owe a special thank you to Geralyn Caradona, Administrative Assistant of the Mathematics Department of Tulane University, who managed to oversee virtually all of the small details of running the conference (even to the extent of relocating the major social gathering on the day it occurred) with such dispatch that the rest of us were able to concentrate on the main order of business, the program.

Michael Main *Austin Melton* *Michael Mislove* *David Schmidt*

February, 1990

Table of Contents

A Generalized Kahn Principle for Abstract Asynchronous Networks

Samson Abramsky

Department of Computing

Imperial College of Science, Technology and Medicine

180 Queen's Gate

London SW7 2BZ

England

Abstract

Our general motivation is to answer the question: "What is a model of concurrent computation?". As a preliminary exercise, we study dataflow networks. We develop a very general notion of model for asynchronous networks. The "Kahn Principle", which states that a network built from functional nodes is the least fixpoint of a system of equations associated with the network, has become a benchmark for the formal study of dataflow networks. We formulate a generalized version of the Kahn Principle, which applies to a large class of non-deterministic systems, in the setting of abstract asynchronous networks; and prove that the Kahn Principle holds under certain natural assumptions on the model. We also show that a class of models, which represent networks that compute over arbitrary event structures, generalizing dataflow networks which compute over streams, satisfy these assumptions.

1 Introduction

There are by now a proliferation of mathematical structures which have been proposed to model concurrent systems. These include synchronization trees [Win85], event structures [Win86], Petri nets [Rei85], failure sets [Hoa85], trace monoids [Maz89], pomsets [Pra82] and many others. One is then led to ask: what general structural conditions should a model of concurrency satisfy? There is an obvious analogy with the λ-calculus, where a consensus on the appropriate notions of model only emerged some time after a number of particular model constructions had been discovered (*cf.* [Bar84]). Indeed, we would like to pose the question:

"What is a model of concurrent computation?"

in the same spirit as the title of Meyer's excellent paper [Mey82].

One important disanalogy with the λ-calculus is that the field of concurrent computation so far lacks a canonical syntax; and at a deeper level, there is as yet no analogue of Church's thesis for concurrent computation. The various formalisms which have been proposed actually draw inspiration from a highly varied phenomenology: synchronous, asynchronous, real-time, dataflow, shared-memory, declarative, object-oriented, systolic, SIMD, neural nets, etc. etc. In these circumstances, some more modest and circumscribed attempts at synthesis seem justified. At the same time, merely finding general definitions which subsume a number of concrete models is not enough; good definitions should show their cutting edge by yielding some non-trivial results.

In the present study, we start from a particular class of concurrent systems, the *non-deterministic dataflow networks* [Par82]. A problem which has established itself as a benchmark for the formal study of such systems is the *Kahn Principle* [Kah74], which states that if a network is composed of functional nodes, its behaviour is captured by the least fixpoint of a system of equations associated with the network in a natural way.

We attempt to formulate a notion of model for such networks in the most general and abstract form which still allows us to prove the Kahn Principle. In this way, we hope both to shed light on the initial motivating question of the axiomatics of process semantics, and to expose the essence of the Kahn Principle. In the course of doing so, we shall attain a level of generality, both as regards the notion of asynchronous network we consider, and the statement of the Kahn Principle, far in excess of anything we have seen in the literature.

The structure of the remainder of the paper is as follows. In section 2, we review some background on domain theory and dataflow networks. Then in section 3 we introduce our general notion of model, state a generalized version of the Kahn Principle, and prove that certain conditions on models are sufficient to imply the Kahn Principle. As far as I know, these are the first results of this form, as opposed to proofs of the Kahn Principle for specific models. Some directions for further research are given in section 4.

2 Background

We begin with a review of some notions in Domain theory; see e.g. [GS89] for further information and motivation.

We write $\mathsf{Fin}(X)$ for the set of finite subsets of a set X; and $A \subseteq^{\mathsf{f}} X$ for the assertion that A is a finite subset of X. A *poset* is a structure $(P, \leqslant)$, where P is a set, and $\leqslant$ a reflexive, transitive, anti-symmetric relation on P. Let $(P, \leqslant)$ be a poset. We write $\downarrow x = \{y \in P \mid y \leqslant x\}$, $\uparrow x = \{y \in P \mid y \geqslant x\}$ for $x \in P$; and $\downarrow X = \bigcup_{x \in X} \downarrow x$, $\uparrow X = \bigcap_{x \in X} \uparrow x$ for $X \subseteq P$. A subset $S \subseteq P$ is *directed* if every finite subset of S has an upper bound in S. A poset is *directed-complete* if every directed subset S has a least upper bound, written $\bigsqcup S$. A cpo (complete partial order) is a directed-complete poset with a least element, written $\bot$. An element $b \in D$ of a cpo $(D, \sqsubseteq)$ is *compact* if whenever $S \subseteq D$ is directed, and $b \sqsubseteq \bigsqcup S$, then $b \sqsubseteq d$ for some $d \in S$. We write $K(D)$ for the set of compact elements of D, and $K(d) = \downarrow d \cap K(D)$ for $d \in D$. A cpo D is *algebraic* if for all $d \in D$, $K(d)$ is directed, and $d = \bigsqcup K(d)$; and ω-algebraic if in addition $K(D)$ is countable. An *ideal* over a poset P is a directed subset $I \subseteq P$ such that $x \leqslant y \in I \Rightarrow x \in I$. The *ideal completion* of a poset P is the set of ideals over P, ordered by inclusion. If P has a least element, this is an algebraic cpo; it is ω-algebraic if P is countable.

A map $f : D \to E$ of cpo's is *continuous* if for every directed subset $S \subseteq D$, $f(\bigsqcup S) = \bigsqcup f(S)$; and *strict* if $f(\bot_D) = \bot_E$. A subset $U \subseteq D$ of a cpo D is *Scott-open* if $U = \uparrow U$, and whenever $\bigsqcup S \in U$ for a directed subset $S \subseteq D$, then $S \cap U \neq \varnothing$. The Scott-open subsets form a topology on D; a function between cpo's is continuous as defined above iff it is continuous in the topological sense with respect to the Scott topology. The Scott-open subsets of an algebraic cpo D are those of the form $\bigcup_{i \in I} \uparrow b_i$, where $b_i \in K(D)$ for all $i \in I$.

We define some standard constructions on cpo's. Given a set X, the algebraic cpo of *streams* over X, $\mathsf{Str}(X)$, is the set of finite and infinite sequences over X, with the prefix ordering. If D, E are cpo's, $[D \to E]$ is the cpo of continuous functions from D to E, with the pointwise ordering; if $\{D_i\}_{i \in I}$ is a family of cpo's, $\prod_{i \in I} D_i$ is the cartesian product cpo, with the componentwise ordering. If $f : D \to D$ is a continuous map on a cpo D, it has a least fixed point, defined by

$$\mathsf{lfp}(f) = \bigsqcup_{k \in \omega} f^k(\bot).$$

We shall assume some small knowledge of category theory in the sequel; suitable references are [ML71,AM75]. We write **Cpo** for the category of cpo's and continuous maps, $\mathbf{Cpo}^s$ for the subcategory of strict continuous maps; and $\omega\mathbf{Alg}$, $\omega\mathbf{Alg}^s$ for the corresponding categories of ω-algebraic cpo's.

We define the *weak covering relation* on a poset $(P, \leqslant)$ by:

$$x \preceq y \overset{\text{def}}{\iff} x \leqslant y \;\&\; \forall z. (x \leqslant z \leqslant y \Rightarrow (x = z \text{ or } y = z))$$

and the *covering relation* by

$$x \prec y \overset{\text{def}}{\iff} x \preceq y \;\&\; x \neq y.$$

The computational intuition behind the covering relation as used in Domain theory is that it represents an atomic computation step, or the occurrence of an atomic event; this idea can be traced back to [KP78].

A *covering sequence* in an algebraic cpo D is a non-empty finite or infinite sequence of compact elements (b_n), such that $b_0 = \bot$, and $b_n \prec b_{n+1}$ for all terms b_n, b_{n+1} in the sequence. A covering sequence can be taken as a representation of $d = \bigsqcup b_n$, which gives a step-by-step description of how it was computed.

Given an algebraic cpo D, we can form the algebraic cpo $\mathcal{C}(D)$ of covering sequences over D, with the prefix ordering. There is a continuous map $\mu : \mathcal{C}(D) \to D$, with $\mu((b_n)) = \bigsqcup b_n$.

Finally, we define the *relative covering relation* in D by:

$$[b, c] \sqsubseteq d \overset{\text{def}}{\iff} b, c \in K(d) \;\&\; b \prec c.$$

We can think of $b \prec c$ as an atomic step at some finite stage in the computation of d.

A *prime event structure* [Win86] is a structure $\mathcal{E} = (E, \leqslant, \mathsf{Con})$, where $(E, \leqslant)$ is a countable poset, and $\mathsf{Con} \subseteq \mathsf{Fin}(E)$ a family of finite subsets of E, satisfying:

- $\forall e \in E. (\downarrow e \text{ is finite})$.
- $\forall e \in E. (\{e\} \in \mathsf{Con})$.
- $A \subseteq B \in \mathsf{Con} \Rightarrow A \in \mathsf{Con}$.
- $A \in \mathsf{Con} \Rightarrow \downarrow A \in \mathsf{Con}$.

We refer to elements of E as *events*, to $\leqslant$ as the *causality* or *enabling* relation, and to Con as the *consistency* predicate. A *configuration* of $\mathcal{E}$ is a set $x \subseteq E$ such that

- $e \leqslant e' \in x \Rightarrow e \in x$
- $A \subseteq^{\mathsf{f}} x \Rightarrow A \in \mathsf{Con}$.

The set $|\mathcal{E}|$ of configurations of $\mathcal{E}$, ordered by inclusion, is an algebraic cpo; the compact elements are the finite configurations. Note that in $|\mathcal{E}|$, $x \prec y$ iff $y \backslash x = \{e\}$ for some $e \in E$; and that if $x \sqsubseteq y$ for compact elements x, y, there is a sequence $e_1, \ldots, e_n$ such that $x = z_0 \prec \cdots \prec z_n = y$, where $z_i = x \cup \{e_1, \ldots, e_i\}$. The algebraic cpo's which arise from prime event structures are characterized in [Win86]; we refer to them as *event domains*. They form quite an extensive class, containing models of type-free and polymorphic lambda calculi (using stable functions), as well as the usual datatypes of functional programming [CGW87].

We now turn to the dataflow model of concurrent computation. Consider a process network, represented by a directed (multi)graph $G = (N, A, s, t)$, where N is the set of nodes, A the set of arcs, and $s, t : A \to N$ are the source and target functions. Each node is labelled with a sequential process, while each arc corresponds to a buffered message channel, which behaves like an unbounded FIFO queue. In addition to the usual sequential constructs, each node n can read from its input channels (those α with $t(\alpha) = n$), and write to its output channels (those α with $s(\alpha) = n$). Although this computational model might be criticised as unrealistic because of the unbounded buffering, this very feature enables a high degree of parallelism, and the model is appealingly simple, and quite close to a number of actually proposed and implemented dataflow languages and architectures [WA85,KLP79,KM77,GGKW84]. Kahn's brilliant insight in his seminal paper [Kah74] was that the behaviour of such networks could be captured denotationally in a very simple and elegant fashion, using some elementary domain theory. The key idea is to model the behaviour of each message channel α, on which atomic values from the set D_α can be transmitted, as a stream from the domain $\mathsf{Str}(D_\alpha)$. Using standard denotational techniques, the behaviour of the process at node n, with input channels $\alpha_1, \ldots, \alpha_k$, and output channnels $\beta_1, \ldots, \beta_l$, can be modelled by a continuous function

$$f : \mathsf{Str}(D_{\alpha_1}) \times \cdots \times \mathsf{Str}(D_{\alpha_k}) \to \mathsf{Str}(D_{\beta_1}) \times \cdots \times \mathsf{Str}(D_{\beta_l}).$$

The behaviour of the whole system can be modelled by setting up a system of equations, one for each channel in the network, of the overall form

$$\mathbf{X} = G(\mathbf{X}),$$

where $G : \prod_\alpha \mathsf{Str}(D_\alpha) \to \prod_\alpha \mathsf{Str}(D_\alpha)$; and solving by taking the least fixed point $\mathsf{lfp}(G) \in \prod_\alpha \mathsf{Str}(D_\alpha)$.

It is worth noting that Kahn never *proved* the coincidence of this denotational semantics with an operational semantics based directly on the computational model sketched above; indeed, he never defined any formal operational semantics for dataflow networks. Nevertheless, no-one has ever seriously doubted the accuracy of his semantics. A number of subsequent attempts have been made to fill this gap in the theory [Fau82,LS88]; it has proved surprisingly difficult to give a clean and elegant account.

In another direction, many attempts have been made to overcome one crucial limitation built into Kahn's framework; namely, the assumption that all processes in the network are deterministic, and hence their behaviour can be described by functions. This limitation must be overcome in order for these networks to be sufficiently expressive to model general-purpose concurrent systems (see e.g. [Hen82,Abr84]). However, as soon as non-deterministic processes are allowed, the denotational description of dataflow networks becomes much more complicated. In fact, naive attempts to extend Kahn's model have been shown to be doomed to failure by certain "anomalies" which were found by Keller [Kel78] and Brock and Ackerman [BA81]. In particular, Brock and Ackerman exhibited a pair of deterministic processes N_1, N_2 with the same Kahn semantics, and a non-deterministic context $C[\cdot]$ such that $C[N_1] \neq C[N_2]$ with respect to the intended operational semantics. The main point of this is to show that in the presence of non-determinism, the behaviour of a system is no longer adequately modelled by a "history tuple" $d \in \prod_\alpha \mathsf{Str}(D_\alpha)$.

Such a tuple records the order in which values are realized on each channel, but fails to record causality relations which may exist between items of data on *different* channels. A number of more detailed models have been proposed which reflect this kind of information. Two in particular have received some attention.

Definition 2.1 *Let S be a set of channel names, where for each $\alpha \in S$, there is a set D_α of atomic data which can be transmitted over α. The domain of linear traces over S, LTr_S, is the stream domain $\mathsf{Str}(E_S)$, where*

$$E_S = \{(\alpha, d) \mid \alpha \in S, d \in D_\alpha\}.$$

The idea is that a linear trace represents a sequential observer's view of a computation in the network, as a sequence of atomic events (α, d)—namely, the production of the atomic value d on the channel α. We can regard linear traces as more detailed—perhaps even *over-specified*—representations of history tuples; indeed, there is an obvious "result" or "output" map $\mu_S : \mathsf{LTr}_S \to \prod_{\alpha \in S} \mathsf{Str}(D_\alpha)$. It is a useful exercise to verify that this is strict and continuous.

Given $S \supseteq T$, we can define a (strict, continuous) *restriction map*, $\rho^S_T : \mathsf{LTr}_S \to \mathsf{LTr}_T$, where $\rho^S_T(s)$ is obtained by deleting all (α, d) from s such that $\alpha \notin T$.

In the linear trace model, a process is modelled by a pair (S, P), where S is the set of channels incident to the process, and $P \subseteq \mathsf{LTr}_S$ describes its (possibly non-deterministic) behaviour. The key definition is that of the operation of *network composition*, which glues together a family of processes along their coincident channels. Let $\{(S_j, P_j)\}_{j \in J}$ be a family of processes; we define $\|_{j \in J} (S_j, P_j) = (S, P)$, where

$$\begin{aligned} S &= \textstyle\bigcup_{j \in J} S_j \\ P &= \{s \in \mathsf{LTr}_S \mid \forall j \in J. (\rho^S_{S_j}(s) \in P_j)\}. \end{aligned}$$

Note that this definition of the behaviour of a net is quite different in form to the Kahn semantics; we have replaced continuous functions by sets of traces, and the iterative construction of a least fixed point by a product-like construction. It thus becomes a matter of some importance to see if this definition actually *coincides* with the Kahn semantics in the case when each node in the network is in fact computing some continuous function. (Of course, we must firstly define what that means in terms of sets of traces). We refer to this task as the proof of the *Kahn Principle* for the linear trace model.

The linear trace model has recently been proved to be *fully abstract* in a certain sense [Jon89]; however, some other models have also received considerable attention, and avoid the apparent over-specification of linear traces. In particular there are the *pomset* models [Pra82], which were inspired by Brock and Ackerman's *scenarios* [BA81]. The idea is to allow partial orders of events, rather than insisting on purely sequential observations.

Definition 2.2 *The domain of partially ordered traces PTr_S is the ideal completion of the finite partially-ordered traces with the prefix ordering, where:*

- *A finite partially-ordered trace is an isomorphism type of finite labelled partial orders* $(V, \leqslant, \ell)$, *where* $\ell : V \rightarrow E_S$, *and for each* $\alpha \in S$, *the subposet*

$$\{v \in V \mid \exists d \in D_\alpha(\ell(v) = (\alpha, d))\}$$

is linearly ordered.

- *The prefix ordering is defined on representatives by:*

$$\begin{aligned}(V, \leqslant, \ell) \sqsubseteq (V', \leqslant', \ell') \quad \stackrel{\text{def}}{\Longleftrightarrow} \quad & V \subseteq V' \;\&\; \leqslant = \leqslant' \cap V^2 \;\&\; \ell = \ell' \restriction V \\ & \&\; v \leqslant' v' \in V \;\Rightarrow\; v \in V.\end{aligned}$$

Note that, if we identify sequences with isomorphism types of labelled *linear* orders, we have the inclusion $\mathsf{LTr}_S \subseteq \mathsf{PTr}_S$. Once again, there is an evident definition of a restriction map $\rho^S_T : \mathsf{PTr}_S \rightarrow \mathsf{PTr}_T$ for $S \supseteq T$, and, by virtue of the stipulation that events at each channel are linearly ordered, a map $\mu_S : \mathsf{PTr}_S \rightarrow \prod_{\alpha \in S} \mathsf{Str}(D_\alpha)$.

We can then define the notion of network composition in the partially ordered trace model in *exactly the same way* as we did for the linear traces, modulo the different notions of "trace" and "restriction"; and formulate the Kahn Principle in exactly the same terms. The main previous work on proving the Kahn Principle for (essentially) the partially ordered trace model is described in [GP87].

Our aim is firstly to extract the essential properties of this situation to arrive at a general notion of model, and then to prove the Kahn principle in this general setting. Apart from yielding the particular results for the linear and partially-ordered trace models for dataflow networks as instances of our general result, there are a number of other insights that we hope this work provides:

- The abstract networks we consider compute over a much broader class of domains than just the stream domains of dataflow—our results apply at least to the event domains.

- The version of the Kahn Principle we formulate and prove in fact applies not only to the deterministic case, but to a broad class of *non-deterministic* networks—namely those in which each node computes one of a *set* of possible continuous functions. This includes for example the so-called "infinity-fair merge", though not the "angelic merge" [PS88]. As far as I know, this major extension to the Kahn Principle is new, even for the specific models described above.

- Although our notion of model is abstracted from the dataflow family, and cannot be claimed to be fully general, we hope it is a useful step along the way to answering the question raised in the opening paragraph, namely: "what is a model of concurrent computation?".

3 Results

3.1 Models

We assume a class Chan of *channel names*, ranged over by α, β, γ. We refer to sets of channels as *sorts*; the class of sorts, partially ordered by inclusion, is denoted by **Sort**. We use S, T, U to range over sorts.

Definition 3.1 *A model $\mathcal{M} = (\mathcal{T}, \mathcal{V}, \mu)$ comprises:*

- *functors $\mathcal{T}, \mathcal{V} : \mathbf{Sort}^{\mathsf{op}} \to \mathbf{Cpo}^s$*
- *a natural transformation $\mu : \mathcal{T} \to \mathcal{V}$*

such that $\mathcal{V}$ preserves limits.

We refer to $\mathcal{T}_S$ as the *traces* of sort S, $\mathcal{V}_S$ as the *values* of sort S, and μ as the *output* or *evaluation* map.

More explicitly, $\mathcal{T}$ assigns to each sort S a cpo $\mathcal{T}_S$, and to each $S \supseteq T$ a strict, continuous *restriction map* $\rho^S_T : \mathcal{T}_S \to \mathcal{T}_T$, such that:

- $S \supseteq T \supseteq U \;\Rightarrow\; \rho^T_U \circ \rho^S_T = \rho^S_U$
- $\rho^S_S = \mathsf{id}_{\mathcal{T}_S}$.

Similarly, $\mathcal{V}$ assigns a cpo $\mathcal{V}_S$ to each sort S. The requirement that $\mathcal{V}$ preserves limits amounts to asking that $\mathcal{V}$ takes *unions* in **Sort** to *products* in $\mathbf{Cpo}^s$. Since each sort is the union of its singletons, this means that if $\mathcal{V}_\alpha$ is the value domain of sort $\{\alpha\}$,

$$\mathcal{V}_S = \prod_{\alpha \in S} \mathcal{V}_\alpha;$$

and that the restriction maps will be the projections onto sub-products: for $S \supseteq T$, $\pi^S_T : \mathcal{V}_S \to \mathcal{V}_T$. Thus $\mathcal{V}$ is completely determined by the $\mathcal{V}_\alpha$.

Finally, for each sort S there is a strict, continuous map $\mu_S : \mathcal{T}_S \to \mathcal{V}_S$, such that for all $S \supseteq T$,

$$\mu_T \circ \rho^S_T = \pi^S_T \circ \mu_S.$$

Notation. We write $\nu^S_T = \mu_T \circ \rho^S_T = \pi^S_T \circ \mu_S$.

Examples

(1). Firstly, from the discussion in the previous Section, it is easy to see that both linear and partially-ordered traces yield examples of models. More precisely, for each channel α fix a set D_α of atomic values; then define $\mathcal{V}_\alpha = \mathsf{Str}(D_\alpha)$, and $\mathcal{T}_S = \mathsf{PTr}_S$ (LTr_S), ρ^S_T, μ_S as in Section 2. The verification of the required functoriality and naturality conditions is straightforward.

(2). We now describe a general class of models. For each channel α, fix an event structure $\mathcal{E}_\alpha = (E_\alpha, \leqslant_\alpha, \mathsf{Con}_\alpha)$. Define $\mathcal{V}_\alpha = |\mathcal{E}_\alpha|$, the domain of configurations over

$\mathcal{E}_\alpha$. For a sort S, we define $\mathcal{E}_S = \prod_{\alpha \in S} \mathcal{E}_\alpha$, where the product of event structures is defined as their *disjoint union* [Win86]: $\mathcal{E}_S = (E_S, \leqslant_S, \mathsf{Con}_S)$, where

$$\begin{array}{rcl} E_S & \stackrel{\text{def}}{=} & \{(\alpha, e) \mid \alpha \in S, e \in E_\alpha\} \\ (\alpha, e) \leqslant_S (\beta, e') & \stackrel{\text{def}}{\iff} & \alpha = \beta \;\&\; e \leqslant_\alpha e' \\ A \in \mathsf{Con}_S & \stackrel{\text{def}}{\iff} & \forall \alpha \in S. (\{e \mid (\alpha, e) \in A\} \in \mathsf{Con}_\alpha). \end{array}$$

We have [Win86]: $|\mathcal{E}_S| \cong \prod_{\alpha \in S} |\mathcal{E}_\alpha|$, and we shall take $\mathcal{V}_S = |\mathcal{E}_S|$. For $S \supseteq T$, the projections $\pi^S_T : |\mathcal{E}_S| \to |\mathcal{E}_T|$ are defined by $\pi^S_T(x) = x \cap E_T$.

In order to define the traces over $\mathcal{E}_S$, we follow the idea that

$$\text{traces} = \text{data} + \text{causality}.$$

Thus a trace is a configuration together with extra information about the order in which data was actually produced in a particular computation, reflecting some causal constraints.

Definition 3.2 *A trace over an event structure $\mathcal{E} = (E, \leqslant, \mathsf{Con})$ is a pair $t = (x_t, \leqslant_t)$, where $x_t \in |\mathcal{E}|$, and $\leqslant_t$ is a partial order on x_t such that:*

- $\forall e \in x_t. (\{e' \in x_t \mid e' \leqslant_t e\}$ *is finite*)
- $(\leqslant \cap\, x_t^2) \subseteq \leqslant_t$.

Traces are partially ordered as follows:

$$t \sqsubseteq t' \stackrel{\text{def}}{\iff} x_t \subseteq x_{t'} \;\&\; \leqslant_t = \leqslant_{t'} \cap\, x_t^2 \;\&\; (e \leqslant_{t'} e' \in x_t \;\Rightarrow\; e \in x_t).$$

Clearly, traces with this ordering form an algebraic cpo $\mathbf{P}\mathcal{E}$. A trace t is *linear* if $\leqslant_t$ is a linear order; the linear traces also form an algebraic cpo, $\mathbf{L}\mathcal{E}$, and $\mathbf{L}\mathcal{E} \subseteq \mathbf{P}\mathcal{E}$. The compact elements of $\mathbf{P}\mathcal{E}$ are those t for which x_t is a finite configuration of $|\mathcal{E}|$. Also, $t \prec u$ in $\mathbf{P}\mathcal{E}$ iff $x_u \setminus x_t = \{e\}$ for some e which is *maximal* in $\leqslant_u$. The following construction on trace domains will be useful. Given $t \in \mathbf{P}\mathcal{E}$, and $X \subseteq x_t$, we define $t{\upharpoonright}X$ by:

$$\begin{array}{rcl} x_{t\upharpoonright X} & = & \{e \in x_t \mid \exists e' \in X. e \leqslant_t e'\} \\ \leqslant_{t\upharpoonright X} & = & \leqslant_t \cap (x_{t\upharpoonright X})^2. \end{array}$$

Clearly $t{\upharpoonright}X$ is a well-defined trace, and $t{\upharpoonright}X \sqsubseteq t$; moreover, $X \subseteq Y \;\Rightarrow\; t{\upharpoonright}x \sqsubseteq t{\upharpoonright}Y$. This construction can also be applied to $\mathbf{L}\mathcal{E}$.

We can now complete the definitions for our two families of models, $\mathcal{M}_\mathbf{P}$ (partially ordered traces over event structures) and $\mathcal{M}_\mathbf{L}$ (the sub-model of linearly ordered traces). The trace domains for $\mathcal{M}_\mathbf{P}$ are defined by $\mathcal{T}_S = \mathbf{P}\mathcal{E}_S$, and for $\mathcal{M}_\mathbf{L}$ by $\mathcal{T}_S = \mathbf{L}\mathcal{E}_S$. The evaluation maps are defined for both by

$$\mu_S(t) = x_t,$$

and the restriction maps by

$$\rho^S_T(t) = (x_t \cap E_T, \leqslant_t \cap E_T^2),$$

for $S \supseteq T$.

The verification that these definitions yield models is straightforward. Note that $\mathcal{M}_{\mathsf{P}}$ and $\mathcal{M}_{\mathsf{L}}$ are really *families* of models, parameterized by the choice of event structures $\mathcal{E}_\alpha$ for each α. Our results will apply to *all* models in these families.

We now show how the concrete dataflow models of (1) are special cases of $\mathcal{M}_{\mathsf{P}}$ and $\mathcal{M}_{\mathsf{L}}$. Fix a set D_α for each channel α, and define an event structure $\mathcal{E}_\alpha$ as follows:

- $E_\alpha = \{(s, sd) \mid s \in D_\alpha^\star, d \in D_\alpha\}$.
- $(s, sd) \leqslant_\alpha (s', s'd') \stackrel{\text{def}}{\iff} sd \sqsubseteq s'd'$.
- $A \in \mathsf{Con} \stackrel{\text{def}}{\iff} \forall (s, sd), (s', s'd') \in A.(sd \sqsubseteq s'd' \text{ or } s'd' \sqsubseteq sd)$.

It can easily be verified that $|\mathcal{E}_\alpha| \cong \mathsf{Str}(D_\alpha)$. Also, we have

Proposition 3.3 *For all sorts S,*

$$\begin{aligned} \mathsf{PTr}_S &\cong \mathsf{P}\mathcal{E}_S \\ \mathsf{LTr}_S &\cong \mathsf{L}\mathcal{E}_S. \end{aligned}$$

PROOF. Given $t \in K(\mathsf{P}\mathcal{E}_S)$, we define a labelled poset $(x_t, \leqslant_t, \ell)$, where

$$\ell((\alpha, (s, sd))) = (\alpha, d).$$

This defines a map $\phi : K(\mathsf{P}\mathcal{E}_S) \to K(\mathsf{PTr}_S)$. (Note that the condition $(\leqslant_S \cap x_t) \subseteq \leqslant_t$ is needed to ensure that α-events are linearly ordered in $\phi(t)$ for each $\alpha \in S$). Now consider a trace in $K(\mathsf{PTr}_S)$ with representative labelled poset $(V, \leqslant, \ell)$. For each $v \in V$, let $\ell(v) = (\alpha, d)$. The set of α-labelled predecessors of v is linearly ordered, say

$$v_1 < \cdots < v_n < v,$$

and hence yields a finite sequence $s = d_1 \cdots d_n \in K(\mathsf{Str}(D_\alpha))$, where $d_i = \mathsf{snd} \circ \ell(v_i)$, $i = 1, \ldots, n$. We can thus define a new labelling function ℓ', which maps v to $(\alpha, (s, sd)) \in E_S$. Note that ℓ' is *injective*, and hence we can dispense with V, and take the induced order on $\ell'(V)$: $\ell'(v) \leqslant' \ell'(v') \stackrel{\text{def}}{\iff} v \leqslant v'$, yielding a trace $(\ell'(V), \leqslant')$ in $\mathsf{P}\mathcal{E}_S$. Thus we obtain a map $\psi : K(\mathsf{PTr}_S) \to K(\mathsf{P}\mathcal{E}_S)$. It is easily checked that ϕ and ψ are monotone and mutually inverse, yielding an order-isomorphism $K(\mathsf{P}\mathcal{E}_S) \cong K(\mathsf{PTr}_S)$, and hence by algebraicity, $\mathsf{P}\mathcal{E}_S \cong \mathsf{PTr}_S$. Finally, ϕ, ψ cut down to an isomorphism $K(\mathsf{L}\mathcal{E}_S) \cong K(\mathsf{LTr}_S)$, and so $\mathsf{L}\mathcal{E}_S \cong \mathsf{LTr}_S$. ∎

One further connection will be useful: the linear traces over an event structure are isomorphic to the covering sequences over its domain of configurations.

Proposition 3.4 *For any event structure $\mathcal{E}$, $\mathsf{L}\mathcal{E} \cong \mathcal{C}(|\mathcal{E}|)$.*

PROOF. From our description of covering relations in event domains, it follows that any covering sequence in $|\mathcal{E}|$ has the form

$$x_0 \prec \cdots x_n \prec \cdots$$

where $x_0 = \varnothing$, $x_{n+1} \setminus x_n = \{e_n\}$ for some $e \in E$. We can then define the linear trace t with $x_t = \bigcup x_n$, $e_n \leqslant_t e_m \Leftrightarrow n \leqslant m$. Conversely, any linear trace must, by countability of E and the well-foundedness property of traces, amount to a (finite or infinite) sequence (e_n), from which we can define a covering sequence (x_n), where $x_n = \{e_j \mid j \leqslant n\}$. The fact that each $x_n \in |\mathcal{E}|$ follows from the conditions on traces. These passages between $\mathsf{L}\mathcal{E}$ and $\mathcal{C}(\mathcal{E})$ are easily checked to be monotone and mutually inverse, establishing the required isomorphism. ∎

For the remainder of this section, we fix a model $\mathcal{M} = (\mathcal{T}, \mathcal{V}, \mu)$.

Definition 3.5 *A process in $\mathcal{M}$ is a pair (S, P), where $P \subseteq \mathcal{T}_S$. Let $\{(S_j, P_j)\}_{j \in J}$ be a family of processes. The network composition of this family is defined by:*

$$\|_{j \in J} (S_j, P_j) = (S, P),$$

where

$$\begin{aligned} S &= \textstyle\bigcup_{j \in J} S_j \\ P &= \{t \in \mathcal{T}_S \mid \forall j \in J. (\rho^S_{S_j}(t) \in P_j)\}. \end{aligned}$$

This definition was predictable from our discussion of concrete dataflow models in the previous section. The next definition is a key one, which answers the question of how to characterize when a process, *qua* set of traces, is computing a function. In fact, we deal with the more general situation when a process is computing any one (non-deterministically chosen) from a *set* of functions.

Definition 3.6 *Let (S, P) be a process, with $S = I \cup O$, and let $F \subseteq [\mathcal{V}_I \to \mathcal{V}_O]$ be a set of continuous functions. We say that (S, P) computes F if for all $t \in \mathcal{T}_S$:*

$$\begin{aligned} & t \in P \iff \exists f \in F: \\ & (1)\ \nu^S_O(t) = f(\nu^S_I(t)) \\ & (2)\ [u, v] \sqsubseteq t \Rightarrow \nu^S_O(v) \sqsubseteq f(\nu^S_I(u)). \end{aligned}$$

Condition (1) in this definition is the obvious stipulation that the overall effect of the trace is to compute an input-output pair in the graph of one of the functions $f \in F$. Condition (2) is more subtle; it insists that the *way* this input-output pair is computed must be "causally consistent", in the sense that for any step $u \prec v$ towards computing t, the output values realized after the step—at v—are no more than what was justified as f applied to the input values available before the step—at u.[1]

As regards the generality conferred by the use of *sets* of functions, consider the following example from dataflow [Par82]: the deterministic merge function

$$\mathsf{dmerge} : \mathsf{Str}(X) \times \mathsf{Str}(X) \times \mathsf{Str}(\{0,1\}) \to \mathsf{Str}(X)$$

which uses an oracle to guide its choices. This satisfies the equations:

$$\begin{aligned} \mathsf{dmerge}(a : x, y, 0 : o) &= a : \mathsf{dmerge}(x, y, o) \\ \mathsf{dmerge}(x, b : y, 1 : o) &= b : \mathsf{dmerge}(x, y, o). \end{aligned}$$

Now for any set of oracles O we can define:

$$F = \{\lambda x, y.\mathsf{dmerge}(x, y, o) \mid o \in O\}.$$

If we take O to be the set of *fair* oracles, *i.e.* infinite binary sequences containing infinitely many zeroes and infinitely many ones, then F corresponds to the "infinity-fair merge" [PS88]; however, note that the "angelic merge" cannot be obtained in this way.

Now let $\{(S_j, P_j)\}_{j \in J}$ be a family of processes, with $(S, P) = \|_{j \in J} (S_j, P_j)$. We say that $\{(S_j, P_j)\}_{j \in J}$ is a *non-deterministic functional network* if the following conditions hold:

1. For all $j \in J$, $S_j = I_j \cup O_j$ and (S_j, P_j) computes $F_j \subseteq [\mathcal{V}_{I_j} \to \mathcal{V}_{O_j}]$.

2. For all $\alpha \in S$, there is exactly one $j \in J$ with $\alpha \in O_j$.

If F_j is a singleton for all $j \in J$, we say that the network is *deterministic*.

Condition (2) is worth some comment. The constraint that each channel has *at most* one producer precludes non-determinism by "short circuit". The requirement that there be *exactly* one producer is a technical convenience; it means that we can avoid considering input channels—*i.e.* those with no producer in the system—separately. Of course, we can still handle input channels, in a "pointwise" fashion; for each given input value, we add a process which behaves like the constant function producing that value on the channel. Indeed, in our approach this is immediately generalized to allow a *set* of values to be produced.

Now we generalize the Kahn semantics for dataflow in the obvious way. For each $f \in \prod_{j \in J} F_j$, we define $G_f : \mathcal{V}_S \to \mathcal{V}_S$ by:

$$G_f = \langle \pi^{O_j}_{\alpha} \circ f_j \circ \pi^{S}_{I_j} \rangle_{\alpha \in S, \alpha \in O_j}.$$

By virtue of condition (2) on the network, there is exactly one component of the tuple defining G_f for each $\alpha \in S$.

---

[1]These conditions were directly inspired by Misra's "limit" and "smoothness" conditions in his notion of *descriptions* [Mis89]; his definition was made in the specific setting of the linear trace domain LTr_S, and in a rather different context.

We say that the network satisfies the *Generalized Kahn Principle* if the following condition holds:

$$\text{(GKP)} \quad \mu_S(P) = \{\mathsf{lfp}(G_f) \mid f \in \prod_{j \in J} F_j\}.$$

We say that $\mathcal{M}$ satisfies the Generalized Kahn Principle if (GKP) holds for every non-deterministic functional network in $\mathcal{M}$. We say that $\mathcal{M}$ satisfies the (ordinary) Kahn Principle if (GKP) holds for every deterministic functional network. Note that in this case, $\prod_{j \in J} F_j$ is a singleton, and hence so is the right-hand side of (GKP).

Our main objective will be to give sufficient conditions on $\mathcal{M}$ to ensure that (GKP) holds. (GKP) states an equality between two sets; it is convenient to consider the two inclusions separately. Firstly, we have

$$(\mathrm{GKP}_s) \quad \mu_S(P) \subseteq \{\mathsf{lfp}(G_f) \mid f \in \prod_{j \in J} F_j\}.$$

This is a *safety* property, since it asserts that every behaviour of the network computes one of the values specified by the (generalized) Kahn semantics. The converse:

$$(\mathrm{GKP}_l) \quad \{\mathsf{lfp}(G_f) \mid f \in \prod_{j \in J} F_j\} \subseteq \mu_S(P)$$

is a *liveness* property, since it asserts that every specified value is realized by some computation.

3.2 Safety

Definition 3.7 *An ω-algebraic cpo is incremental if whenever $b \sqsubseteq c$ in $K(D)$, there is a finite covering sequence*

$$b = b_0 \prec \cdots \prec b_n = c.$$

A strict, continuous function $f : D \to E$ on incremental domains is an incremental morphism if:

- *f weakly preserves relative covers:*

$$[b, c] \sqsubseteq d \Rightarrow [f(b), f(c)] \sqsubseteq f(d) \text{ or } f(b) = f(c) \in K(d).$$

- *f lifts relative covers:*

$$[b', c'] \sqsubseteq d' = f(d) \Rightarrow \exists b, c.([b, c] \sqsubseteq d \ \& \ f(b) = b', f(c) = c').$$

Incremental domains and morphisms form a category **IncDom**. We say that a functor $F : \mathbf{C} \to \mathbf{Cpo}^s$ is incremental if it factors through the inclusion $\mathbf{IncDom} \hookrightarrow \mathbf{Cpo}^s$, and that a model $\mathcal{M} = (\mathcal{T}, \mathcal{V}, \mu)$ is incremental if $\mathcal{T}$ is.

Note that all event domains, and all ideal completions of countable posets satisfying both the ascending and descending chain conditions, are incremental. The reason for our terminology is that incremental domains are precisely the specialization to posets of the incremental categories introduced in [GJ88].

Proposition 3.8 *$\mathcal{M}_{\mathsf{P}}$ and $\mathcal{M}_{\mathsf{L}}$ are incremental.*

PROOF. We have already observed that the domains $\mathsf{P}\mathcal{E}_S$, $\mathsf{L}\mathcal{E}_S$ are incremental. The fact the restriction maps weakly preserve relative covers follows easily from the definitions. We must verify the lifting property. We give the argument for $\mathcal{M}_{\mathsf{P}}$ only. Suppose then that $S \supseteq T$, $[u', v'] \sqsubseteq t'$ in $\mathsf{P}\mathcal{E}_T$, and $\rho^S_T(t) = t'$. We define $v = t \restriction x_{v'}$. Since $x_{v'} \subseteq x_{t'} \subseteq x_t$, this is well-defined, and yields $v \sqsubseteq t$. Let $w = \rho^S_T(v)$. Since $x_{v'} \subseteq x_v$, $x_{v'} \subseteq x_w$. For the converse, suppose $e \in x_w$. This implies that $e \in E_T$, and that for some $e' \in x_{v'}$, $e \leqslant_t e'$. But this implies $e \leqslant_{t'} e'$, since $\rho^S_T(t) = t'$, and hence $e \in x_{v'}$, since $v' \sqsubseteq t'$ and $e' \in x_{v'}$. Thus $x_w = x_{v'}$. The same reasoning shows that $\leqslant_w = \leqslant_{v'}$, and so $w = v'$.

To define u, recall that $u' \prec v'$ iff $x_{v'} \backslash x_{u'} = \{e\}$ for some $e \in E_T$ which is maximal in $\leqslant_{v'}$. But then e must also be maximal with respect to $\leqslant_v$, since otherwise we would have $e <_v e' \in x_{v'}$, which would imply $e <_{v'} e'$, contradicting $<_{v'}$-maximality of e. Thus if we define $x_u = x_v \setminus \{e\}$, $\leqslant_u = \leqslant_v \cap x_u^2$, we see that $v \restriction x_u = (x_u, \leqslant_u)$. Clearly $u \prec v$; and if $w = \rho^S_T(u)$,

$$x_w = x_u \cap E_T = (x_v \setminus \{e\}) \cap E_T = (x_v \cap E_T) \setminus \{e\} = x_{v'} \setminus \{e\} = x_{u'}.$$

Similarly $\leqslant_w = \leqslant_{u'}$, yielding $\rho^S_T(u) = u'$, and the proof is complete. ∎

Our main objective in the remainder of this subsection is to prove:

Theorem 3.9 *If $\mathcal{M}$ is incremental, it satisfies* (GKP$_s$).

Our strategy is to use incrementality of the restriction maps to move between local conditions expressing the functional behaviour of the nodes and global conditions expressing the functional behaviour of the whole network.

Lemma 3.10 *Let (S, P) be a non-deterministic functional process computing F, where $S = I \cup O$. For all $t \in P$ computing $f \in F$, and $u \sqsubseteq t$:*

$$\nu^S_O(u) \sqsubseteq f(\nu^S_I(u)).$$

PROOF. Suppose firstly that u is compact. Either $u = t$, in which case the conclusion follows directly from the first condition for $t \in P$, or by incrementality of $\mathcal{T}_S$, for some compact v, $[u, v] \sqsubseteq t$. Applying the second condition for $t \in P$,

$$\nu^S_O(u) \sqsubseteq \nu^S_O(v) \sqsubseteq f(\nu^S_I(u)).$$

The general result follows from this special case, since

$$\nu^S_O(u) = \bigsqcup_{v \in K(u)} \nu^S_O(v) \sqsubseteq \bigsqcup_{v \in K(u)} f(\nu^S_I(v)) = f(\nu^S_I(u)). \quad ∎$$

Lemma 3.11 *Let $\{(S_j, P_j)\}_{j \in J}$ be a non-deterministic functional network computing F_j at each $j \in J$, where $S_j = I_j \cup O_j$. Let $(S, P) = \|_{j \in J} (S_j, P_j)$. Then for all $t \in \mathcal{T}_S$:*

$$t \in P \Leftrightarrow \forall j \in J. \exists f_j \in F_j.$$

- $\nu^S_{O_j}(t) = f_j(\nu^S_{I_j}(t))$ (1)
- $[u, v] \sqsubseteq t \Rightarrow \nu^S_{O_j}(v) \sqsubseteq f_j(\nu^S_{I_j}(u)$ (2)

PROOF. We shall write $t_j = \rho^S_{S_j}(t)$ for $t \in \mathcal{T}_S$. By definition of network composition,

$$\begin{array}{rcll} t \in P & \Leftrightarrow & \forall j \in J.\, t_j \in P_j & \\ & \Leftrightarrow & \forall j \in J.\, \exists f_j \in F_j. & \\ & \bullet & \nu^{S_j}_{O_j}(t) = f_j(\nu^{S_j}_{I_j}(t_j)) & (1') \\ & \bullet & [u_j, v_j] \sqsubseteq t_j \Rightarrow \nu^{S_j}_{O_j}(v) \sqsubseteq f_j(\nu^{S_j}_{I_j}(u_j)) & (2') \end{array}$$

Now it suffices to show that for all $t \in \mathcal{T}_S$, $j \in J$, $f_j \in F_j$: (1) $\Longleftrightarrow$ (1′) and (2) $\Longleftrightarrow$ (2′). The equivalence of (1) and (1′) follows from the functoriality of ρ. To show that (2′) implies (2), we use the fact that ρ weakly preserves covers. Suppose $[u, v] \sqsubseteq t$. If $u_j = v_j$, we can apply Lemma 3.10 to get (2); if $u_j \prec v_j$, we can apply (2′). Finally, we show that (2) implies (2′). Suppose $[u', v'] \sqsubseteq t_j$. Since ρ lifts covers, for some $u, v \in \mathcal{T}_S$,

$$\rho^S_{S_j}(u) = u', \rho^S_{S_j}(v) = v', \ \& \ [u, v] \sqsubseteq t.$$

We can now apply (2) to get (2′), as required. ▮

As an immediate Corollary of Lemma 3.11, we obtain:

Proposition 3.12 *With notation as in Lemma 3.11:*

$$\begin{array}{rcll} t \in P & \Longleftrightarrow & \exists f \in \prod_{j \in J} F_j. & \\ & \bullet & \mu_S(t) = G_f(\mu_S(t)) & (1) \\ & \bullet & [u, v] \sqsubseteq t \Rightarrow \mu_S(v) \sqsubseteq G_f(\mu_S(u)) & (2) \end{array}$$

PROOF OF THEOREM 3.9. With notation as in Lemma 3.11, suppose $t \in P$. Applying Proposition 3.12 (1), for some $f \in \prod_{j \in J} F_j$, $\mu_S(t) = G_f(\mu_S(t))$, whence $\mathsf{lfp}(G_f) \sqsubseteq \mu_S(t)$. To show that $\mu_S(t) \sqsubseteq \mathsf{lfp}(G_f)$, let (t_k) be a covering sequence for t, which must exist by incrementality of $\mathcal{T}_S$; we show by induction on k that:

$$\forall k \in \omega.(\mu_S(t_k) \sqsubseteq G^k_f(\bot)).$$

The base case follows from the strictness of μ_S. For the inductive step,

$$\begin{array}{rcll} \mu_S(t_{k+1}) & \sqsubseteq & G_f(\mu_S(t_k)) & \text{Proposition 3.12 (2)} \\ & \sqsubseteq & G_f(G^k_f(\bot)) & \text{by induction hypothesis.} \end{array}$$ ▮

3.3 Liveness

Consider an algebraic domain D, and a chain of compact elements $C = (b_k)$ in D, with $\bigsqcup b_k = d$. We can consider C as a (partial) specification of a particular way of computing d, which induces a causality relation on compact approximations of d, as follows. Define $\|\cdot\|_C : K(d) \to \omega$ by

$$\|b\|_C = \min\{k \mid b \sqsubseteq b_k\}.$$

Now we can define:

$$b <_C c \overset{\text{def}}{\iff} \|b\|_C < \|c\|_C,$$

for $b, c \in K(d)$.

Now let t be a trace in $\mathcal{T}_S$, with $\mu_S(t) = d \in \mathcal{V}_S$. We can define a relation $<_t$ on $K(d)$ which reflects the causal constraints on how d can be realized introduced by t:

$$b <_t c \overset{\text{def}}{\iff} \text{for every covering sequence } (t_k) \text{ for } t:$$
$$\min\{k \mid b \sqsubseteq \mu_S(t_k)\} < \min\{k \mid c \sqsubseteq \mu_S(t_k)\}.$$

Definition 3.13 *Let $\mathcal{M} = (\mathcal{T}, \mathcal{V}, \mu)$ be an incremental model in which each value domain $\mathcal{V}_S$ is ω-algebraic. $\mathcal{M}$ is causally expressive if for every sort S, $d \in \mathcal{V}_S$, and chain of compact elements $C = (b_k)$ with $\bigsqcup b_k = d$, there exists $t \in \mathcal{T}_S$ such that:*

- $\mu_S(t) = d$
- $<_t \supseteq <_C$.

Proposition 3.14 *$\mathcal{M}_{\mathbf{P}}$ and $\mathcal{M}_{\mathbf{L}}$ are causally expressive.*

PROOF. Since $\mathcal{M}_{\mathbf{L}}$ is a sub-model of $\mathcal{M}_{\mathbf{P}}$, it suffices to prove causal expressiveness for $\mathcal{M}_{\mathbf{L}}$. Suppose then that a compact chain $C = (b_n)$ in $\mathcal{E}_S$ is given, with $\bigsqcup b_n = d$. Since $\mathcal{E}_S$ is incremental, C can be refined into a covering sequence C'; clearly $<_{C'} \supseteq <_C$. Now let t be the trace in $\mathbf{L}\mathcal{E}_S$ corresponding to C' under the isomorphism of Proposition 3.4. We note the general fact that for *any* algebraic cpo D, and covering sequence (c_n) in D, there is a *unique* covering sequence for (c_n) in $\mathcal{C}(D)$; a consequence of this is that $\mathcal{C}(\mathcal{C}(D)) \cong \mathcal{C}(D)$. If follows that $<_t = <_{C'} \supseteq <_C$, as required. ∎

We shall need a technical lemma about fixpoints in ω-algebraic cpo's. This was conjectured by the author, and proved under the hypothesis that the domain is SFP. The ingenious proof of the general result is due to Achim Jung (personal communication); it is reproduced here with his kind permission.

Lemma 3.15 (Jung) *Let D be an ω-algebraic cpo, and $f : D \to D$ a continuous function. There exists a chain (b_n) of compact elements in D such that:*

1. $b_0 = \bot$
2. $\forall n.\, b_{n+1} \sqsubseteq f(b_n)$
3. $\bigsqcup b_n = \mathsf{lfp}(f)$.

PROOF. For each $f^n(\bot)$ we choose a chain of compact elements (c^n_m) with least upper bound $f^n(\bot)$. By taking a diagonal sequence we find a chain (c_n) with the property $c^{n'}_{m'} \sqsubseteq c_n \sqsubseteq f^n(\bot)$ for all $n', m' \leqslant n$. The least upper bound of this chain is equal to $\mathsf{lfp}(f)$. Let $C_n = \uparrow c_n$.

We shall define the required sequence (b_n) inductively, to satisfy the following properties:

1. $b_n \sqsubseteq f(b_{n-1}), \quad n \geqslant 1$

2. $b_n \sqsubseteq f^n(\bot), \quad n \geqslant 0$

3. $b_n \in O_n = \bigcap_{m\in\omega, 0\leqslant 2m\leqslant n} g^{-n+2m}(C_{n-m}), \quad n \geqslant 0.$

For $n = 2k$, the last property implies in particular that $b_n \in C_k$, and together with (2) this ensures that the limit of the b_n is the least fixed point of f.

Let $b_0 = \bot$. Then (2) is obviously satisfied, and (3) evaluates to

$$O_0 = f^0(C_0) = C_0 = \uparrow c_0 = \uparrow\bot = D,$$

and is satisfied too.

Given $b_0, \ldots, b_n$ we find b_{n+1} as follows. First note that $b_n \sqsubseteq f(b_{n-1}) \sqsubseteq f(b_n)$ by (1) (for $n = 0$ this is trivially satisfied); and that $f(b_n) \sqsubseteq f^{n+1}(\bot)$ by (2). We shall select b_{n+1} below $f(b_n)$ and above b_n, so (1) and (2) will be satisfied. As for (3), we calculate:

$$\begin{aligned}
&b_n \in O_n \;\Rightarrow\; f(b_n) \in f(O_n)\\
&\subseteq \textstyle\bigcap_{0\leqslant 2m\leqslant n} f^{-n+2m+1}(C_{n-m})\\
&= \textstyle\bigcap_{2\leqslant 2m+2\leqslant n+2} f^{-n-1+(2m+2)}(C_{n+1-(m+1)})\\
&= \textstyle\bigcap_{2\leqslant 2m'\leqslant n+2} f^{-n-1+2m'}(C_{n+1-m'})\\
&\subseteq \textstyle\bigcap_{2\leqslant 2m'\leqslant n+1} f^{-n-1+2m'}(C_{n+1-m'}).
\end{aligned}$$

Note that $f^{n+1}(\bot)$ is contained in C_{n+1}, so we have

$$\bot \in f^{-n-1}(f^{n+1}(\bot)) \subseteq f^{-n-1}(C_{n+1}),$$

which tells us that $f^{-n-1}(C_{n+1}) = D$. So

$$f(b_n) \in \bigcap_{0\leqslant 2m'\leqslant n+1} f^{-n-1+2m'}(C_{n+1-m'}) = O_{n+1}.$$

Since O_{n+1} is Scott-open, it contains a compact element below $f(b_n)$; let b_{n+1} be such an element above b_n. ∎

Theorem 3.16 *If $\mathcal{M}$ is causally expressive, it satisfies* (GKP$_l$).

PROOF. We adopt the same notation as in Lemma 3.11. Suppose $f \in \prod_{j\in J} F_j$. We must show that for some $t \in P$, $\mu_S(t) = \mathsf{lfp}(G_f)$. We apply Lemma 3.15 to obtain a chain of compact elements $C = (b_k)$ with $\bigsqcup b_k = \mathsf{lfp}(G_f)$, $b_0 = \bot$, and $b_{k+1} \sqsubseteq G_f(b_k)$ for all k. Since $\mathcal{M}$ is causally expressive, for some $t \in \mathcal{T}_S$, $\mu_S(t) = \bigsqcup b_k = \mathsf{lfp}(G_f)$, and $<_t \supseteq <_C$. It remains to show that $t \in P$. By Proposition 3.12, it suffices to show that for all $[u, v] \sqsubseteq t$, $\mu_S(v) \sqsubseteq G_f(\mu_S(u))$, which in turn is equivalent to:

$$\forall b \in K(\mathcal{V}_S).\,(b \sqsubseteq \mu_S(v) \;\Rightarrow\; b \sqsubseteq G_f(\mu_S(u))).$$

Suppose then that $b \sqsubseteq \mu_S(v) \sqsubseteq \mu_S(t) = \mathsf{lfp}(G_f)$. Since b is compact, $b \sqsubseteq b_k$ for some k. If $b = \bot$ we are done; otherwise, for some k, $b \sqsubseteq b_{k+1}$, $b \not\sqsubseteq b_k$. This implies $b_k <_C b$, and hence $b_k <_t b$. By incrementality of $\mathcal{T}_S$, we can find a covering sequence (t_k) for t with $u = t_n$, $v = t_{n+1}$ for some n. But since $b \sqsubseteq \mu_S(v)$ and $b_k <_t b$, this implies $b_k \sqsubseteq \mu_S(u)$, and hence

$$b \sqsubseteq b_{k+1} \sqsubseteq G_f(b_k) \sqsubseteq G_f(\mu_S(u)),$$

as required. ▮

As an immediate Corollary of Propositions 3.8 and 3.14 and Theorems 3.9 and 3.16, we obtain:

Theorem 3.17 *$\mathcal{M}_{\mathbf{P}}$ and $\mathcal{M}_{\mathbf{L}}$ satisfy (GKP).*

4 Concluding Remarks

The results in this paper are of a preliminary nature. Even within the asynchronous network model, there are a number of interesting topics for further investigation. These include the characterisation of models in terms of properties of *extensionality* and *expressive completeness*; and connections with *full abstraction*. Also, it would be of interest to specify a uniform operational semantics for our general class of models $\mathcal{M}_{\mathbf{P}}$, and to prove some correspondence results. A good basis for this should be given by [Cur86]. It would also be interesting to formulate a notion of *continuous* (*e.g.* probabilistic) computation in a network, replacing algebraic domains by continuous ones. Much of the theory developed here should generalize; note in particular that Lemma 3.15 is valid for ω-continuous cpo's, replacing "compact" by "relatively compact". Beyond asynchronous networks, we would like to give a general notion of model in categorical terms, which would subsume a wide range of concurrency formalisms, including process algebras and Petri nets, as well as dataflow. The ideas of [Win88] should be relevant here.

Acknowledgements. I would like to thank Jay Misra for providing the initial stimulus to this work by sending me his paper [Mis89]; much of the present paper can be seen as an attempt to understand some of his ideas in a general setting. I would also like to thank Achim Jung for helpful discussions while the ideas developed, and for proving Lemma 3.15; and Mike Mislove for inviting me to the 1989 MFPS conference, where I presented this material at an informal "pre-meeting"; and for inviting me to submit this paper to the Proceedings. My thanks also to my hosts at the University of Pennsylvania for providing such a stimulating and friendly environment during my visit in the first half of 1989; the Nuffield Foundation, for their support in the form of a Science Research Fellowship for 1988–89; and the U.K. SERC and U.S.A. NSF for their financial support.

References

[Abr84] S. Abramsky. Reasoning about concurrent systems: a functional approach. In F. Chambers, D. Duce, and G. Jones, editors, *Distributed*

Computing, volume 20 of *APIC Studies in Data Processing*, pages 307–319. Academic Press, 1984.

[AM75] M. A. Arbib and E. Manes. *Arrows, Structures and Functors: the Categorical Imperative.* Academic Press, 1975.

[BA81] J. D. Brock and W. B. Ackerman. Scenarios: a model of nondeterminate computation. In *Formalization of Programming Concepts*, pages 252–259. Springer-Verlag, 1981. Lecture Notes in Computer Science Vol. 107.

[Bar84] H. Barendregt. *The Lambda Calculus: Its Syntax and Semantics.* North-Holland, revised edition, 1984.

[CGW87] T. Coquand, C. Gunter, and G. Winskel. dI-domains as a model of polymorphism. In *Third Workshop on the Mathematical Foundations of Programming Language Semantics*, pages 344–363. Springer-Verlag, 1987.

[Cur86] Pierre-Louis Curien. *Categorical Combinators, Sequential Algorithms and Functional Programming.* Pitman, 1986.

[Fau82] A. Faustini. *The equivalence of a denotational and an operational semantics for pure dataflow.* PhD thesis, University of Warwick, 1982.

[GGKW84] J. Glauert, J. Gurd, C. Kirkham, and I. Watson. The dataflow approach. In *Distributed Computing*, volume 20 of *APIC Studies in Data Processing*, pages 1–53. Academic Press, 1984.

[GJ88] C. Gunter and A. Jung. Coherence and consistency in domains. In *Third Annual Symposium on Logic in Computer Science*, pages 309–317. Computer Society Press of the IEEE, 1988.

[GP87] H. Gaifman and V. R. Pratt. Partial order models of concurrency and the computation of functions. In *Symposium on Logic in Computer Science*, pages 72–85. Computer Society Press of the IEEE, 1987.

[GS89] C. Gunter and D. S. Scott. Semantic domains. Technical Report MS-CIS-89-16, University of Pennsylvania, Department of Computer and Information Science, 1989.

[Hen82] P. Henderson. Purely functional operating systems. In J. Darlington, P. Henderson, and D. Turner, editors, *Functional Programming.* Cambridge University Press, 1982.

[Hoa85] C. A. R. Hoare. *Communicating Sequential Processes.* Prentice Hall International, 1985.

[Jon89] B. Jonsson. A fully abstract trace model for dataflow networks. In *Sixteenth Annual ACM Symposium on Principles of Programming Languages*, pages 155–165, 1989.

[Kah74] G. Kahn. The semantics of a simple language for parallel programming. In J. L. Rosenfeld, editor, *Information Processing 74*, pages 471–475, Amsterdam, 1974. North Holland.

[Kel78] R. M. Keller. Denotational models for parallel programs with indeterminate operators. In E. J. Neuhold, editor, *Formal Description of Programming Concepts*, pages 337–366. North Holland, 1978.

[KLP79] R. M. Keller, G. Lindstrom, and S. Patil. A loosely-coupled applicative multiprocessing system. In *AFIPS Conference Proceedings 46*, pages 613–622, 1979.

[KM77] G. Kahn and D. B. MacQueen. Coroutines and networks of parallel processes. In B. Gilchrist, editor, *Information Processing 77*, pages 993–998, Amsterdam, 1977. North Holland.

[KP78] G. Kahn and G. Plotkin. Domaines concrets. Technical Report 336, IRIA-Laboria, 1978.

[LS88] N. A. Lynch and E. W. Stark. A proof of the Kahn principle for input/output automata. *Information and Computation*, 1988. To appear.

[Maz89] A. Mazurkiewicz. Basic notions of trace theory. In J. W. de Bakker, W.-P. de Roever, and G. Rozenberg, editors, *Linear Time, Branching Time and Partial Order in Logics and Models for Concurrency*, pages 285–363. Springer-Verlag, 1989. Lecture Notes in Computer Science Vol. 354.

[Mey82] Albert Meyer. What is a model of the lambda calculus? *Information and Control*, 52:87–122, 1982.

[Mis89] J. Misra. Equational reasoning about nondeterministic processes. Department of Computer Science, The University of Texas at Austin, 1989.

[ML71] S. Mac Lane. *Categories for the Working Mathematician*. Springer-Verlag, Berlin, 1971.

[Par82] D. Park. The "fairness" problem and non-deterministic computing networks. In *Foundations of Computer Science IV Part 2*, volume 159 of *Mathematical Centre Tracts*, pages 133–161. Centrum voor Wiskunde en Informatica, 1982.

[Pra82] V. R. Pratt. On the composition of processes. In *Ninth Annual ACM Symposium on Principles of Programming Languages*, 1982.

[PS88] P. Panangaden and E. W. Stark. Computation, residuals and the power of indeterminacy. In *Automata, Languages and Programming*, pages 439–454. Springer-Verlag, 1988. Lecture Notes in Computer Science Vol. 317.

[Rei85] W. Reisig. *Petri Nets*, volume 4 of *EATCS Monographs on Theoretical Computer Science*. Springer-Verlag, 1985.

[WA85] W. W. Wadge and E. A. Ashcroft. *Lucid, the Dataflow Programming Language*, volume 22 of *APIC Studies in Data Processing*. Academic Press, 1985.

[Win85] G. Winskel. Synchronisation trees. *Theoretical Computer Science*, May 1985.

[Win86] G. Winskel. Event structures. In W. Brauer, W. Reisig, and G. Rozenberg, editors, *Petri Nets: Applications and Relationships to other Models of Concurrency*. Springer-Verlag, 1986. Lecture Notes in Computer Science Vol. 255.

[Win88] G. Winskel. A category of labelled Petri nets and compositional proof system. In *Third Annual Symposium on Logic in Computer Science*, pages 142–154. Computer Society Press of the IEEE, 1988.

Operations on Records

(Summary[1])

Luca Cardelli
Digital Equipment Corporation
Systems Research Center

John C. Mitchell
Department of Computer Science
Stanford University

Abstract

We define a simple collection of operations for creating and manipulating record structures, where records are intended as finite associations of values to labels. A second-order type system over these operations supports both subtyping and polymorphism. We provide typechecking algorithms and limited semantic models.

Our approach unifies and extends previous notions of records, bounded quantification, record extension, and parametrization by row-variables. The general aim is to provide foundations for concepts found in object-oriented languages, within a framework based on typed lambda-calculus.

1. Introduction

Object-oriented programming is based on record structures (called *objects*) intended as named collections of values (*attributes*) and functions (*methods*). Collections of objects form *classes*. A *subclass* relation is defined on classes with the intention that methods work "appropriately" on all members belonging to the subclasses of a given class. This property is important in software engineering because it permits after-the-fact extensions of systems by subclasses, without requiring modifications to the systems themselves.

The first object-oriented language, Simula67, and most of the more recent ones (see references) are typed by using simple extensions of the type rules for Pascal-like languages. These extensions mainly involve a notion of *subtyping*. In addition to subtyping, we are interested here in more powerful type systems that smoothly incorporate *parametric polymorphism*.

Type systems for record structures have recently received much attention. They provide foundations for typing in object-oriented languages, data base languages, and their extensions. In [Cardelli 84&88] the basic notions of record types, as intended here, were defined in the context of a first-order type system

[1] *Full paper to appear in Mathematical Structures in Computer Science.*

for fixed-size records. Then Wand [Wand 87] introduced the concept of *row-variables* while trying to solve the type inference problem for records; this led to a system with extensible records and limited second-order typing. His system was later refined and shown to have principal types in [Jategaonkar Mitchell 88], [Rémy 89], and again in [Wand 89]. The resulting system provides a flexible integration of record types and Milner-style type inference [Milner 78].

Meanwhile [Cardelli Wegner 85] defined a full second-order extension of the system with fixed-size records, based on techniques from [Mitchell 84]. In that system, a program can work polymorphically over all the subtypes *B* of a given record type *A*, and it can preserve the "unknown" fields (the ones in *B* but not in *A*) of record parameters from input to output. However, some natural functions are not expressible. For example, by the nature of fixed-size records there is no way to add a field to a record and preserve all its unknown fields. Less obviously, a function that updates a record field, in the purely applicative sense of making a modified copy of it, is forced to remove all the unknown fields from the result. Imperative update also requires a careful typing analysis.

In this paper we describe a second-order type system that incorporates extensible records and solves the problem of expressing the natural functions mentioned above. We believe this second-order approach makes the presentation of record types more natural. The general idea is to extend a standard second-order (or even higher-order) type system with a notion of subtyping at all types. Record types are then introduced as specialized type constructions with some specialized subtyping rules. These new constructions interact well with the rest of the system. For example, row-variables fall out naturally from second-order type variables, and contravariance of function spaces and universal quantifiers mixes well with record subtyping.

In moving to second-order typing we give up the principal type property of weaker type systems, in exchange for some additional expressiveness. But most importantly for us, we gain some perspective on the space of possible operations on records and record types, unencumbered (at least temporarily) by questions about type inference. Since it is not clear yet where the bounds of expressiveness may lie, this perspective should prove useful for comparisons and further understanding.

The first part of the paper is informal and introduces the main concepts and problems by means of examples. Then we formalize our intuitions by a collection of type rules. In this summary of our work, we briefly describe a normalization procedure for record types, and show soundness of the rules with respect to a simple semantics for the pure calculus of records. Applications and extensions of the basic calculus are described in the full paper.

2. Informal development

Before looking at a formal system, we describe informally the desired operations on records and we justify the rules that are expected to hold. The final formal system is rather subtle, so these explanations should be useful in understanding it.

We also give simple examples of how records and their operations can be used in the context of object-oriented languages.

2.1 Record values

A *record value* is intended to represent, in some intuitive semantic sense, a finite map from labels to values where the values may belong to different types. Syntactically, a record value is a collection of *fields*, where each field is a labeled value. To capture the notion of a map, the labels in a given record must be distinct. Hence the labels can be used to identify the fields, and the fields can be taken to be unordered. This is the notation we use:

$\langle\rangle$	the empty record.
$\langle x=3, y=true\rangle$	a record with two fields, labeled x and y, equivalent to $\langle y=true, x=3\rangle$.

There are three basic operations on record values.

• *Extension* $\langle r \mid x=a\rangle$; adds a field of label x and value a to a record r, provided a field of label x is not already present. (This condition will be enforced statically.) We write $\langle r \mid x=a \mid y=b\rangle$ for $\langle\langle r \mid x=a\rangle \mid y=b\rangle$.

• *Restriction* $r\backslash x$; removes the field of label x, if any, from the record r. We write $r\backslash xy$ for $r\backslash x\backslash y$.

• *Extraction* $r.x$; extracts the value corresponding to the label x from the record r, provided a field having that label is present. (This condition will be enforced statically.)

We have chosen these three operations because they seem to be the fundamental constituents of more complex operations. An alternative, considered in [Wand 87], would be to replace extension and restriction by a single operation that either modifies or adds a field of label x, depending on whether another field of label x is already present. In our system, the extension operation is not required to check whether a new field is already present in a record: its absence is guaranteed statically. The restriction operation has the task of removing unwanted fields and fulfilling that guarantee. This separation of tasks has advantages for efficiency, and for static error detection since fields cannot be overwritten unintentionally by extension alone. Based on a comparison between the systems of [Wand 87] and [Jategaonkar Mitchell 88], it also seems possible that a reasonable fragment of our language will have a practical type inference algorithm.

Here are some simple examples. The symbol $\leftrightarrow$ (value equivalence) means that two expressions denote the same value.

$\langle\langle\rangle \mid x{=}3\rangle$	$\leftrightarrow$	$\langle x{=}3\rangle$	extension	
$\langle\langle x{=}3\rangle \mid y{=}true\rangle$	$\leftrightarrow$	$\langle x{=}3, y{=}true\rangle$		
$\langle x{=}3, y{=}true\rangle\backslash y$	$\leftrightarrow$	$\langle x{=}3\rangle$	restriction	(cancelling y)
$\langle x{=}3, y{=}true\rangle\backslash z$	$\leftrightarrow$	$\langle x{=}3, y{=}true\rangle$		(no effect)
$\langle x{=}3, y{=}true\rangle.x$	$\leftrightarrow$	3	extraction	

$\langle\langle x{=}3\rangle \mid x{=}4\rangle$	invalid extension
$\langle x{=}3\rangle.y$	invalid extraction

Some useful derived operators can be defined in terms of the ones above.
- *Renaming* $r[x{\leftarrow}y] =_{\text{def}} \langle r\backslash x \mid y{=}r.x\rangle$: changes the name of a record field.
- *Overriding* $\langle r \leftarrow x{=}a\rangle =_{\text{def}} \langle r\backslash x \mid x{=}a\rangle$: if x is present in r, overriding replaces its value with one of a possibly unrelated type, otherwise extends r (compare with [Wand 89]). Given adequate type restrictions, this can be seen as an updating operator, or a method overriding operator. We write $\langle r \leftarrow x{=}a \leftarrow y{=}b\rangle$ for $\langle\langle r \leftarrow x{=}a\rangle \leftarrow y{=}b\rangle$.

Obviously, all records can be constructed from the empty record using extension operations. In fact, in the formal presentation of the calculus, we regard the syntax for a record of many fields as an abbreviation for iterated extensions of the empty record, e.g.:

$$\langle x{=}3\rangle \quad =_{\text{def}} \quad \langle\langle\rangle \mid x{=}3\rangle$$
$$\langle x{=}3, y{=}true\rangle \quad =_{\text{def}} \quad \langle\langle\langle\rangle \mid x{=}3\rangle \mid y{=}true\rangle$$

This definition allows us to express the fundamental properties of records in terms of combinations of simple operators of fixed arity, as opposed to n-ary operators. Hence, we never have to use schemas with ellipses, such as $\langle x_1{=}a_1, \ldots, x_n{=}a_n\rangle$, in our formal treatment.

Since $r\backslash x \leftrightarrow r$ whenever r lacks a field of label x, we can formulate the definition above using any of the following expressions:

$$\langle\langle\rangle \mid x{=}3 \mid y{=}true\rangle \;\leftrightarrow\; \langle\langle\langle\rangle\backslash x \mid x{=}3\rangle\backslash y \mid y{=}true\rangle \;\leftrightarrow\; \langle\langle\rangle \leftarrow x{=}3 \leftarrow y{=}true\rangle$$

The latter forms match better a similar definition for record types, given next.

2.2 Record types

In describing operations on record values we made positive assumptions of the form "a field of label *x* *must* occur in record *r*" and negative assumptions of the form "a field of label *x* *must not* occur in record *r*".

These constraints will be verified statically by embedding them in a type system, hence *record types* will convey both positive and negative information. Positive information describes the fields that members of a record type *must* have. (Members may have additional fields.) Negative information describes the fields the members of that type *must not* have. (Members may lack additional fields.)

Note that both positive and negative information expresses constraints, hence increasing either kind of information will lead to smaller sets of values. The smallest amount of information is expressed by the record type with no fields, $\langle\!\langle\rangle\!\rangle$, which therefore denotes the collection of all records, since all records have at least no fields and lack at least no fields. This type is called the *total* record type.

$\langle\!\langle\rangle\!\rangle$	the type of all records. Contains, e.g.: $\langle\rangle$, $\langle x{=}3\rangle$.
$\langle\!\langle\rangle\!\rangle\backslash x$	the type of all records which lack fields of label *x*. E.g.: $\langle\rangle$, $\langle y{=}true\rangle$, but not $\langle x{=}3\rangle$.
$\langle\!\langle x{:}Int, y{:}Bool\rangle\!\rangle$	the type of all records which have *at least* fields of labels *x* and *y*, with values of types *Int* and *Bool*. E.g.: $\langle x{=}3, y{=}true\rangle$, $\langle x{=}3, y{=}true, z{=}\text{"str"}\rangle$, but not $\langle x{=}3, y{=}4\rangle$, $\langle x{=}3\rangle$.
$\langle\!\langle x{:}Int\rangle\!\rangle\backslash y$	the type of all records which have *at least* a field of label *x* and type *Int*, and no field of label *y*. E.g. $\langle x{=}3, z{=}\text{"str"}\rangle$, but not $\langle x{=}3, y{=}true\rangle$.

Hence a record type is characterized by a finite collection of (*positive*) *type fields* (i.e. labeled types) and *negative type fields* (i.e. labels)[2]. We often simply say "fields" for "type fields". The positive fields must have distinct labels and are unordered. Negative fields are also unordered. We have assumed so far that types are normalized so that positive and negative labels are distinct, otherwise positive and negative fields may cancel, as described shortly.

[2] In this section we consider only *ground* record types, i.e., those containing no record type variables.

As with record values, we have three basic operations on record types.

• *Extension* $\langle\!\langle R \mid x{:}A \rangle\!\rangle$: This type denotes the collection obtained from R by adding x fields with values in A in all possible ways (provided that none of the elements of R has x fields). More precisely, this is the collection of those records $\langle r \mid x{=}a \rangle$ such that r is in R and a is in A, provided that a positive type field x is not already present in R. (This condition will be enforced statically.) We sometimes write $\langle\!\langle R \mid x{:}A \mid y{:}B \rangle\!\rangle$ for $\langle\!\langle \langle\!\langle R \mid x{:}A \rangle\!\rangle \mid y{:}B \rangle\!\rangle$.

• *Restriction* $R\backslash x$: this type denotes the collection obtained from R by removing the field x (if any) from all its elements. More precisely, this is the collection of those records $r\backslash x$ such that r is in R. We write $R\backslash xy$ for $R\backslash x\backslash y$.

• *Extraction* $R.x$: this type denotes the type associated with label x in R, provided R has such a positive field. (This condition will be enforced statically.)

Again, derived operators can be defined in terms of the ones above.

• *Renaming* $R[x{\leftarrow}y] =_{\text{def}} \langle\!\langle R\backslash x \mid y{=}R.x \rangle\!\rangle$: changes the name of a record type field.

• *Overriding* $\langle\!\langle R \leftarrow x{:}A \rangle\!\rangle =_{\text{def}} \langle\!\langle R\backslash x \mid x{:}A \rangle\!\rangle$: if a type field x is present in R, overriding replaces it with a field x of type A, otherwise extends R. Given adequate type restrictions, this can be used to override a method type in a class signature (i.e. record type) with a more specialized one, to produce a subclass signature.

The crucial formal difference between these operators on types and the similar ones on values is that type restrictions do not cancel as easily, for example: $\langle\!\langle\rangle\!\rangle\backslash y \neq \langle\!\langle\rangle\!\rangle$, $\langle\!\langle x{:}A \rangle\!\rangle\backslash y \neq \langle\!\langle x{:}A \rangle\!\rangle$, etc., since $\langle\!\langle\rangle\!\rangle\backslash y$ is a smaller set than $\langle\!\langle\rangle\!\rangle$. As a consequence, one must always make a type restriction before making a type extension, as can be seen in the examples below, because the extension operator needs proof that the extension label is missing. The symbol $\leftrightarrow$ (type equivalence) means also that two type expressions denote the same type.

$\langle\!\langle \langle\!\langle\rangle\!\rangle\backslash x \mid x{:}Int \rangle\!\rangle$	$\leftrightarrow$	$\langle\!\langle x{:}Int \rangle\!\rangle$	extension	
$\langle\!\langle \langle\!\langle x{:}Int \rangle\!\rangle\backslash y \mid y{:}Bool \rangle\!\rangle$	$\leftrightarrow$	$\langle\!\langle x{:}Int, y{:}Bool \rangle\!\rangle$		
$\langle\!\langle x{:}Int, y{:}Bool \rangle\!\rangle\backslash y$	$\leftrightarrow$	$\langle\!\langle x{:}Int \rangle\!\rangle\backslash y$	restriction	(cancelling y)
$\langle\!\langle x{:}Int, y{:}Bool \rangle\!\rangle\backslash z$	$\leftrightarrow$	$\langle\!\langle x{:}Int, y{:}Bool \rangle\!\rangle\backslash z$		(no effect on x,y)
$\langle\!\langle x{:}Int, y{:}Bool \rangle\!\rangle.x$	$\leftrightarrow$	Int	extraction	

$\langle\!\langle \langle\!\langle\rangle\!\rangle \mid x{:}Int \rangle\!\rangle$	invalid extension
$\langle\!\langle \langle\!\langle x{:}Int \rangle\!\rangle \mid x{:}Int \rangle\!\rangle$	invalid extension
$\langle\!\langle x{:}Int \rangle\!\rangle.y$	invalid extraction

It helps to read these examples in terms of the collections they represent. For example, the first example for restriction says that if we take the collection of records that have x and y (and possibly more) fields, and remove the y field from all the elements in the collection, then we obtain the collection of records that have an x field (and possibly more fields) but no y field. In particular, we do not obtain the collection of records that have x and possibly more fields, because those would include y.

The way positive and negative information is formally manipulated is easier to understand if we regard record types as abbreviations, as we did for record values, e.g.:

$$\langle\!\langle x{:}Int \rangle\!\rangle \quad =_{\text{def}} \quad \langle\!\langle \langle\!\langle\rangle\!\rangle\backslash x \mid x{:}Int \rangle\!\rangle$$
$$\langle\!\langle x{:}Int, y{:}Bool \rangle\!\rangle \quad =_{\text{def}} \quad \langle\!\langle \langle\!\langle \langle\!\langle\rangle\!\rangle\backslash x \mid x{:}Int \rangle\!\rangle\backslash y \mid y{:}Bool \rangle\!\rangle$$

Then, when considering $\langle\!\langle y{:}Bool \rangle\!\rangle\backslash y$ we actually have the expansion $\langle\!\langle \langle\!\langle\rangle\!\rangle\backslash y \mid y{:}Bool \rangle\!\rangle\backslash y$. If we allow the outside positive and negative y labels to cancel, we are still left with $\langle\!\langle\rangle\!\rangle\backslash y$. In other words, the inner y restriction reminds us that y fields have been eliminated.

2.3 Record value variables

Now that we have a first understanding of record types, we can introduce record value variables which are declared to have some record type. For example, $r{:}\langle\!\langle\rangle\!\rangle\backslash y$ means that r must not have a field y, and $r{:}\langle\!\langle x{:}A \rangle\!\rangle$ means that r must have a field x of type A. The well-formed record expressions can now be formulated more precisely:

$\langle r \mid x{=}a \rangle$	where $r{:}\langle\!\langle\rangle\!\rangle\backslash x$
$r\backslash x$	where $r{:}\langle\!\langle\rangle\!\rangle$
$r.x$	where $r{:}\langle\!\langle x{:}A \rangle\!\rangle$ for some A

Record value variables can now be used to write function abstractions. Here we have a function that increments a field of a record, and adds another field to it:

$$\begin{array}{l} let\ f(r{:}\ \langle\!\langle x{:}Int \rangle\!\rangle\backslash y) : \langle\!\langle x{:}Int, y{:}Int \rangle\!\rangle = \\ \quad \langle r \leftarrow x{=}r.x{+}1 \mid y{=}0 \rangle \end{array}$$

This function requires an argument with a field x and no field y; it has type:

$$f : \langle\!\langle x{:}Int \rangle\!\rangle\backslash y \rightarrow \langle\!\langle x{:}Int, y{:}Int \rangle\!\rangle$$

and can be used as follows:

$$f(\langle x{=}3\rangle) \quad \leftrightarrow \quad \langle x{=}4, y{=}0\rangle : \langle\!\langle x{:}Int, y{:}Int\rangle\!\rangle$$
$$f(\langle x{=}3, z{=}true\rangle) \quad \leftrightarrow \quad \langle x{=}4, y{=}0, z{=}true\rangle : \langle\!\langle x{:}Int, y{:}Int\rangle\!\rangle$$

The first application uses the non-trivial fact that $\langle x{=}3\rangle : \langle\!\langle x{:}Int\rangle\!\rangle\backslash y$. We could also have matched the parameter type precisely by $f(\langle x{=}3\rangle\backslash y)$, which is of course equivalent. The second application is noticeable for several reasons. First, it uses the non-trivial fact that $\langle x{=}3, z{=}true\rangle : \langle\!\langle x{:}Int\rangle\!\rangle\backslash y$. Second, the "extra" field z is preserved in the result value, because of the way f is defined. Third, the "extra" field z is not preserved in the result type, because f has a fixed result type; we shall come back to this problem.

2.4 Record type variables

In the previous section we introduced record value variables, and we used record types to impose restrictions on the values which could be bound to such variables. Now we want to introduce record type variables in order to write programs that are polymorphic over a collection of record types. We similarly need to express restrictions on the admissible types that these variables can be bound to; these restrictions are written as subtype specifications.

To write subtype specifications, we use a predicate $A{<:}B$ meaning that A is a *subtype* of B: in other words, every value of A is also a value of B. The typing rule based on this condition is called *subsumption*, and will play a central role in the formal system.

Using subtype assumptions, we can better formulate the restrictions on the record type operators:

$\langle\!\langle R \mid x{:}A\rangle\!\rangle$	where $R <: \langle\!\langle\rangle\!\rangle\backslash x$
$R\backslash x$	where $R <: \langle\!\langle\rangle\!\rangle$
$R.x$	where $R <: \langle\!\langle x{:}A\rangle\!\rangle$ for some A

We may now write a polymorphic version of the function f of the previous section:

$$let\ f(R{<:}\langle\!\langle x{:}Int\rangle\!\rangle\backslash y)(r{:}R) : \langle\!\langle R \mid y{:}Int\rangle\!\rangle =$$
$$\langle r \leftarrow x{=}r.x{+}1 \mid y{=}0\rangle$$

This function expects first a type parameter R which must be a subtype of $\langle\!\langle x{:}Int\rangle\!\rangle\backslash y$, and then an actual value parameter of type R. An example application is:

$$f(\langle\!\langle x{:}Int, z{:}Bool\rangle\!\rangle\backslash y)(\langle x{=}3, z{=}true\rangle) \leftrightarrow$$
$$\langle x{=}4, y{=}0, z{=}true\rangle : \langle\!\langle x{:}Int, y{:}Int, z{:}Bool\rangle\!\rangle$$

First, note that R is bound to $\langle\!\langle x{:}Int,\ z{:}Bool \rangle\!\rangle \backslash y$, which is a subtype of $\langle\!\langle x{:}Int \rangle\!\rangle \backslash y$ as required. Second, $\langle x{=}3,\ z{=}true \rangle$ has type $\langle\!\langle x{:}Int,\ z{:}Bool \rangle\!\rangle \backslash y$ as required. Third, the result type, obtained by instantiating R, is $\langle\!\langle \langle\!\langle x{:}Int,\ z{:}Bool \rangle\!\rangle \backslash y \mid y{:}Int \rangle\!\rangle$, which is the same as $\langle\!\langle x{:}Int,\ y{:}Int,\ z{:}Bool \rangle\!\rangle$ by definition. Finally, note that the "extra" field z has not been forgotten in the result type this time, because all the "extra" fields are carried over from input to output type by the type variable R. This is the advantage of writing f in polymorphic style.

What is the type of f then? We cannot write this type with simple function arrows, because we have a free variable R to bind. Moreover, we want to mark the precise location where this binding occurs, because this permits more types to be expressed. Hence, we use an explicit *bounded universal quantifier*:

$$f : \forall (R <: \langle\!\langle x{:}Int \rangle\!\rangle \backslash y)\ R \rightarrow \langle\!\langle R \mid y{:}Int \rangle\!\rangle$$

This reads rather naturally: "for all types R which are subtypes of $\langle\!\langle x{:}Int \rangle\!\rangle \backslash y$, f is a function form R to $\langle\!\langle R \mid y{:}Int \rangle\!\rangle$". (The scope of a quantifier extends to the right as much as possible.)

2.5 Subtype hierarchies

Our operations on record types and record values make it easy to define new types and values by *reusing* previously defined types and values.

For example, we want to express the subtype hierarchy shown in the diagram below, where various entities can have a combination of coordinates x and y, radius r, and color c.

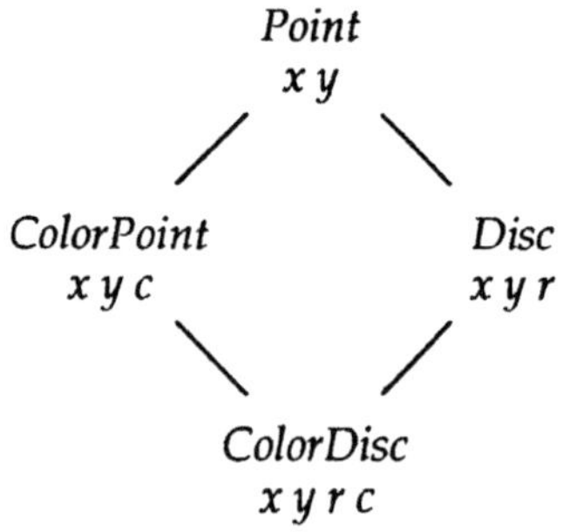

First, we could define each type independently:

let Point = $\langle\!\langle x{:}Real,\ y{:}Real \rangle\!\rangle$
let ColorPoint = $\langle\!\langle x{:}Real,\ y{:}Real,\ c{:}Color \rangle\!\rangle$

$let\ Disc = \langle\!\langle x{:}Real, y{:}Real, r{:}Real\rangle\!\rangle$
$let\ ColorDisc = \langle\!\langle x{:}Real, y{:}Real, r{:}Real, c{:}Color\rangle\!\rangle$

But these explicit definitions do not scale up easily to large hierarchies; it is much more convenient to define each type in terms of previous ones, e.g:

$let\ Point = \langle\!\langle x{:}Real, y{:}Real\rangle\!\rangle$
$let\ ColorPoint = \langle\!\langle Point \leftarrow c{:}Color\rangle\!\rangle$
$let\ Disc = \langle\!\langle Point \leftarrow r{:}Real\rangle\!\rangle$
$let\ ColorDisc = \langle\!\langle ColorPoint \leftarrow r{:}Real\rangle\!\rangle$

Note that $\langle\!\langle Point \mid c{:}Color\rangle\!\rangle$ would not be well-formed here, since members of *Point* may have a *c* label.

Similarly, record values can be defined by reusing available values:

$let\ p{:}Point = \langle x{=}3, y{=}4\rangle$
$let\ cp{:}ColorPoint = \langle p \leftarrow c{=}green\rangle$
$let\ cd{:}ColorDisc = \langle cp \leftarrow r{=}1\rangle$
$let\ d{:}Disc = cd\backslash c$

We should notice here that the subtyping relation depends only on the structure of the types, and not on how the types are named or constructed. Similarly, record values belong to record types uniquely based on their structure, independently of how they are declared or constructed.

Another observation, which we already made in a more abstract context, is that $Point\backslash r <: Point$ since *Point* does not contain *r*, but $Point\backslash y$ is incomparable with *Point* since *Point* requires *y:Int* while $Point\backslash y$ forbids it.

2.6 The update problem

The type system for records we have described in the previous sections was initially motivated by a single example which involves typing an update function. Here updating is intended in the functional sense of creating a copy of a record with a modified field, but the discussion is also relevant to imperative updating.

The problem is to define a function that updates a field of a record and returns the new record; the type of this function should be such that when an argument of the function has a subtype of the expected input type, the result has a related subtype. That is, no type information regarding additional fields should be lost in updating. (We have already seen that bounded quantification can be useful in this respect.)

It is pretty clear what the body of such a function should look like; for example for an input r and a boolean field b which has to be negated, we would write:

$$\langle r \leftarrow b{=}not(r.b)\rangle \qquad \text{(an abbreviation for } \langle r\backslash b \mid b{=}not(r.b)\rangle\text{)}$$

The overriding operator here preserves the additional fields of r.

One might expect the following typing, which seems to preserve subtype information as desired:

$$\begin{aligned}&let\ update(R<:\langle\!\langle b{:}Bool\rangle\!\rangle)(r{:}R){:}\ R =\\&\quad \langle r \leftarrow b{=}not(r.b)\rangle\end{aligned}$$

In words, we expect *update* to be a function from R to R, for any subtype R of $\langle\!\langle b{:}Bool\rangle\!\rangle$. But this typing is not derivable from our rules and, worse, it is semantically unsound. To see this, assume we have a type $True <: Bool$ with unique element *true*, as follows[3]:

$$\begin{aligned}&true : True <: Bool\\&not : Bool \rightarrow Bool \qquad \text{(alternatively, } not : \forall(A<:Bool)A\rightarrow Bool\text{)}\end{aligned}$$

$$update(\langle\!\langle b{:}True\rangle\!\rangle)(\langle b{=}true\rangle) \;\leftrightarrow\; \langle b{=}false\rangle : \langle\!\langle b{:}True\rangle\!\rangle$$

This use of *update* produces an obviously incorrect result type. In general, a function with result type R has a fixed range; it cannot restrict its output to an arbitrary subtype of R, even when this subtype is given as a parameter.

To avoid this problem, we must update the result type as well as the result. The correct typing comes naturally from typechecking the body of *update* according to the rules for each construct involved; note how the shape of the result type matches the shape of the body of the function:

$$\begin{aligned}&let\ update(R<:\langle\!\langle b{:}Bool\rangle\!\rangle)(r{:}R){:}\ \langle\!\langle R\leftarrow b{:}Bool\rangle\!\rangle =\\&\quad \langle r \leftarrow b{=}not(r.b)\rangle\end{aligned}$$

$$\begin{aligned}&update(\langle\!\langle b{:}True\rangle\!\rangle)(\langle b{=}true\rangle) \;\leftrightarrow\\&\quad \langle b{=}false\rangle : \;(\langle\!\langle \langle\!\langle b{:}True\rangle\!\rangle \leftarrow b{:}Bool\rangle\!\rangle \leftrightarrow \langle\!\langle b{:}Bool\rangle\!\rangle)\end{aligned}$$

The outcome is that the overriding operator on types, which involves manipulation of negative information, is necessary to express the type of update functions. Bounded quantification by itself is not sufficient.

The type $\forall(B<:A)\ B \rightarrow B$ turns out to contain only the identity function on

[3]Although the singleton type *True* may seem artificial, this argument can be reformulated with any proper inclusion between two types.

A in many natural semantic models, such as [Bruce Longo 88]. For example take $A=Int$ and let the subranges $[n..m]$ be subtypes of Int. Then any function of type $\forall(B<:Int)\ B \rightarrow B$ can be instantiated to $[n..n] \rightarrow [n..n]$, hence it must be the identity on $[n..n]$ for any n, and hence the identity over all of Int.

A further complication manifests itself when updating acts deep in a structure, because then we have to preserve type information with subtyping occurring at multiple levels. Here is the body of a function that negates the $s.a.b$ field of a record s of type $\langle\!\langle a{:}\langle\!\langle b{:}Bool\rangle\!\rangle\rangle\!\rangle$:

$$\langle s\leftarrow a{=}\langle s.a\leftarrow b{=}not(s.a.b)\rangle\rangle$$

The following is a correct typing which does not lose information on subtypes (simpler typings would). Here we need to introduce an additional type parameter in order to use two type variables in the result type and to avoid two possible ways of losing type information:

$$\begin{array}{l} let\ deepUpdate(R<:\langle\!\langle b{:}Bool\rangle\!\rangle)(S<:\langle\!\langle a{:}R\rangle\!\rangle)(s{:}S){:}\ \langle\!\langle S\leftarrow a{:}\langle\!\langle R\leftarrow b{:}Bool\rangle\!\rangle\rangle\!\rangle = \\ \quad \langle s\leftarrow a{=}\langle s.a\leftarrow b{=}not(s.a.b)\rangle\rangle \end{array}$$

Of course this is rather clumsy; we need one additional type parameter for each additional depth level of updating. Fortunately, we can avoid the extra type parameters by using *extraction* types $S.a$. Again, the following typing comes naturally from typechecking the body of *deepUpdate* according to the rules for each construct:

$$\begin{array}{l} let\ deepUpdate(S<:\langle\!\langle a{:}\langle\!\langle b{:}Bool\rangle\!\rangle\rangle\!\rangle)(s{:}S){:}\ \langle\!\langle S\leftarrow a{:}\langle\!\langle S.a\leftarrow b{:}Bool\rangle\!\rangle\rangle\!\rangle = \\ \quad \langle s\leftarrow a{=}\langle s.a\leftarrow b{=}not(s.a.b)\rangle\rangle \end{array}$$

The output type is still complex (it could be inferred) but the input is more natural. Here is a use of this function:

$$\begin{array}{l} deepUpdate(\langle\!\langle a{:}\langle\!\langle b{:}True,\ c{:}C\rangle\!\rangle,\ d{:}D\rangle\!\rangle)(\langle a{=}\langle b{=}true,\ c{:}v\rangle,\ d{:}w\rangle) \ \leftrightarrow \\ \quad \langle a{=}\langle b{=}true,\ c{:}v\rangle,\ d{:}w\rangle : \langle\!\langle a{:}\langle\!\langle b{:}Bool,\ c{:}C\rangle\!\rangle,\ d{:}D\rangle\!\rangle \end{array}$$

Here we have provided an argument type that is a subtype of $\langle\!\langle a{:}\langle\!\langle b{:}Bool\rangle\!\rangle\rangle\!\rangle$ in "all possible ways".

Finally, we should remark that the complexity of the update problem seems to manifests itself only in the functional case, while simpler solutions are available in the imperative case. Simpler type systems for records, such as the one in [Cardelli Wegner 85], may be adequate for imperative languages when properly extended with imperative constructs, as sketched below.

The imperative updating operator := has the additional constraint that the new record should have the same type as the old record, since intuitively updating is done "in place". This requirement produces something very similar to the typing we have initially shown to be unsound. Here assignable fields are identified by *var*:

let update(R<:⟪var b:Bool⟫)(r:R): R =
r.b := not(r.b)

Soundness is then recovered by requiring that assignable fields be both covariant and contravariant. Hence, *True* <: *Bool* does not imply ⟪*var b:True*⟫ <: ⟪*var b:Bool*⟫, thereby blocking the counterexamples to soundness.

Imperative update, with the natural requirement of not changing the type of a record, leads to simpler typing. However, this approach does not completely solve the problem we have discussed in this section. Imperative update alone does not provide the functionality of polymorphically extending existing records; when this is added, all the problems discussed above about functional update resurface.

3. Formal development

Now that we have acquired some intuitions, we can discuss the formal type inference rules in detail. We first define judgment forms and environment structures. Then we look at inference rules individually, and we analyze their properties. Finally, we provide a set-theoretical semantics for the pure calculus of records.

3.1 Judgments and inferences

A *judgment* is an inductively defined predicate between environments, value terms and type terms. The following judgments are used in formalizing our system:

$\vdash E\ env$	E is an environment
$E \vdash A\ type$	A is a type
$E \vdash A <: B$	A is a subtype of B
$E \vdash a : A$	a has type A
$E \vdash A \leftrightarrow B$	equivalent types
$E \vdash a \leftrightarrow b : A$	equivalent values of type A

The formal system is given by a set of *inference rules* below, each expressed as a finite set of *antecedent* judgments and side conditions (above a horizontal line) and a single *conclusion* judgment (below the line). Most inference rules are actually *rule schemas*, where meta-variables must be instantiated to obtain concrete inferences. For typographical reasons, we write the side conditions for these schemas as part of the antecedent.

3.2 Environments

An environment E is a finite sequence of (a) unconstrained type variables, (b) type variables constrained to be subtypes of a given type, and (c) value variables associated with their type.

We use $dom(E)$ for the set of type and value variables defined in an environment.

(ENV1) $\dfrac{}{\vdash \emptyset\ env}$ (ENV2) $\dfrac{X \notin dom(E)}{\vdash E, X\ env}$ (ENV3) $\dfrac{E \vdash A\ type \quad X \notin dom(E)}{\vdash E, X{<:}A\ env}$ (ENV4) $\dfrac{E \vdash A\ type \quad x \notin dom(E)}{\vdash E, x{:}A\ env}$

Hence, a legal environment is obtained by starting with the empty environment $\emptyset$ and extending it with a finite set of *assumptions* for type and value variables. Note that the assumptions involve distinct variables; we could perhaps allow multiple assumptions (e.g., $\emptyset, X{<:}A, X{<:}B$) but this would push us into the more general discipline of *conjunctive types*.

Assumptions about variables can then be extracted from well-formed environments:

(VAR1) $\dfrac{\vdash E,X,E'\ env}{E,X,E' \vdash X\ type}$ (VAR2) $\dfrac{\vdash E,X{<:}A,E'\ env}{E,X{<:}A,E' \vdash X\ type}$ (VAR3) $\dfrac{\vdash E,X{<:}A,E'\ env}{E,X{<:}A,E' \vdash X{<:}A}$ (VAR4) $\dfrac{\vdash E,x{:}A,E'\ env}{E,x{:}A,E' \vdash x{:}A}$

All legal inferences take place in (well-formed) environments. All judgments are recursively defined in terms of other judgments. For example, above we have used the typing judgment $E \vdash A\ type$ in constructing environments; vice versa, well-formed environments are involved in constructing types.

We now consider the remaining judgments in turn.

3.3 Record type formation

The following collection of rules determines when record types are well-formed. There is some interdependence between this section and the following ones, since equivalence rules have assumption that involve subtyping, which is discussed later. Fortunately, these assumptions are fairly simple, so a full understanding of the subtype relation is not required at this point.

(F1)
$$\frac{\vdash E\ env}{E \vdash \langle\!\langle\rangle\!\rangle\ type}$$

(F2)
$$\frac{E \vdash R <: \langle\!\langle\rangle\!\rangle \backslash x \quad E \vdash A\ type}{E \vdash \langle\!\langle R \mid x{:}A \rangle\!\rangle\ type}$$

(F3)
$$\frac{E \vdash R <: \langle\!\langle\rangle\!\rangle}{E \vdash R \backslash x\ type}$$

(F4)
$$\frac{E \vdash R <: \langle\!\langle S \mid x{:}A \rangle\!\rangle <: \langle\!\langle\rangle\!\rangle}{E \vdash R.x\ type}$$

As shown above, and already discussed informally, the legal record types are: the type of all records, $\langle\!\langle\rangle\!\rangle$; a record type variable X, (because of (VAR3) in the previous section); an extension $\langle\!\langle R \mid x{:}A \rangle\!\rangle$ of a record type R, provided R does not have x; and a restriction $R \backslash x$ of a record type R. Moreover, extracting a component $R.x$ of a record type R that has a label x, produces a legal type.

In general, if R does not have x, then R will be a subtype of the type $\langle\!\langle\rangle\!\rangle \backslash x$ of all records without x. This explains the hypothesis of rule (F2). In rule (F4) we use $R <: \langle\!\langle S \mid x{:}A \rangle\!\rangle$ to guarantee that every record in R has an x field.

3.4 Record type equivalence

When are two record types equivalent? We discuss here the formal rules for answering such a question. Type equivalence, as a relation, is reflexive (over well-formed expressions), symmetric, and transitive; it is denoted by the symbol $\leftrightarrow$. Substituting two equivalent types in a third type should produce an equivalent result; this is called the *congruence* property, and requires a number of rules to be fully formalized. We now consider, by cases, the equivalence of extended, restricted and extracted record types.

Two extended record types are equivalent if we can reorder their fields to make them identical (or, recursively, equivalent). This simple fact is expressed by the following rule. A number of applications of this rule, and of the congruence property, may be necessary to adequately reorder the fields of a record type.

(TE1)
$$\frac{E \vdash R <: \langle\!\langle\rangle\!\rangle \backslash xy \quad E \vdash A,B\ type \quad x \neq y}{E \vdash \langle\!\langle\langle\!\langle R \mid x{:}A \rangle\!\rangle \mid y{:}B \rangle\!\rangle \leftrightarrow \langle\!\langle\langle\!\langle R \mid y{:}B \rangle\!\rangle \mid x{:}A \rangle\!\rangle}$$

Similarly, we can reorder restrictions. Moreover, a double restriction $R \backslash xx$ reduces to $R \backslash x$. This fact is expressed in slightly more general form below, since the assumption that R does not have x is sufficient to deduce that $R \backslash x$ is the same as R:

(TE2)

$$\frac{E \vdash R <: \langle\!\langle\rangle\!\rangle \backslash x}{E \vdash R \backslash x \leftrightarrow R}$$

(TE3)

$$\frac{E \vdash R <: \langle\!\langle\rangle\!\rangle}{E \vdash R \backslash xy \leftrightarrow R \backslash yx}$$

The most interesting rules concern the distribution of restriction over extension. An outside restriction and inner extension of the same variable can cancel each other. Otherwise, a restriction can be pushed inside or outside of an extension of a different variable.

(TE5)

$$\frac{E \vdash R <: \langle\!\langle\rangle\!\rangle \backslash x \quad E \vdash A \ \mathit{type}}{E \vdash \langle\!\langle R \mid x{:}A \rangle\!\rangle \backslash x \leftrightarrow R}$$

(TE6)

$$\frac{E \vdash R <: \langle\!\langle\rangle\!\rangle \backslash x \quad E \vdash A \ \mathit{type} \quad x \neq y}{E \vdash \langle\!\langle R \mid x{:}A \rangle\!\rangle \backslash y \leftrightarrow \langle\!\langle R \backslash y \mid x{:}A \rangle\!\rangle}$$

Note however that in a situation like $\langle\!\langle R \backslash x \mid x{:}A \rangle\!\rangle$ no cancellation or swap can occur. The inner restriction may be needed to guarantee that the extension is sensible, and so neither is redundant.

Finally, a record extraction is equivalent to the extracted type:

(TE7)

$$\frac{E \vdash R <: \langle\!\langle\rangle\!\rangle \backslash x \quad E \vdash A \ \mathit{type}}{E \vdash \langle\!\langle R \mid x{:}A \rangle\!\rangle .x \leftrightarrow A}$$

(TE8)

$$\frac{E \vdash R <: \langle\!\langle S \mid y{:}B \rangle\!\rangle \backslash x <: \langle\!\langle\rangle\!\rangle \quad E \vdash A \ \mathit{type} \quad x \neq y}{E \vdash \langle\!\langle R \mid x{:}A \rangle\!\rangle .y \leftrightarrow R.y}$$

(TE4)

$$\frac{E \vdash R <: \langle\!\langle S \mid y{:}B \rangle\!\rangle <: \langle\!\langle\rangle\!\rangle \quad x \neq y}{E \vdash R \backslash x.y \leftrightarrow R.y}$$

These equivalence rules can be given a direction and interpreted as rewrite rules producing a normal form for record types; normalization is investigated in a later section.

3.5 Record subtyping

We have seen that subtyping is central to the notion of abstracting over record type variables, and we have intuitively justified some of the valid subtype assertions. In this section we take a more rigorous look at the subtype relation.

Subtyping should at least be a pre-order: a reflexive and transitive relation. Given a substitutive type equivalence relation $\leftrightarrow$, such as the one discussed in the previous section, we require:

(G1)

$$\frac{E \vdash A \leftrightarrow B}{E \vdash A <: B}$$

(G2)

$$\frac{E \vdash A <: B \quad E \vdash B <: C}{E \vdash A <: C}$$

Reflexivity is a special case of (G1).

It would be natural to require subtyping to be anti-symmetric, hence obtaining a partial order. A reasonable semantics of subtyping will in fact construct such a partial order. However, it might be too strong to require anti-symmetry as a type rule. In some systems anti-symmetry may introduce obscure ways of proving type equivalence, while in other systems it may be provable from the other rules. Moreover, anti-symmetry does not seem very useful for typechecking, hence we do not include it.

The basic intuition about subtyping is that it behaves much like the subset relation; this is expressed by the *subsumption* rule, which claims that if $A<:B$ and a is an element of A, then a is also an element of B.

(G3)

$$\frac{E \vdash a{:}A \quad E \vdash A <: B}{E \vdash a : B}$$

We feel strongly that subsumption should be included in the type system, since this rule gives object-oriented programming much of its flavor. One should not be satisfied, for programming purposes, with emulating subsumption by explicit coercions. The latter technique is interesting and adequate for providing semantics to a language with subsumption [Breazu-Tannen Coquand Gunter Scedrov 89] [Curien Ghelli 89], but even then it would seem more satisfactory to exhibit a model that satisfies subsumption directly.

Combining (G1) and (G3) we obtain another standard type rule:

$$\frac{E \vdash a{:}A \quad E \vdash A \leftrightarrow B}{E \vdash a : B}$$

This rule is normally taken as primitive, but here it is derived.

We are now ready to talk about subtyping between record types. It helps if we break this problem into pieces and ask what are the subtypes of: (1) the total record type $\langle\!\langle\rangle\!\rangle$, (2) an extended record type $\langle\!\langle R \mid x{:}A \rangle\!\rangle$, (3) a restricted record type $R \backslash x$, and (4) a record type extraction $R.x$.

Case (1). Every record type should be a subtype of the total record type. Hence, we have three subcases: (1a) the total record type is of course a subtype of itself, and this is simply a consequence of (G1); (1b) any well-formed extended record type is a subtype of $\langle\!\langle\rangle\!\rangle$; and (1c) any well-formed restricted record type is a subtype of $\langle\!\langle\rangle\!\rangle$. Hence we have the following rules corresponding to 1b and 1c respectively:

(S1)

$$\frac{E \vdash R <: \langle\!\langle\rangle\!\rangle \backslash x \quad E \vdash A \ type}{E \vdash \langle\!\langle R \mid x{:}A \rangle\!\rangle <: \langle\!\langle\rangle\!\rangle}$$

(S2)

$$\frac{E \vdash R <: \langle\!\langle\rangle\!\rangle}{E \vdash R \backslash x <: \langle\!\langle\rangle\!\rangle}$$

Case (2). A subtype of an extended record type will be another extended record type, provided all respective components are in the subtype relation:

(S3)

$$\frac{E \vdash R <: S <: \langle\!\langle\rangle\!\rangle \backslash x \quad E \vdash A <: B}{E \vdash \langle\!\langle R \mid x{:}A \rangle\!\rangle <: \langle\!\langle S \mid x{:}B \rangle\!\rangle}$$

The condition $A <: B$ says that we can produce a subtype by weakening the type of a given field. The condition $R <: S$ tells us that we can produce a subtype either (a) by weakening other fields inductively, because of (S3) itself, or (b) by requiring the presence of additional components, because of (S1), or (c) by requiring the absence of additional components, for example y, because from (S2) we are able to derive $\langle\!\langle\rangle\!\rangle \backslash yx <: \langle\!\langle\rangle\!\rangle \backslash x$.

Case (3). The subtype rule for restricted types is semantically straightforward: if every r in R occurs in S, then every $r \backslash x$ in $R \backslash x$ occurs in $S \backslash x$:

(S4)

$$\frac{E \vdash R <: S <: \langle\!\langle\rangle\!\rangle}{E \vdash R \backslash x <: S \backslash x}$$

Case (4). We have to consider the subtypes of record type extractions; that is situations of the form $R.x <: T.x$, or more generally $R.x <: A$ under an assumption $R <: \langle\!\langle S \mid x{:}B \rangle\!\rangle$. If R can be converted to the form $R = \langle\!\langle R' \mid x{:}A \rangle\!\rangle$, then the extraction $R.x$ simplifies and no special rule is required to deduce $R.x <: A$. But if R is a type variable, for example, the following rule is necessary:

(S5)

$$\frac{E \vdash R <: \langle\!\langle S \mid x{:}A \rangle\!\rangle <: \langle\!\langle\rangle\!\rangle}{E \vdash R.x <: A}$$

This says that if R has an x field of type A, then $R.x$ is a subtype of A (and possibly equal to A).

Finally, there is a another subtyping rule that we must consider. If every record r in R has an x field, then any such r is described also by the type $\langle\!\langle R\backslash x \mid x{:}R.x\rangle\!\rangle$, since $r\backslash x$ is described by $R\backslash x$ and the x field of r is described by $R.x$. Therefore we have the following inclusion:

(S6)
$$\frac{E \vdash R <: \langle\!\langle S \mid x{:}A\rangle\!\rangle <: \langle\!\langle\rangle\!\rangle}{E \vdash R <: \langle\!\langle R\backslash x \mid x{:}R.x\rangle\!\rangle}$$

The inverse inclusion is not necessarily valid, although it might seem natural to require it as we shall see later.

The rule (S6) can be used in the following derivation, which provides a "symmetrical" version of (S5) as a derived rule:

$$\begin{array}{ll} & E \vdash R <: S <: \langle\!\langle T \mid x{:}A\rangle\!\rangle <: \langle\!\langle\rangle\!\rangle \\ \hline \text{(S6)} & E \vdash S <: \langle\!\langle S\backslash x \mid x{:}S.x\rangle\!\rangle \\ \hline \text{(G2)} & E \vdash R <: \langle\!\langle S\backslash x \mid x{:}S.x\rangle\!\rangle \\ \hline \text{(S5)} & E \vdash R.x <: S.x \end{array}$$

In absence of (S6), the derived rule above would have to be taken as primitive, replacing (S5).

3.6 Record typing and equivalence

Now that we have seen the rules for type equivalence and subtyping, the rules for record values follow rather naturally. The only subtle point is about the empty record. We must be able to assign it a type which lacks any given set of labels. This is obtained by repeatedly applying the following two rules:

(I1)
$$\frac{\vdash E\ env}{E \vdash \langle\rangle : \langle\!\langle\rangle\!\rangle}$$

(I2)
$$\frac{E \vdash \langle\rangle : R <: \langle\!\langle\rangle\!\rangle}{E \vdash \langle\rangle : R\backslash x}$$

The remaining constructions on record values are typed by the corresponding constructions on record types, given the appropriate assumptions:

(I3)
$$\frac{E \vdash r{:}R <: \langle\!\langle\rangle\!\rangle\backslash x \quad E \vdash a{:}A}{E \vdash \langle r \mid x{=}a\rangle : \langle\!\langle R \mid x{:}A\rangle\!\rangle}$$

(E1)
$$\frac{E \vdash r{:}R <: \langle\!\langle\rangle\!\rangle}{E \vdash r\backslash x : R\backslash x}$$

(E2)
$$\frac{E \vdash r{:}\langle\!\langle R \mid x{:}A\rangle\!\rangle <: \langle\!\langle\rangle\!\rangle}{E \vdash r.x : A}$$

As we did in the previous section, we can use the rule (S6) to derive a "symmetrical" version of (I2):

$$\cfrac{\cfrac{\cfrac{E \vdash r{:}R{<:}⟪S \mid x{:}A⟫{<:}⟪⟫}{E \vdash R{<:}⟪R\backslash x \mid x{:}R.x⟫}\ \text{(S6)}}{E \vdash r{:}⟪R\backslash x \mid x{:}R.x⟫}\ \text{(G3)}}{E \vdash r.x : R.x}\ \text{(E2)}$$

Finally, we have to examine the rules for record value equivalence. These rules are formally very similar to the ones already discussed for record type equivalence; record extensions can be permuted, record components can be extracted, and restrictions can be permuted and pushed inside extensions, sometimes cancelling each other.

The main formal difference between these and the rules for types is that we equate $⟨⟩\backslash x \leftrightarrow ⟨⟩$. Hence, restriction can always be completely eliminated from variable-free records.

Because of the formal similarity we omit a detailed discussion; the complete set of rules for our type system follows in the next section.

3.7 Type rules

We can now summarize and complete the rules for record types and values, along with selected auxiliary rules. These rules are designed to be immersed in a second-order λ-calculus with bounded quantification (see [Cardelli Wegner 85]), and possibly with recursive values and types.

We only list the names of the rules that have already been discussed.

Environments

(ENV1)...(ENV4), (VAR1)...(VAR4)

General properties of <: and ↔

(G1)...(G3)

(G4)
$$\frac{E \vdash A \leftrightarrow B}{E \vdash B \leftrightarrow A}$$

(G5)
$$\frac{E \vdash A \leftrightarrow B \quad E \vdash B \leftrightarrow C}{E \vdash A \leftrightarrow C}$$

(G6)
$$\frac{E \vdash a \leftrightarrow b : A}{E \vdash b \leftrightarrow a : A}$$

(G7)
$$\frac{E \vdash a \leftrightarrow b : A \quad E \vdash b \leftrightarrow c : A}{E \vdash a \leftrightarrow c : A}$$

Formation

(F1)...(F4)

Subtyping

(S1)...(S6)

Introduction/Elimination

(I1)...(I3), (E1), (E2)

Type Congruence

(TC1)
$$\frac{\vdash E\ env}{E \vdash \langle\!\langle\rangle\!\rangle \leftrightarrow \langle\!\langle\rangle\!\rangle}$$

(TC2)
$$\frac{E \vdash X\ type}{E \vdash X \leftrightarrow X}$$

(TC3)
$$\frac{E \vdash R \leftrightarrow S <: \langle\!\langle\rangle\!\rangle\backslash x \quad E \vdash A \leftrightarrow B}{E \vdash \langle\!\langle R \mid x{:}A\rangle\!\rangle \leftrightarrow \langle\!\langle S \mid x{:}B\rangle\!\rangle}$$

(TC4)
$$\frac{E \vdash R \leftrightarrow S <: \langle\!\langle\rangle\!\rangle}{E \vdash R\backslash x \leftrightarrow S\backslash x}$$

(TC5)
$$\frac{E \vdash R \leftrightarrow S <: \langle\!\langle T \mid x{:}A\rangle\!\rangle <: \langle\!\langle\rangle\!\rangle}{E \vdash R.x \leftrightarrow S.x}$$

Type Equivalence

(TE1)...(TE8)

Value Congruence

(VC1a)
$$\frac{\vdash E\ env}{E \vdash \langle\rangle \leftrightarrow \langle\rangle : \langle\!\langle\rangle\!\rangle}$$

(VC2)
$$\frac{E \vdash x : A}{E \vdash x \leftrightarrow x : A}$$

(VC3)
$$\frac{E \vdash r \leftrightarrow s : R <: \langle\!\langle\rangle\!\rangle\backslash x \quad E \vdash a \leftrightarrow b : A}{E \vdash \langle r \mid x{=}a\rangle \leftrightarrow \langle s \mid x{=}b\rangle : \langle\!\langle R \mid x{:}A\rangle\!\rangle}$$

(VC4)
$$\frac{E \vdash r \leftrightarrow s : R <: \langle\!\langle\rangle\!\rangle}{E \vdash r\backslash x \leftrightarrow s\backslash x : R\backslash x}$$

(VC5)
$$\frac{E \vdash r \leftrightarrow s : R <: \langle\!\langle S \mid x{:}A\rangle\!\rangle <: \langle\!\langle\rangle\!\rangle}{E \vdash r.x \leftrightarrow s.x : R.x}$$

Value Equivalence

(VE1)
$$\frac{E \vdash r{:}R <: \langle\!\langle\rangle\!\rangle\backslash xy \quad E \vdash a{:}A \quad E \vdash b{:}B \quad x{\neq}y}{E \vdash \langle\langle r \mid x{=}a\rangle \mid y{=}b\rangle \leftrightarrow \langle\langle r \mid y{=}b\rangle \mid x{=}a\rangle : \langle\!\langle\langle\!\langle R \mid x{:}A\rangle\!\rangle \mid y{:}B\rangle\!\rangle}$$

(VE2)
$$\frac{\vdash E\ env}{E \vdash \langle\rangle\backslash x \leftrightarrow \langle\rangle : \langle\!\langle\rangle\!\rangle}$$

(VE3)
$$\frac{E \vdash r{:}R <: \langle\!\langle\rangle\!\rangle\backslash x}{E \vdash r\backslash x \leftrightarrow r : R}$$

(VE4)
$$\frac{E \vdash r{:}R <: \langle\!\langle\rangle\!\rangle}{E \vdash r\backslash xy \leftrightarrow r\backslash yx : R\backslash xy}$$

(VE5)
$$\frac{E \vdash r{:}\langle\!\langle R \mid x{:}A\rangle\!\rangle <: \langle\!\langle\rangle\!\rangle \quad x{\neq}y}{E \vdash r\backslash y.x \leftrightarrow r.x : A}$$

(VE6)

$$\frac{E \vdash r{:}R{<:}\langle\!\langle\rangle\!\rangle\backslash x \quad E \vdash a{:}A}{E \vdash \langle r \mid x{=}a\rangle\backslash x \leftrightarrow r : R}$$

(VE7)

$$\frac{E \vdash r{:}R{<:}\langle\!\langle\rangle\!\rangle\backslash x \quad E \vdash a{:}A \quad x{\neq}y}{E \vdash \langle r \mid x{=}a\rangle\backslash y \leftrightarrow \langle r\backslash y \mid x{=}a\rangle : \langle\!\langle R \mid x{:}A\rangle\!\rangle\backslash y}$$

(VE8)

$$\frac{E \vdash r{:}R{<:}\langle\!\langle\rangle\!\rangle\backslash x \quad E \vdash a{:}A}{E \vdash \langle r \mid x{=}a\rangle.x \leftrightarrow a : A}$$

(VE9)

$$\frac{E \vdash r{:}\langle\!\langle R \mid y{:}B\rangle\!\rangle\backslash x{<:}\langle\!\langle\rangle\!\rangle \quad E \vdash a{:}A \quad x{\neq}y}{E \vdash \langle r \mid x{=}a\rangle.y \leftrightarrow r.y : B}$$

(VE10)

$$\frac{E \vdash r{:}R{<:}\langle\!\langle S \mid x{:}A\rangle\!\rangle{<:}\langle\!\langle\rangle\!\rangle}{E \vdash r \leftrightarrow \langle r\backslash x \mid x{=}r.x\rangle : R}$$

Special rules

In the following sections we discuss the rules (VC1b) and (TE9) below; these are valid only with respect to particular semantic interpretations.

(VC1b)

$$\frac{E \vdash r{:}\langle\!\langle\rangle\!\rangle \quad E \vdash s{:}\langle\!\langle\rangle\!\rangle}{E \vdash r \leftrightarrow s : \langle\!\langle\rangle\!\rangle}$$

(TE9)

$$\frac{E \vdash R{<:}\langle\!\langle S \mid x{:}A\rangle\!\rangle{<:}\langle\!\langle\rangle\!\rangle}{E \vdash R \leftrightarrow \langle\!\langle R\backslash x \mid x{:}R.x\rangle\!\rangle}$$

In presence of (TE9), the rule (S6) is redundant, and the rules (TC5) and (VC5) are implied by the simpler (TC5b) and (VC5b) below.

(TC5b)

$$\frac{E \vdash R \leftrightarrow \langle\!\langle S \mid x{:}A\rangle\!\rangle{<:}\langle\!\langle\rangle\!\rangle}{E \vdash R.x \leftrightarrow A}$$

(VC5b)

$$\frac{E \vdash r \leftrightarrow s : \langle\!\langle R \mid x{:}A\rangle\!\rangle{<:}\langle\!\langle\rangle\!\rangle}{E \vdash r.x \leftrightarrow s.x : A}$$

Properties

Lemma 3.7.1:
(1) If $E \vdash A\ type$, then $\vdash E\ env$.
(2) If $E \vdash A <: B$, then $\vdash E\ env$.

Lemma 3.7.2:
(1) If $E \vdash A \leftrightarrow B$, then $E \vdash A\ type$ and $E \vdash B\ type$.
(2) If $E \vdash A <: B$, then $E \vdash A\ type$ and $E \vdash B\ type$.

3.8 Semantics of the pure calculus of records

Our stated intent is to define a second-order type system for record structures. However, models of such a system are rather complex, and outside the scope of this paper.

In this section we provide a simple set-theoretical model of the pure calculus of records, without any additional functional or polymorphic structure. The intent here is to show the plausibility of the inference rules for records, by proving their soundness with respect to a natural model.

This model is natural because it embodies the strong set-theoretical intuitions of subtyping seen as a subset relation, and of records seen as finite tuples. Although this model does not extend to more complex language features, it exhibits the kind of simple-minded but (usually) sound reasoning that guides the design and implementation of object-oriented languages.

Syntax

We start with the language implied by the type rules of section 3.7. Since no basic non-record values are expressible in this calculus, we must make some arbitrary choices to get started. To this end, we will consider an extension of the pure calculus with any collection $G_1, G_2, \ldots$ of basic (ground) type symbols and an arbitrary collection of subtype relations $G_i <: G_j$ between them. To incorporate these new symbols into the calculus, we add the following two rules (which preserve lemmas 3.7.1 and 3.7.2):

$$\frac{\vdash E\ env}{E \vdash G_i\ type} \qquad \frac{\vdash E\ env}{E \vdash G_i <: G_j} \quad \text{(as appropriate)}$$

For simplicity we do not introduce value constants, instead we work with environments containing assumptions of the form $k : G_i$.

We will now construct a model of the extended calculus.

Semantic domains

In the following, we rely largely on context to distinguish between syntactic expressions and semantic expressions, and we often identify terms with their denotations.

We start by choosing some fixed set of labels L, and a collection of sets $\mathcal{G}_1$, $\mathcal{G}_2, \ldots$ corresponding to the type symbols $G_1, G_2, \ldots$ such that $\mathcal{G}_i \subseteq \mathcal{G}_j$ if $G_i <: G_j$ is a subtyping axiom.

For simplicity, we assume that no element of any $\mathcal{G}_i$ is a finite partial function on L (i.e. a record, as we shall see shortly). This assumption is useful when we define the subtype relations of section 3.9.

Since $⟪⟫$ serves as a type of all records, we will need some value space closed under record formation. This property may be accomplished by regarding records as finite functions from L to values, and using *ranked* values with rank $< \omega$. We use $A \rightarrow_{\text{fin}} B$ for the set of partial functions from A to B with finite domain, $f(x)\uparrow$ to indicate that the partial function f is undefined at x, and $f(x)\downarrow$ to indicate that f is defined at x.

Define set $\mathcal{R}_i$ of records of rank i, and set $\mathcal{V}_i$ of values of rank i, as follows:

$$\mathcal{V}_0 = \cup_j \mathcal{G}_j \qquad \mathcal{V}_{i+1} = \mathcal{R}_i \cup \mathcal{V}_i$$
$$\mathcal{R}_0 = L \rightharpoonup_{\text{fin}} \mathcal{V}_0 \qquad \mathcal{R}_{i+1} = L \rightharpoonup_{\text{fin}} \mathcal{V}_{i+1}$$

$\mathcal{R} = \cup_{i<\omega} \mathcal{R}_i$ the set of *records*

$\mathcal{V} = \cup_{i<\omega} \mathcal{V}_i$ the set of *values*

The essential properties of this construction are summarized by the relationship:

$$\mathcal{R} = (L \rightharpoonup_{\text{fin}} \mathcal{V}) \subseteq \mathcal{V}$$

It is clear by construction that $\mathcal{R}_i \subseteq \mathcal{V}_{i+1}$ and so $\mathcal{R} \subseteq \mathcal{V}$. To see that $\mathcal{R} = L \rightharpoonup_{\text{fin}} \mathcal{V}$, we first show that $L \rightharpoonup_{\text{fin}} \mathcal{V} \subseteq \mathcal{R}$. If $r \in L \rightharpoonup_{\text{fin}} \mathcal{V}$, then since $dom(r)$ is finite there is some i with $range(r) \subseteq \mathcal{V}_i$; hence $r \in \mathcal{R}_i \subseteq \mathcal{R}$. The converse follows from the fact that if $r \in \mathcal{R}$, then $r \in \mathcal{R}_i = (L \rightharpoonup_{\text{fin}} \mathcal{V}_i) \subseteq L \rightharpoonup_{\text{fin}} \mathcal{V}$.

We now summarize the notation used to describe the semantic interpretation of syntactic constants and operators:

$\varnothing \quad = \quad \lambda y \in L. \uparrow$

$r\text{-}x \quad =_{\text{def}} \quad \lambda y \in L.$ *if* $y=x$ *then* $\uparrow$ *else* $r(y)$
provided $r \in \mathcal{R}$ and $x \in L$

$r[x=a] \quad =_{\text{def}} \quad \lambda y \in L.$ *if* $y=x$ *then* a *else* $r(y)$
provided $r \in \mathcal{R}$, $x \in L$, $a \in \mathcal{V}$, and $x \notin dom(r)$.

$r(x)$ is well-defined,
provided $r \in \mathcal{R}$, $x \in L$, and $x \in dom(r)$.

Lemma 3.8.1:

(1) The empty record $\varnothing$ is an element of $\mathcal{R}$.
(2) For any $r \in \mathcal{R}$ we have $r\text{-}x \in \mathcal{R}$.
(3) If $r \in \mathcal{R}$ is not defined on x, then for any $a \in \mathcal{V}$ we have $r[x=a] \in \mathcal{R}$.
(4) If $r \in \mathcal{R}$ is defined on x, then $r(x) \in \mathcal{V}$.

Types and type operations

Types are interpreted as subsets of our global value set; hence we have a type of all values, and a type of all records. Subtyping is interpreted as set inclusion.

We introduce the following notation for operations on record types:

$R\text{-}x$	$=_{def}$	$\{r\text{-}x \mid r \in R\}$ if $R \subseteq \mathcal{R}$
$R[x{:}A]$	$=_{def}$	$\{r[x{=}a] \mid r \in R, a \in A\}$ if $R \subseteq \mathcal{R}\text{-}x$ (R undefined on x) and $A \subseteq \mathcal{V}$
$R(x)$	$=_{def}$	$\{r(x) \mid r \in R\}$ if $R \subseteq S[x{:}A]$ for some $S \subseteq \mathcal{R}$ and $A \subseteq \mathcal{V}$

Lemma 3.8.2:

Under the conditions stated above, the sets $R\text{-}x$ and $R[x{:}A]$ are subsets of $\mathcal{R}$, and the sets $R(x)$ are subsets of $\mathcal{V}$.

Interpretation of judgments

An *assignment* ρ is a partial map from type variables to subsets of $\mathcal{V}$, and from ordinary variables to elements of $\mathcal{V}$. We say that an assignment ρ *satisfies* an environment E if the following conditions are satisfied:

If X in E,	then $\rho(X) \subseteq \mathcal{V}$
If $X <: A$ in E,	then $\rho(X) \subseteq A_\rho \subseteq \mathcal{V}$
If $x : A$ in E,	then $\rho(x) \in A_\rho \subseteq \mathcal{V}$

where A_ρ is the type defined by A under the assignment ρ. Similarly, by a_ρ we indicate the value of a term a under an assignment ρ for its free variables.

The judgments of our system are interpreted as follows.

$\vdash E\ env$	≈	for every initial segment $E',X{<:}A$ or $E',x{:}A$ of E, if ρ satisfies E' then $A_\rho \subseteq \mathcal{V}$.
$E \vdash A\ type$	≈	$A_\rho \subseteq \mathcal{V}$, for every ρ satisfying E.
$E \vdash A <: B$	≈	$A_\rho \subseteq B_\rho \subseteq \mathcal{V}$, for every ρ satisfying E.
$E \vdash A \leftrightarrow B$	≈	$A_\rho = B_\rho \subseteq \mathcal{V}$, for every ρ satisfying E.
$E \vdash a : A$	≈	$a_\rho \in A_\rho \subseteq \mathcal{V}$, for every ρ satisfying E.
$E \vdash a \leftrightarrow b : A$	≈	$a_\rho = b_\rho \in A_\rho \subseteq \mathcal{V}$, for every ρ satisfying E.

Type and value expressions are interpreted using:

⟪⟫	≈	$\mathcal{R}$
$R \backslash x$	≈	$R\text{-}x$
⟪$R \mid x{:}A$⟫	≈	$R[x{:}A]$
$R.x$	≈	$R(x)$

$$
\begin{array}{lcl}
\langle\rangle & \approx & \varnothing \\
r\backslash x & \approx & r\text{-}x \\
\langle r \mid x{=}a\rangle & \approx & r[x{=}a] \\
r.x & \approx & r(x)
\end{array}
$$

Soundness

Finally, we can show that this semantics satisfies the type rules. More precisely, we consider the system *S1* consisting of all the rules listed in section 3.7, except for the special rules (VC1b) and (TE9).

Theorem 3.8.3 (soundness):
The inference rules of system *S1* are sound with respect to the interpretation of judgments given in this section.

3.9 Other semantic constructions

The type equivalence rule below seems very natural semantically. It also simplifies the types associated with the override operation, and has application to extensional models studied in the next section.

(TE9)

$$
\frac{E \vdash R <: \langle\!\langle S \mid x{:}A \rangle\!\rangle <: \langle\!\langle\rangle\!\rangle}{E \vdash R \leftrightarrow \langle\!\langle R\backslash x \mid x{:}R.x \rangle\!\rangle}
$$

In the simple model described in section 3.8, it is easy to see that if $R \subseteq \langle\!\langle x{:}A \rangle\!\rangle$, then, as required by (S6):

$$R \subsetneq \langle\!\langle R\backslash x \mid x{:}R.x \rangle\!\rangle$$

The reason is that every record r in R has an x component $r(x) \in R(x)$, and remaining components $r\text{-}x$ in $R\text{-}x$. However, it is not necessarily true that every combination of $r\text{-}x$ from $R\text{-}x$ and $r(x)$ from $R(x)$ occur together in a single record in R. For example, the set of records:

$$R = \{\langle x{=}1, y{=}true\rangle, \langle x{=}0, y{=}false\rangle\}$$

is clearly a subset of $\langle\!\langle x{:}Int \rangle\!\rangle$. However, $R \neq \langle\!\langle R\backslash x \mid x{:}R.x \rangle\!\rangle$ since the records $\langle x{=}1, y{=}false\rangle$ and $\langle x{=}0, y{=}true\rangle$ do not appear in R. In category-theoretic terms, the equation $R = \langle\!\langle R\backslash x \mid x{:}R.x \rangle\!\rangle$ says that R is the product of $R\backslash x$ and $R.x$.

In the full paper we present a variant of the construction of section 3.8 in which rule (TE9) is sound. Since we are ultimately interested in polymorphism and bounded quantification, we construct a model with $R = \langle\!\langle R\backslash x \mid x{:}R.x \rangle\!\rangle$ for every semantic type R with $R.x$ defined. The construction uses the same collection of values as before, but allows only certain subsets of $\mathcal{V}$ as types. In this way we eliminate sets of records which violate (TE9).

Another construction arises from the following inference rule, which gives us an extensional equality between records:

(VC1b)

$$\frac{E \vdash r{:}\langle\!\langle\rangle\!\rangle \quad E \vdash s{:}\langle\!\langle\rangle\!\rangle}{E \vdash r \leftrightarrow s : \langle\!\langle\rangle\!\rangle}$$

The intuitive explanation of this rule is that if r and s both belong to $\langle\!\langle\rangle\!\rangle$, then r and s are indistinguishable. In fact, assume r and s differ at some label x. We cannot use $r.x$ or $s.x$ to distinguish them since neither is well-typed; if we use $r\backslash x$ or $s\backslash x$ then we simply remove the difference.

In addition to giving us more equations between records of type $\langle\!\langle\rangle\!\rangle$, rule (VC1b) implies the following extensionality property: for any $r,s : \langle\!\langle x_1{:}A_1 , \ldots , x_k{:}A_k \rangle\!\rangle$, we have $r \leftrightarrow s : \langle\!\langle x_1{:}A_1 , \ldots , x_k{:}A_k \rangle\!\rangle$ iff $r.x_i {\leftrightarrow} s.x_i : A_i$ for $i = 1 \ldots k$. The straightforward proof of this uses $r\backslash x_1 \ldots x_k \leftrightarrow s\backslash x_1 \ldots x_k : \langle\!\langle\rangle\!\rangle$ and the value congruence rules.

In the full paper, we construct a model of the pure record calculus satisfying (TE9) and (VC1b). In this construction, (TE9) is essential; we do not know how to construct an extensional model satisfying (VC1b) without requiring that record types satisfy $R = \langle\!\langle R\backslash x \mid x{:}R.x \rangle\!\rangle$. The main use of (TE9) lies in showing that if R is a record type with extensional equality, then both $R\text{-}x$ and $R(x)$, when defined, are extensional record types.

3.10 Normalization and decidability

Even though the basic ideas behind the record calculus are relatively simple, the formal system has quite a few rules. As a consequence, it is not easy to see, by inspection, how we could determine whether a supposed type A is well-formed, or whether a record expression has type R.

In this section, we outline a proof that all of the basic properties of the calculus are decidable, using relatively natural algorithms. In the process, we show that every type expression has a unique normal form (modulo permuting the order of fields) and every typable record expression has a *principal type* in each suitable environment.

The first properties we consider are deciding whether a supposed environment E is well-formed and whether a given A is a well-formed type expression in E. A quick glance at the formation rules shows that in order to determine whether a type is well-formed we must be able to decide the following apparently simple properties; assuming $E \vdash R$ *type* is derivable, we want to know whether $E \vdash R <: \langle\!\langle\rangle\!\rangle \backslash x$ and whether there exist S and A such that $E \vdash R <: \langle\!\langle S \mid x{:}A \rangle\!\rangle$. Therefore, we consider these first. Once we develop a simple method for these, it is easy to check whether a type or environment is well-formed.

For each derivable $E \vdash R$ *type*, we define a labeled tree *Tree*($E \vdash R$ *type*) with:

edges: labeled by field names
vertices: labeled by finite sets of field names

If v is a vertex in *Tree*($E \vdash R$ *type*), we call the finite set of field names at v the *absent set at* v.

Intuitively, if $p = x_1 x_2 \ldots x_k$ is a path from the root of *Tree*($E \vdash R$ *type*) and $N = \{y_1, y_2, \ldots, y_l\}$ is the absent set of the vertex designated by this path, then:

$$E \vdash (..(R.x_1).x_2 \ldots).x_k \ \textit{type}$$
$$E \vdash (..(R.x_1).x_2 \ldots).x_k <: \langle\!\langle\rangle\!\rangle \backslash y_1 y_2 \ldots y_l$$

A convenient notational shorthand is to write $R.p$ for $(..(R.x_1).x_2 \ldots).x_k$, where p is the path $p = x_1 x_2 \ldots x_k$. If $p = \varepsilon$ is the empty path, then we may write $R.\varepsilon$ for R. If e is an edge leading from the root of a tree to the root of some subtree, we call e a *root edge*.

The inductive definition of *Tree*($E \vdash R$ *type*) is given in the full paper.

Lemma 3.10.1:

Suppose $E \vdash R$ *type* and let $T =$ *Tree*($E \vdash R$ *type*).
(1) If p is a path in T, then $E \vdash R.p$ *type*.
(2) If x is in the absent set of T at position p, then $E \vdash R.p <: \langle\!\langle\rangle\!\rangle \backslash x$.

Lemma 3.10.2:

Suppose $E \vdash R$ *type* and let $T =$ *Tree*($E \vdash R$ *type*).
There is a semantic model $\mathcal{M}$ and assignment ρ such that:
(1) If p is a sequence of labels which is not a path in T, then there is some record r in R_ρ with $r.p$ undefined.
(2) If p is a path in T with x absent from every record in $(R.p)_\rho$, then x is in the absent set of T at the vertex located at p.

By constructing trees of absent sets, it is relatively easy to decide whether a purported environment or type expression is well-formed. The basic idea is simply to check whether $\vdash E$ *env* or $E \vdash R$ *type* by reading the environment and formation rules backwards. This gives us mutually recursive procedures which rely on *Tree*($E \vdash R$ *type*) in checking the hypotheses of (F2) and (F4).

Theorem 3.10.3:

Given environment E and expression A, there are mutually recursive procedures which decide whether $\vdash E$ *env* and $E \vdash A$ *type.*

The next problems to consider are, given well-formed types $E \vdash A$ *type* and $E \vdash B$ *type,* whether $E \vdash A \leftrightarrow B$ or $E \vdash A <: B$. Since type equality may be used to prove subtyping assertions, both depend on our choice of type equality rules. For definiteness, let us assume we have (TE9). Similar results seem to hold without (TE9), but we have not checked the details.

Theorem 3.10.4:

Given $E \vdash A$ *type* and $E \vdash B$ *type,* there are straightforward algorithms to determine whether $E \vdash A \leftrightarrow B$ or $E \vdash A <: B$. Moreover, the proof rules are semantically complete for deducing type equality and subtype assertions.

The final algorithmic problem is, given $E \vdash R$ *type* and an expression r, determine whether $E \vdash r{:}R$.

Since we can decide whether one type is a subtype of another, it suffices to compute a minimal type S with $E \vdash r{:}S$ and check whether $E \vdash S <: R$.

However, most record expressions do not have a minimal type. This stems from the fact that for any sequence $x_1 \ldots x_k$ of labels, we have $\langle\rangle : \langle\!\langle\rangle\!\rangle \backslash x_1 \ldots x_k$, and we can always obtain a smaller type by adding more labels. To get around this problem, we use *type schemas* that contain sequence variables. We show that each typable record expression r has a scheme S such that every type for r is a supertype of some instance of S. This allows us to test whether a record expression has any given type.

Theorem 3.10.5:

There is an algorithm PTS(E,r) such that, given $\vdash E$ *env* and an expression r, if $E \vdash r{:}R$ then $PTS(E,r)$ succeeds, producing S with $E \vdash S' <: R$ for some instance S' of S. Otherwise, $PTS(E,r)$ fails. Furthermore, given $S = PTS(E,r)$ and $E \vdash R$ *type,* it is easy to compute the smallest instance S' of S such that if any instance is a subtype of R, then $E \vdash S' <: R$.

This concludes our investigation of decidability properties. We leave extensions of these properties to functions and polymorphism for further work.

4. Conclusions

We have investigated a theory of record operations in presence of type variables and subtyping. The intent is to embed this record calculus in a polymorphic λ-calculus, thus providing a full second-order theory of record structures and their types. Although we have not investigated the type inference problem for this calculus, we have provided typechecking and subtyping algorithms. We have also presented several models of the basic record calculus; a full second-order model is left for future work.

The result is a very flexible system for typing programs that manipulate records. In particular, polymorphism and subtyping are incorporated in full generality. We expect that this theory will be useful in analyzing fundamental aspects of object-oriented programming.

Acknowledgements

We would like to acknowledge G. Longo and E. Moggi, for several clarifying discussions.

References

[Breazu-Tannen Coquand Gunter Scedrov 89] V.Breazu-Tannen, T.Coquand, C.Gunter, A.Scedrov: ***Inheritance and explicit coercion***, Proc. of the Fourth Annual Symposium on Logic in Computer Science, 1989.

[Bruce Longo 88] K.B.Bruce, G.Longo: ***Modest models for inheritance and explicit polymorphism***, Proc. of the Third Annual Symposium on Logic in Computer Science, 1988.

[Bruce Meyer Mitchell 89] K.B.Bruce, A.R.Meyer, J,C.Mitchell: ***The semantics of second order lambda calculus***, Information and Computation, 1989 (to appear).

[Cardelli Donahue Glassman Jordan Kalsow Nelson 88] L.Cardelli, J.Donahue, L.Glassman, M.Jordan, B.Kalsow, G.Nelson: ***Modula-3 report***, Research Report n.31, DEC Systems Research Center, Sep. 1988.

[Cardelli 84&88] L.Cardelli: ***A semantics of multiple inheritance***, in Information and Computation 76, pp 138-164, 1988. (First appeared in Semantics of Data Types, G.Kahn, D.B.MacQueen and G.Plotkin Ed. Lecture Notes in Computer Science n.173, Springer-Verlag 1984.)

[Cardelli Wegner 85] L.Cardelli, P.Wegner: ***On understanding types, data abstraction and polymorphism***, Computing Surveys, Vol 17 n. 4, pp 471-522, December 1985.

[Curien Ghelli 89] P.-L.Curien, G.Ghelli: ***Coherence of subsumption***, to appear.

[Dahl Nygaard 66] O.Dahl, K.Nygaard: ***Simula, an Algol-based simulation language***, Communications of the ACM, Vol 9, pp. 671-678, 1966.

[Girard 71] J-Y.Girard: ***Une extension de l'interprétation de Gödel à l'analyse, et son application à l'élimination des coupures dans l'analyse et la théorie des types***, Proceedings of the second Scandinavian logic symposium, J.E.Fenstad Ed. pp. 63-92, North-Holland, 1971.

[Girard 72] J-Y.Girard: ***Interprétation fonctionelle et élimination des coupures dans l'arithmétique d'ordre supérieur,*** Thèse de doctorat d'état, University of Paris, 1972.

[Jategaonkar Mitchell 88] L.A.Jategaonkar, J.C.Mitchell: ***ML with extended pattern matching and subtypes,*** Proc. of the ACM Conference on Lisp and Functional Programming, pp.198-211, 1988.

[Longo Moggi 88] G.Longo, E.Moggi: ***Constructive natural deduction and its 'ω-set' interpretation,*** Report CMU-CS-88-131, CMU, Dept. of Computer Science, 1988.

[Meyer 88] B.Meyer: ***Object-oriented software construction,*** Prentice Hall, 1988.

[Milner 78] R.Milner: ***A theory of type polymorphism in programming,*** Journal of Computer and System Science 17, pp. 348-375, 1978.

[Mitchell 84] J.C.Mitchell: ***Coercion and type inference,*** Proc. of the 11th ACM Symposium on Principles of Programming Languages, pp.175-185, 1984.

[Mitchell 86] J.C.Mitchell: ***A type inference approach to reduction properties and semantics of polymorphic expressions,*** Proc. Symposium on Lisp and Functional Programming, pp.308-319, 1986. (Revised version to appear in Logic Foundations of Functional Programming, ed. G. Huet, Addison-Wesley, 1989.)

[Mitchell 90] J.C.Mitchell: ***Type systems for programming languages,*** in Handbook of Theoretical Computer Science, ed. J. van Leeuwen et al. North Holland, 1990 (to appear).

[Ohori Buneman Breazu-Tannen 88] A.Ohori, P.Buneman, V.Breazu-Tannen: ***Database programming in Machiavelli - a polymorphic languaage with static type inference,*** Report MS-CIS-88-103, University of Pennsylvania, Computer and Information Science Dept., 1988.

[Ohori Buneman 88] A.Ohori, P.Buneman: ***Type inference in a database programming language,*** Proc. of the ACM Conference on LISP and Functional Programming, pp.174-183, Snowbird, Utah, 1988.

[Rémy 89] D. Rémy: ***Typechecking records and variants in a natural extension of ML,*** Proc. of the 16th ACM Symposium on Principles of Programming Languages, pp.77-88, 1989.

[Reynolds 74] J.C.Reynolds: ***Towards a theory of type structure,*** in Colloquium sur la programmation pp. 408-423, Springer-Verlag Lecture Notes in Computer Science, n.19, 1974.

[Schaffert Cooper Bullis Kilian Wilpolt 86] C.Schaffert, T.Cooper, B.Bullis, M.Kilian, C.Wilpolt: ***An introduction to Trellis/Owl,*** Proc. OOPSLA'86.

[Stroustrup 86] B.Stroustrup: ***The C++ programming language,*** Addison-Wesley 1986.

[Wand 87] M.Wand: ***Complete Type Inference for Simple Objects,*** Proc. of the Second Annual Symposium on Logic in Computer Science, June 1987, Cornell University.

[Wand 89] M.Wand: ***Type inference for record concatenation and multiple inheritance,*** Proc. of the Fourth Annual Symposium on Logic in Computer Science, pp. 92-97, 1989.

Connections between a Concrete and an Abstract Model of Concurrent Systems

Eugene W. Stark*

Department of Computer Science

State University of New York at Stony Brook

Stony Brook, NY 11794 USA

Abstract

We define a concrete operational model of concurrent systems, called *trace automata*. For such automata, there is a natural notion of *permutation equivalence* of computation sequences, which holds between two computation sequences precisely when they represent two interleaved views of the "same concurrent computation." Alternatively, permutation equivalence can be characterized in terms of a *residual operation* on transitions of the automaton, and many interesting properties of concurrent computations can be expressed with the help of this operation. In particular, concurrent computations, ordered by "prefix," form a Scott domain whose structure we characterize up to isomorphism.

By axiomatizing the properties of the residual operation, we obtain a more abstract formulation of automata, which we call *concurrent transition systems* (CTS's). By exploiting a correspondence between concurrent alphabets and certain CTS's, we are able to use the rich algebraic structure of CTS's to obtain results in trace theory. Finally, we connect CTS's and trace automata by obtaining a characterization of those CTS's that correspond in a natural way to trace automata, and we show how the correspondence suggests an interesting notion of morphism of trace automata.

1 Introduction

Labeled transition systems (LTS's) have been used frequently as an operational semantics of concurrent processes. In typical formulations, an LTS is a tuple $A = (E, Q, T, *)$, where E is a set of *events*, Q is a set of *states*, $* \in Q$ is a distinguished *start state*, and $T \subseteq Q \times E \times Q$ is a set of *transitions*, which represent potential computation steps. Although useful for many applications, LTS's are not ideally suited as a model of concurrency, since they contain no mathematical

*Research supported in part by NSF Grant CCR-8702247.

structure with which concurrency can be represented and reasoned about directly. Instead, concurrency in computations must be represented somewhat artificially by interleaving, and reasoning about concurrency requires that we make use of auxiliary information (*e.g.* which pairs of transitions "commute") not explicitly formalized in the LTS model.

In an effort to get explicit concurrency information into the definition of a transition system, we might introduce a symmetric, irreflexive *concurrency relation* $\|$ on E, and require that the transitions respect this concurrency information in a suitable sense. Although there is some flexibility in the exact sense in which concurrency is to be respected by the transitions, the end result is essentially the class of *trace automata* which we define below. This kind of automaton, which arises naturally in the study of trace theory [1,11], has been the subject of investigation by Bednarczyk [2], Kwiatkowska [8], and Shields [13], and has been used by the author to study nondeterministic dataflow networks [12,17]. The familiar mapping that takes a finite computation sequence of an automaton to the string it generates is now replaced by a monotone mapping from the prefix-ordered set of finite computation sequences to the prefix-ordered set of *traces*, which are equivalence classes of strings modulo a congruence relation induced by the concurrency relation.

We can actually go a bit further than this. If we regard computation sequences having the same trace as "equivalent interleaved views" of a single concurrent computation, and we factor the poset of computation sequences by this equivalence, then the trace mapping becomes an isomorphism between the resulting poset of "finite concurrent computations" and the "trace language" generated by the automaton, where the latter is viewed as a subset of the prefix-ordered set of all traces. Ideal completion of the poset of finite concurrent computations results in a Scott domain containing both finite and infinite concurrent computations. The domain of concurrent computations is much more interesting than the poset of finite computation sequences, since concurrency is reflected in the former through the existence of nontrivial upper bounds. Since our goal is to make concurrency explicit, one might argue that concurrent computations, rather than computation sequences, ought to be the main focus of attention.

One of our main results is the following characterization of the structure of the domains of concurrent computations of trace automata:

> The domain of concurrent computations of a trace automaton is isomorphic to a normal subdomain U of the domain $\bar{E}$ of traces generated by the event set E and concurrency relation $\|$, where the inclusion of U in $\bar{E}$ preserves prime intervals. Conversely, if U is a normal subdomain of a domain $\bar{E}$ of traces, such that the inclusion of U in $\bar{E}$ preserves prime intervals, then U is isomorphic to the domain of concurrent computations of a trace automaton.

The proof of this characterization theorem uses in an essential way the observation that equivalence of computations can be described, independently of trace theory, using the concept of a "residual operation" on computation sequences. Intuitively, taking the residual of a computation sequence γ "after" a computation sequence δ corresponds to "cancelling from γ the greatest common prefix, up to concurrency, of γ and δ." The computation sequences γ and δ are equivalent precisely when each is completely cancelled by the other. Residuals have previously been used by Lévy [9] in the study of the λ-calculus, to define the notion of strongly equivalent reductions. The same ideas can also be applied [3,4,7] to the study of recursive programs and left-linear term-rewriting systems without critical pairs. In that work, residuals are used to keep track of what happens to one redex in a term while other redexes are contracted. Our use here is analogous: the residual operation allows us to keep track of what happens to one enabled transition in a system while other concurrent transitions are executed. Boudol and Castellani [5] have also exploited the use of residuals and permutation equivalence in reasoning about concurrency.

By axiomatizing the properties of a residual operation necessary to obtain the equivalence relation on computation sequences, we arrive at the definition of *concurrent transition systems* (CTS's) [15,16]. The defining axioms generate a rich algebraic theory, which we have found to be of use in the study of concurrent systems [12,15,17]. A suitable definition of morphism makes the class of all CTS's into a category **CTS**, which has small limits, small coproducts, small filtered colimits, and is cartesian closed. Moreover, many interesting constructions on automata have universal or couniversal characterizations either in **CTS** or in functor categories built from it. Included among these constructions are those that extract the computational behavior of a CTS.

In this paper, we give the details of the story outlined above. Our goal is to motivate explicitly the connection, between the concrete, easily understood trace automaton model, and the more abstract concurrent transition systems which the author has described elsewhere [15,16]. To complete this connection between abstract and concrete, we exhibit properties that characterize up to isomorphism those CTS's that are derived from "event automata."

2 Trace Automata

In this paper, all sets whose cardinality is left unspecified are assumed to be at most countable.

A *concurrent alphabet* is a set E, equipped with a symmetric, irreflexive binary relation $\|_E$, called the *concurrency relation.*

A *trace automaton* (henceforth simply "automaton") is a tuple $A = (E, Q, T)$, where

- E is a concurrent alphabet, whose elements are called *events*. We assume that E does not contain the special symbol ϵ, called the *identity event.*

- Q is a set of *states.*

- $T \subseteq Q \times (E \cup \{\epsilon\}) \times Q$ is a set of *transitions*. We usually write $t : q \xrightarrow{a} r$, or just $q \xrightarrow{a} r$, to denote a transition $t = (q, a, r)$ in T.

These data are required to satisfy the following conditions:

(Identity) $q \xrightarrow{\epsilon} r$ iff $q = r$.

(Disambiguation) If $q \xrightarrow{a} r$ and $q \xrightarrow{a} r'$, then $r = r'$.

(Commutativity) For all states q and events a, b, if $a \|_E b$, $q \xrightarrow{a} r$, and $q \xrightarrow{b} s$, then for some state p there exist transitions $s \xrightarrow{a} p$ and $r \xrightarrow{b} p$.

A trace automaton *with start state* is a tuple $(E, Q, T, *)$, where (E, Q, T) is a trace automaton, and $* \in Q$ is a distinguished state.

Intuitively, if $a \in E$, then a transition $q \xrightarrow{a} r$ represents a potential computation step of A in which event a occurs and the state changes from q to r. *Identity transitions* $\mathrm{id}_q = (q \xrightarrow{\epsilon} q)$ do not represent steps of A. Rather, these transitions play a purely technical role, which will become evident when we define the notion of a "residual operation" below. We say that event $a \in E$ is *enabled* in state q if there exists a transition $q \xrightarrow{a} r$. By the disambiguation condition, if $q \xrightarrow{a} r$, then r is uniquely determined by q and a. If $t : q \xrightarrow{a} r$, then q is called the *domain* $\mathrm{dom}(t)$ of t and r is called the *codomain* $\mathrm{cod}(t)$ of t. Transitions t and u are called *coinitial* if $\mathrm{dom}(t) = \mathrm{dom}(u)$.

A *finite computation sequence* for an automaton is a finite sequence γ of non-identity transitions of the form:

$$q_0 \xrightarrow{a_1} q_1 \xrightarrow{a_2} \cdots \xrightarrow{a_n} q_n.$$

The number n is called the *length* $|\gamma|$ of γ. By convention, we regard an identity transition id_q. as identical to the computation sequence of length zero from state q. An *infinite computation sequence* is an infinite sequence of non-identity transitions:

$$q_0 \xrightarrow{a_1} q_1 \xrightarrow{a_2} \cdots.$$

We extend notation and terminology for transitions to computation sequences, so that if γ is a computation sequence, then the *domain* $\mathrm{dom}(\gamma)$ of γ is the state q_0, and if γ is finite, then the *codomain* $\mathrm{cod}(\gamma)$ of γ is the state q_n. We write $\gamma : q \to r$ to assert that γ is a finite computation sequence with domain q and codomain r. A computation sequence γ is *initial* if $\mathrm{dom}(\gamma)$ is the distinguished start state $*$. If $\gamma : q \to r$ and $\delta : q' \to r'$ are finite computation sequences, then γ and δ are called *composable* if $q' = r$, and we define their *composition* to be the finite computation sequence $\gamma\delta : q \to r'$, obtained by concatenating γ and δ and identifying $\mathrm{cod}(\gamma)$ with $\mathrm{dom}(\delta)$. The operation of composition of finite computation sequences is associative, and identity transitions (computation sequences of length 0) behave as units for it. A finite computation sequence γ is a *prefix* of a computation sequence δ, and we write $\gamma \leq \delta$, iff there exists a computation sequence ξ with $\gamma\xi = \delta$.

Trace automata (with start state) are nearly identical to the "forward stable asynchronous systems" of Bednarczyk [2]; the difference being that we retain his forward stability axiom (our commutativity axiom), but we omit his axiom stating that if $q \xrightarrow{a} r$, $r \xrightarrow{b} s$, and $a \|_E b$, then for some state p there exist transitions $q \xrightarrow{b} p$ and $p \xrightarrow{a} s$. Although Bednarczyk seems to treat this axiom as more fundamental than forward stability, it seems overly restrictive, and we shall see that much can be done without it.

Trace automata can be used as the basis for an operational model of nondeterministic dataflow networks, which consist of a collection of concurrently and asynchronously executing processes that communicate by passing "value tokens" over named "ports." This is done by introducing additional structure to distinguish between "input events," "output events," and "internal events," and then requiring that input and output events have a specific form that reflects the port structure. We give below some of the definitions to illustrate how this is done. The reader wishing further discussion is referred to [10,12,15,17].

An *input/output* automaton is a triple (A, X, Y), where $A = (E, Q, T, *)$ is a trace automaton with start state, and X, Y are disjoint subsets of E, called the sets of *input events* and *output events*, respectively. Elements of $E \setminus (X \cup Y)$ are called *internal events*. The following property is required to hold:

(Receptivity) For all states q and input events a, event a is enabled in state q.

A *port automaton* is an input/output automaton (A, X, Y) equipped with a set V of *values*, a set I of *input ports*, and a set O of *output ports*, such that $X = I \times V$, $Y = O \times V$, and such that whenever (p, v) and (p', v') are events in $X \cup Y$, then $(p, v) \|_E (p', v')$ iff $p = p'$.

Two particularly well-behaved classes of input/output automata are the "monotone" automata and the "determinate" automata. An input/output automaton is *monotone* if the following additional property holds:

(Monotonicity) $a \|_E b$ whenever $a \in X$ and $b \in E \setminus X$.

Intuitively, the monotonicity property states that the arrival of input cannot disable any enabled output or internal transitions, since if b is an output event enabled in state q, and if a is an arbitrary input event, then $a \|_E b$ by monotonicity, $q \xrightarrow{a} r$ for some r by receptivity, hence b is enabled in state r by commutativity. An automaton is *determinate* if it satisfies the following condition:

(Determinacy) Suppose $q \xrightarrow{b} r$ and $q \xrightarrow{c} s$, where b and c are distinct non-input events. Then $b \|_E c$.

It can be shown [15,16] that determinate input/output automata have functional input/output behavior.

2.1 Concurrent Computations

A partially ordered set $(D, \sqsubseteq)$ is *consistently complete* if each pair of elements of D that have an upper bound, have a least upper bound. A (Scott) *domain* is an ω-algebraic, consistently complete CPO $D = (D, \sqsubseteq, \perp)$. A domain D is *finitary* if for all finite (=isolated=compact) elements $d \in D$ the set $\{d' \in D : d' \sqsubseteq d\}$ is finite. An *atom* of D is a minimal non-$\perp$ element of D. If D and E are domains, then a monotone map $f : D \to E$ is *continuous* if it preserves suprema of ω-chains, *strict* if $f(\perp_D) = \perp_E$, and *additive* if whenever d, d' are consistent elements of D, then $f(d), f(d')$ are consistent elements of E, and $f(d \sqcup d') = f(d) \sqcup f(d')$. The map f *reflects consistency* if whenever $f(d)$ and $f(d')$ are consistent elements of E, then d and d' are consistent elements of D.

The set of all finite and infinite initial computation sequences of a trace automaton with start state $A = (E, Q, T, *)$, forms a domain when equipped with the prefix ordering $\leq$. This domain is of limited utility, since it does not take into account any concurrency information. However, there is a natural notion of "permutation equivalence" of computations of A, which captures the idea of "equivalent interleaved views" of the same concurrent computation. By factoring the domain of initial computation sequences by permutation equivalence we obtain a more useful and interesting domain of "concurrent computations."

Formally, define permutation equivalence to be the least congruence $\sim$, respecting concatenation, on the set of finite computation sequences of A such that:

- Computation sequences $q \xrightarrow{a} r \xrightarrow{b} p$ and $q \xrightarrow{b} s \xrightarrow{a} p$ are $\sim$-related if $a \|_E b$.

Closely related to permutation equivalence is the *permutation preorder* relation $\precsim$ on finite computation sequences of A, which is defined to be the transitive closure of $(\leq \cup \sim)$. It is not difficult to see that $\gamma \sim \delta$ iff $\gamma \precsim \delta$ and $\delta \precsim \gamma$.

Permutation preorder extends in a straightforward way to infinite computation sequences as well: if γ' and δ' are coinitial finite or infinite computation sequences, then define $\gamma' \precsim \delta'$ to hold iff for every finite $\gamma \leq \gamma'$ there exists a finite $\delta \leq \delta'$, such that $\gamma \precsim \delta$. We may then extend permutation equivalence to infinite computation sequences by defining $\gamma' \sim \delta'$ iff $\gamma' \precsim \delta'$ and $\delta' \precsim \gamma'$.

2.1.1 Permutation Preorder and Traces

Because the concurrency in a trace automaton is completely determined by the concurrency relation on events, we can describe permutation preorder as the preorder induced by a certain mapping from computation sequences to a domain of "traces." To state this formally, we need some basic definitions from trace theory [1,2,11].

Suppose E is a concurrent alphabet. Let E^* denote the free monoid generated by E, then there is a least congruence $\sim_E$ on E^* such that $a \|_E b$ implies $ab \sim_E ba$ for all $a, b \in E$. The quotient $E^* / \sim_E$ is the *free partially commutative monoid* generated by E, and its elements are called *traces*. We use ϵ to denote the monoid identity, and if $x \in E^*$, then we use $[x]$ to denote the corresponding element of $E^* / \sim_E$.

Define the relation $\sqsubseteq$ on the monoid $E^* /{\sim_E}$ by: $[x] \sqsubseteq [y]$ iff $\exists[z]([x][z] = [y])$. It is not difficult to show that $\sqsubseteq$ is a partial order, with ϵ as a least element. Let $\bar{E}$ denote the ideal completion of this partial order, then $\bar{E}$ is an algebraic CPO whose finite elements are the principal ideals generated by elements of $E^* /{\sim_E}$. We call $\bar{E}$ the *domain of traces generated by* the concurrent alphabet E. (This terminology is justified by Lemma 2.3, which shows that $\bar{E}$ is consistently complete, hence a domain. For the moment, we only need the fact that $\bar{E}$ is an algebraic CPO.) Notice that since the finite elements of $\bar{E}$ are in bijective correspondence with the elements of $E^* /{\sim_E}$, they inherit the monoid operation of $E^* /{\sim_E}$, with the least element of $\bar{E}$ as the monoid identity. In the sequel, we identify elements of $E^* /{\sim}$ with the corresponding finite elements of $\bar{E}$.

Now, suppose

$$\gamma = q_0 \xrightarrow{a_1} q_1 \xrightarrow{a_2} \dots$$

is a finite or infinite computation sequence of an automaton $A = (E, Q, T)$. Define the *trace* of γ to be the element $\text{tr}(\gamma) = \bigsqcup_{k \geq 0} [a_1 a_2 \dots a_k]$ of the domain of traces $\bar{E}$. Obvious consequences of this definition are: (1) $\text{tr}(\text{id}_q) = \epsilon$, (2) if γ and δ are composable finite computation sequences, then $\text{tr}(\gamma\delta) = \text{tr}(\gamma)\text{tr}(\delta)$, and (3) the map tr is continuous, with respect to the prefix ordering $\leq$ on computation sequences and the ordering $\sqsubseteq$ on traces.

Theorem 1 *Suppose* γ *and* δ *are coinitial computation sequences. Then* $\gamma \precsim \delta$ *holds iff* $\text{tr}(\gamma) \sqsubseteq \text{tr}(\delta)$.

Proof – We first prove the result for the special case that γ and δ are finite. Suppose $\gamma \precsim \delta$. Then there exists a finite sequence

$$\gamma = \xi_0, \xi_1, \dots, \xi_n = \delta,$$

such that for each k with $0 \leq k < n$, one of the following two relationships holds:

1. $\xi_k \leq \xi_{k+1}$.

2. ξ_{k+1} is obtained from ξ_k by replacing a subsequence of the form $q \xrightarrow{a} r \xrightarrow{b} p$, where $a \|_E b$, by a sequence $q \xrightarrow{b} s \xrightarrow{a} p$.

In case (1), $\text{tr}(\xi_k) \sqsubseteq \text{tr}(\xi_{k+1})$, and in case (2), $\text{tr}(\xi_k) = \text{tr}(\xi_{k+1})$. Hence $\text{tr}(\gamma) \sqsubseteq \text{tr}(\delta)$ by reflexivity and transitivity of $\sqsubseteq$.

Conversely, suppose $\text{tr}(\gamma) \sqsubseteq \text{tr}(\delta)$. Let x and y be the sequences of events appearing in γ and δ, respectively. Then $[x] = \text{tr}(\gamma) \sqsubseteq \text{tr}(\delta) = [y]$, so by definition of $\sqsubseteq$ there exists z such that $xz \sim y$. It follows that the string y can be transformed into the string xz by a finite sequence of steps in which adjacent pairs of concurrent symbols are permuted. By performing the same sequence of permutation steps starting with δ, we obtain a proof that $\delta \sim_E \gamma\xi$ for some computation sequence ξ.

We now extend the result to include infinite computation sequences. Suppose $\gamma' \precsim \delta'$, where γ' and δ' are arbitrary. Given any finite $[x] \sqsubseteq \text{tr}(\gamma')$, by the continuity of the map tr with respect to the prefix ordering $\leq$, there exists a finite $\gamma \leq \gamma'$ such that $[x] \sqsubseteq \mathrm{tr}(\gamma)$. Choose a finite $\delta \leq \delta'$ such that $\gamma \precsim \delta$, then $\mathrm{tr}(\gamma) \sqsubseteq \mathrm{tr}(\delta)$ by the finite case of the theorem. Thus, for each finite $[x] \sqsubseteq \mathrm{tr}(\gamma')$ there exists a finite $[y] \sqsubseteq \mathrm{tr}(\delta')$ such that $[x] \sqsubseteq [y]$. By the fact that the CPO $\bar{E}$ is algebraic, it follows that $\mathrm{tr}(\gamma') \sqsubseteq \mathrm{tr}(\delta')$.

Conversely, suppose $\mathrm{tr}(\gamma') \sqsubseteq \mathrm{tr}(\delta')$. Given a finite $\gamma \leq \gamma'$, let $[x] = \mathrm{tr}(\gamma)$. Then $[x] \sqsubseteq \mathrm{tr}(\gamma')$ so by algebraicity of $\bar{E}$ we may choose a finite $[y] \sqsubseteq \mathrm{tr}(\delta')$ such that $[x] \sqsubseteq [y]$. By continuity of tr with respect to the prefix ordering $\leq$ on computation sequences, there exists a finite $\delta \leq \delta'$ such that $[y] \sqsubseteq \mathrm{tr}(\delta)$. But then $\gamma \precsim \delta$ by the finite case of the theorem. ∎

2.1.2 Permutation Preorder and Residuals

The permutation preorder can also be characterized in a much different, and ultimately more useful way, using the notion of the "residual" of one finite computation sequence "after" another. Residuals, previously used for the λ-calculus and term-rewriting systems [3,4,7,9], have been shown in [12,15,16] to be extremely useful in reasoning about concurrent systems. Here, we formalize the notion of residual as a partial binary operation $\uparrow$ on finite computation sequences, such that $\gamma \uparrow \delta$ (read γ "after" δ) is defined, and γ and δ are said to be *consistent*, exactly when γ and δ are coinitial finite computation sequences that could both be part of the "same concurrent computation." In general, δ will then contain some transitions that "overlap" with γ and some transitions that are "concurrent" with γ, and the residual $\gamma \uparrow \delta$ is defined to be what is left of γ after the part of δ that overlaps with it has been "cancelled." The residual $\gamma \uparrow \delta$ is undefined when γ contains some nondeterministic choice that is incompatible with a choice made in δ. In this case we say that γ and δ *conflict*. Observe that we distinguish between two types of choice that may be represented in an automaton: *concurrent* choice, in which events a, b with $a \| b$ are both enabled in the same state q, and *nondeterministic* choice, in which a and b are both enabled in state q but we do not have $a \| b$.

We first define the residual operation for coinitial computation sequences $\gamma : q \to r$ and $\delta : q \to s$ of length ≤ 1. There are three cases:

1. If $\gamma = \mathrm{id}_q$, then $\gamma \uparrow \delta = \mathrm{id}_s$ and $\delta \uparrow \gamma = \delta$.

2. If γ is a non-identity transition $q \xrightarrow{a} r$, and δ is a non-identity transition $q \xrightarrow{a} s$, then $r = s$ by the disambiguation condition. Define $\gamma \uparrow \delta = \mathrm{id}_s = \mathrm{id}_r = \delta \uparrow \gamma$.

3. If γ is a non-identity transition $q \xrightarrow{a} r$, and δ is a non-identity transition $q \xrightarrow{b} s$, where $a \neq b$, then $\gamma \uparrow \delta$ and $\delta \uparrow \gamma$ are defined iff $a \|_E b$. In this case, the commutativity property implies there must exist transitions $s \xrightarrow{a} p$ and $r \xrightarrow{b} p$, which we take as $\gamma \uparrow \delta$ and $\delta \uparrow \gamma$, respectively.

Lemma 2.1 *The operation $\uparrow$ has the following properties, where γ, δ, and ξ denote computation sequences of length ≤ 1.*

1. *If* $\gamma \uparrow \delta$ *is defined, then so is* $\delta \uparrow \gamma$, *and we have* $\mathrm{dom}(\gamma \uparrow \delta) = \mathrm{cod}(\delta)$, $\mathrm{dom}(\delta \uparrow \gamma) = \mathrm{cod}(\gamma)$, *and* $\mathrm{cod}(\gamma \uparrow \delta) = \mathrm{cod}(\delta \uparrow \gamma)$.
2. *For all* $\gamma : q \to r$, (a) $\mathrm{id}_q \uparrow \gamma = \mathrm{id}_r$; (b) $\gamma \uparrow \mathrm{id}_q = \gamma$; *and* (c) $\gamma \uparrow \gamma = \mathrm{id}_r$.
3. *For all coinitial* γ, δ, ξ, $(\xi \uparrow \gamma) \uparrow (\delta \uparrow \gamma) = (\xi \uparrow \delta) \uparrow (\gamma \uparrow \delta)$, *whenever either side is defined.*
4. *For all coinitial* $\gamma : q \to r$ *and* $\delta : q \to s$, *if* $\gamma \uparrow \delta = \mathrm{id}_s$ *and* $\delta \uparrow \gamma = \mathrm{id}_r$, *then* $\gamma = \delta$.

Moreover, $\uparrow$ *extends uniquely to an operation on finite computation sequences, in such a way that properties (1)-(3) are preserved and such that the following additional identities hold whenever either side is defined:*

5. $$\begin{aligned} \gamma \uparrow \delta\xi &= (\gamma \uparrow \delta) \uparrow \xi \\ \delta\xi \uparrow \gamma &= (\delta \uparrow \gamma)(\xi \uparrow (\gamma \uparrow \delta)). \end{aligned}$$

Proof – (1), (2), and (4) are obvious from the definitions. (3) is proved by a straightforward case analysis on γ, δ, and ξ (see [12]).

The extension of $\uparrow$ to finite computation sequences is done using (5) as a recursive definition. Verification that the resulting extension has properties (1), (2), and (3) is then done by induction on the lengths of the computation sequences involved. Property (4) does not necessarily hold for the extension; for example, because if $\gamma = q \xrightarrow{a} r \xrightarrow{b} p$ and $\delta = q \xrightarrow{b} s \xrightarrow{a} p$ with $a \|_E b$, then $\gamma \uparrow \delta = \mathrm{id}_p = \delta \uparrow \gamma$, but $\gamma \neq \delta$. ∎

We now derive the connection between residuals and permutation preorder. If γ and δ are coinitial, then define $\gamma \precsim \delta$ iff $\gamma \uparrow \delta$ is an identity.

Lemma 2.2 *The relation* $\precsim$ *is a preorder.*

Proof – Reflexivity holds because $\gamma \uparrow \gamma = \mathrm{id}_q$, where $q = \mathrm{cod}(\gamma)$. To show transitivity, suppose $\gamma \precsim \delta$ and $\delta \precsim \xi$. Then $\gamma \uparrow \delta$ is an identity, so $(\gamma \uparrow \delta) \uparrow (\xi \uparrow \delta)$ is an identity. Since $(\gamma \uparrow \delta) \uparrow (\xi \uparrow \delta) = (\gamma \uparrow \xi) \uparrow (\delta \uparrow \xi)$, it follows that $(\gamma \uparrow \xi) \uparrow (\delta \uparrow \xi)$ is an identity. But $\delta \uparrow \xi$ is an identity because $\delta \precsim \xi$, hence $\gamma \uparrow \xi$ is an identity. ∎

Theorem 2 *Suppose* γ *and* δ *are coinitial finite computation sequences. Then* $\gamma \sqsubseteq \delta$ *iff* $\gamma \precsim \delta$.

Proof – Suppose $\gamma \sqsubseteq \delta$. Then there exists a finite sequence

$$\gamma = \xi_0, \xi_1, \ldots, \xi_n = \delta,$$

such that for each k with $0 \leq k < n$, one of the following two relationships holds:

1. $\xi_k \leq \xi_{k+1}$.

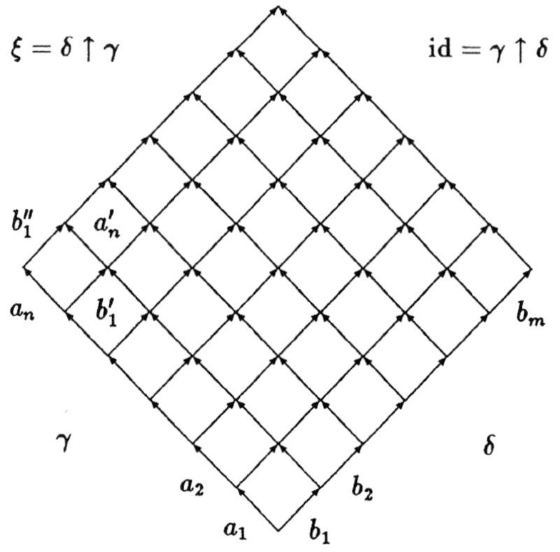

Figure 1: Residuals and Permutation Preorder

2. ξ_{k+1} is obtained from ξ_k by replacing a subsequence of the form $q \xrightarrow{a} r \xrightarrow{b} p$, where $a \|_E b$, by a sequence $q \xrightarrow{b} s \xrightarrow{a} p$.

In case (1), $\xi_{k+1} = \xi_k \zeta$ for some ζ, so by Lemma 2.1, $\xi_k \uparrow \xi_{k+1} = (\xi_k \uparrow \xi_k) \uparrow \zeta$, which is an identity, hence $\xi_k \precsim \xi_{k+1}$. In case (2), $\xi_k = \zeta t u' \zeta'$ and $\xi_{k+1} = \zeta u t' \zeta'$, where t is the transition $q \xrightarrow{a} r$, u is the transition $q \xrightarrow{b} s$, $t' = t \uparrow u$, and $u' = u \uparrow t$. From this, a straightforward calculation using Lemma 2.1 shows that $\xi_k \uparrow \xi_{k+1}$ is an identity, so $\xi_k \precsim \xi_{k+1}$. Since $\xi_k \precsim \xi_{k+1}$ both in case (1) and in case (2), the result $\gamma \precsim \delta$ follows by the reflexivity and transitivity of $\precsim$.

Conversely, suppose $\gamma \precsim \delta$. Let $\xi = \delta \uparrow \gamma$, then γ, δ, and ξ are related as depicted in Figure 1, where the sides of each small diamond are single transitions, and the apex of each small diamond is obtained by applying the residual operation to the transitions forming its base. From this diagram, we may read off a proof that $\gamma\xi \sim \delta$. This is done by starting with the computation sequence $\gamma\xi$ represented by the path around the left-hand side of the large diamond, and, beginning with the leftmost small diamond in the diagram, successively replacing the pairs of transitions forming the left-hand sides of the small diamonds by the pairs of transitions forming the right-hand sides. For example, in the first step we would replace $a_n b_1''$ by $b_1' a_n'$. Eventually, these replacement steps transform the sequence $\gamma\xi$ into the computation sequence represented by the path around the right-hand side of the large diamond; that is, into δ. ∎

Having characterized permutation preorder in terms of residuals, we may now obtain a great deal of information about the structure of the set of computation sequences, under the permutation preorder. The main result is Theorem 4 below. I do not know how to prove such a strong result without using residuals.

Lemma 2.3 *Suppose γ and δ are coinitial finite computation sequences. Then γ and δ have an upper bound with respect to $\precsim$ iff $\gamma \uparrow \delta$ is defined. Moreover, if $\gamma \uparrow \delta$ is defined, then γ and δ have a supremum with respect to $\precsim$, given by $\gamma(\delta \uparrow \gamma)$ or $\delta(\gamma \uparrow \delta)$, which are permutation equivalent.*

Proof – If γ and δ have an upper bound ξ with respect to $\precsim$, then $\gamma \uparrow \xi$ and $(\gamma \uparrow \xi) \uparrow (\delta \uparrow \xi)$ are identities. But $(\gamma \uparrow \xi) \uparrow (\delta \uparrow \xi) = (\gamma \uparrow \delta) \uparrow (\xi \uparrow \delta)$, hence $\gamma \uparrow \delta$ is defined.

Conversely, if $\gamma \uparrow \delta$ is defined, then since $\gamma \uparrow \gamma(\delta \uparrow \gamma) = (\gamma \uparrow \gamma) \uparrow (\delta \uparrow \gamma)$ and $\delta \uparrow \gamma(\delta \uparrow \gamma) = (\delta \uparrow \gamma) \uparrow (\delta \uparrow \gamma)$, both of which are identities, it is clear that $\gamma(\delta \uparrow \gamma)$ is a $\precsim$-upper bound of γ and δ. Suppose ξ is any $\precsim$-upper bound of γ and δ. Then

$$\gamma(\delta \uparrow \gamma) \uparrow \xi = (\gamma \uparrow \xi)((\delta \uparrow \gamma) \uparrow (\xi \uparrow \gamma)) = (\gamma \uparrow \xi)((\delta \uparrow \xi) \uparrow (\gamma \uparrow \xi)),$$

which is an identity, so $\gamma(\delta \uparrow \gamma) \precsim \xi$. ∎

Theorem 3 *Suppose $A = (E, Q, T)$ is a trace automaton, and $q \in Q$. Then the set of permutation equivalence classes of computation sequences from state q, equipped with the partial order induced by $\precsim$, is a domain, whose finite elements are exactly the equivalence classes of finite computation sequences.*

Proof – The set of finite computation sequences from state q is countable, and from Lemma 2.3 we know that whenever γ and δ have an upper bound with respect to $\precsim$, then they have a supremum with respect to $\precsim$. Then the ideal completion $\mathcal{I}$ of the preorder $\precsim$ is a domain whose finite elements are exactly the principal ideals. Let h be the map that takes each $\precsim$-equivalence class $[\gamma]$ to the set $h([\gamma]) = \{\delta : \delta \text{ finite}, \delta \precsim \gamma\}$. The set $h([\gamma])$ is obviously nonempty and downward-closed. It is also directed, because if $\delta \in h([\gamma])$ and $\delta' \in h([\gamma])$, then δ and δ' have a $\precsim$-supremum, which must also be in $h([\gamma])$. Thus, $h([\gamma])$ is an ideal of $\precsim$.

We claim that the map h is an order-isomorphism, from the partially ordered set of permutation equivalence classes of computation sequences from state q, to $\mathcal{I}$. Since h takes each equivalence class $[\gamma]$ with γ finite to the principal ideal generated by γ, we will then have the desired result. Obviously h satisfies $h([\gamma]) \subseteq h([\delta])$ iff $\gamma \precsim \delta$. Note that h is injective, because if $[\gamma] \neq [\delta]$ then either γ has a finite prefix γ' such that $\gamma' \notin h([\delta])$, or else δ has a finite prefix δ' such that $\delta' \notin h([\gamma])$. To complete the proof, we must show that h is also surjective; that is, every $\precsim$-ideal Γ of the set of finite computation sequences is $h([\gamma])$ for some computation sequence γ.

Suppose $\Gamma \in \mathcal{I}$. We first observe that Γ is at most countable (because the set of all finite computation sequences is countable), hence has an enumeration (perhaps with repetition) $\delta_0, \delta_1, \ldots$. Next, we inductively construct a sequence $\gamma_0 \leq \gamma_1 \leq \ldots$ of elements of Γ, forming a chain under the prefix ordering, such that $\delta_k \precsim \gamma_{k+1}$ for all $k \geq 0$. For the basis step, let $\gamma_0 = \mathrm{id}_q$, which is in Γ because Γ is an ideal. For the induction step, suppose $\gamma_k \in \Gamma$ has been defined for some $k \geq 0$. Since $\delta_k, \gamma_k \in \Gamma$, and Γ is directed, it follows by Lemma 2.3 that δ_k and γ_k are consistent. Define $\gamma_{k+1} = \gamma_k(\delta_k \uparrow \gamma_k)$. Clearly, γ_k is a prefix of γ_{k+1}. Since γ_{k+1} is a $\precsim$-supremum of $\{\gamma_k, \delta_k\} \subseteq \Gamma$, and since the ideal Γ is closed under suprema of finite subsets, it follows that $\gamma_{k+1} \in \Gamma$. Also, $\delta_k \precsim \gamma_{k+1}$, since $\delta_k \uparrow \gamma_{k+1} = (\delta_k \uparrow \gamma_k) \uparrow (\delta_k \uparrow \gamma_k) = \mathrm{id}$.

Let γ be the supremum of the chain $\gamma_0 \leq \gamma_1 \leq \ldots$ with respect to the prefix ordering. We claim that $h([\gamma]) = \Gamma$. Clearly, if $\delta \in \Gamma$, then $\delta = \delta_k$ for some $k \geq 0$, hence $\delta \precsim \gamma_{k+1}$. This shows $\Gamma \subseteq h([\gamma])$. Conversely, if δ is a finite computation sequence with $\delta \precsim \gamma$, then $\delta \precsim \xi$ for some finite prefix ξ of γ. But this means $\delta \precsim \gamma_k$ for some $k \geq 0$, hence $h([\gamma]) \subseteq \Gamma$ because $\gamma_k \in \Gamma$ and Γ is an ideal. ∎

To proceed further, we need some additional information about the structure of the domain of traces $\bar{E}$ generated by a concurrent alphabet E. Define a *trace domain* to be a structure $(D, \sqsubseteq, \perp, \cdot)$, where $(D, \sqsubseteq, \perp)$ is a finitary domain with least element $\perp$, and if D° denotes the set of finite elements of D, then $(D^\circ, \cdot, \perp)$ is a monoid. In addition, the following are required to hold:

1. For all $x, y \in D^\circ$, we have $x \sqsubseteq y$ iff there exists $z \in D^\circ$ with $xz = y$.

2. For all $x, y, z \in D^\circ$, if $xy = xz$ then $y = z$.

3. For all distinct atoms $a, b \in D$, we have a, b consistent iff $ab = ba$, and then $a \sqcup b = ab = ba$.

On any trace domain D, we may define a partial binary operation $\setminus$ on D°, such that $x \setminus y$ is defined iff x and y are consistent, and if so, then $x \setminus y$ is the unique z such that $yz = x \sqcup y$. Obviously the domain of definition of $\setminus$ is symmetric, and $x = y$ iff $x \setminus y = \epsilon = y \setminus x$.

Lemma 2.4 *Suppose $(D, \sqsubseteq, \perp, \cdot)$ is a trace domain. Then the following identities hold for finite x, y, z, whenever either side is defined:*

1. $x(y \sqcup z) = xy \sqcup xz$.
2. $z \setminus xy = (z \setminus x) \setminus y$.
3. $xy \setminus z = (x \setminus z)(y \setminus (z \setminus x))$.

Proof – Omitted. ∎

We will prove the following lemma in Section 3.5 below.

Lemma 2.5 *Suppose E is a concurrent alphabet. Then $(\bar{E}, \sqsubseteq, \epsilon, \cdot)$ is a trace domain. Conversely, a structure $(D, \sqsubseteq, \perp, \cdot)$ is a trace domain iff it is isomorphic to $(\bar{E}, \sqsubseteq, \epsilon, \cdot)$ for some concurrent alphabet E.*

Lemma 2.6 *Suppose γ and δ are coinitial finite computation sequences of a trace automaton. Then γ and δ are consistent iff* $\mathrm{tr}(\gamma)$ *and* $\mathrm{tr}(\delta)$ *are consistent, in which case* $\mathrm{tr}(\gamma \uparrow \delta) = \mathrm{tr}(\gamma) \setminus \mathrm{tr}(\delta)$.

Proof – The proof is by induction on the pair of lengths $(|\gamma|, |\delta|)$. For the special cases: $|\gamma| = 0$ or $|\delta| = 0$, the result is trivial. For the basis case of $|\gamma| = |\delta| = 1$ we have that γ is a single transition $t : q \xrightarrow{a} r$ and δ is a single transition $u : q \xrightarrow{b} s$, so $\mathrm{tr}(\gamma) = [a]$ and $\mathrm{tr}(\delta) = [b]$. Then γ and δ are consistent iff one of the following cases occurs:

1. $a = b$.

2. $a \neq b$ and $a \|_E b$.

By Lemma 2.5, traces $[a]$ and $[b]$ are consistent iff one of these two cases occurs. In case (1), traces $[a]$ and $[b]$ are equal, and $\mathrm{tr}(\gamma \uparrow \delta) = \epsilon = \mathrm{tr}(\gamma) \setminus \mathrm{tr}(\delta)$. In case (2), $[a] \setminus [b] = [a]$, hence $\mathrm{tr}(\gamma \uparrow \delta) = [a] = \mathrm{tr}(\gamma) \setminus \mathrm{tr}(\delta)$.

For the induction step, suppose $|\gamma| > 1$ and $|\delta| \geq 1$. (We omit the case $|\gamma| \geq 1$ and $|\delta| > 1$, which is symmetric.) Then $\gamma = \xi\zeta$, where $|\xi| \geq 1$ and $|\zeta| \geq 1$. Suppose $\mathrm{tr}(\xi) = [x]$, $\mathrm{tr}(\zeta) = [y]$, and $\mathrm{tr}(\delta) = [z]$. By Lemma 2.1, γ and δ are consistent iff ξ and δ are consistent and also $\delta \uparrow \xi$ and ζ are consistent. If γ and δ are consistent, then

$$\begin{aligned} \gamma \uparrow \delta &= (\xi \uparrow \delta)(\zeta \uparrow (\delta \uparrow \xi)) \\ \delta \uparrow \gamma &= (\delta \uparrow \xi) \uparrow \zeta. \end{aligned}$$

Applying the induction hypothesis and Lemma 2.4, we see that

$$\begin{aligned} \mathrm{tr}(\gamma \uparrow \delta) &= (\mathrm{tr}(\xi) \uparrow \mathrm{tr}(\delta))(\mathrm{tr}(\zeta) \uparrow (\mathrm{tr}(\delta) \uparrow \mathrm{tr}(\xi))) = [x][y] \setminus [z] = \mathrm{tr}(\gamma) \setminus \mathrm{tr}(\delta) \\ \mathrm{tr}(\delta \uparrow \gamma) &= (\mathrm{tr}(\delta) \uparrow \mathrm{tr}(\xi)) \uparrow \mathrm{tr}(\zeta) = [z] \setminus [x][y] = \mathrm{tr}(\delta) \setminus \mathrm{tr}(\gamma). \end{aligned}$$

Conversely, suppose $\mathrm{tr}(\gamma)$ and $\mathrm{tr}(\delta)$ are consistent. Since then $\mathrm{tr}(\xi)$ and $\mathrm{tr}(\delta)$ are consistent, by the induction hypothesis we know that ξ and δ are consistent, $\mathrm{tr}(\delta \uparrow \xi) = \mathrm{tr}(\delta) \setminus \mathrm{tr}(\xi)$, and $\mathrm{tr}(\xi \uparrow \delta) = \mathrm{tr}(\xi) \setminus \mathrm{tr}(\delta)$. Now, since $\mathrm{tr}(\gamma)$ and $\mathrm{tr}(\delta)$ are consistent, and $\mathrm{tr}(\xi) \sqsubseteq \mathrm{tr}(\gamma)$, it follows that $\mathrm{tr}(\gamma)$ and $\mathrm{tr}(\xi) \sqcup \mathrm{tr}(\delta)$ are consistent. Moreover, $\mathrm{tr}(\gamma) = \mathrm{tr}(\xi)\mathrm{tr}(\zeta)$ and $\mathrm{tr}(\delta) \sqcup \mathrm{tr}(\xi) = \mathrm{tr}(\xi)(\mathrm{tr}(\delta) \setminus \mathrm{tr}(\xi))$, so it follows that $\mathrm{tr}(\zeta)$ and $\mathrm{tr}(\delta) \setminus \mathrm{tr}(\xi)$ are consistent. Applying the induction hypothesis once again shows that ζ and $\delta \setminus \xi$ are consistent. Since ξ and δ are consistent, and $\delta \uparrow \xi$ and ζ are consistent, it follows that γ and δ are consistent, as was to be shown. ∎

A *subdomain* of a domain D is a subset U of D, which is a domain under the restriction of the ordering on D, and is such that the inclusion of U in D is strict and continuous. A subdomain U of D is *normal* if for all $d \in D$, the set $\{e \in U : e \sqsubseteq d\}$ is directed.

Lemma 2.7 *Domain D' is isomorphic to a normal subdomain of D iff there exists a strict, additive, continuous injection $f : D' \to D$ that also reflects consistency.*

Proof – Omitted. ∎

An *interval* in a domain D is a pair (d, d') of elements of D, with $d \sqsubseteq d'$. A *prime* interval is an interval (d, d') such that $d \sqsubset d'$, and there exists no $d'' \in D$ with $d \sqsubset d'' \sqsubset d'$.

Theorem 4 *Suppose D is the domain of permutation equivalence classes of initial computation sequences of a trace automaton with start state $A = (E, Q, T, *)$. Then D is isomorphic to a normal subdomain U of $\bar{E}$, such that the inclusion of U in $\bar{E}$ preserves prime intervals. Conversely, if U is a normal subdomain of $\bar{E}$, such that the inclusion of U in $\bar{E}$ preserves prime intervals, then U is isomorphic to the domain of permutation equivalence classes of initial computation sequences of a trace automaton with start state.*

Proof – Suppose domain D is the domain of permutation equivalence classes $[\gamma]$ of initial computation sequences γ of a trace automaton with start state $A = (E, Q, T, *)$. Each computation sequence γ of A determines a trace $\mathrm{tr}(\gamma) \in \bar{E}$. By Lemma 2.6, if $\gamma \precsim \delta$, then $\mathrm{tr}(\gamma) \setminus \mathrm{tr}(\delta) = \epsilon$, so $\mathrm{tr}(\gamma) \sqsubseteq \mathrm{tr}(\delta)$. It follows that the mapping tr from computation sequences to traces induces a monotone injection (which we also denote by tr) from D to $\bar{E}$. Clearly, the map tr is strict. It also preserves prime intervals, because an interval $([\gamma], [\delta])$ in D is prime iff $\delta \sim \gamma t$ for some nonidentity transition t, and an interval $([x], [y])$ in $\bar{E}$ is prime iff $y \sim xe$ for some $e \in E$.

To see that tr is continuous, it suffices to show that it preserves suprema of $\precsim$-chains. Suppose $\gamma_0 \precsim \gamma_1 \precsim \ldots$ is a chain with supremum γ. By a dovetailing argument using ω-algebraicity and consistent completeness, we may construct a prefix-ordered chain of finite computation sequences $\delta_0 \leq \delta_1 \leq \ldots$ with the following properties:

1. For all $j \geq 0$, there exists $k \geq 0$ such that $\delta_j \precsim \gamma_k$.

2. For all $k \geq 0$, and all finite $\xi \precsim \gamma_k$, there exists $j \geq 0$ such that $\xi \precsim \delta_j$.

It follows that $\bigsqcup_k \mathrm{tr}(\delta_k) = \bigsqcup_k \mathrm{tr}(\gamma_k)$, and if δ is the $\leq$-supremum of $\delta_0 \leq \delta_1 \leq \ldots$, then $\delta \sim \gamma$. Since it is obvious from the definition that tr is continuous with respect to the prefix ordering $\leq$, we may then conclude that $\mathrm{tr}(\gamma) = \mathrm{tr}(\delta) = \bigsqcup_k \mathrm{tr}(\delta_k) = \bigsqcup_k \mathrm{tr}(\gamma_k)$.

To see that tr is additive, suppose γ and δ are consistent finite computation sequences. Then $\xi = \gamma(\delta \uparrow \gamma)$ is a $\precsim$-supremum of γ and δ. Moreover, $\mathrm{tr}(\xi) = \mathrm{tr}(\gamma)(\mathrm{tr}(\delta) \setminus \mathrm{tr}(\gamma)) = \mathrm{tr}(\gamma) \sqcup \mathrm{tr}(\delta)$ by Lemmas 2.4 and 2.5. Finally, if $\mathrm{tr}(\gamma)$ and $\mathrm{tr}(\delta)$ are consistent, then γ and δ are consistent by Lemma 2.6, so tr reflects consistency. Since the map tr is a strict, additive, continuous injection that reflects consistency, it is a isomorphism from D to a normal subdomain of $\bar{E}$ by Lemma 2.7.

Conversely, suppose U is a normal subdomain of $\bar{E}$ for some concurrent alphabet E, such that the inclusion of U in $\bar{E}$ preserves prime intervals. Let $A = (E, U^\circ, T, \epsilon)$, where U° is the set of finite elements of U, and where T contains id_ϵ all identity transitions $\mathrm{id}_{[x]}$ with $[x] \in U^\circ$, and all transitions $[x] \xrightarrow{a} [xa]$ where both $[x]$ and $[xa] \in U^\circ$. It is easy to check that A satisfies the conditions for an automaton. If γ is an initial computation sequence of A, then $\mathrm{tr}(\delta)$ is a finite element of U for each finite prefix δ of γ, so $\mathrm{tr}(\gamma) \in U$ by continuity of tr. Conversely, if $[x] \in U$, then we may choose a chain $[x_0] \sqsubseteq [x_1] \sqsubseteq \ldots$ of finite elements of U such that $\bigsqcup_k [x_k] = [x]$.

Since the inclusion of U in $\bar{E}$ preserves prime intervals, and the domain $\bar{E}$ is finitary, we may assume without loss of generality that for each $k \geq 0$, either $x_{k+1} = x_k$ or else $x_{k+1} = x_k a$ for some $a \in E$. It is then a simple matter to construct a corresponding initial computation sequence γ of A, such that $\mathrm{tr}(\gamma) = [x]$. ∎

It can be shown (see [14]), that a domain D is isomorphic to a normal subdomain U of a domain of traces $\bar{E}$, where the inclusion of U in $\bar{E}$ preserves prime intervals, if and only if D is an *event domain* [6,18]. Event domains, in turn, are those that are isomorphic to the domains of "configurations" of an "event structure" determined by an enabling relation and a binary conflict relation. Thus, trace automata generate the same class of domains as event structures.

3 Concurrent Transition Systems

The notion of the residual of one computation sequence after another was crucial in the results of the previous section. The properties of the residual operation that were used in the proofs can be axiomatized, and the result is *concurrent transition systems* [15,16]. As we have shown in these papers and in [12,17], many interesting properties of automata can be established from these axioms. An advantage of the more abstract, axiomatic formulation is that there is an obvious way in which concurrent transition systems can be made into a category, which we call **CTS**. The category **CTS** has a surprisingly rich structure, and categorical constructions in it and related functor categories correspond directly to many natural constructions we wish to perform on automata, including the extraction of their computational behavior. After investigating the properties of **CTS**, we shall see that a certain mapping from trace automata to **CTS** suggests an interesting way to make trace automata into a category.

An *abstract automaton* is a structure $A = (Q, T, \mathrm{dom}, \mathrm{cod}, \mathrm{id})$, where

- Q is a set of *states*.
- T is a set of *transitions*.
- $\mathrm{dom}, \mathrm{cod} : T \to Q$ are functions that map each transition to its *domain* and *codomain*, respectively.
- $\mathrm{id} : Q \to T$ maps each state q in Q to a distinguished *identity transition* $\mathrm{id}_q \in T$, such that $\mathrm{dom}(\mathrm{id}_q) = \mathrm{cod}(\mathrm{id}_q) = q$.

Let $\mathrm{Coin}(A)$ denote the set of all coinitial pairs of transitions of A. A *residual operation* on an automaton $A = (Q, T, \mathrm{dom}, \mathrm{cod}, \mathrm{id})$ is a partial function $\uparrow$: $\mathrm{Coin}(A) \to T$, such that the following conditions hold (we write $t \uparrow u$ instead of $\uparrow(t, u)$, and we read it as "t *after* u"):

1. If $t \uparrow u$ is defined, then so is $u \uparrow t$, and then $\mathrm{dom}(t \uparrow u) = \mathrm{cod}(u)$, $\mathrm{dom}(u \uparrow t) = \mathrm{cod}(t)$, and $\mathrm{cod}(t \uparrow u) = \mathrm{cod}(u \uparrow t)$.

2. For all $t : q \to r$ in T, *(a)* $\mathrm{id}_q \uparrow t = \mathrm{id}_r$; *(b)* $t \uparrow \mathrm{id}_q = t$; and *(c)* $t \uparrow t = \mathrm{id}_r$.
3. For all coinitial $t, u, v \in T$, $(v \uparrow t) \uparrow (u \uparrow t) = (v \uparrow u) \uparrow (t \uparrow u)$, whenever either side is defined.

A residual operation is *extensional* if it satisfies the additional condition:

4. For all coinitial transitions t, u, if $t \uparrow u$ and $u \uparrow t$ are both identities, then $t = u$.

A *concurrent transition system* (CTS) is a pair $C = (A, \uparrow)$, where A is an abstract automaton and $\uparrow$ is an extensional residual operation on A. If $\uparrow$ is a residual operation, but not necessarily extensional, then C is called a *pre-CTS*. Coinitial transitions t, u of a pre-CTS are called *consistent* if $t \uparrow u$ is defined (equivalently, if $u \uparrow t$ is defined), otherwise they are called *conflicting*. We think of consistent transitions as representing actions that might arise in a single concurrent computation, and of conflicting transitions as representing incompatible nondeterministic choices. If t and u are consistent, then we think of $t \uparrow u$ as what remains of the transition t after the transition u has occurred.

The results of the previous section allow us to associate with each trace automaton $A = (E, Q, T)$ a CTS C_A whose states, transitions, and identities are those of A, and whose residual operation is defined as in Section 2.1.2. It is also interesting to observe that a concurrent alphabet E can be viewed as the set of nonidentity transitions of a one-state CTS, such that $a, b \in E$ are consistent exactly when either $a = b$ or $a \|_E b$, and in the latter case $a \uparrow b = a$ and $b \uparrow a = b$. We shall put this observation to good use in Section 3.5.

Suppose $A = (Q, T, \mathrm{dom}, \mathrm{cod}, \mathrm{id})$ and $A' = (Q', T', \mathrm{dom}', \mathrm{cod}', \mathrm{id}')$ are automata. Then a *morphism* from A to A' is a pair of maps $\rho = (\rho_o, \rho_a)$, where $\rho_o : Q \to Q'$ and $\rho_a : T \to T'$, such that $\mathrm{dom}' \circ \rho_a = \rho_o \circ \mathrm{dom}$, $\mathrm{cod}' \circ \rho_a = \rho_o \circ \mathrm{cod}$, and $\mathrm{id}' \circ \rho_o = \rho_a \circ \mathrm{id}$. In the sequel, we will drop the notational distinction between ρ_o and ρ_a, writing simply ρ in both cases. Let **Auto** denote the category of automata and their morphisms.

A *morphism* from a pre-CTS $B = (A, \uparrow)$ to a pre-CTS $B' = (A', \uparrow')$ is a morphism $\rho : A \to A'$ of the underlying automata, with the following additional property:

- If t, u are consistent transitions of B, then $\rho(t)$ and $\rho(u)$ are consistent transitions of B', and $\rho(t \uparrow u) = \rho(t) \uparrow' \rho(u)$.

The class of all pre-CTS's forms a category **PCTS**, when equipped with the CTS-morphisms as arrows. Let **CTS** denote the full subcategory of **PCTS** whose objects are the CTS's.

The following result shows that there is a universal way to obtain a CTS from a pre-CTS.

Lemma 3.1 *Suppose B is a pre-CTS. Then there exists a CTS $B^\flat$, and a morphism $\flat_B : B \to B^\flat$, with the following property: if C is any CTS equipped with a morphism $\rho : B \to C$, then there exists a unique morphism $\rho^\flat : B^\flat \to C$ with $\rho = \rho^\flat \circ \flat_B$.*

Proof – Suppose $B = (A, \uparrow)$. Define a binary relation $\sim$ on transitions of B by: $t \sim u$ iff $t \uparrow u = \mathrm{id}$ and $u \uparrow t = \mathrm{id}$. Let $A^\flat$ be the automaton whose states are those of A, but whose transitions are $\sim$-equivalence classes $[t]$ of transitions of A, with domain, codomain, and identities defined in the obvious way. Define $\uparrow^\flat$ by: $[t] \uparrow^\flat [u] = [t \uparrow u]$, and let $B^\flat = (A^\flat, \uparrow^\flat)$. Define $\flat_B : B \to B^\flat$ to take each transition t of A to the corresponding $\sim$-equivalence class $[t]$, which is a transition of $A^\flat$. One may now verify that $B^\flat$ and $\flat_B$ are well-defined and have the required properties. ∎

Corollary 3.1 **CTS** *is a reflective subcategory of* **PCTS**.

Proof – Lemma 3.1 shows that the inclusion of **CTS** in **PCTS** has a left adjoint, whose object map takes B to $B^\flat$. ∎

There is also a universal way to lift an automaton to a CTS.

Lemma 3.2 *Suppose A is an automaton. Then there exists a CTS $A^\sharp = (A, \uparrow)$ with the following property: Given a CTS $C' = (A', \uparrow')$, every morphism $\rho : A \to A'$ in* **Auto** *is also a morphism $\rho : A^\sharp \to C'$ in* **CTS**.

Proof – Given an automaton A, let $t \uparrow u$ be defined for coinitial transitions t, u of A iff either $t = u$ or else either t or u is an identity. If $t = u$ or $t = \mathrm{id}$, then $t \uparrow u = \mathrm{id}$, and if u is an identity, then $t \uparrow u = t$. Let $A^\sharp = (A, \uparrow)$, then it is easy to see that $A^\sharp$ has the required property. ∎

Corollary 3.2 **Auto** *is isomorphic to a coreflective subcategory of* **CTS**.

Proof – Lemma 3.2 shows that the forgetful functor Auto : **CTS** $\to$ **Auto** has a left adjoint, whose object map takes A to $A^\sharp$, and which is full and faithful. ∎

3.1 Basic Consequences of the CTS Axioms

Define a relation $\preceq$ on the transitions of a CTS by: $t \preceq u$ iff t, u are coinitial and $t \uparrow u = \mathrm{id}$. We say that a transition v is a *join* of the coinitial transitions t, u if $t \preceq v$, $u \preceq v$, $v \uparrow t = u \uparrow t$, and $v \uparrow u = t \uparrow u$. If t and u are composable, then we say that a transition v is a *composite* of t, u if $t \uparrow v = \mathrm{id}$ and $v \uparrow t = u$.

The following results are proved in [16]. The reader may enjoy working out the proofs, since they are good examples of how to work with residuals.

Lemma 3.3 *Suppose C is a CTS.*

1. *The relation $\preceq$ is a partial order.*
2. *If a composite of t and u exists, then it is unique. (We denote it by tu.)*
3. *Suppose t and v are coinitial, and the composite tu of t and u exists.*
   (a) $v \uparrow tu = (v \uparrow t) \uparrow u$.

   (b) $tu \uparrow v = (t \uparrow v)(u \uparrow (v \uparrow t))$.

4. *Composition obeys the following laws:*
   (a) *For all* t, $t = \mathrm{id}\, t = t\, \mathrm{id}$.
   (b) i. *If* tu *and* $(tu)v$ *exist, then* uv *and* $t(uv)$ *exist, and* $(tu)v = t(uv)$.
       ii. *If* tu, uv, *and* $t(uv)$ *exist, then* $(tu)v$ *exists and* $(tu)v = t(uv)$.
   (c) *If* tu *and* tv *exist, and* $tu = tv$, *then* $u = v$.
5. *A transition* v *is a join of* t *and* u *iff* $v = t(u \uparrow t)$. *Hence a join of* t *and* u, *if it exists, is unique. (We denote it by* $t \vee u$.*) Moreover, if* $t \vee u$ *exists, then it is the least upper bound of* t *and* u *under* $\preceq$.

Lemma 3.4 *Suppose* $\rho : C \to C'$ *is a morphism. Then*

1. $\rho(tu) = \rho(t)\rho(u)$ *whenever* tu *exists.*
2. $\rho(t \vee u) = \rho(t) \vee \rho(u)$ *whenever* $t \vee u$ *exists.*

3.2 Completion of a CTS

A CTS is called *complete* if every composable pair of transitions has a composite. Complete CTS's play a fundamental role in the process of extracting the computational behavior of a CTS. In particular, each CTS C freely generates a *completion* C^*, whose states are the same as those of C, and whose transitions are certain equivalence classes of finite sequences of transitions of C. In case C is the CTS C_A corresponding to a trace automaton A, then the transitions of C_A^* are precisely the permutation equivalence classes of finite computation sequences of A.

Lemma 3.5 *Suppose* C *is a CTS. Then there exists a complete CTS* C^*, *and a morphism* $\natural_C : C \to C^*$, *with the following property: if* C' *is any complete CTS equipped with a morphism* $\rho : C \to C'$, *then there exists a unique morphism* $\rho^* : C^* \to C'$ *such that* $\rho = \rho^* \circ \natural_C$.

Proof – Suppose $C = (A, \uparrow)$ is a CTS. Let A^* be the automaton whose states are those of A, and whose transitions are the finite computation sequences of A, with the computation sequences of length 0 as the identity transitions. If we identify each nonidentity transition of A with the corresponding computation sequence of length 1, then the residual operation $\uparrow$ on A extends uniquely to a residual operation $\uparrow^*$ on A^*, such that the following hold whenever either side is defined:

$$\xi \uparrow^* \gamma\delta = (\xi \uparrow^* \gamma) \uparrow^* \delta, \qquad \gamma\delta \uparrow^* \xi = (\gamma \uparrow^* \xi)(\delta \uparrow^* (\xi \uparrow^* \gamma)).$$

Then $B = (A^*, \uparrow^*)$ is a pre-CTS. Moreover, the map $\mu : C \to B$ that takes each state of C to the same state of B and each nonidentity transition of C to the corresponding computation sequence of length 1, is a morphism. Define $C^* = B^\flat$

and define $\natural_C : C \to C^*$ by: $\natural_C = \flat_B \circ \mu$. One may now verify that C^* and $\natural_C$ have the stated properties. ∎

Let **CCTS** denote the full subcategory of **CTS** having the complete CTS's as objects.

Corollary 3.3 **CCTS** *is a reflective subcategory of* **CTS**.

Proof – Lemma 3.5 shows that the inclusion of **CCTS** in **CTS** has a left adjoint, whose object map takes C to C^*. ∎

Theorem 5 *If C_A is the CTS associated with a trace automaton A, then up to isomorphism, C_A^* is the CTS whose states are the states of A and whose transitions are the permutation equivalence classes of finite computation sequences of A, with residual given by Lemma 2.1.*

Proof – By the construction in Lemma 3.5, C_A^* has as states the states of A and as transitions the $\sim$-equivalence classes of finite computation sequences of A, where $\gamma \sim \delta$ holds iff $\gamma \uparrow^* \delta = \mathrm{id}$ and $\delta \uparrow^* \gamma = \mathrm{id}$. By Theorem 2, $\sim$ is nothing more than permutation equivalence. ∎

In later sections, we need a characterization of the complete CTS's that was proved in [16]. Define a *computation category* to be a small category C with the following properties:

1. Every arrow is an epimorphism.

2. The only isomorphisms are identities.

3. C has *bounded pushouts*: whenever t, u are coinitial arrows of C such that $tv = uw$ for some arrows v, w, then t and u have a pushout.

For the statement of the next result, we observe that a category is nothing more than a pair $(A, \cdot)$, where A is an automaton and $\cdot$ is an associative composition on the transitions of A, having the identities of A as units.

Lemma 3.6 *Suppose $C = (A, \uparrow)$ is a complete CTS, and let $\cdot$ denote the composition operation of C. Then $C' = (A, \cdot)$ is a computation category. Conversely, suppose $C' = (A, \cdot)$ is a computation category. For coinitial arrows t, u of A, let $t \uparrow u$ be defined iff $tv = uw$ for some arrows v, w, in which case let $t \uparrow u$ be the side opposite t in a pushout square with t, u as its base. Then $C = (A, \uparrow)$ is a complete CTS, whose composition operation coincides with that of C.*

Proof – Omitted. ∎

3.3 Computation Diagrams

In this section we generalize to CTS's the notions of "computation tree" and "computation" for ordinary transition systems. Given a CTS C, the set of all transitions of its completion C^* from a designated state $*$, forms the set of states of another complete CTS D, which we call the *complete computation diagram* of C with respect to $*$. Transitions of D represent prefix relationships between concurrent computations; accordingly, there is at most one transition between any two states of D.

Formally, a *pointed CTS* is a pair $(C, *)$, where C is a CTS, and $*$ is a distinguished state of C, called the *start state*. Let $\mathbf{CTS}_*$ (resp. $\mathbf{CCTS}_*$) denote the category of pointed CTS's (resp. pointed, complete CTS's) and morphisms that preserve the start state.

A *complete computation diagram* is a pointed CTS $(D, \bot)$, such that D is complete, and for each state q of D, there is a unique transition $0_q : \bot \to q$ in D. The $\preceq$ partial order on transitions of D determines a corresponding ordering $\preceq$ on states of D, defined by: $q \preceq r$ iff $0_q \preceq 0_r$. Then the set of states of D, partially ordered by $\preceq$, has $\bot$ as a least element, and by Lemma 3.6 has the property that any pair of states with an upper bound, has a least upper bound.

There is a universal way to obtain a complete computation diagram from a pointed, complete CTS.

Lemma 3.7 *Suppose $(C, *)$ is a pointed, complete CTS. Then there exists a complete computation diagram* $\mathrm{CDiag}(C, *) = (D, \bot)$, *and a* $\mathbf{CTS}_*$*-morphism*

$$\varepsilon : \mathrm{CDiag}(C, *) \to (C, *),$$

with the following property: if $(D', \bot')$ is any other complete computation diagram, equipped with a morphism $\rho : (D', \bot') \to (C, *)$, then there exists a unique $\mathbf{CTS}_*$*-morphism $\rho^\dagger : (D', \bot') \to \mathrm{CDiag}(C, *)$ such that $\rho = \varepsilon \circ \rho^\dagger$.*

Proof – Given $(C, *)$, let D be the complete CTS whose states are all transitions t of C such that $\mathrm{dom}(t) = *$, and in which there is a unique transition from t to u iff $t \preceq u$. Let $\bot = \mathrm{id}_*$. Let $\varepsilon : (D, \bot) \to (C, *)$ take each state t of D to the state $\mathrm{cod}(t)$ of C, and each transition from t to u in D to the transition $u \uparrow t$ in C. One may now easily check that $(D, \bot)$ and ε have the required properties. ∎

Let **CDiag** denote the full subcategory of $\mathbf{CTS}_*$ whose objects are the complete computation diagrams.

Theorem 6 **CDiag** *is coreflective in* $\mathbf{CCTS}_*$.

Proof – Lemma 3.7 shows that the inclusion of **CDiag** in $\mathbf{CCTS}_*$ has a right adjoint, whose object map takes $(C, *)$ to $\mathrm{CDiag}(C, *)$. ∎

Theorem 7 *Suppose $(C_A, *)$ is the pointed CTS determined by a trace automaton A with start state, and let $(D, \perp) = \mathrm{CDiag}(C_A, *)$ be its complete computation diagram. Then up to isomorphism, D is the poset of finite initial concurrent computations of A.*

Proof – Obvious from Theorem 5 and the construction of $\mathrm{CDiag}(C_A, *)$ given in the proof of Lemma 3.7. ▮

Noting that the domain of all computations of A may be obtained by ideal completion of the poset of finite computations, we make the following general definition:

- A *computation* of a pointed CTS $(C, *)$ is an ideal of its complete computation diagram. A computation is *finite* if it is a principal ideal, otherwise it is *infinite*.

3.4 The Category CTS

The category **CTS** has a great deal of structure making it suitable for use in constructing models of concurrent systems.

Theorem 8 *The category* **CTS***:*

1. *has equalizers and small products, hence all small limits.*
2. *has small coproducts.*
3. *is cartesian closed.*
4. *has small filtered colimits.*

Proof – The proofs of (1) and (2) use the obvious constructions. Assertion (3) is proved in [16].

To show (4), let D be a small filtered category, and let $L : D \to \mathbf{CTS}$ be a functor. Our objective is to construct a colimit of L. Let V denote the set of objects of D, and let E denote the set of its arrows. For each $i \in V$, let $C_i = (A_i, \uparrow_i)$ denote the CTS Li, where $A_i = (Q_i, T_i, \mathrm{dom}_i, \mathrm{cod}_i, \mathrm{id}_i)$. Define T to be the disjoint union $\coprod_{i \in V} T_i$. Let the relation $\sim$ on T be defined as follows: $t \sim u$ iff there exist arrows $f : i \to k$ and $g : j \to k$ in D such that $t \in T_i$, $u \in T_j$, and $Lf(t) = Lg(u)$. One may now show that $\sim$ is an equivalence relation, and then construct a CTS C having the equivalence classes of $\sim$ as its transitions. The CTS C is the base of a colimiting cone over L. ▮

3.5 Application to Trace Theory

In this section, we show how the theory of CTS's can be used to obtain results in trace theory. Typically, proofs in trace theory make use of the representation of traces by their "dependency graphs" [1,11]. In contrast, proofs using CTS theory involve residuals. One of our goals is to prove Lemma 2.5, which we have already used in the proof of Theorem 4.

We first establish a correspondence between concurrent alphabets and certain CTS's. This correspondence extends to yield a correspondence between free partially commutative monoids and certain complete CTS's.

Lemma 3.8 *Suppose E is a concurrent alphabet. Then E is the set of nonidentity transitions of a one-state CTS C_E, in which $a \uparrow b$ is defined for $a, b \in E$ precisely when $a \|_E b$, and in that case $a \uparrow b = a$. Conversely, suppose C is a one-state CTS in which $a \uparrow b = a$ whenever a and b are consistent, distinct, nonidentity transitions. Let E_C be the concurrent alphabet whose elements are the nonidentity transitions of C, with $a \|_{E_C} b$ defined to hold iff a and b are consistent and distinct. Then $E_{C_E} = E$ and $C_{E_C} \simeq C$.*

Proof – Straightforward. ∎

Lemma 3.9 *Suppose C is a one-state complete CTS. Then $x \uparrow y$ is defined iff x and y have an upper bound with respect to the ordering $\preceq$. Moreover, the transitions of C, equipped with the identity and composition of C, form a monoid* Mon(C) *with the following properties:*

1. *For all $x, y, z \in \mathrm{Mon}(C)$, if $xy = xz$ then $y = z$.*
2. *Define $x \sqsubseteq y$ iff $\exists z(xz = y)$. Then $\sqsubseteq$ is a consistently complete partial order with the identity transition of C as a least element.*

Proof – Immediate from Lemma 3.6. ∎

From Lemma 3.9, it follows that the residual operation on C can be recovered from the monoid Mon(C), because $x \uparrow y$ is defined iff x and y have an upper bound with respect to $\sqsubseteq$ (which is the same relation as $\preceq$), in which case $x \uparrow y$ is the unique z such that $yz = x \sqcup y$.

Lemma 3.10 *Suppose E is a concurrent alphabet. Then* $\mathrm{Mon}(C_E) \simeq E^* /\!\sim$.

Proof – The transitions of C_E^* are permutation equivalence classes $[\gamma]$ of finite computation sequences γ of C_E. Define a map μ from $\mathrm{Mon}(C_E^*)$ to $E^* /\!\sim$ by: $\mu([\gamma]) = \mathrm{tr}(\gamma)$. Note that γ is well-defined, because if $[\gamma] = [\delta]$, then $\mathrm{tr}(\gamma) = \mathrm{tr}(\delta)$ by Theorem 1. Also, μ is a monoid homomorphism, because $\mu(\mathrm{id}) = \epsilon$ and $\mu([\gamma][\delta]) = \mu([\gamma\delta]) = \mathrm{tr}(\gamma\delta) = \mathrm{tr}(\gamma)\mathrm{tr}(\delta) = \mu([\gamma])\mu([\delta])$. The map μ is injective, because if $\mu([\gamma]) = \mu([\delta])$, then $\mathrm{tr}(\gamma) = \mathrm{tr}(\delta)$ hence $[\gamma] = [\delta]$ by Theorem 1. Finally, μ is surjective, because if $[x] \in E^* /\!\sim$, then we may construct a computation sequence γ of C_E, such that the sequence of events appearing in γ is x, hence such that $\mathrm{tr}(\gamma) = [x]$. Since μ is a bijective monoid homomorphism, it is an isomorphism. ∎

We are now equipped to prove Lemma 2.5.

Proof of Lemma 2.5 – ($\Rightarrow$) Suppose E is a concurrent alphabet. By definition of $\bar{E}$, the set $\bar{E}^{\circ}$ of finite elements of $\bar{E}$ is the set of elements of a monoid isomorphic to the free partially commutative monoid $(E^* /\!\sim_E, \cdot, \epsilon)$. By Lemma 3.10, this monoid is isomorphic to the monoid $\mathrm{Mon}(C_E^*)$. Since the residual operation of C_E^* can be recovered from $\mathrm{Mon}(C_E^*)$ it makes sense to use this operation to reason about $\mathrm{Mon}(C_E^*)$, hence about $E^* /\!\sim$.

Now, by Lemma 3.9, the relation $\sqsubseteq$ on $\mathrm{Mon}(C_E^*)$ is a consistently complete partial order with ϵ as a least element. Since the ideal completion of a consistently complete partial order with a least element is a domain, it follows that $\bar{E}$ is a domain. We observe that $\bar{E}$ is finitary, because if $[x]$ is a finite element of $\bar{E}$, then $|\{[y] : [y] \sqsubseteq [x]\}|$ is bounded by the number of prefixes of permutations of x, which is finite.

Property (1) of a trace domain holds for $\bar{E}$ by definition of $\sqsubseteq$. Property (2) of a trace domain is immediate from Lemma 3.9. It remains to verify property (3) of a trace domain. The atoms of $E^* /\!\sim$ are precisely the $\sim$-equivalence classes $[a]$ of elements a of E. Suppose $[a]$ and $[b]$ are distinct atoms. By Lemma 3.9, $[a]$ and $[b]$ are consistent iff $[a] \uparrow [b]$ is defined. But by definition of C_E, $[a] \uparrow [b]$ is defined for distinct $[a], [b]$ iff $a \| _E b$, hence iff $[a][b] = [b][a]$. In that case, $[a] \sqcup [b] = [a]([b] \uparrow [a]) = [a][b] = [b][a]$, by definition of $\uparrow$ on C_E.

($\Leftarrow$) Conversely, suppose $(D, \sqsubseteq, \cdot, \bot)$ is a trace domain. Define a concurrent alphabet E whose elements are the atoms of D, and which has $a \| _E b$ iff $a \neq b$ and $[a][b] = [b][a]$. By property (3) of a trace domain, the monoid homomorphism that takes each element $[a_1 \dots a_n]$ of E^* (with $a_1, \dots, a_n \in E$) to the corresponding finite element $a_1 \dots a_n$ of D respects $\sim$, thus induces a monoid homomorphism from $(E^* /\!\sim) \to D^{\circ}$, and this monoid homomorphism extends uniquely to a continuous map $\mu : \bar{E} \to D$. Property (1) of a trace domain, together with the fact that D is finitary, implies that every element of D° factors via $\cdot$ into a finite sequence of atoms. Thus, μ is surjective.

It remains to be shown that μ is injective. It suffices to show that μ is injective when restricted to $E^* /\!\sim$, since then the injectiveness of μ on all of $\bar{E}$ will follow by algebraicity. It is easy to see that $\mu(x) = \bot$ iff $x = \epsilon$. A straightforward argument by induction on the length of a factorization into atoms shows that $[x] \setminus [y]$ is defined in $\bar{E}$ iff $\mu([x]) \setminus \mu([y])$ is defined in D, and then $\mu([x] \setminus [y]) = \mu([x]) \setminus \mu([y])$. Hence, $\mu([x]) = \mu([y])$ iff $\mu([x]) \setminus \mu([y]) = \bot = \mu([y]) \setminus \mu([x])$, which holds iff $\mu([x] \setminus [y]) = \epsilon = \mu([y] \setminus [x])$, that is, iff $[x] \setminus [y] = \epsilon = [y] \setminus [x]$. Thus, μ is injective, hence is an isomorphism. ∎

The correspondence between concurrent alphabets and one-state CTS's suggests a way to make the class of concurrent alphabets into a category. Formally, if E and F are concurrent alphabets, then define a *strong morphism* from E to F to be an ϵ-preserving map $\mu : (E \cup \{\epsilon\}) \to (F \cup \{\epsilon\})$, such that $a \| _E b$ implies $\mu(a) \neq \mu(b)$. Let **SAlph** denote the category of concurrent alphabets and strong morphisms.

Lemma 3.11 *The map that takes a concurrent alphabet E to the CTS C_E extends to an equivalence of* **SAlph** *to the full subcategory of* **CTS** *whose objects are the* *one-state CTS's C such that $a \uparrow b = a$ whenever a and b are distinct consistent nonidentity transitions of C.*

Proof – Omitted. ∎

The category **SAlph** is not particularly interesting. However, there is another way to map concurrent alphabets to one-state CTS's. This mapping leads to an alternative notion of "weak morphism" of concurrent alphabets. The resulting category **Alph** of concurrent alphabets and weak morphisms is also equivalent to a full subcategory of **CTS**, and is much more interesting than **SAlph**.

Formally, if E is a concurrent alphabet, then call a subset U of E *commuting* if $a \|_E b$ whenever a and b are distinct elements of E. Let $\mathrm{Com}(E)$ denote the set of all finite commuting subsets of E. A *weak morphism* from E to F is a function $\mu : \mathrm{Com}(E) \to \mathrm{Com}(F)$ such that

1. $\mu(\emptyset) = \emptyset$.

2. If $U \cup V \in \mathrm{Com}(E)$, then $\mu(U) \cup \mu(V) \in \mathrm{Com}(F)$, and $\mu(U \setminus V) = \mu(U) \setminus \mu(V)$.

Here the symbol $\setminus$ denotes set difference. Let **Alph** denote the category of concurrent alphabets and weak morphisms.

A CTS C is called *join-complete* if every consistent coinitial pair of transitions of C has a join. The *join-completion* C is the sub-CTS $\widehat{C}$ of C^* whose transitions are precisely those transitions of C^* that can be expressed as a finite join $t_1 \vee \ldots \vee t_n$, where $t_1, \ldots, t_n$ are transitions of C.

Theorem 9 *The map that takes a concurrent alphabet E to the CTS $\widehat{C}_E$ extends to an equivalence of* **Alph** *to the full subcategory of* **CTS** *whose objects are the one-state, join-complete CTS's C with the following properties:*

1. *$t \uparrow u = t$ whenever t and u are distinct, consistent, $\preceq$-minimal nonidentity transitions.*

2. *Every nonidentity transition t of C can be expressed as a join $t_1 \vee \ldots \vee t_n$ of a nonempty set of $\preceq$-minimal nonidentity transitions.*

Proof – Omitted. ∎

We do not have space here to develop in detail the features of the category **Alph**. We merely note that it can be shown that **Alph** has binary products and coproducts, which correspond to intuitively appealing notions of "concurrent product" and "nondeterministic sum," respectively.

3.6 Characterization of Trace CTS's

Notably absent from the CTS definition are any sort of "concreteness" axioms that would have as a consequence, for example, a theorem stating that every transition factors into a finite sequence of $\preceq$-minimal transitions. Part of the reason we have not included such axioms is that we can often do without them, using instead a principle of "computational induction" that arises out of the fact that C^* is freely generated by C. However, another reason we have not given any concreteness axioms is that we are still looking for attractive axioms that are satisfied by the CTS's in some subcategory of **CTS** with sufficient completeness properties. The axioms we have been able to discover are either too weak in the sense that the CTS's satisfying them are not very concrete, or else they are too strong in the sense that the resulting subcategory does not admit countable products.

As an example of what we are able to do, we obtain properties that characterize the CTS's obtained from trace automata. The result is patterned after Winskel's characterization of the domains of configurations of event structures defined by an enabling relation and a binary conflict relation [6,18].

Define a CTS C to be *atomic* if the following holds:

- $t \uparrow v = u \uparrow v$ implies either $t = u$, $t = v$, or $u = v$, whenever t, u, and v are transitions of C, with t, v consistent and u, v consistent.

This is a very strong concreteness axiom which ensures that no nontrivial $\preceq$-relationships hold between transitions.

Lemma 3.12 *Suppose C is an atomic CTS. If t and u are coinitial transitions of C such that $t \preceq u$, then either $t = u$ or else $t = \mathrm{id}$.*

Proof – If $t \preceq u$, then $t \uparrow u = \mathrm{id}$, so $t \uparrow u = \mathrm{id} \uparrow u$. By the atomicity property, either $t = \mathrm{id}$, $u = \mathrm{id}$, or $t = u$. But $u = \mathrm{id}$ implies $t = u = \mathrm{id}$, so either $t = \mathrm{id}$ or $t = u$. ∎

If C is an atomic CTS, then we may define $\equiv$ to be the least equivalence relation on transitions of C such that $t \equiv t \uparrow u$ holds whenever t is an arbitrary transition and transition $u \neq t$ is a nonidentity transition consistent with t.

Theorem 10 *A CTS C is isomorphic to the CTS C_A associated with a trace automaton A iff the following conditions hold:*

1. *C is atomic.*
2. *For all transitions t, t' of C, if $t \equiv t'$ and $\mathrm{dom}(t) = \mathrm{dom}(t')$, then $t = t'$.*
3. *For all transitions $t \equiv t'$ and $u \equiv u'$, if the transitions t, u are consistent, and the transitions t', u' are coinitial, then t', u' are consistent as well.*

Proof – If C_A is the CTS associated with a trace automaton A, then it is a straightforward application of the definition of the residual operation on A to see that C_A has properties (1)-(3).

Conversely, suppose C is a CTS with the stated properties. Define the *events* of C to be the $\equiv$-equivalence classes $[t]$ of nonidentity transitions t of C. Let E be the set of all events of C, with concurrency relation $\|_E$ defined as follows: $[t]\|_E[u]$

iff there exist $t' \equiv t$ and $u' \equiv u$ such that t', u' are consistent and $t' \neq u'$. Clearly, the relation $\|_E$ is symmetric, and hypothesis (2) implies that it is irreflexive.

Define $A = (E, Q, T)$, where T contains all transitions $q \xrightarrow{\epsilon} q$ and all transitions $q \xrightarrow{[t]} r$ such that $t : q \to r$ is a nonidentity transition of C. One may now verify that A is an automaton, with $A \simeq C_A$. ∎

The equivalence, discussed in Section 3.5, from the category **Alph** of concurrent alphabets and weak morphisms to a full subcategory of **CTS**, suggests an interesting way to make the class of trace automata into a category **TrAuto**.

Formally, suppose $A = (E, Q, T)$ and $A' = (E', Q', T')$ are trace automata. Define a *morphism* from A to A' to be a pair (ρ_e, ρ_s), where $\rho_e : E \to E'$ is a weak morphism of concurrent alphabets, and $\rho_s : Q \to Q'$ is a function, such that the following holds:

- Suppose $q \xrightarrow{e} r \in T$, with $e \neq \epsilon$. Then for every enumeration $\{e'_1, \ldots, e'_n\}$ of $\rho_e(\{e\})$, there exists a (necessarily unique) finite computation sequence

$$\rho_s(q) = r'_0 \xrightarrow{e'_1} r'_1 \xrightarrow{e'_2} \ldots \xrightarrow{e'_n} r'_n = \rho_s(r)$$

  of A'.

Let **TrAuto** denote the category of trace automata and their morphisms.

The map that takes a trace automaton A to the CTS $\widehat{C}_A$ extends to a functor AuCts : **TrAuto** $\to$ **CTS**. This functor does not yield an equivalence with a full subcategory of **CTS**, because AuCts(A) does not contain sufficient information to recover the concurrent alphabet of A up to isomorphism. Nevertheless, the category **TrAuto** seems quite interesting, and further study of it and its relationship to **CTS** seems worthwhile.

4 Conclusion

We have seen that by using the notion of a concurrent alphabet to introduce concurrency information into ordinary nondeterministic transition systems, we obtain a model of concurrent computation having a great deal of algebraic structure. The essential features of this structure can be expressed nicely with the help of the notion of the residual of one transition after another. The resulting algebra of residuals can be used to obtain useful insights into the capabilities and limitations of concurrency. In particular, the characterization of the domain of concurrent computations given by Theorem 4 leads at once to interesting characterizations (see [17]) of the input/output relations that are computable by various classes of concurrent automata.

References

[1] I. J. Aalbersberg and G. Rozenberg. Theory of traces. *Theoretical Computer Science*, 60(1):1–82, 1988.

[2] M. Bednarczyk. *Categories of Asynchronous Systems*. PhD thesis, University of Sussex, October 1987.

[3] G. Berry and J.-J. Lévy. Minimal and optimal computations of recursive programs. *Journal of the ACM*, 26(1):148–175, January 1979.

[4] G. Boudol. Computational semantics of term rewriting systems. In M. Nivat and J. Reynolds, editors, *Algebraic Methods in Semantics*, pages 169–236, Cambridge University Press. 1985.

[5] G. Boudol and I. Castellani. A non-interleaving semantics for CCS based on proved transitions. *Fundamenta Informaticae*, XI:433–452, 1988.

[6] P.-L. Curien. *Categorical Combinators, Sequential Algorithms, and Functional Programming*. *Research Notes in Theoretical Computer Science*, Pitman, London, 1986.

[7] G. Huet. Formal structures for computation and deduction (first edition). May 1986. Unpublished manuscript. INRIA, France.

[8] M. Kwiatkowska. *Categories of Asynchronous Systems*. PhD thesis, University of Leicester, May 1989.

[9] J.-J. Lévy. *Réductions Correctes et Optimales dans le Lambda Calcul*. PhD thesis, Université Paris VII, 1978.

[10] N. A. Lynch and E. W. Stark. A proof of the Kahn principle for input/output automata. *Information and Computation*, 82(1):81–92, July 1989.

[11] A. Mazurkiewicz. Trace theory. In *Advanced Course on Petri Nets*, GMD, Bad Honnef, September 1986.

[12] P. Panangaden and E. W. Stark. Computations, residuals, and the power of indeterminacy. In *Automata, Languages, and Programming*, pages 439–454, Springer-Verlag. Volume 317 of *Lecture Notes in Computer Science*, 1988.

[13] M. W. Shields. Deterministic asynchronous automata. In *Formal Methods in Programming*, North-Holland. 1985.

[14] E. W. Stark. Compositional relational semantics for indeterminate dataflow networks. In *Category Theory and Computer Science*, pages 52–74, Springer-Verlag. Volume 389 of *Lecture Notes in Computer Science*, Manchester, U. K., 1989.

[15] E. W. Stark. Concurrent transition system semantics of process networks. In *Fourteenth ACM Symposium on Principles of Programming Languages*, pages 199–210, January 1987.

[16] E. W. Stark. Concurrent transition systems. *Theoretical Computer Science*, 64:221–269, 1989.

[17] E. W. Stark. On the relations computed by a class of concurrent automata. In *Seventeenth Annual ACM Symposium on Principles of Programming Languages*, January 1990.

[18] G. Winskel. *Events in Computation*. PhD thesis, University of Edinburgh, 1980.

A hierarchy of domains for real-time distributed computing

G.M. Reed[1]

Oxford University Computing Laboratory
7-11 Keble Road, Oxford OX1 3QD, U.K.

ABSTRACT. *Together with A.W. Roscoe, the author has earlier presented two models (the Timed Stability Model and the Timed Failures-Stability Model) offering timed versions of Hoare's CSP. In this paper, the author outlines a hierarchy of untimed and timed models for CSP which includes the two above, and which allows one to reason about concurrent processes in a uniform fashion with the minimum of complexity. This hierarchy supports timewise refinement of specifications and the development of powerful proof rules for verification.*

1 Introduction

Programming languages which involve some form of parallelism are becoming prominent as computer science seeks to take advantage of the opportunities of distributed computing. At present, there remain serious difficulties with our efforts to reason effectively about the behaviour of such systems. Concurrent execution introduces nondeterminism and such undesirable behaviour as deadlock and starvation, and it creates the crucial need for formal proof systems to understand such pathological behaviour. These proof systems require in turn the development of formal semantic models to establish their consistency and completeness, and to assist in achieving correct designs and implementations. Among the major obstacles to the full exploitation of parallel languages are the absence of standard semantic models, and consequently the lack of a consistent calculus for the rigorous specification and verification of concurrent programs.

In the case of sequential programming languages, it is well understood that most programs can be taken to denote input-output functions or state-transformations and that the logical systems for specification and verification can be transparently based on the standard denotational semantics for the language in question. However,

[1]The work reported in this paper was supported by the U.S. Office of Naval Research.

there is no agreement on an accepted method for assigning meanings to concurrent programs. Many different semantic models have been proposed. Each of the models attempted to describe effectively a particular aspect of the complex behaviour of concurrent programs. Hence, in a given model, it may be relatively easy to reason about one type of semantic property, but difficult or impossible to reason about others. Furthermore, it is difficult to establish the relative consistency between the various existing models of a given language, since each model may be based on different mathematical or operational structures. Thus, it would be extremely beneficial at this point to isolate a single structure on which to base a uniform theory for generating a hierarchy of models for a common language capable of expressing the full complexity of distributed computing.

One major goal of such a theory would be the generation of successful models for real-time distributed computing. Although widely used throughout the world in such critical applications as aviation and nuclear power, real-time programming is a poorly understood discipline. The solutions to current problems involving sequential real-time systems will be most difficult and will take many years of work. Furthermore, the complexity of these problems will only intensify as we implement distributed real-time systems with nondeterministic behaviour. It is imperative that we begin now to develop the formal models on which the eventual solutions must be based.

The objective of this paper is to outline a uniform mathematical theory of real-time distributed computing as described above within the context of the parallel language CSP (Communicating Sequential Processes) [H,1985]. This language, which was initially described by Hoare in [H,1978] has become a major tool for the analysis of structuring methods and proof systems involving parallelism. For example, the Ada programming language can be said to be "CSP-like" in the sense that concurrently active processes interact by some form of synchronized communication: the so-called "handshake" in CSP and the "rendezvous" in Ada. Furthermore, Occam [Occ,1984], the parallel language designed for the parallel commercial computer, the INMOS Transputer, was specifically based on CSP concepts. The theoretical parallel languages CCS and SCCS ([M,1980] and [M,1983]) are even more closely related to CSP. There are, by now, hundreds of research papers in the literature concerning this family of languages. Hence a uniform theory for the definition, specification, and verification of CSP processes would be a valuable contribution towards a consensus on the formal foundations of concurrency.

The unifying mathematical structure used in our hierarchy is that of a complete ultrametric space. Real time gives a particularly natural measure for comparing processes: we can think of two processes as being t-alike if they are indistinguishable up to time t. This notion is easily formalised as a metric over the space of processes which provides a natural fixed point theory, seemingly with few of the disadvantages of the traditional ways of defining fixed points in untimed models. In particular, we are able to deal effectively for the first time with the problems of unbounded nondeterminism, infinite hiding, and the subtle relationship between divergence and deadlock.

The relationships between the models in our hierarchy are based on the key concept of stability. In untimed CSP, it is only necessary to know that a given process can or cannot diverge after engaging in a trace s; in the timed models, it is necessary to know (if the process cannot diverge after s) *when* it will again be ready to respond to the environment. This analysis leads us to consider the untimed Divergence Models ([Ros,1982], [B,1983], [OH,1983], and [BR,1985]) as providing discrete information for a given trace s ("0" cannot diverge, "∞" can diverge), and our corresponding timed models as providing continuous information ($\alpha \in [0, \infty]$ such that the process is guaranteed to be stable within α time after engaging in s). Our topological models rely on this notion of **stability**, which is the dual of divergence.

We begin our hierarchy of CSP models with the topological *Trace Model* (M_T) from [Ros,1982]. We then reformulate the untimed Divergence and Failures Models into topological terms. First, we convert the Divergence Model from [OH,1983] into the *Stability Model* (M_S). This model distinguishes between deadlock and divergence in a manner that can be readily extended to the consideration of timed processes. Processes are identified with a set of ordered pairs (s, α), where $\alpha = 0$, if the process cannot diverge after engaging in the trace s and (s, ∞) otherwise. Next, we construct a topological *Failures Model* (M_F), where a process is identified with a set of ordered pairs (s, X) such that X represents the set of alphabet events which the process can refuse after engaging in s. We then merge the two models into the *Failures-Stability Model* (M_{FS}), where a process is identified with a set of ordered three tuples (s, α, X). Having developed a uniform hierarchy of untimed models, we then construct step-by-step their timed equivalents, the *Timed Trace Model* (TM_T), the *Timed Stability Model* (TM_S), and the *Timed Failures-Stability Model* (TM_{FS}). These models are based on timed traces, continuous stability values, and timed refusals. We also consider a "half-way" model, the *(Untimed Failures)-(Timed Stability) Model* (TM^*_{FS}), which provides a relatively simple and useful model for reasoning about an important class of timed processes.

The fact that our models are complete metric spaces and all recursions are contraction mappings make them natural vehicles for correctness proofs using the form of recursion induction described in [Ros,1982]. (A predicate that represents a non-empty closed subset and which is preserved by a recursion must contain the unique fixed point.) The introduction of stability seems to enhance the range of useful predicates which represent closed sets, since it (to a limited extent) allows us to look into the future. It is also our topological structure which lets us have the choice of infinite hiding, infinite alphabet renaming, unbounded nondeterminism, or the non-equivalence of deadlock and divergence as we wish. Our basic models are flexible with respect to these and many other issues which are predetermined in the partial order models.

Finally, we formulate a hierarchical structure via projection mappings from the more complex models to the simpler ones. It is the uniformity of behaviour after a process has become stable that allows these projection mappings to preserve information. In particular, since the liveness properties predicted by the Timed Failures-Stability Model for a given process can be inferred from the time of stability on, we can often exploit this fact by reasoning in the simpler (Untimed-Failures)-(Timed Stability) Model. Indeed we have given case studies whereby the design of quite complicated timed processes can be started in the simple Traces Model and then moved gradually up the hierarchy to the Timed Failures-Stability Model, where at each step the specification and verification techniques of the relevant model are appropriate to the complexity of the design decision.

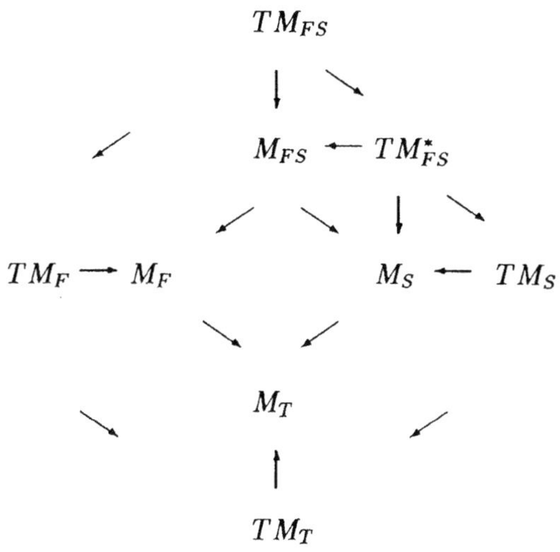

1.1 Organization of the paper

Due to the scope of the hierarchy outlined above, we will restrict our attention in this paper to the models M_{FS}, TM^*_{FS}, and TM_{FS}. Although some material is repeated for clarity, the reader should refer to [RR,1986] and [RR,1987] for further background. A complete discussion of the hierarchy with proofs of the relevant assertions about the models as well as case studies of its use in specification and verification can be found in [Re,1988].

In section 2, we summarize material from [RR,1986] and [RR,1987] concerning the objectives and assumptions behind the view of real time expressed in this paper. In section 3, we introduce the above three models, and discuss the mathematical subtlities in their treatment of deadlock, divergence, and nondeterminism. In section 4, we define the projection mappings between the models. We explore where such mappings are continuous (in the topological sense) and under what conditions they preserve determinism. In section 5, we develop specification and verification techniques based upon our hierarchical approach, whereby the untimed and timed models are simultaneously employed to reason about concurrent processes with a minimum of complexity. Finally, in section 6, we draw conclusions about our work and make comparisons with other work in the area.

2 Time and topology

2.1 Abstract syntax for CSP and TCSP (Timed Communicating Sequential Processes)

We shall essentially extend the abstract syntax for untimed CSP from [BHR,1984], [BR,1985] (with the addition of $\perp$, the diverging process which engages in no event visible to the environment). We use P, Q, R to range over syntactic processes; a, b over the alphabet Σ; X, Y over subsets of Σ; f over the set of functions from Σ to Σ; and F over "appropriate" compositions of our syntactic operators.

The basic requirement for analysing real-time programming languages is the ability to model time-outs and interrupts. This can be accomplished in CSP simply by the addition of a process *WAIT* t for each real number $t \geq 0$: the process which engages in no visible event to the environment and which terminates successfully after t units of time. Intuitively, *SKIP* should coincide with *WAIT* 0.

TCSP

$$P ::= \perp \mid STOP \mid SKIP \mid WAIT\ t \mid (a \rightarrow P) \mid P \Box Q \mid P \sqcap Q \mid P \parallel Q \mid$$
$$P\ {}_X\!\parallel_Y Q \mid P \mid\mid\mid Q \mid P;Q \mid P \setminus X \mid f^{-1}(P) \mid f(P) \mid \mu p.F(p)$$

2.2 Objectives

Our objective in [RR,1987] was the construction of a comprehensive timed CSP model (TM_{FS}), which satisfied the following:

1) **Continuous with respect to time.** The time domain should consist of all nonnegative real numbers, and there should be no lower bound on the time difference between consecutive observable events from two processes operating asynchronously in parallel.

2) **Realistic.** A given process should engage in only finitely many events in a bounded period of time.

3) **Continuous and distributive with respect to semantic operators.** All semantic operators should be continuous, and all the basic operators as defined in [BHR,1984], except recursion, should distribute over nondeterministic choice.

4) **Of verifiable design.** The model should provide a basis for the definition, specification, and verification of time critical processes with an adequate treatment of nondeterminism, which assists in avoidance of deadlock and divergence.

5) **Compatible.** The model and its associated proof systems should be a "natural" extension of untimed models and proof systems, and there should be well-understood links between the various models. The model should contain

the timed equivalents of those CSP constructs modelled in [BHR,1984], and [BR,1985].

6) **Adaptable.** The model should be flexible with respect to such issues as the equivalence of deadlock and divergence, infinite choice, infinite hiding, and unbounded nondeterminism. That is, the basic model should be adaptable to suit the design decisions made on a wide variety of issues including those mentioned above, and the consequences of such decisions should be well-understood.

The crucial element in achieving a CSP model satisfying the above requirements proved to be in making the subtle distinction between deadlock and divergence in timed processes. Previous constructions of timed CSP models have either relied on unrealistic (in the sense defined above) processes to make this distinction [J,1982], or else by design have not distinguished between these two concepts [KSRGAK]. For compatibility with the untimed models, we must be able to make this distinction. Furthermore, we believe that it is beneficial to distinguish between a property that can be established in finite time (deadlock) and one that cannot (divergence).

In [RR,1986], [RR,1987], Bill Roscoe and the author proposed a theory of communicating sequential processes based on an underlying topological structure. This choice was in a sense imposed upon us by our considerations of time and divergence. Most of the previous untimed CSP models have been based on domains of complete partial orders, and different models often have incompatible orderings. As we shall see in this paper, our complete metric space structure allows us to develop an unexpectedly elegant hierarchy of CSP models, the largest of which meets all our objectives.

2.3 Timing postulates

Let us present our basic assumptions about timing in a distributed system.

(1) **A global clock.** We assume that all events recorded by processes within the system relate to a *conceptual* global clock.

(2) **A system delay constant.** As previously indicated, we realistically postulate that a process can engage in only finitely many events in a bounded period of time. The structure of our timed models allows several parameters by which to ensure adherence with this postulate. In the current presentation, for simplicity we assume the existence of a single delay constant δ such that:

a) For each $a \in \Sigma$ and each process P, the process $(a \rightarrow P)$ is ready to engage in P only after a delay of time δ from participation in the event a.

b) A given recursive process is only ready to engage in an observable event after a delay of δ time from making a recursive call.

(3) **Hiding.** We wish $(a \rightarrow P)$ to denote the process that is willing at *any* time to engage in the event a and then to behave like the process P. Clearly, if $P = a \rightarrow P$,

we then wish $P \setminus a = \bot$. However, consider $P = a \rightarrow STOP$ (the process that is willing to engage in a at any time ≥ 0 and then to deadlock). What do we wish $P \setminus a$ to denote?

The basic concept of CSP is that a process is willing to make a communication *at the discretion* of the environment. By hiding, we remove external control. Hence, any time a process is willing to engage in an internal action, it is permitted to do so. Thus, we assume that each hidden event has taken place as soon as such event was possible.

Our intuitive model of hiding is that of placing a given process within a box in which all the events to be hidden are constantly on offer, and then concealing all the hidden events within the box from the environment.

In the above example, we would wish:

$$(a \rightarrow STOP) \setminus a = WAIT\ \delta; STOP$$

In order to model this idea of an event occurring as soon as it becomes available, we will record (either explicitly or implicitly) not only those times at which events are available, but also those at which they can *become* available.

(4) **Timed stability.** We model a timed CSP process in TM_{FS} as a specified set of ordered 3-tuples $(s, \alpha, \aleph)$, where s is a *timed trace* of the process, $\aleph$ is a timed refusal of the process (a subset of Σ in which the process can fail to engage over a specified time interval), and α is the time at which the process is guaranteed to be stable after the "observation" of s and $\aleph$. If $(s, \alpha, \aleph)$ is in the process P and $\alpha < \infty$, then the next observable event in the life of the process following s may occur at any time on or after time α at the discretion of the environment, and the set of possible next events must be the same at *all* such times after stability. Clearly no event can *become* available after α.

We think of timed stability as a red light on the outside of a process which goes off when the process can make no more internal progress.

(5) **Termination and sequential composition.** In the model TM_{FS}, the sequential composition operator treats the termination of its first argument very much as a hidden event. That is, we postulate in $P; Q$ the process P *must* terminate as soon as it can not refuse to do so. Thus, we assume that participation in the "hidden" event $\surd$ has taken place as soon as such participation was possible.

For example, consider:

$$((a \rightarrow STOP) \Box WAIT\ 1); b \rightarrow STOP$$

$$(WAIT\ 1 \mid\mid\mid WAIT\ 2); STOP$$

The first process is prepared to participate in the event a for the first unit of time and then deadlock, or to wait one unit of time and then participate in the event b and then deadlock; after time 1, it is no longer able to participate in the event a. The second process is simply equivalent to $WAIT\ 1; STOP$.

As we shall see in section 5, it is the above assumption about termination that will allow us to model interrupts in Timed CSP.

3 The models M_{FS}, TM^*_{FS}, and TM_{FS}

3.1 The Failures-Stability Model (M_{FS})

In formulating our untimed models of CSP, we will take the view that such models are to represent only that information which can be inferred from our ultimate timed model when the timing of our observations are ignored and only the ordering of events is preserved. Unavoidably, this will lead us to a pessimistic analysis of untimed information. For example, if a timed process is *not homogeneous* (i.e., if it is capable of performing the same sequence of alphabet events at different times with different future behaviours), we will be limited in our knowledge of the process inferred from untimed traces. Hence, we do not demand that these models fulfill completely our operational concept of CSP, only that they are consistent with the more informative timed models doing so. As we shall see, our non-strict treatment of divergence will mean that certain laws fail in the untimed models simply because we have lost the ability to distinguish between observations based on their relative timings.

3.1.1 Notation

$STAB = \{0, \infty\}$, the *stability set for* M_{FS}

For $S \subseteq \Sigma^* \times STAB \times P(\Sigma)$:

$$Traces(S) = \{s \mid \exists \alpha \in STAB,\ X \in P(\Sigma) \text{ such that } (s, \alpha, X) \in S\}$$

$$Stab(S) = \{(s, \alpha) \mid \exists X \in P(\Sigma) \text{ such that } (s, \alpha, X) \in S\}$$

$$Fail(S) = \{(s, X) \mid \exists \alpha \in STAB \text{ such that } (s, \alpha, X) \in S\}$$

$$SUP(S) = \{(s, \alpha, X) \mid (s, X) \in Fail(S) \wedge \alpha = sup\{\beta \mid (s, \beta) \in Stab(S)\}\}$$

For $s, u \in \Sigma^*$ and $X \in P(\Sigma)$, we define $\#s$ to be the length of s, $s \upharpoonright X$ to be the maximal subsequence w of s such that each member of w is contained in X, and $Merge(s, u)$ to be the set of all traces obtained by interleaving u and s.

3.1.2 The evaluation domain M_{FS}

We formally define M_{FS} to be those subsets S of $\Sigma^* \times STAB \times P(\Sigma)$ satisfying:

1. $\langle\rangle \in Traces(S)$
2. $s.w \in Traces(S) \Rightarrow s \in Traces(S)$
3. $(s,\alpha),(s,\alpha') \in Stab(S) \Rightarrow \alpha = \alpha'$
4. $(s,\alpha,X) \in S \wedge Y \subseteq X \Rightarrow (s,\alpha,Y) \in S$
5. $(s,\alpha,X) \in S \wedge (\forall a \in Y,\ s.\langle a\rangle \notin Traces(S)) \Rightarrow (s,\alpha,X \cup Y) \in S$

3.1.3 The complete metric on M_{FS}

If $S \in M_{FS}$, we define

$$\begin{aligned} S(n) &= \{(s,\alpha,X) \in S \mid \#s < n\} \\ &\quad \cup(s,\alpha,X) \mid \#s = n \wedge s \in Traces(S)\} \end{aligned}$$

The complete metric on M_{FS} is defined:

$$d(S_1,S_2) = inf\{2^{-n} \mid S_1(n) = S_2(n)\}$$

3.1.4 The semantic function $\mathcal{E}$

We now define the semantic function $\mathcal{E} : CSP \to M_{FS}$.

$$\begin{aligned}
\mathcal{E}[\![\perp]\!] &= \{(\langle\rangle,\infty,X) \mid X \in P(\Sigma)\} \\
\mathcal{E}[\![STOP]\!] &= \{(\langle\rangle,0,X)\} \mid X \in P(\Sigma)\} \\
\mathcal{E}[\![SKIP]\!] &= \{(\langle\rangle,0,X) \mid \surd \notin X\} \cup \{(\langle\surd\rangle,0,X) \mid X \in P(\Sigma)\} \\
\mathcal{E}[\![a \to P]\!] &= \{(\langle\rangle,0,X) \mid a \notin X\} \\
&\quad \cup\{(\langle a\rangle.s,\alpha,X) \mid (s,\alpha,X) \in \mathcal{E}[\![P]\!]\} \\
\mathcal{E}[\![P \Box Q]\!] &= SUP(\{(\langle\rangle,\alpha,X) \mid (\langle\rangle,\alpha) \in Stab(\mathcal{E}[\![P]\!]) \cup Stab(\mathcal{E}[\![Q]\!]) \\
&\qquad \wedge(\langle\rangle,X) \in Fail(\mathcal{E}[\![P]\!] \cap Fail(\mathcal{E}[\![Q]\!])\} \\
&\qquad \cup\{(s,\alpha,X) \in \mathcal{E}[\![P]\!] \cup \mathcal{E}[\![Q]\!] \mid s \neq \langle\rangle\}) \\
\mathcal{E}[\![P \sqcap Q]\!] &= SUP(\mathcal{E}[\![P]\!] \cup \mathcal{E}[\![Q]\!])
\end{aligned}$$

$$
\begin{aligned}
\mathcal{E}[\![P \| Q]\!] &= SUP(\{s, max\{\alpha_P, \alpha_Q\}, X_P \cup X_Q) \mid (s, \alpha_P, X_P) \in \mathcal{E}[\![P]\!] \\
&\qquad \wedge (s, \alpha_Q, X_Q) \in \mathcal{E}[\![Q]\!]\}) \\
\mathcal{E}[\![P\ {}_X\|_Y Q]\!] &= \{(s, max\{\alpha_P, \alpha_Q\}, Z_P \cup Z_Q \cup Z) \mid \\
&\qquad s \upharpoonright (X \cup Y) = s \ \wedge\ Z_P \subseteq X \ \wedge\ Z_Q \subseteq Y \\
&\qquad \wedge\ Z \subseteq (\Sigma - (X \cup Y)) \ \wedge\ (s \upharpoonright X, \alpha_P, Z_P) \in \mathcal{E}[\![P]\!] \\
&\qquad \wedge\ (s \upharpoonright Y, \alpha_Q, Z_Q) \in \mathcal{E}[\![Q]\!]\} \\
\mathcal{E}[\![P \,|||\, Q]\!] &= SUP(\{(s, max\{\alpha_P, \alpha_Q\}, X) \mid \exists (u, \alpha_P, X) \in \mathcal{E}[\![P]\!], \\
&\qquad (v, \alpha_Q, X) \in \mathcal{E}[\![Q]\!] \ \textit{such that}\ s \in Merge(u, v)\}) \\
\mathcal{E}[\![P ; Q]\!] &= SUP(\{(s, \alpha, X) \in \mathcal{E}[\![P]\!] \mid \surd \notin s\ \wedge (s, \alpha, X \cup \{\surd\}) \in \mathcal{E}[\![P]\!]\} \\
&\qquad \cup \{(s.w, \alpha, X) \mid \surd \notin s\ \wedge\ s.\langle \surd \rangle \in Traces(\mathcal{E}[\![P]\!]) \\
&\qquad \wedge (w, \alpha, X) \in \mathcal{E}[\![Q]\!]\}) \\
\mathcal{E}[\![P \setminus X]\!] &= SUP(\{(s \setminus X, \alpha, Y) \mid (s, \alpha, X \cup Y) \in \mathcal{E}[\![P]\!]\} \\
&\qquad \cup \{(s, \infty, Y) \mid \forall n \geq \#s, \exists w_n \in Traces(\mathcal{E}[\![P]\!]) \ \textit{such that} \\
&\qquad\quad w_n < w_{n+1}\ \wedge\ s = w_n \setminus X\}) \\
\mathcal{E}[\![f^{-1}(P)]\!] &= \{(s, \alpha, X) \mid (f(s), \alpha, f(X)) \in \mathcal{E}[\![P]\!]\} \\
\mathcal{E}[\![f(P)]\!] &= SUP(\{(f(s), \alpha, X) \mid (s, \alpha, f^{-1}(X)) \in \mathcal{E}[\![P]\!]\}) \\
\mathcal{E}[\![\mu P.F(P)]\!] &= \text{The unique fixed point of the contraction mapping } C \text{ on } M_{FS} \\
&\qquad \text{represented by } F.
\end{aligned}
$$

Note. Although each mapping formed by the composition of CSP operators represents a continuous mapping with respect to the complete partial order of [BHR,1984] and [BR,1985], it is not the case that each such mapping represents a contraction mapping (or even a continuous mapping) with respect to the complete metric d on M_{FS}.

3.1.5 Remarks on the Failures-Stability Model

Our Failures-Stability Model supports a formal treatment of deadlock, divergence, and nondeterminism over the full range of untimed CSP processes (with the restrictions to recursion imposed by our demand for contraction mappings).

However, the topological Failures-Stability Model differs in several ways from the partial order Failures-Divergences Model of [Br,1985]. We gain the ability to allow infinite hiding, to distinguish between divergence and the "possibility" of divergence, and to model unbounded nondeterminism. Due to our topological structure we lose the continuity of the hiding operator, and due to our view of the model as containing only partial (i.e., untimed information) we lose the law that $(P \setminus a) \setminus b = (P \setminus b) \setminus a$.

Discontinuity of the hiding operator

Consider P, Q, and P_n where:

$$
\begin{aligned}
Q &= b \to Q & P_0 &= Q \\
P &= a \to P & \forall n \geq 1,\ P_n &= a \to P_{n-1}
\end{aligned}
$$

Clearly, in the Failures-Stability Model semantics, we have $\lim_{n\to\infty} P_n = P$, and $P \setminus a = \perp \neq STOP$. However $\lim_{n\to\infty}(P_n \setminus a) = Q \neq (P \setminus a)$. Hence, we have distinguished between divergence and deadlock, but the hiding operator is not continuous in our topological model. However, as noted in [RR,1986] and [RR,1987], such continuity is reclaimed in the timed models.

Distinction between divergence and the possibility of divergence

Note that the Failures-Stability Model also differs from the Failures-Divergences Model [BR,1985] in that $\Box$ is not strict with respect to $\perp$. In fact, it differs from all previously mentioned CSP models relevant to divergence in that $(s, \infty) \in Stab(P)$ does not imply that $(s.w, \infty) \in Stab(P)$ for all traces w. That is, just because a process may diverge after engaging in a given trace, it does not mean that some time later after extending the trace, the process might not again become stable. For example, let $P = a \to P$ and consider the process $R = (b \to (P \setminus a)) \Box (b \to (b \to STOP))$. Both $(\langle b \rangle, \infty)$ and $(\langle bb \rangle, 0) \in Stab(R)$. If our only observation is $\langle b \rangle$, we must assume the worst; however, once we observe $\langle bb \rangle$, we know that we are safe. Although it is possible to modify our model to conform to the other untimed models in this regard, we choose to allow the finer distinction of CSP processes made possible by the topological structure of our evaluation domain.

These distinctions, which will occur in all our models, are not possible in the models of [HBR,1982], [OH,1983], [BHR,1984], and [BR,1985] precisely because of their use of least fixed points to define recursions. In any model where the partial order is based on nondeterminism or definedness ($P \sqsubseteq Q$ iff Q is more deterministic than P), the least fixed point of $\mu P.P$ (operationally, a simply diverging process) is the most nondeterministic process. Hence, one is thus forced to identify the diverging process with one that can do anything (including diverge). This is closely related to the philosophy of the Smyth powerdomain (the powerdomain of dæmonic nondeterminism).

The complete partial order models essentially state that we cannot specify anything about a process' behaviour after the possibility of divergence. This is tantamount to saying that we will never be prepared to accept, for any practical purpose, a process that can diverge. Some authors have disliked being forced to take this very strict view, and would have preferred a theory more like that of our models. This has lead them to use alternative fixed point theories such as *optimal* fixed points [Br,1982] (usually using more than one partial order). By using complete metric space domains, we are able to achieve the same effect more easily (although with restrictions on possible recursions in the untimed models).

Compatibility with the Laws of [BR,1985]

Throughout the paper, we will use the 31 laws of [BR,1985] as a reasonable test for compatibility of our models with existing CSP theory.

The Failures-Stability Model satisfies all the laws of [BR,1985] with the exception of $(P \setminus a) \setminus b = (P \setminus b) \setminus a$. The failure of this law is the price we pay in our untimed model for the finer distinctions mentioned above.

Consider:

$$\begin{array}{rcll} P & = & (c & \rightarrow \; SKIP \\ & & \Box & \\ & & a & \rightarrow \; P \,;\, b \rightarrow SKIP) \end{array}$$

The process P is prepared for all $n \geq 0$ to engage in n consecutive "a's", then to perform a single "c", then to engage in n consecutive "b's".

The untimed complete partial order models are able both to distinguish between deadlock and divergence and to have $(P \setminus a) \setminus b = (P \setminus b) \setminus a$. They do so by equating both sides to divergence. However, we wish to distinguish between $\perp$, the unstable process which engages in no visible event, and $(P \setminus b) \setminus a$, a process which not only is capable of engaging in the event c but of then becoming stable.

In the timed models, we will regain the above law. Since *time* cannot be hidden, after hiding a, the time at which the c occurs will allow us to deduce the associated finite stability value when b is hidden. The problem in the untimed model is simply a lack of information.

Nondeterminism

A process P in M_{FS} is *deterministic* provided

$$(s, X) \in Fail(P) \;\wedge\; a \in X \Rightarrow s.\langle a \rangle \notin Traces(P).$$

A process in M_F is *nondeterministic* if it fails to be deterministic.

Note that unlike the complete partial order models, $\perp$ is deterministic in our model. Observe also that the property of being deterministic is a continuous specification on M_{FS} in the sense of [Ros,1982]. That is, if a process fails to be deterministic, it must do so on a trace of finite length. Hence the limit of deterministic processes must also be deterministic, and the set of all deterministic processes is a closed subset of (M_{FS}, d). Thus, we have that $\mu P.F(P)$ is deterministic provided $\forall Q \in M_F$, (Q deterministic $\Rightarrow C(Q)$ deterministic), where C is the contraction mapping represented by F.

Unbounded nondeterminism

The reader will note that the Axiom of Bounded Nondeterminism from the Failures-Divergences Model in [BR,1985] is not included in our Model:

$$(\forall Y \in p(X),\; (s, Y) \in S) \;\Rightarrow\; (s, X) \in S$$

This condition does not hold in the topological Failures-Divergences Model. For example, let $N \subseteq \Sigma$ and consider a process which is prepared to accept either a b and then output the number 1 or to accept an a and behave like the process which is prepared to accept either a b and then output the number 2 or to accept an a and behave like the process which is prepared to $\cdots$.

Such a process can be defined by use of infinite mutual recursion.

$$\begin{aligned} P_n &= (a \rightarrow P_{n+1} \\ & \quad \Box \\ & \quad b \rightarrow (n \rightarrow STOP)) \end{aligned}$$

Clearly, if P_0 is such a process, $\{(\langle b \rangle, X) \mid X \in p(\Sigma)\} \subseteq Fail(P_0 \setminus a)$. However, $(\langle b \rangle, \Sigma) \notin Fail(P_0 \setminus a)$. (Note $P_0 \setminus a = \bot$ in the Failures-Divergences Model of [BR,1985].)

The above axiom is needed in the complete partial order models to ensure the existence of limits. This is not the case in the topological models. By dropping the axiom from our domain definition, we gain the ability to hide infinite subsets of Σ. Such hiding is not possible in the complete partial order models precisely because infinite hiding would lead to a violation of the axiom.

For example, let Q denote the process which is prepared to input any odd integer n and then to participate in the event b and then to output $n+1$. (See [H,1985] for the obvious generalization to our syntax.)

$$Q = n : Odds \rightarrow (b \rightarrow (n+1 \rightarrow STOP))$$

Now, Q is certainly definable in the complete partial order Failures-Divergences Model as well as the topological Failures-Stability Model. However, $\{(b, X) \mid X \in p(Evens)\} \subseteq Fail(Q \setminus Odds)$, but $(b, Evens) \notin Fail(Q \setminus Odds)$. Hence, $Q \setminus Odds$ is not allowed under the above axiom.

Perhaps surprisingly, we will find that in the Timed Failures-Stability Model we are again free to introduce timed equivalents of the above axiom. We will have our choice: (the axiom + finite hiding) or (not the axiom and infinite hiding). Again, the topological structure allows more flexibility in our design decisions.

However, note that we have introduced problematic behaviour in our untimed model in that it is possible to obtain unbounded nondeterminism from only *finite* hiding. For example, consider the process $P_0 \setminus a$. After communicating a b, this process can apparently choose to communicate any natural number, but cannot refuse them all.

3.2 The (Untimed Failures)-(Timed Stability) Model (TM_{FS}^*)

Now, from [Re,1988], we combine the untimed Failures Model with the Timed Stability Model. Although the resulting (Untimed Failures)-(Timed Stability) Model fails to adequately describe certain of our operational intuition about the behaviour of timed processes, it serves as a useful and relatively simple model for a wide class of timed processes.

3.2.1 Notation

We need to distinguish between the times when a process can communicate an event, and the times at which it can become ready to communicate it. As we shall see in the Timed Failures-Stability Model, one way to record such information is via the subtle use of timed failures. However, without that additional structure, we will for the time being need two ways to record each event that might occur: for $a \in \Sigma$, a will denote the communication of a at any time, but we will use the special notation $\hat{a}$ to denote communication of a the moment it becomes available. We will denote the set $\Sigma \cup \{\hat{a} \mid a \in \Sigma\}$ by $\hat{\Sigma}$; e will be a typical element of $\hat{\Sigma}$. A *timed event* is an ordered pair (t, e), where e is the communication and $t \in [0, \infty)$ is the time at which it occurs. The set $[0, \infty) \times \hat{\Sigma}$ of all timed events is denoted $T\hat{\Sigma}$, and $[0, \infty) \times \Sigma$ is denoted $T\Sigma$. Two sets of *timed traces* are

$$(T\Sigma)^*_{\leq} = \{s \in (T\Sigma)^* \mid \text{if } (t, a) \text{ precedes } (t', a') \text{ in } s, \text{ then } t \leq t'\}.$$

$$(T\hat{\Sigma})^*_{\leq} = \{s \in (T\hat{\Sigma})^* \mid \text{if } (t, e) \text{ precedes } (t', e') \text{ in } s, \text{ then } t \leq t'\}.$$

Note $(T\Sigma)^*_{\leq} \subseteq (T\hat{\Sigma})^*_{\leq}$.

If $s \in (T\hat{\Sigma})^*_{\leq}$, we define (i) $\#s$ to be the length (i.e., number of events) of s; (ii) $\tilde{s}$ to be the sequence where any $\hat{}$'s have been removed from the communications in s: thus, if $s = \langle (t, a), (t', \hat{b}) \rangle$, $\tilde{s} = \langle (t, a), (t', b) \rangle$; and (iii) $\Sigma(s)$ to be the set of all communication events appearing in $\tilde{s}$ (i.e., the second components of all timed communications in $\tilde{s}$). If $s, w \in (T\hat{\Sigma})^*_{\leq}$, we define $s \cong w$ if and only if s is a permutation of w. If $S \subseteq (T\hat{\Sigma})^*_{\leq}$, then $\Sigma(S)$ denotes $\bigcup\{\Sigma(s) \mid s \in S\}$.

Let $begin(s)$ and $end(s)$ denote respectively the earliest and latest times of any of the timed events in s. (For completeness we define $begin(\langle\rangle) = \infty$ and $end(\langle\rangle) = 0$.)

If $X \subseteq \Sigma$, $s \upharpoonright X$ is the maximal subsequence w of s such that $\Sigma(\tilde{w}) \subseteq X$. $s \setminus X = s \upharpoonright (\Sigma - X)$. If $t \in [0, \infty)$, $s \upharpoonright t$ is the subsequence of s consisting of all those events which occur no later than t. If $s = \langle\rangle$ and $t \in [0, \infty)$, define $s + t = s$, and if $t \in [-begin(s), \infty)$ and $s = \langle (t_0, e_0), (t_1, e_1), \ldots, (t_n, e_n) \rangle$, define

$$s + t = \langle (t_0 + t, e_0), (t_1 + t, e_1), \ldots, (t_n + t, e_n) \rangle .$$

If $s, w \in (T\hat{\Sigma})^*_{\leq}$, $Tmerge(s, w)$ is defined to be the set of all traces in $(T\hat{\Sigma})^*_{\leq}$ obtained by interleaving s and w.

Let $TSTAB = [0, \infty] = [0, \infty) \cup \{\infty\}$. This is the set of all "timed stability values".

For $S \subseteq (T\hat{\Sigma})^*_{\leq} \times TSTAB \times P(\Sigma)$:

$$\begin{aligned}
Traces(S) &= \{s \mid \exists \alpha \in TSTAB,\ X \in P(\Sigma) \textit{ such that } (s, \alpha, X) \in S\} \\
Stab(S) &= \{(s, \alpha) \mid \exists X \in P(\Sigma) \textit{ such that } (s, \alpha, X) \in S\} \\
Fail(S) &= \{(s, X) \mid \exists \alpha \in TSTAB \textit{ such that } (s, \alpha, X) \in S\} \\
SUP(S) &= \{(s, \alpha, X) \mid s \in Traces(S) \\
&\qquad \wedge\ \alpha = sup\{\beta \mid \exists (w, \beta) \in Stab(S) \textit{ such that } \tilde{s} = \tilde{w}\} \\
&\qquad \wedge\ (((s, X) \in Fail(S)) \vee (\alpha = \infty \wedge X \in P(\Sigma)))\ \} \\
CL_{\cong}(S) &= \{(s, \alpha, X) \mid \exists (w, \alpha, X) \in S \textit{ such that } s \cong w\}
\end{aligned}$$

3.2.2 The evaluation domain TM^*_{FS}

We formally define TM^*_{FS} to be those subsets S of $(T\hat{\Sigma})^*_{\leq} \times TSTAB \times P(\Sigma)$ satisfying:

1. $\langle\rangle \in Traces(S)$
2. $s.w \in Traces(S) \ \Rightarrow\ s \in Traces(S)$
3. $(s, \alpha, X) \in S \ \Rightarrow\ (\tilde{s}, \alpha, X) \in S$
4. $(s, \alpha, X) \in S\ \wedge\ s \cong w \ \Rightarrow\ (w, \alpha, X) \in S$
5. $s.\langle(t, a)\rangle \in Traces(S) \ \Rightarrow\ \exists t' \leq t \textit{ such that } (s \upharpoonright t').\langle(t', \hat{a})\rangle \in Traces(S)\ \wedge$ $(t' \leq t'' < t \Rightarrow (s \upharpoonright t'').\langle(t'', a)\rangle \in Traces(S)\ )$
6. $\forall t \in [0, \infty),\ \exists n(t) \in \mathbf{N} \textit{ such that } \forall s \in Traces(S), (end(s) \leq t \Rightarrow \#s \leq n(t))$
7. $(s, \alpha), (s, \alpha') \in Stab(S) \ \Rightarrow\ \alpha = \alpha'$
8. $(s, \alpha) \in Stab(S) \ \Rightarrow\ end(s) \leq \alpha$
9. $(s, \alpha) \in Stab(S) \wedge s.\langle(t, \hat{a})\rangle \in Traces(S) \ \Rightarrow\ t \leq \alpha$
10. $(s, \alpha) \in Stab(S) \ \Rightarrow$ if $t > \alpha$, $t' \geq \alpha$, $a \in \Sigma$ and $w \in (T\hat{\Sigma})^*_{\leq}$ is such that $w = \langle(t, a)\rangle.w'$, then $(s.w, \alpha', X) \in S \Rightarrow (s.(w + (t' - t)), \gamma, X) \in S$, where $\gamma \geq \alpha' + (t' - t)$
11. $(s, \alpha, X) \in S\ \wedge\ Y \subseteq X \ \Rightarrow\ (s, \alpha, Y) \in S$
12. $(s, \alpha, X) \in S \wedge \exists Y \in P(\Sigma) \textit{ such that } (\forall a \in Y,$ $\exists t \geq \alpha \textit{ such that } (s.\langle(t, a)\rangle, \emptyset) \notin Fail(S)) \ \Rightarrow\ (s, \alpha, X \cup Y) \in S$
13. $(s, \infty) \in Stab(S)\ \wedge\ X \in P(\Sigma) \Rightarrow (s, \infty, X) \in S$

3.2.3 The complete metric on TM^*_{FS}

If $S \in TM^*_{FS}$ and $t \in [0,\infty)$, we define

$$\begin{aligned} S(t) \;=\; & \{(s,\alpha,X) \in S \mid \alpha < t\} \\ & \cup\{(s,\infty,X) \mid end(s) < t \\ & \quad \wedge \exists \alpha \geq t \text{ such that } (s,\alpha) \in Stab(S) \ \wedge\ X \in P(\Sigma)\}. \end{aligned}$$

The complete metric on TM^*_{FS} is defined:

$$d(S_1,S_2) \;=\; inf\{2^{-t} \mid S_1(t) = S_2(t)\}$$

3.2.4 The semantic function $\mathcal{E}^*_T$

We now define the semantic function $\mathcal{E}^*_T : TCSP \to TM^*_{FS}$.

$$\mathcal{E}^*_T[\![\bot]\!] \;=\; \{(\langle\rangle,\infty,X) \mid X \in P(\Sigma)\}$$

$$\mathcal{E}^*_T[\![STOP]\!] \;=\; \{(\langle\rangle,0,X) \mid X \in P(\Sigma)\}$$

$$\begin{aligned} \mathcal{E}^*_T[\![SKIP]\!] \;=\; & \{(\langle\rangle,0,X) \mid \surd \notin X\} \\ & \cup\{(\langle(0,\hat{\surd})\rangle,0,X) \mid X \in P(\Sigma)\} \\ & \cup\{(\langle(t,\surd)\rangle,t,X) \mid t \geq 0 \ \wedge\ \surd \notin X\} \end{aligned}$$

$$\begin{aligned} \mathcal{E}^*_T[\![WAIT\,t]\!] \;=\; & \{(\langle\rangle,t,X) \mid \surd \notin X\} \\ & \cup\{(\langle(t,\hat{\surd})\rangle,t,X) \mid X \in P(\Sigma)\} \\ & \cup\{(\langle(t',\surd)\rangle,t',X) \mid t' \geq t \ \wedge\ X \in P(\Sigma)\} \end{aligned}$$

$$\begin{aligned} \mathcal{E}^*_T[\![a \to P]\!] \;=\; & \{(\langle\rangle,0,X) \mid a \notin X\} \\ & \cup\{(\langle(0,\hat{a})\rangle.(s+\delta),\alpha+\delta,X) \mid (s,\alpha,X) \in \mathcal{E}^*_T[\![P]\!]\} \\ & \cup\{(\langle(t,a)\rangle.(s+(t+\delta)),\alpha+t+\delta,X) \mid t \geq 0 \\ & \quad \wedge (s,\alpha,X) \in \mathcal{E}^*_T[\![P]\!]\} \end{aligned}$$

$$\begin{aligned} \mathcal{E}^*_T[\![P \Box Q]\!] \;=\; & SUP(\{(\langle\rangle,\alpha,X) \mid (\langle\rangle,\alpha) \in Stab(\mathcal{E}^*_T[\![P]\!] \cup \mathcal{E}^*_T[\![Q]\!]) \\ & \qquad \wedge (\langle\rangle,X) \in Fail(\mathcal{E}^*_T[\![P]\!]) \cap Fail(\mathcal{E}^*_T[\![Q]\!])\} \\ & \qquad \cup\{(s,\alpha,X) \in \mathcal{E}^*_T[\![P]\!] \cup \mathcal{E}^*_T[\![Q]\!] \mid s \neq \langle\rangle\}) \end{aligned}$$

$$\mathcal{E}^*_T[\![P \sqcap Q]\!] \;=\; SUP(\mathcal{E}^*_T[\![P]\!] \cup \mathcal{E}^*_T[\![Q]\!])$$

$$\begin{aligned} \mathcal{E}^*_T[\![P \| Q]\!] \;=\; & SUP(\{((s_P \vee s_Q), max\{\alpha_P,\alpha_Q\}, X_P \cup X_Q) \mid \\ & \qquad (s_P,\alpha_P,X_P) \in \mathcal{E}^*_T[\![P]\!] \ \wedge (s_Q,\alpha_Q,X_Q) \in \mathcal{E}^*_T[\![Q]\!] \wedge \tilde{s}_P = \tilde{s}_Q\}) \end{aligned}$$

$$\begin{aligned} \mathcal{E}^*_T[\![P\ {}_X\|_Y Q]\!] \;=\; & \{(s, max\{\alpha_P,\alpha_Q\}, Z_P \cup Z_Q \cup Z) \mid \exists (s_P,\alpha_P,Z_P) \in \mathcal{E}^*_T[\![P]\!] \\ & \wedge (s_Q,\alpha_Q,Z_Q) \in \mathcal{E}^*_T[\![Q]\!] \text{ with } Z_P \subseteq X \wedge Z_Q \subseteq Y \text{ such that} \\ & s \in (s_P\ {}_X\|_Y\, s_Q) \wedge Z \subseteq (\Sigma - (X \cup Y))\} \\ & \text{where} \\ & v\ {}_X\|_Y\, w = \{s \in (T\Sigma)^*_{\leq} \mid s\restriction(X \cup Y) = s \ \wedge\ \tilde{s}\restriction X = \tilde{v} \ \wedge\ \tilde{s}\restriction Y = \tilde{w} \ \wedge \\ & \qquad s\restriction X - Y = v\restriction X - Y \ \wedge\ s\restriction Y - X = w\restriction Y - X \ \wedge \\ & \qquad s\restriction X \cap Y = (v\restriction X \cap Y \vee w\restriction Y \cap X)\}. \end{aligned}$$

$$\mathcal{E}_T^*[\![P \,|||\, Q]\!] = SUP(\{(s, max\{\alpha_P, \alpha_Q\}, X) \mid \exists (u, \alpha_P, X) \in \mathcal{E}_T^*[\![P]\!] \wedge (v, \alpha_Q, X) \in \mathcal{E}_T^*[\![Q]\!] \textit{ such that } s \in Tmerge(u,v)\})$$

$$\mathcal{E}_T^*[\![P; Q]\!] = CL_{\cong}(SUP(\{(s,\alpha,X) \mid (s,\alpha,X \cup \{\checkmark\}) \in \mathcal{E}_T^*[\![P]\!] \wedge \checkmark \notin \Sigma(s)\} \cup \{(s.(w+t), \alpha + t, X) \mid s.\langle(t,\hat{\checkmark})\rangle \in Traces(\mathcal{E}_T^*[\![P]\!]) \wedge \checkmark \notin \Sigma(s) \wedge (w,\alpha,X) \in \mathcal{E}_T^*[\![Q]\!]\}))$$

$$\mathcal{E}_T^*[\![P \setminus X]\!] = \{(s \setminus X, \alpha, Y) \mid s \text{ is } X\text{-active in } \mathcal{E}_T^*[\![P]\!] \wedge \alpha = sup\{\beta \mid \exists (w,\beta) \in \mathcal{E}_T^*[\![P]\!] \textit{ such that } w \text{ is } X\text{-active} \wedge w \setminus X = s \setminus X\} \wedge (\alpha < \infty \wedge (s, X \cup Y) \in Fail(\mathcal{E}_T^*[\![P]\!]) \vee (\alpha = \infty \wedge Y \in P(\Sigma)))\}$$

where s is X-active provided s contains no element of the form (t,a) for $a \in X$ (all communications in X are in the form $\hat{a}$).

$$\mathcal{E}_T^*[\![f^{-1}(P)]\!] = \{(s,\alpha,X) \mid (f(s),\alpha,f(X)) \in \mathcal{E}_T^*[\![P]\!]\}$$

$$\mathcal{E}_T^*[\![f(P)]\!] = SUP(\{(f(s),\alpha,X) \mid (s,\alpha,f^{-1}(X)) \in \mathcal{E}_T^*[\![P]\!]\})$$

$\mathcal{E}_T^*[\![\mu P.F(P)]\!]$ = The unique fixed point of the contraction mapping $\hat{C}(Q) = C(WAIT\delta; Q)$, where C is the mapping on TM_{FS}^* represented by F.

Note. Due to our introduction of the system delay constant δ, we regain the property that all mappings formed by composition of the operators represent non-expanding (therefore continuous) mappings with respect to the complete metric space d.

3.2.5 Remarks on the (Untimed Failures)-(Timed Stability) Model

The ability to consider only "untimed" refusals is due to our concept of stability; once a process has become stable, its possibilities for participation in subsequent events must be the same for all future times. Hence, in our present model, we are actually restricting our concern to refusals which occur *after* stability. As a consequence, we are forced to a treatment of divergence which is essentially different from the other models in that we must now allow arbitrary refusals on a trace having infinite stability value. This closure property is accomplished in our semantics by the SUP operator. (A somewhat unfortunate result is the law $P \Box \bot = P \sqcap \bot$.)

A process P in TM_{FS}^* is *deterministic* provided

$$(s,X) \in P \wedge a \in X \Rightarrow s.\langle(t,a)\rangle \notin Traces(P).$$

A process in M_F is *nondeterministic* if it fails to be deterministic.

Note that as in the untimed Failures-Stability Model, $\bot$ is deterministic in the (Untimed Failures)-(Timed Stability) Model. However the property of being deterministic is not a continuous specification on TM_{FS}^*. In the timed models, continuous specifications are those based on predicates such that the failure of a process to satisfy the predicate can be decided in *finite time*.

For example, $\forall n > 0$, let $P_n = (a \rightarrow STOP)\Box(WAIT\ n)$. Each P_n is deterministic. But $\lim_{n\rightarrow\infty} P_n = (a \rightarrow STOP)\Box\bot$, and $(a \rightarrow STOP)\Box\bot$ is not deterministic.

The above issues are not crucial to the applications of the (Untimed Failures)-(Timed Stability) Model, since the model is only suitable for consideration of stable processes (i.e., processes P such that $(s, \alpha) \in Stab(P) \Rightarrow \alpha < \infty$). (see section 4)

Clearly, a vending machine which alternated between offering chocolates and biscuits at one hour intervals would never be stable. Nevertheless, such processes are certainly of practical concern. We will need the more complex Timed Failures-Stability Model to deal with such processes.

However, the (Untimed Failures)-(Timed Stability) Model is a useful model for specifying, implementing, and verifying the correctness of timed processes in which the only choices involved are between a set of events, *not* between waiting and participation. For example, suppose we wish to specify a timed vending machine which is capable of an unbounded number of transactions and which (1) does not give out more chocolates and biscuits than it receives payment for, (2) on becoming stable, cannot refuse to offer either a chocolate or a biscuit if it has received more payments than the number of chocolates and biscuits it has given out, (3) if it has given out exactly as many chocolates and biscuits as it has received payment for it cannot refuse a further coin when it becomes stable, (4) cannot engage in any action for 4 seconds after payment, and (5) is initially stable and never waits more than 4 seconds after any action before becoming stable again.

The Specification. $\forall Q \in TM^*_{FS}$, Q <u>sat</u> $S = (S_1 \wedge S_2 \wedge S_3 \wedge S_4 \wedge S_5)$ where

$$S_1)\ s \in Traces(Q) \Rightarrow \Sigma(s) \subseteq \{5p, choc, bisc\} \\ \wedge\ \#(s\restriction\{choc\}) + \#(s\restriction\{bisc\}) \leq \#(s\restriction\{5p\})$$

$$S_2)\ ((s, X) \in Fail(Q)\ \wedge\ \#(s\restriction\{choc\}) + \#(s\restriction\{bisc\}) < \#(s\restriction\{5p\})) \\ \Rightarrow \{choc, bisc\} \cap X = \emptyset$$

$$S_3)\ ((s, X) \in Fail(Q)\ \wedge\ \#(s\restriction\{choc\}) + \#(s\restriction\{bisc\}) = \#(s\restriction\{5p\})) \Rightarrow 5p \notin X$$

$$S_4)\ s.\langle(t, 5p)(t', a)\rangle \in Traces(Q) \Rightarrow t + 4 \leq t'$$

$$S_5)\ (\langle\rangle, 0) \in Stab(Q)\ \wedge\ (s.\langle(t, a)\rangle,\ \alpha) \in Stab(Q) \Rightarrow \alpha \leq t + 4$$

Consider the timed vending machine $TV^*_\Box$, where

$$TV^*_\Box \ = \ 5p \ \rightarrow \ (WAIT\ (4 - \delta);\ (choc \rightarrow TV^*_\Box \ \Box \ bisc \rightarrow TV^*_\Box))$$

To show that $TV^*_\Box$ satisfies the above specification S, we need to show that fix($\hat{C}$) <u>sat</u> S, where

$$\hat{C}(Q) \ = \ 5p \ \rightarrow \ (WAIT\ (4 - \delta);\ (choc \rightarrow Q \ \Box \ bisc \rightarrow Q))$$

Now, we can use a timed theory of recursion induction to reduce the verification of the above process to the verification of each component specification. Under such a theory (see section 5.5), it follows that each of S_1,S_2,S_4, and S_5 is continuous and satisfiable by *STOP*, and S_3 is continuous and satisfiable by $(5p \rightarrow STOP)$. Hence, we need only to show for i =1,5 and $\forall Q$, $(Q \ \underline{\text{sat}}\ S_i) \Rightarrow (\hat{C}(Q)\ \underline{\text{sat}}\ S_i)$. ($S_4$ will require that $\delta \leq 1$.) We can conclude $TV_{\Box}^{*} = \text{fix}(\hat{C})\ \underline{\text{sat}}\ S$.

Compatibility with the Laws of [BR,1985]

All but 3 of the 31 laws of [BR,1985] hold in the (Untimed Failures)-(Timed Stability) Model. These three are:

$$\begin{aligned} P \parallel STOP &= STOP \quad \text{if } P \neq \bot \\ &= \bot \qquad\quad\ \text{if } P = \bot \end{aligned}$$

$$(a \rightarrow P) \,|||\, (b \rightarrow Q) = (a \rightarrow (P \,|||\, (b \rightarrow Q)))\Box(b \rightarrow ((a \rightarrow P) \,|||\, Q))$$

$$\begin{aligned} (a \rightarrow P) \setminus b &= (a \rightarrow P \setminus b) \quad \text{if } a \neq b \\ &= P \setminus b \qquad\qquad \text{if } a = b \end{aligned}$$

The failure of the first and third laws simply reflects the passage of time (for example, $WAITn \parallel STOP = WAITn; STOP$). The failure of the second law reflects our use of the delay constant δ to implement our view of realism: two process in parallel can run faster than a sequential process.

It is easily seen that hiding is now a continuous operator (see [R,1988] for a proof in the Timed Failures-Stability Model). Also, as desired in section 2.3, we now have $(a \rightarrow STOP) \setminus a = WAIT\ \delta; STOP$.

3.3 Hiding and choice in Timed CSP

As mentioned above, the (Untimed Failures)-(Timed Stability) Model can not deal with certain of our operational assumptions about the behaviour of timed processes. In particular, although it is adequate to model the choice between participation in different events at a given time, it does not have the capacity to model our intuitive notion of the choice between participation and waiting. This inadequacy is especially evident in applications of the hiding and sequential composition operators, which have the effect of pre-empting certain traces from occurring.

Let us postulate the effect of the environment being given the choice of participating in a given process or of waiting. For example, $P = ((a \rightarrow STOP)\Box\, WAIT\ 1)$ offers the environment the initial choice of participating in the event a or of terminating successfully after 1 second. Again, what do we wish $P \setminus a$ to denote? Since we have consistently assumed that the hidden event takes place as soon as possible, we would expect (under the assumption that $\delta < 1$):

$$((a \rightarrow STOP)\Box\, WAIT\ 1) \setminus a = WAIT\ \delta; STOP$$

Similarly, we would wish:

$$(((a \to SKIP) \Box WAIT\, 1)\ ;\ b \to STOP) \setminus a \quad = \quad WAIT\, \delta\ ;\ b \to STOP$$
$$(((a \to STOP) \Box WAIT\, 1)\ ;\ b \to STOP) \setminus a \quad = \quad WAIT\, \delta\ ;\ STOP$$

Clearly, none of the above relationships hold in the (Untimed Failures)-(Timed Stability) Model. However, they all do hold in the Timed Failure-Stability Model TM_{FS}. Hence, when defining $P \setminus X$ in the context of $\Box$, we must be able to exclude some traces in P from consideration based on information about their possible refusals prior to stability.

Note also that

$$P = ((a \to STOP) \Box WAIT\ 1); b \to STOP$$

does not behave as we postulated in assumption (5) from section 2.3. Since $Traces(P)$ includes $\langle(t, a)\rangle$ for $t > 1$, the process $(a \to STOP) \Box WAIT\ 1$ was not terminated at time 1 by the sequential operator as we would expect in a timed failures model. In fact, the situation is even more complicated. Consider:

$$P_1 \quad = \quad ((a \to STOP) \Box (b \to STOP)) \sqcap (a \to c \to STOP)$$
$$P_2 \quad = \quad ((a \to c \to STOP) \Box (b \to STOP)) \sqcap (a \to STOP)$$

In the (Untimed Failures)-(Timed Stability) Model, as well as in the untimed Failures-Stability Model, $P_1 = P_2$. Such processes would seem free of our current concern since they do not involve either hiding or delays. However, let

$$Q = (WAIT\, 1 \Box (b \to STOP))\ ;\ a \to c \to STOP$$

Operationally, as indicated in our assumption (5) from section 2.3, we would expect:

$$(P_1 \parallel Q) \setminus b \neq (P_2 \parallel Q) \setminus b$$

In particular, we would expect:

$$\langle(1, a)(1 + \delta, c)\rangle \in Traces((P_1 \| Q) \setminus b) \ \text{ but } \ \langle(1, a)(1 + \delta, c)\rangle \notin Traces((P_2 \| Q) \setminus b)$$

Hence, in our final Timed Failures-Stability Model, we must distinguish between processes such as P_1 and P_2. Note that it will be impossible to make such a distinction based on what a process can refuse <u>after</u> a given state has been achieved. Hence, it will be necessary not only to record refusals on a given trace prior to stability, but also to record what refusals were involved in the state changes which led to the final state witnessed by the trace. This is a *crucial* issue in achieving a successful semantics for real-time parallel languages. As we will see, the distributivity of the hiding operator over $\sqcap$ depends on the subtle resolution of this issue.

3.4 The Timed Failures-Stability Model (TM_{FS}) [RR,1987]

Finally, we are ready to present our Timed Failures-Stability Model which meets all our objectives. Note there are subtle changes in certain of the domain axioms as presented in [RR,1986] and [RR,1987]. These changes were brought about by understanding gained during the development of our hierarchy of untimed and timed models. The definitive version of the Timed Failures-Stability Model will appear in a forthcoming journal article by the author and Bill Roscoe.

Unfortunately to define the Timed Failures-Stability Model, it is not simply a matter of doing the "obvious thing" by combining timed stabilities and timed failures. We must be a bit careful.

Consider the processes:

$$P = WAIT\ 1; a \rightarrow STOP$$

$$Q = (b \rightarrow STOP \;\Box\; WAIT\ 1; a \rightarrow (WAIT\ 1; STOP))$$

As previously mentioned, to achieve our desired operational view of allowing a process the choice between waiting and participation, we must allow certain traces to be pre-empted by hidden events. Hence, although $\langle(1,a)\rangle$ is a trace in $P \sqcap Q$, we would expect it to be pre-empted in $(Q \setminus b)$ but not $(P \setminus b)$. However, if this is so and we calculate stability values in the usual manner, we would have:

$$(\langle(1,a)\rangle, 1+\delta) \in Stab((P \setminus b) \sqcap (Q \setminus b))$$
$$(\langle(1,a)\rangle, 2+\delta) \in Stab((P \sqcap Q) \setminus b)$$

Thus, we are in danger of losing the distributivity of the hiding operator over nondeterministic choice. Fortunately, we need only to keep track of stability values during state transitions in the same manner as we must record past refusal behaviour, i.e., stability values become associated to each $(s, \aleph)$.

3.4.1 Notation

I: $TINT$	$= \{\, [l(I), r(I)) \mid 0 \leq l(I) < r(I) < \infty\}$	(Time Intervals)
T: $RTOK$	$= \{I \times X \mid I \in TINT \wedge X \in P(\Sigma)\}$	(Refusal Tokens)
$\aleph$: $RSET$	$= \{\bigcup Z \mid Z \subseteq RTOK \wedge Z \text{ finite}\}$	(Refusal Sets)

1) $\forall \aleph \in RSET$,

$$\begin{aligned}
\Sigma(\aleph) &= \{a \in \Sigma \mid \exists t \in [0,\infty) \textit{ such that } (t,a) \in \aleph\} \\
I(\aleph) &= \{t \in [0,\infty) \mid \exists a \in \Sigma \textit{ such that } (t,a) \in \aleph\} \\
begin(\aleph) &= inf(I(\aleph)),\ \forall \aleph \neq \emptyset \\
end(\aleph) &= sup(I(\aleph)),\ \forall \aleph \neq \emptyset \\
begin(\aleph) &= \infty,\ \text{for } \aleph = \emptyset \\
end(\aleph) &= 0,\ \text{for } \aleph = \emptyset \\
\forall t \geq -begin(\aleph),\ \aleph + t &= \{(t'+t,a) \mid (t',a) \in \aleph\},\ \text{for } \aleph \neq \emptyset \\
\aleph + t &= \aleph,\ \text{for } \aleph = \emptyset \\
\forall t \in [0,\infty),\ \aleph \upharpoonright t &= \aleph \cap ([0,t) \times \Sigma).
\end{aligned}$$

2) $\forall S \subseteq (T\Sigma)^*_{\leq} \times TSTAB \times RSET$,

$$\begin{aligned}
Traces(S) &= \{s \mid \exists \alpha \in TSTAB,\ \aleph \in RSET \textit{ such that } (s,\alpha,\aleph) \in S\} \\
Stab(S) &= \{(s,\alpha) \mid \exists \aleph \in RSET \textit{ such that } (s,\alpha,\aleph) \in S\} \\
Fail(S) &= \{(s,\aleph) \mid \exists \alpha \in TSTAB \textit{ such that } (s,\alpha,\aleph) \in S\} \\
\underline{SUP}(S) &= \{(s,\alpha,\aleph) \mid (s,\aleph) \in Fail(S) \\
&\qquad \wedge\ \alpha = sup\{\beta \mid (s,\beta,\aleph) \in S\}\} \\
CL_{\cong}(S) &= \{(s,\alpha,\aleph) \mid \exists (w,\alpha,\aleph) \in S \textit{ such that } s \cong w\}
\end{aligned}$$

3.4.2 The evaluation domain TM_{FS}

We formally define TM_{FS} to be those subsets S of $(T\Sigma)^*_{\leq} \times TSTAB \times RSET$ satisfying:

1. $\langle\rangle \in Traces(S)$
2. $(s.w,\aleph) \in Fail(S) \ \Rightarrow\ (s, \aleph \upharpoonright begin(w)) \in Fail(S)$
3. $(s,\alpha,\aleph) \in S\ \wedge\ s \cong w\ \Rightarrow\ (w,\alpha,\aleph) \in S$
4. $(s,\aleph) \in Fail(S) \wedge t \geq 0 \ \Rightarrow\ \exists \aleph' \in RSET \textit{ such that } \aleph \subseteq \aleph' \ \wedge\ (s,\aleph') \in Fail(S)\ \wedge$ $(t' \leq t\ \wedge\ (t',a) \notin \aleph') \Rightarrow (s \upharpoonright t'.\langle(t',a)\rangle, \aleph' \upharpoonright t') \in Fail(S)$
5. $t \in [0,\infty) \Rightarrow \exists n(t) \in \mathbf{N}$ such that $\forall s \in Traces(S)$, $end(s) \leq t \Rightarrow \#s \leq n(t)$
6. $(s,\alpha,\aleph),(s,\beta,\aleph) \in S \Rightarrow\ \alpha = \beta$
7. $(s,\alpha,\aleph) \in S\ \Rightarrow\ end(s) \leq \alpha$

8. $(s, \alpha, \aleph) \in S \ \wedge \ (s.\langle(t, a)\rangle, \aleph) \in Fail(S) \ \wedge \ t > t' \geq \alpha \ \wedge \ t \geq end(\aleph) \ \Rightarrow (t', a) \notin \aleph$

9. $(s, \alpha, \aleph) \in S \ \Rightarrow$ if $t > \alpha$, $t' \geq \alpha$, $a \in \Sigma$ and
   $w \in (T\hat{\Sigma})^*_{\leq}$ is such that $w = \langle(t, a)\rangle.w'$, then
   $(s.w, \alpha', \aleph') \in S \ \wedge \ \aleph \subseteq \aleph' \upharpoonright t \ \Rightarrow$
   $\exists \gamma \geq \alpha' + (t' - t)$ *such that*
   $(s.(w + (t' - t)), \gamma, \aleph_1 \cup \aleph_2 \cup (\aleph_3 + (t' - t))) \in S$,
   where $\aleph_1 = \aleph' \upharpoonright \alpha$, $\aleph_2 = [\alpha, t') \times \Sigma(\aleph' \cap ([\alpha, t) \times \Sigma))$,
   and $\aleph_3 = \aleph' \cap ([t, \infty) \times \Sigma)$.

10. $(s, \alpha, \aleph) \in S \wedge \aleph' \in RSET$ such that $\aleph' \subsetneq \aleph$
    $\Rightarrow \ \exists \alpha' \geq \alpha$ such that $(s, \alpha', \aleph') \in S$

11. $(s.w, \alpha, \aleph) \in S \wedge \aleph' \in RSET$ is such that $end(s) \leq begin(\aleph') \ \wedge$
    $end(\aleph') \leq begin(w) \ \wedge \ (\forall (t, a) \in \aleph', (s.\langle(t, a)\rangle, \aleph \upharpoonright t) \notin Fail(S))$
    $\Rightarrow \ (s.w, \alpha, \aleph \cup \aleph') \in S$

12. $(s, \alpha, \aleph) \in S \Rightarrow$
    $(\forall I \in TINT, \ \ I \subseteq [\alpha, \infty) \Rightarrow (s, \alpha, \aleph \cup (I \times \Sigma(\aleph \cap ([\alpha, \infty) \times \Sigma)))) \in S)$

Although some of these axioms appear complex, each reflects a simple healthiness property. Now that we have reached our final model, let us give a brief intuitive explanation of each axiom.

1. Every process has initially done nothing at all.

2. If a process has been observed to communicate $s.w$ while refusing $\aleph$ then at the time when the first event of w occurred the pair $(s, \aleph \upharpoonright begin(w))$ had been observed.

3. Traces which are equivalent (i.e., are the same except for the permutation of events happening at the same times) are interchangeable.

4. Given a trace/refusal pair $(s, \aleph)$ and a time t, there exists a single refusal $\aleph'$ in $RSET$ containing $\aleph$ such that $(s, \aleph')$ is also a "state" of the process, and all timed events (up to time t) not in $\aleph'$ are possible communications for the process consistent with that state. This axiom is a finitary axiom since $\aleph'$, being the union of finitely many products of timed intervals and event sets, ensures that only finitely many changes in refusal behaviour can occur by time t. It is needed to establish links with the "hatted" models (see [R,1988]), and it is necessary for our assumption that hidden events must occur as soon as possible. Unfortunately, its existence makes impossible the introduction of general infinite choice (see 3.4.5 below).

5. The process cannot perform an infinite number of events in a finite time.

6. There is only one stability value for each trace/refusal pair: the least time by which we can guarantee stability after the given observation.

7. The time of stability is always after the end of the trace.

8. A stable process cannot communicate an event which it has been seen to refuse since stability.

9. After stability the same set of events is available at all times. Furthermore the behaviour of a process after such an event does not depend on the exact time at which it was executed. Thus the trace w and the corresponding part of the refusal may be translated so as to make the first event of w now occur at time t'. The stability value γ corresponding to the translated behaviour may, in general, be greater than the obvious value because the translated behaviour may in some circumstances be possible for other reasons.

10. If a process has been observed to communicate s while refusing $\aleph$ then it can communicate the same trace while refusing any subset of $\aleph$. This simply reflects the fact that the environment might offer it less and so have less refused. However, because less has been observed, the stability value can, in general, be greater.

11. Any set of impossible events *must* be refused if offered. Such observation does not give any extra information to the observer, so the stability value is not affected.

12. Something that is refused at one time on or after stability is refused at all such times.

Note. In both axioms 8 and 9, we carefully distinguish (via t' and t) between events *at* stability and events *after* stability. This is a necessary distinction. For example, the process $P = (a \rightarrow STOP \Box WAIT\ 1); b \rightarrow STOP$ will become stable on the pair $(\langle\rangle, [0,1) \times (\Sigma - \{a\}))$ at time 1; however $\langle(1,a)\rangle \in Traces(P)$ but $\forall t > 1, \langle(t,a)\rangle \notin Traces(P)$. Essentially, we are introducing nondeterminism at the time of stability in such processes. This is an unavoidable consequence of using $\surd$ as a termination *event*; a similar situation occurs in the untimed models of [BR,1985] with $(a \rightarrow STOP \Box SKIP); b \rightarrow STOP$.

3.4.3 The complete metric on TM_{FS}

If $S \in TM_{FS}$ an $t \in [0,\infty)$, we define

$$\begin{aligned} S(t) &= \{(s,\alpha,\aleph) \in S \mid \alpha < t \wedge end(\aleph) < t\} \\ &\quad \cup\{(s,\infty,\aleph) \mid end(s) < t \wedge end(\aleph) < t \wedge \exists \alpha \geq t \textit{ such that } (s,\alpha,\aleph) \in S\}. \end{aligned}$$

The complete metric on TM_{FS} is defined:

$$d(S_1,S_2) = inf\{2^{-t} \mid S_1(t) = S_2(t)\}$$

3.4.4 The Semantic function $\mathcal{E}_T$

We now define the semantic function $\mathcal{E}_T : TCSP \to TM_{FS}$.

$$\mathcal{E}_T[\![\bot]\!] = \{(\langle\rangle, \infty, \aleph) \mid \aleph \in RSET\}$$

$$\mathcal{E}_T[\![STOP]\!] = \{(\langle\rangle, 0, \aleph) \mid \aleph \in RSET\}$$

$$\begin{aligned}\mathcal{E}_T[\![SKIP]\!] = {} & \{(\langle\rangle, 0, \aleph) \mid \surd \notin \Sigma(\aleph)\} \\ & \cup\{(\langle(t,\surd)\rangle, t, \aleph_1 \cup \aleph_2) \mid t \geq 0 \wedge (I(\aleph_1) \subseteq [0,t) \wedge \surd \notin \Sigma(\aleph_1)) \\ & \wedge I(\aleph_2) \subseteq [t, \infty)\}\end{aligned}$$

$$\begin{aligned}\mathcal{E}_T[\![WAIT\, t]\!] = {} & \{(\langle\rangle, t, \aleph) \mid \aleph \cap ([t,\infty) \times \{\surd\}) = \emptyset\} \\ & \cup\{(\langle(t',\surd)\rangle, t', \aleph_1 \cup \aleph_2 \cup \aleph_3) \mid t' \geq t \wedge I(\aleph_1) \subseteq [0,t) \\ & \wedge(I(\aleph_2) \subseteq [t,t') \wedge \surd \notin \Sigma(\aleph_2)) \wedge I(\aleph_3) \subseteq [t', \infty)\}\end{aligned}$$

$$\begin{aligned}\mathcal{E}_T[\![a \to P]\!] = {} & \{(\langle\rangle, 0, \aleph) \mid a \notin \Sigma(\aleph)\} \\ & \cup\{(\langle(t,a)\rangle.(s + (t+\delta)), \alpha + t + \delta, \aleph_1 \cup \aleph_2 \cup (\aleph_3 + (t+\delta))) \mid t \geq 0 \\ & \wedge(I(\aleph_1) \subseteq [0,t) \wedge a \notin \Sigma(\aleph_1)) \wedge I(\aleph_2) \subseteq [t, t+\delta) \\ & \wedge(s, \alpha, \aleph_3) \in \mathcal{E}_T[\![P]\!]\}\end{aligned}$$

$$\begin{aligned}\mathcal{E}_T[\![P \Box Q]\!] = {} & \underline{SUP}(\{(\langle\rangle, max\{\alpha_P, \alpha_Q\}, \aleph) \mid (\langle\rangle, \alpha_P, \aleph) \in \mathcal{E}_T[\![P]\!] \\ & \quad \wedge (\langle\rangle, \alpha_Q, \aleph) \in \mathcal{E}_T[\![Q]\!]\} \\ & \quad \cup\{(s, \alpha, \aleph) \mid s \neq \langle\rangle \ \wedge \ (s, \alpha, \aleph) \in \mathcal{E}_T[\![P]\!] \cup \mathcal{E}_T[\![Q]\!] \\ & \quad \wedge (\langle\rangle, \aleph \restriction begin(s)) \in Fail(\mathcal{E}_T[\![P]\!]) \cap Fail(\mathcal{E}_T[\![Q]\!])\})\end{aligned}$$

$$\mathcal{E}_T[\![P \sqcap Q]\!] = \underline{SUP}(\mathcal{E}_T[\![P]\!] \cup \mathcal{E}_T[\![Q]\!])$$

$$\begin{aligned}\mathcal{E}_T[\![P \| Q]\!] = {} & \underline{SUP}(\{(s, max\{\alpha_P, \alpha_Q\}, \aleph_P \cup \aleph_Q) \mid (s, \alpha_P, \aleph_P) \in \mathcal{E}_T[\![P]\!] \\ & \quad \wedge(s, \alpha_Q, \aleph_Q) \in \mathcal{E}_T[\![Q]\!]\})\end{aligned}$$

$$\begin{aligned}\mathcal{E}_T[\![P\ {}_X\|_Y Q]\!] = {} & \{(s, max\{\alpha_P, \alpha_Q\}, \aleph_P \cup \aleph_Q \cup \aleph_Z) \mid \exists (s_P, \alpha_P, \aleph_P) \in \mathcal{E}_T[\![P]\!], \\ & (s_Q, \alpha_Q, \aleph_Q) \in \mathcal{E}_T[\![Q]\!] \text{ with } \Sigma(\aleph_P) \subseteq X \wedge \Sigma(\aleph_Q) \subseteq Y \ such\ that \\ & s \in (s_P\ {}_X\|_Y\, s_Q) \wedge \Sigma(\aleph_Z) \subseteq (\Sigma - (X \cup Y))\} \\ & \text{where} \\ & v\ {}_X\|_Y\, w = \{s \in (T\Sigma)^*_{\leq} \mid s \restriction (X \cup Y) = s \ \wedge \ s \restriction X = v \ \wedge \ s \restriction Y = w\}\end{aligned}$$

$$\begin{aligned}\mathcal{E}_T[\![P \,|||\, Q]\!] = {} & \underline{SUP}(\{(s, max\{\alpha_P, \alpha_Q\}, \aleph) \mid \exists (u, \alpha_P, \aleph) \in \mathcal{E}_T[\![P]\!] \\ & \quad \wedge (v, \alpha_Q, \aleph) \in \mathcal{E}_T[\![Q]\!] \ such\ that\ s \in Tmerge(u, v)\})\end{aligned}$$

$$\begin{aligned}\mathcal{E}_T[\![P; Q]\!] = {} & CL_{\cong}(\underline{SUP}(\{(s, \alpha, \aleph) \mid \surd \notin \Sigma(s) \ \wedge \ \forall I \in TINT \\ & \quad (s, \alpha, \aleph \cup (I \times \{\surd\})) \in \mathcal{E}_T[\![P]\!]\} \\ & \quad \cup\{(s.(w+t), \alpha + t, \aleph_1 \cup (\aleph_2 + t)) \mid \ \surd \notin \Sigma(s) \\ & \quad \wedge\ end(\aleph_1) \leq t \\ & \quad \wedge (s.\langle(t,\surd)\rangle, \aleph_1 \cup ([0,t) \times \{\surd\})) \in Fail(\mathcal{E}_T[\![P]\!]) \\ & \quad \wedge (w, \alpha, \aleph_2) \in \mathcal{E}_T[\![Q]\!]\}))\end{aligned}$$

$$\mathcal{E}_T[\![P \setminus X]\!] = \underline{SUP}(\{s \setminus X, \beta, \aleph) \mid \exists \alpha \geq \beta \geq end(s) \text{ such that } (s, \alpha, \aleph \cup ([0, max\{\beta, end(\aleph)\}) \times X)) \in \mathcal{E}_T[\![P]\!]\})$$

$$\mathcal{E}_T[\![f^{-1}(P)]\!] = \{(s, \alpha, \aleph) \mid (f(s), \alpha, f(\aleph)) \in \mathcal{E}_T[\![P]\!]\}$$

$$\mathcal{E}_T[\![f(P)]\!] = \underline{SUP}(\{(f(s), \alpha, \aleph) \mid (s, \alpha, f^{-1}(\aleph)) \in \mathcal{E}_T[\![P]\!]\})$$

$\mathcal{E}_T[\![\mu p.F(p)]\!]$ = The unique fixed point of the contraction mapping $\hat{C}(Q) = C(WAIT\,\delta; Q)$, where C is the mapping on TM_{FS} represented by F.

3.4.5 Remarks on the Timed Failures-Stability Model

The Timed Failures-Stability Model meets all our objectives. It is realistic and continuous with respect to time; it has continuous operators which (except for recursion) distribute over nondeterministic choice; and it is compatible with the semantics and proof systems of the untimed models. Furthermore, it allows for distinction between divergence and the possibility of divergence, and it offers the option either to have infinite hiding or to restrict unbounded nondeterminism.

The Timed Failures-Stability Model satisfies 27 of the 31 laws of [BR,1985]. It fails the same three as did the (Untimed Failures)-(Timed Stability) Model for the same reasons.

In addition, as discovered by Steve Schneider, it fails the law $P \sqcap (Q \Box R) = (P \sqcap Q) \Box (P \sqcap R)$. For example,

$$(a \rightarrow STOP) \sqcap ((b \rightarrow STOP) \Box (c \rightarrow STOP))$$
$$\neq$$
$$((a \rightarrow STOP) \sqcap (b \rightarrow STOP)) \Box ((a \rightarrow STOP) \sqcap (c \rightarrow STOP))$$

Clearly, $(\langle(1, b)\rangle, [0, 1) \times \{c\})$ is in the failures of the second process but not in the failures of the first. Observe that indeed the two processes **are** operationally different in this respect, since as noted, timed state is determined by past refusal behaviour as well as future.

Domain axiom 4 and infinite choice

Consider the following possible addition to our semantics:

$$\mathcal{E}_T[\![\Box_{n \in \mathbb{N}} P_n]\!] = \underline{SUP}(\{(\langle\rangle, \alpha, \aleph) \mid (\langle\rangle, \alpha, \aleph) \in \cup_{n \in \mathbb{N}} \mathcal{E}_T[\![P_n]\!] \wedge (\langle\rangle, \aleph) \in \cap_{n \in \mathbb{N}} Fail(\mathcal{E}_T[\![P_n]\!])\} \cup \{(s, \alpha, \aleph) \mid s \neq \langle\rangle \wedge (s, \alpha, \aleph) \in \cup_{n \in \mathbb{N}} \mathcal{E}_T[\![P_n]\!] \wedge (\langle\rangle, \aleph \restriction begin(s)) \in \cap_{n \in \mathbb{N}} Fail(\mathcal{E}_T[\![P_n]\!])\})$$

Such a construct is not consistent in TM_{FS}, since it violates domain axiom 4. For example, consider

$$P = \Box_{n \in \mathbb{N}^+} ((n \rightarrow STOP \mid\mid\mid WAIT\ (1 - 1/n)); STOP).$$

Suppose axiom 4 holds, and apply it with $(\langle\rangle, \emptyset) \in Fail(P)$ and $1 > 0$ to produce the required $\aleph'$. Then $(t < 1 \ \wedge \ (t,n) \notin \aleph') \Rightarrow (\langle(t,n)\rangle, \aleph' \upharpoonright t) \in Fail(P)$. Hence, $\forall n \in \mathbf{N}^+$, $([1-n^{-1},1) \times \{n\}) \subseteq \aleph'$ and $([0,1-n^{-1}) \times \{n\}) \cap \aleph' = \emptyset$. However, there can be no such $\aleph' \in RSET$, which is a contradiction.

Another example of the failure of the above construct (as pointed out by Steve Schneider) is $P \setminus a$, where

$$P = \Box_{n \in \mathbf{N}^+}(WAIT\ (1/n); a \rightarrow STOP).$$

Even if we delete axiom 4, we would still have a problem. Although P would satisfy all axioms except axiom 4, $P \setminus a$ would violate axiom 11. Here we see that a general deterministic operator is not consistent with our basic assumption that hidden events must occur as soon as possible, since in $P \setminus a$ there is no such time.

[Note that the above concern is not a problem in introducing an infinite $\sqcap$ choice operator However the introduction of such an operator is valid only if there exists a function $n(t)$ from domain axiom 5 which is uniform over the set of argument processes.]

While not having a general infinite choice operator is not a serious loss (since such choice is not implementable anyway), it would be undesirable not to have a timed equivalent of $a : X \rightarrow P(a)$. (The process that is willing to accept any event a from the (possibly infinite) set X and then to behave like the process $P(a)$). Fortunately, we are able to have such an equivalent that is consistent with axiom 4.

$$\begin{aligned}
\mathcal{E}_T[\![a : X \rightarrow P(a)]\!] \ = \ & \underline{SUP}(\{(\langle\rangle, 0, \aleph) \mid X \cap \Sigma(\aleph) = \emptyset\} \\
& \cup \\
& \{(\langle(t,a)\rangle.(s + (t+\delta)), \alpha + (t+\delta), \aleph_1 \cup \aleph_2 \cup (\aleph_3 + (t+\delta))) \mid \\
& \ t \geq 0 \ \wedge \ (I(\aleph_1) \subseteq [0,t) \wedge X \cap \Sigma(\aleph_1) = \emptyset) \\
& \ \wedge I(\aleph_2) \subseteq [t, t+\delta) \wedge (s, \alpha, \aleph_3) \in \mathcal{E}_T[\![P(a)]\!]\})
\end{aligned}$$

Nondeterminism

A process P in TM_{FS} is *deterministic* provided

$$(s, \aleph) \in Fail(P) \ \wedge \ (t,a) \in \aleph \Rightarrow s.\langle(t,a)\rangle \notin Traces(P).$$

A process in M_{FS} is *nondeterministic* if it fails to be deterministic.

Note that as in the untimed Failures-Stability Model, $\perp$ is deterministic in the Timed Failures-Stability Model. Also the property of being deterministic is again a continuous specification on TM_{FS}, and the set of all deterministic processes is a closed subset of (TM_{FS}, d). Thus, we have again that $\mu P.F(P)$ is deterministic provided $\forall Q \in TM_{FS}$, (Q deterministic $\Rightarrow C(Q)$ deterministic), where C is the contraction mapping represented by F (see section 5.5).

Unbounded nondeterminism and infinite hiding

We are now free to introduce an Axiom of Bounded Nondeterminism:

$$I \in TINT \wedge X \in P(\Sigma) \text{ such that } (\forall Y \in p(X), (s, \aleph \cup (I \times Y)) \in Fail(S) \Rightarrow (s, \aleph \cup (I \times X))) \in Fail(S)$$

Since, the convergence in our complete metric space is independent of nondeterminism, we have the choice whether or not to have such an axiom.

Note that the introduction of the above axiom does not allow infinite hiding. For example, as in section 2.2.3, let

$$Q = n : Odds \rightarrow (b \rightarrow (n+1 \rightarrow STOP))$$

Now, Q is definable in the complete partial order failures-divergence model as well as the our current timed model. However, $Q \setminus Odds$ is not allowed under the above axiom, since $\{(\langle(\delta, b)\rangle, [2\delta, 1+2\delta) \times X) \mid X \in p(Evens)\} \subseteq Fail(Q) \setminus Odds$, but $(\langle(\delta, b)\rangle, [2\delta, 1+2\delta) \times Evens) \notin Q \setminus Odds$.

In the model as formulated, infinite hiding is allowed.

Given that in the timed model we have chosen not to ignore what might happen after possible divergence, it might appear that it is possible (as in the untimed model) to obtain unbounded nondeterminism from only finite hiding. For example, consider $P_0 \setminus a$ from section 2.2.3 where

$$\begin{aligned} P_n \quad = \quad & (a \rightarrow P_{n+1} \\ & \Box \\ & b \rightarrow (n \rightarrow STOP)) \end{aligned}$$

As previously indicated, $P_0 \setminus a$ would be problematic in an untimed model. However it does not defy the axiom now because only finitely much nondeterminism is exhibited up to any finite time (as only finitely many hidden 'a's will have occurred).

It will be consistent to bring in an axiom of bounded nondeterminism if, and only if, we do not want to model any operator which, like infinite hiding, has the potential of introducing infinitely many choices in a finite time.

Distributivity of hiding over nondeterministic choice

A proof of this law can be found in [R,1988]. Here, we simply illustrate the semantics of TM_{FS} by considering the example from the introduction to this section:

$$P = WAIT\ 1; a \rightarrow STOP$$

$$\begin{aligned} Q \quad = \quad & (b \rightarrow STOP \\ & \Box \\ & WAIT\ 1; a \rightarrow (WAIT\ 1; STOP)) \end{aligned}$$

Indeed, $(\langle(1, a)\rangle, 1+\delta) \in Stab((P \setminus b) \sqcap (Q \setminus b))$ and $(\langle(1, a)\rangle, 2+\delta) \in Stab((P \sqcap Q) \setminus b)$.

However, $(\langle(1, a)\rangle, \alpha, \aleph) \in (P \setminus b) \sqcap (Q \setminus b) \Leftrightarrow (\langle(1, a)\rangle, \alpha, \aleph) \in (P \sqcap Q) \setminus b$.

Note:

$(\langle(1,a)\rangle, 1+\delta, [0,1) \times \{b\}) \in (P \setminus b)$ and $(\langle(1,a)\rangle, [0,1) \times \{b\}) \notin Fail(Q)$
$\Rightarrow (\langle(1,a)\rangle, 1+\delta, [0,1) \times \{b\}) \in (P \setminus b) \sqcap (Q \setminus b)$

$(\langle(1,a)\rangle, 2+\delta, [0,1) \times \{a\}) \in (Q \setminus b)$ and
$(\langle(1,a)\rangle, 1+\delta, [0,1) \times \{a\}) \in (P \setminus b) \Rightarrow (\langle(1,a)\rangle, 2+\delta, [0,1) \times \{a\}) \in (P \setminus b) \sqcap (Q \setminus b)$

4 The Hierarchy H of CSP Models

Let us now consider the relationships between the various untimed and timed models of CSP which we have introduced. We wish to show that there is a natural hierarchy formed by these models, and that this hierarchy can be exploited in the specification and verification of CSP processes.

4.1 The projection mappings

4.1.1 Notation

Let $strip_t : (T\hat{\Sigma})^*_{\leq} \to \Sigma^*$, where

$$(s = \langle\rangle) \Rightarrow strip_t(s) = \langle\rangle \in \Sigma^*$$
$$(\tilde{s} = \langle(t_0, a_0)(t_1, a_1)...(t_n, a_n)\rangle) \Rightarrow strip_t(s) = \langle a_0 a_1 ... a_n\rangle \in \Sigma^*.$$

$\forall \alpha \in TSTAB$, let

$$\overline{\alpha} = \infty \quad \text{if } \alpha = \infty$$
$$\overline{\alpha} = 0 \quad \text{if } \alpha < \infty.$$

4.1.2 $TM_{FS} \to TM^*_{FS}$

Define $\pi : TM_{FS} \to TM^*_{FS}$ as follows:

Let $P \in TM_{FS}$. For each $s \in (T\hat{\Sigma})^*_{\leq}$, let

$$P[s] = \{(\tilde{s}, \beta, \aleph) \in P \mid s = v_1.\langle(t, \hat{a})\rangle.v_2 \wedge t > 0 \Rightarrow \exists t' < t \text{ such that } [t', t) \times \{a\} \subseteq \aleph\}.$$

$$\begin{aligned}
\pi(P) \;=\; & \{(s, \alpha, X) \in (T\hat{\Sigma})^*_{\leq} \times TSTAB \times P(\Sigma) \mid \\
& \quad (i)\; P[s] \neq \emptyset \\
& \quad \wedge \\
& \quad (ii)\; \alpha = sup\{\beta \mid (\tilde{s}, \beta, \aleph) \in P[s]\} \\
& \quad \wedge \\
& \quad (iii)\; \alpha < \infty \wedge \exists (\tilde{s}, \beta, \aleph) \in P[s] \text{ such that } X = \Sigma(\aleph \cap [\alpha, \infty) \times \Sigma) \\
& \quad \vee \\
& \quad \alpha = \infty \wedge X \in P(\Sigma)\}
\end{aligned}$$

4.1.3 $TM_{FS} \rightarrow M_{FS}$

$\pi : TM_{FS} \longrightarrow M_{FS}$, where $\forall P \in TM_{FS}$,

$$\begin{aligned}\pi(P) \;=\; & SUP(\{(strip_t(s), \overline{\alpha}, X) \mid (s, \alpha, \emptyset) \in P \,\wedge\, \exists(s, \aleph) \in Fail(P) \text{ such that} \\ & \quad \alpha \leq begin(\aleph) < \infty \,\wedge\, \Sigma(\aleph) = X \\ & \quad \vee \\ & \quad \alpha = \infty \,\wedge\, X \in P(\Sigma)\})\end{aligned}$$

4.1.4 $TM_{FS}^{*} \rightarrow M_{FS}$

$\pi : TM_{FS}^{*} \longrightarrow M_{FS}$, where $\forall P \in TM_{FS}^{*}$,

$$\pi(P) \;=\; SUP(\{(strip_t(s), \overline{\alpha}, X) \mid (s, \alpha, X) \in P\})$$

Remark. All of the mappings in our hierarchy commute except those involving $\pi : TM_F \longrightarrow M_F$. Without stability, we lose the uniformity of our hierarchical relationships between the timed and untimed models.

4.2 Continuity of the projection mappings

All of the mappings defined in the previous section between timed models are non-expanding (hence continuous) mappings between the two metric spaces involved. Also, those mappings defined between two untimed models are easily seen to be non-expanding.

However, in general, the projection mapping between a timed model and an untimed model is not continuous. For example, in any of the timed models, let $P = \perp$ and $\forall n > 0$, let $P_n = WAIT\ n;\ a \longrightarrow STOP$; it follows that $\lim_{n \to \infty} P_n = P$. Now, any of the projection mappings to an untimed model will not support $\lim_{n \to \infty} \pi(P_n) = \pi(P)$.

As noted previously, the link between the untimed models and the timed models is our concept of stability. We can not expect our projection mappings to be continuous at "unstable" processes, nor can we expect continuity in the absence of a stability concept such as in TM_T and TM_F.

In the timed stability models $(TM_S, TM_{FS}^{*}, TM_{FS})$, let us define a process P to be *stable* provided $\forall (s, \alpha, \emptyset) \in P$, (or in the case of TM_S, $(s, \alpha) \in P$), $\alpha < \infty$.

A reasonable conjecture would be that, in each of the three timed stability models, a projection mapping to an untimed model would be continuous on the subset of all stable processes. Unfortunately, it is not quite that simple.

Consider:

$$P = \Box_{t \in [0,\infty)} (t \rightarrow WAIT\ t; STOP)$$

$$\forall n > 0,\ P_n = \Box_{t \in [0,n)} (t \rightarrow WAIT\ t; STOP)$$

$$\Box$$

$$\Box_{t \in [n,\infty)} (t \rightarrow WAIT\ t; a \rightarrow STOP)$$

Given the obvious meaning in the various timed models by extending our semantics to include (a limited) infinite choice, the above processes are contained in the present mathematical domains. Furthermore, each of these processes is stable. However, it follows that once again $\lim_{n \to \infty} P_n = P$, but the projection mappings to an untimed model will not support $\lim_{n \to \infty} \pi(P_n) = \pi(P)$.

In fact, we can even produce the same effect with choice on a finite alphabet over a finite interval of time.

Define:

$$(t, a) \rightarrow P = WAIT\ t; ((c \rightarrow STOP) \Box (a \rightarrow SKIP)) \setminus c\,; P$$
$$(\text{where } c \neq a)$$

$(t, a) \rightarrow P$ is the process that can nondeterministically choose to communicate an "a" only at time t and then behave like the process P; it is stable on the empty trace in time $t + \delta$.

Now, let

$$P = \sqcap_{t \in (0,1)} (t, a) \rightarrow (WAIT\ t^{-1}; STOP)$$

$$\forall n > 0,\ P_n = \sqcap_{t \in [n^{-1},1)} (t, a) \rightarrow (WAIT\ t^{-1}; STOP)$$

$$\sqcap$$

$$\sqcap_{t \in (0,n^{-1})} (t, a) \rightarrow (WAIT\ t^{-1}; b \rightarrow STOP).$$

Uniform stability

The problem with the above processes is that there is no bound on the stability over a given time period. Let us define a process P to be *uniformly stable* provided

$$\forall t \in [0, \infty),\ \exists \alpha < \infty \ \textit{such that}\ ((s, \beta) \in Stab(P)\ \wedge\ end(s) \leq t) \Rightarrow \beta \leq \alpha).$$

It follows that if π is a projection mapping from one of the timed stability models to an untimed model and P is a uniformly stable process in the timed model, then π is continuous at P. (see [R,1988])

In fact, each process in a timed stability model is in the closure of the set of all uniformly stable processes in that model.

For example, let $P \in TM_{FS}$ and $\forall n > 0$, let

$$\begin{aligned} P_n &= \{(s, \alpha, \aleph) \in P \mid end(s) < n \ \wedge \ \alpha < n\} \\ &\cup \{(s, n, \aleph \cup \aleph') \mid \exists (s, \beta, \aleph) \in P \ such\ that \\ &\quad end(s) < n \ \wedge \ end(\aleph) < n \ \wedge \ \beta \geq n \ \wedge \ I(\aleph') \subseteq [n, \infty)\} \end{aligned}$$

It follows immediately that $\lim_{n \to \infty} P_n = P$. Hence, $\pi : TM_{FS} \to M_{FS}$ is continuous at each point of a *dense* subset of TM_{FS}. (Recall that a subset of a topological space is dense in the space provided each point of the space is either in the given subset or else is a limit point of elements of the subset.)

4.3 Preservation of determinism by the projection mappings

The mapping from TM_{FS} to TM_F and the mapping from M_{FS} to M_F preserve determinism. However, the situation is more complicated for other mappings involving the timed models. Again we must rely on the concept of stability to derive any information about the preservation of determinism for mappings between a timed model and an untimed model.

In general, the projection mappings are pessimistic with respect to determinism.

For example, consider $\pi : TM_{FS} \to TM^*_{FS}$.

$$\begin{aligned} P \;=\; (a \;&\to\; STOP \\ &\Box \\ &\bot) \end{aligned}$$

Obviously, P is deterministic in TM_{FS} but $\pi(P)$ is not deterministic in TM^*_{FS}.

$$((\langle\rangle, \{a\}) \in Fail(\pi(P)) \text{ and } \langle(0, a)\rangle \in Traces(\pi(P))$$

We might conjecture that $\pi(P)$ would be deterministic in TM^*_{FS} if P were stable and deterministic in TM_{FS}. To find a counterexample, we must produce a stable process in TM_{FS} which can participate in an event prior to stability on a given trace, but which cannot participate in that event on or after stability. Recall that the obvious candidates for a counterexample, processes such as

$$P = (WAIT\ 1 \Box (a \to SKIP)); STOP$$

are not deterministic in TM_{FS} since $(\langle\rangle, [1, 2) \times \{a\}) \in Fail(P)$ and $\langle(1, a)\rangle \in Traces(P)$.

Although not definable under our current syntax, consider:

$$\begin{aligned} Q &= \{(\langle\rangle, 1, \aleph) \mid \aleph \cap ([0,1) \times \{a\}) = \emptyset\} \\ &\cup \{(\langle(t,a)\rangle, t+\delta, \aleph) \mid 0 \le t < 1\} \end{aligned}$$

It follows that Q is in the mathematical domain TM_{FS}, and that Q is uniformly stable and deterministic. However, $(\langle\rangle, \{a\}) \in Fail(\pi(Q))$ and $\langle(0,a)\rangle \in Traces(\pi(Q))$. Hence, $\pi(Q)$ is not deterministic in TM^*_{FS}.

Note that if we had introduced general infinite choice, Q would be definable by:

$$Q = f(\Box_{n \in \mathbf{N}^+} (n \rightarrow STOP \mid\mid\mid WAIT\ (1 - n^{-1}); STOP\ )$$

$$\text{where } \forall x \in (\Sigma - \mathbf{N}),\ f(x) = x \text{ and } \forall x \in \mathbf{N},\ f(x) = a$$

We could axiomatize against processes such as Q, however such a process might in fact be highly desirable. Thus, we choose the option to add them to our syntax at a later date. Hence, let us simply distinguish between two types of deterministic processes in TM_{FS}.

Define a deterministic process $P \in TM_{FS}$ to be *strongly deterministic* provided

$((s, \alpha, \emptyset) \in P\ \wedge\ (s, \aleph) \in Fail(P)\ \wedge\ (t, a) \in \aleph)\ \wedge\ t \ge \alpha)$
$\Rightarrow (s.\langle(t', a)\rangle \notin Traces(P),\ \forall t' \ge end(s)).$

It is now easily seen that

$$P \text{ strongly deterministic in } TM_{FS} \text{ and } P \text{ stable } \Rightarrow \pi(P) \text{ deterministic in } TM^*_{FS}.$$

For mappings from the timed stability models to the untimed models, the preservation of determinism fails for more fundamental reasons. Consider $\pi^* : TM^*_{FS} \rightarrow M_{FS}$.

$$Q = ((a \rightarrow STOP) \mid\mid\mid WAIT\ 1);\ WAIT\ 1; a \rightarrow STOP$$

Again, Q is deterministic in TM^*_{FS} but $\pi^*(Q)$ is not deterministic in M_{FS}. $((\langle a\rangle, \{a\}) \in Fail(\pi^*(Q)$ and $\langle aa\rangle \in Traces(\pi^*(Q))$

Note that the process Q above is certainly not homogeneous in the sense we have informally discussed; $(\langle(2,a)\rangle, \{a\}) \in Fail(Q)$ but $(\langle(0,a)\rangle, \{a\}) \notin Fail(Q)$. Hence, there are two traces with the same sequence of alphabet events, but having different behaviours on the next action.

Let us now formally define the concept of homogeneity for timed processes in TM_{FS} and TM^*_{FS}.

P is *homogeneous* in TM_{FS} provided

1) P is stable
2) $((s, \alpha, \emptyset) \in P\ \wedge\ (w, \beta, \emptyset) \in P\ \wedge\ strip_t(s) = strip_t(w)\ \wedge$
   $(s, \aleph) \in Fail(P)$ *such that* $begin(\aleph) \ge \alpha)$
   $\Rightarrow ((w, \beta, \aleph') \in P$ where $\aleph' = \aleph + (\beta - \alpha))$
   $\wedge\ (s.\langle(t,a)\rangle \in Traces(P)\ \wedge\ t > \alpha \Rightarrow w.\langle(t + (\beta - \alpha), a)\rangle \in Traces(P))$

P is *homogeneous* in TM^*_{FS} provided

1) P is stable
2) $((s, \alpha, X) \in P \ \wedge \ (w, \beta) \in Stab(P) \ \wedge \ strip_t(s) = strip_t(w))$
$\Rightarrow (w, \beta, X) \in P$
$\wedge \ (s.\langle(t, a)\rangle \in Traces(P) \ \wedge \ t > \alpha \Rightarrow w.\langle(t + (\beta - \alpha), a)\rangle \in Traces(P))$

With these concepts, we have the following useful and straightforward results:

P strongly deterministic and stable in TM_{FS}
$\Rightarrow \pi(P)$ deterministic and stable in TM^*_{FS}

P homogeneous in TM_{FS}
$\Rightarrow \pi(P)$ homogeneous in TM^*_{FS}

P strongly deterministic and homogeneous in TM_{FS}
$\Rightarrow \pi(P)$ deterministic and homogeneous in TM^*_{FS}

P deterministic and homogeneous in TM^*_{FS}
$\Rightarrow \pi^*(P)$ deterministic in M_{FS}

Note. The above definition of homogeneity was derived for the application at hand. We suspect that in a future more detailed analysis of the projection mappings, the concept would need to be refined.

(For example, $strip_t(s) = strip_t(w) \Rightarrow (P \ after \ s = P \ after \ w)$, where $P \ after \ s$ denotes the behaviour of P after P has engaged in the trace s.)

The original definition would then become a more restrictive property.

5 A case study of timed specification and verification

Consider the following informal specification of a timed vending machine $V_\#$:

(1) $V_\#$ is capable of an unbounded number of transactions and engages only visibly in the events - chocolates, toffees, coffee, tea, and $5p$.

(2) $V_\#$ does not give out more products than it receives payment for.

(3) $V_\#$ is always operating between the peak demand hours of 6:00 AM to noon, (being prepared to engage in a transaction within four seconds),but if more than one hour passes without payment outside these hours, it switches off to be automatically restarted the next morning.

(4) $V_\#$ never takes more than one $5p$ more in a given hour than the number of products it has given out. (If the offer of a product is rejected for longer than an hour after payment, the offer is withdrawn.)

(5) Upon receiving a payment, $V_{\#}$ cannot engage in any event for four seconds, after which during the next fifty-nine minutes and fifty-six seconds, it cannot (i) refuse to offer a chocolate, or a toffee, nor (ii) refuse to offer coffee during the day, nor (iii) refuse to offer tea during the night.

(6) Upon giving out a product within an hour of the last payment, after a delay of four seconds, $V_{\#}$ can not refuse to accept payment during the remainder of that hour.

We wish to utilize the hierarchy H of CSP models in a typical design process for such a machine.

5.1 A "real-world" evolution of the design

(i) The initial decision is made to design a fair, no-loss machine having as possible products, chocolates, toffee, tea and coffee, and capable of an unbounded number of transactions.

Upon taking this decision, the design team divides into two groups to consider the marketing and implementation restraints.

(ii) Marketing makes the decision definitely to offer the customer a choice of a chocolate or a toffee and at least one of coffee or tea, with the latter option to be decided by more market research.

(iii) Implementation decides the machine should be a stable process. Furthermore, it should be initially in a stable state whereby it is prepared to engage in any appropriate action, it should not engage in any action for 4 seconds after payment while the product is being prepared, and it should never wait more than 4 seconds after any action before becoming stable again.

(iv) A design is produced to incorporate both the marketing and implementation decisions to date.

(v) Implementation decides that operating the machine is not cost-efficient if more than one hour passes from either start-up or the delivery of a product to the receipt of the next payment. Hence, in such an event, the machine is to be automatically switched off.

(vi) Marketing (at a very late date in the process) decides that the machine should dispense coffee only during the day and tea only during the night.

(vii) After production of the machine, implementation belatedly realizes that they actually meant to specify that no more than one hour passes without payment. (What if a customer simply did not accept the product for which he had made payment?) Hence, an adaptation is necessary.

(viii) Finally, after testing the machine, it is decided that the machine should be modified to automatically restart each morning at six o'clock and that it should remain in operation during the peak hours of 6:00AM to noon regardless of frequency of use.

5.2 Time-wise refinement

Now, let us take the above decisions one at a time. We wish in each case to accomplish our design in the least complex of the models which will support the relevant constraints. (We assume throughout that $\delta < 1$ and that time is measured in seconds.)

In [Re,1988], we take the above design step-by-step through our hierarchy, giving at each stage a specification and implementation appropriate to the complexity involved.

We start with decision (i) in M_T. Since (ii) requires nondeterminism, we move to M_F. In (iii), we need the timing constraints from TM_S. In (iv), we unify the design so far in TM^*_{FS}. Decision (v) requires the choice between events and participation (i.e., time-outs), hence we need TM_{FS}. Decisions (vi), (vii), (viii) increase the complexity by requiring non-stable processes and interrupts.

Here, we simply give the final specification in TM_{FS} and outline how the implementations develop.

5.3 The specification of $V_\#$ in TM_{FS}

The Specification. $\forall Q \in TM_{FS}$, Q $\underline{\text{sat}}$ $S^{\#}_{TM_{FS}}$ provided

1) $s \in Traces(Q) \Rightarrow$
$(i)\ \Sigma(s) \subseteq \{choc, toff, tea, coff, 5p,\} \ \wedge$
$(ii)\ \#(s\upharpoonright\{choc, toff, tea, coff\}) \leq \#(s\upharpoonright\{5p\}) \ \wedge$
$(iii)\ s = u.v.w \ \wedge\ 0 \leq end(v) - begin(v) \leq 3600$
$\Rightarrow \#(v\upharpoonright\{5p\}) - 1 \leq \#(v\upharpoonright\{choc, toff, tea, coff\})$

2) $s.\langle(t,a)\rangle \in Traces(Q) \Rightarrow$
$(i)\ n\ even\ \wedge\ t \in (43200n, 43200(n+1)) \Rightarrow (a \neq tea)$
$(ii)\ n\ odd\ \wedge\ t \in (43200n, 43200(n+1)) \Rightarrow (a \neq coff)$
$(iii)\ (\exists n\ such\ that\ t \in [86400n, 86400(n+1)+25200]) \vee (t \leq end(s\upharpoonright\{5p\})+3600)$

3) $(s.\langle(t,5p)\rangle, \aleph) \in Fail(Q) \Rightarrow$
$(i)\ \aleph \cap ([end(s)+4, \infty) \cap [t, t+3600)) \times \{choc, toff\} = \emptyset$
$(ii)\ n$ even $\Rightarrow$
$coff \notin \Sigma(\aleph \cap ([end(s)+4, \infty) \cap [43200n, 43200(n+1)) \cap [t, t+3600)) \times \Sigma)$
$(iii)\ n$ odd $\Rightarrow$
$tea \notin \Sigma(\aleph \cap ([end(s)+4, \infty) \cap [43200n, 43200(n+1)) \cap [t, t+3600)) \times \Sigma)$

4) $(s, \aleph) \in Fail(Q)\ \wedge\ (s = \langle\rangle \vee s = w.\langle(t,a)\rangle$ where $a \neq 5p) \Rightarrow$
$5p \notin \Sigma(\aleph \cap ([end(s)+4, end(s\upharpoonright\{5p\})+3600)) \times \Sigma)$

5) $(s, \aleph) \in Fail(Q)\ \wedge\ n \geq 0\ \wedge\ end(s\upharpoonright\{5p\}) + 3600 \leq t$
$\wedge\ t \in [86400n, 86400n + 21600) \Rightarrow$
$\forall t' > t+4,\ ([t,t') \times \{5p\}) \not\subseteq \aleph$

6) $s.\langle(t,5p)(t',a)\rangle \in Traces(Q) \Rightarrow t + 4 \leq t'$
$\wedge\ (t + 3600 < t' \Rightarrow a = 5p)$

5.4 Implementations in TM_{FS}

An implementation of (iv)

Although as stated above, it is only necessary to consider decision (iv) in the context of TM^*_{FS}, we give the following implementation in TM_{FS} (which remains a syntactic solution in TM^*_{FS}). Actually, it is easily seen that $\pi : TM_{FS} \to TM^*_{FS}$ preserves the relationship sat with respect to appropriate specifications of (iv) in the two models.

$$
\begin{array}{rcll}
V_{TM_{FS}} & = & 5p \to & WAIT\ 4 - \delta;\ ((choc \to V_{TM_{FS}} \\
& & & \qquad \Box \\
& & & \qquad toff \to V_{TM_{FS}}) \\
& & & \qquad \Box \\
& & & \qquad (tea \to V_{TM_{FS}} \\
& & & \qquad \sqcap \\
& & & \qquad coff \to V_{TM_{FS}}))
\end{array}
$$

An implementation of (v)

$$
V'_0 = \left(\begin{array}{ll}
5p \to WAIT\ 4 - \delta; & ((choc \to V'_0 \\
& \Box \\
& toff \to V'_0) \\
& \Box \\
& (tea \to V'_0 \\
& \sqcap \\
& coff \to V'_0)) \\
\Box & \\
WAIT\ (3600 - 2\delta) &
\end{array} \right);\ SKIP
$$

$$V' = WAIT\ 2\delta; V'_0$$

An implementation of (vi)

$$Q'_1 = coff \to Q'_1$$

$$Q'_2 = tea \to Q'_2$$

$$V'_1 = (Q'_1 \mid\mid\mid WAIT\ 43200); (Q'_2 \mid\mid\mid WAIT\ 43200); V'_1$$

Let $X = \{coff, tea, \surd\}$ and $Y = \{5p, choc, toff, coff, tea, \surd\}$.

$$V'' = (V_1' \,|||\, SKIP)\ {}_X\|_Y (WAIT\ 2\delta; V_2)$$

An implementation of (vii)

$$V_3 = ((5p \rightarrow V_3) \,\Box\, (WAIT\ (3600 - 2\delta))); SKIP$$

Let $X' = \{5p, choc, toff, coff, tea, \surd\}$ and $Y' = \{5p, \surd\}$.

$$\begin{aligned} V''' &= (V'' \,|||\, SKIP)\ {}_{X'}\|_{Y'} (WAIT\ 2\delta; V_3) \\ &= (((V_1' \,|||\, SKIP)\ {}_X\|_Y V_2) \,|||\, SKIP)\ {}_{X'}\|_{Y'} (WAIT\ 2\delta; V_3) \end{aligned}$$

An implementation of $V_\#$

$$R = restart \rightarrow R$$

$$R' = (R \,|||\, WAIT\ 21600); WAIT\ 64800; R'$$

$$V_4 = (V''' \,|||\, R); restart \rightarrow V_4$$

Let $X'' = \{choc, toff, tea, coff, 5p, restart\}$ and $Y = \{restart\}$.

$$V_\# = (V_4\ {}_{X''}\|_{Y''} R') \setminus \{restart\}$$

5.5 Verification in H

The theory of recursion induction for the Trace Model outlined in [Ros,1982] was extended to the Timed Stability Model in [RR,1987]. The theory also extends to the other timed models in our hierarchy. The essential concept is that specifications based on predicates being true for all the appropriate combinations of traces, refusals and stability values, will always be continuous in these models; a process which fails a given predicate of this type must do so in *finite* time.

A *specification* S on TM_{FS} is a mapping from the complete metric space TM_{FS} to $\{T, F\}$. We say it is *continuous* if the set $\{P \mid S(P) = T\}$ is closed. A specification S is *satisfiable* provided there exists $Q \in TM_{FS}$ such that $S(Q) = T$ or (Q $\underline{\text{sat}}$ S). Now, where applicable we denote by $Pred_S$, the predicate on traces, refusals, and stability values, such that $(S(Q) \Leftrightarrow Pred_S(s, \alpha, \aleph),\ \forall(s, \alpha, \aleph) \in Q)$.

Theorem 5.5.1 If $\hat{C} : TM_S \to TM_{FS}$ is a contraction mapping and S is a continuous, satisfiable specification, then if $(\forall Q \in TM_{FS},\ Q\ \underline{\text{sat}}\ S \Rightarrow \hat{C}(Q)\ \underline{\text{sat}}\ S)$, then $fix(\hat{C})$ $\underline{\text{sat}}$ S.

Theorem 5.5.2 Suppose S is a specification on TM_{FS} such that $\forall Q \in TM_{FS}$, $(\neg S(Q) \Rightarrow \exists t \in [0, \infty)$ *such that* $\forall Q' \in TM_{FS}$ *satisfying* $Q'(t) = Q(t),\ \neg S(Q'))$. Then S is continuous.

The proofs of the above theorems are direct translations of the corresponding theorems for the Trace Model [Ros,1982], where the metric based on "agreeing up to time t" replaces the metric based on "agreeing on all traces up to length n".

5.6 Verification in TM_{FS} that the timed vending machine $V_{\#}$ meets its specification

Each of the six specifications given in the section 5.3 is easily seen to be continuous. Hence, in each case, verification of the given implementation can be accomplished by appeal to the above theory of recursion induction. For brevity, we consider only the the much simpler process $V_{TM_{FS}}$ and its specification.

We restate the specification at stage (iv) in greater detail.

Recall, we wished to specify a timed vending machine which s capable of an unbounded number of transactions and which (i) engages only in the events - chocolates, toffees, coffee, tea, and $5p$, (ii) does not give out more products than it receives payment for, (iii) never takes more than one $5p$ more than the number of products it has given out, (iv) is initially stable and never waits more than 4 seconds after any transaction before becoming stable again, (v) upon receiving more payments than the number of products it has given out, it cannot refuse to offer a chocolate, or a toffee, and at least one of coffee and tea, after a wait of four seconds during which time it can engage in no event, and (vii) upon giving out as many products as it has received payments for, it can not refuse to accept payment when it becomes stable.

<u>The specification in TM_{FS}</u>

$Pred_1$) $(((\langle\rangle,\alpha,\emptyset) \in Q \Rightarrow \alpha = 0) \ \wedge \ ((s.\langle(t,a)\rangle,\alpha,\emptyset) \in Q \Rightarrow \alpha \leq t+4)$

$Pred_2$) $s \in Traces(Q) \ \Rightarrow \ \Sigma(s) \subseteq \{choc, toff, tea, coff, 5p\}$
$\wedge \ \#(s\restriction\{5p\}) - 1 \leq \#(s\restriction\{choc, toff, tea, coff\}) \leq \#(s\restriction\{5p\})$

$Pred_3$) $(s,\alpha,\aleph) \in Q \ \wedge \ \#(s\restriction\{5p\}) > \#(s\restriction\{choc, toff, tea, coff\}) \Rightarrow$
$\{choc, toff\} \cap \Sigma(\aleph \cap ([\alpha,\infty) \times \Sigma)) = \emptyset \ \wedge \ \{tea, coff\} \not\subseteq \Sigma(\aleph \cap ([\alpha,\infty) \times \Sigma))$

$Pred_4$) $(s,\alpha,\aleph) \in Q \ \wedge \ \#(s\restriction\{5p\}) = \#(s\restriction\{choc, toff, tea, coff\}) \Rightarrow$
$5p \notin \Sigma(\aleph \cap ([\alpha,\infty) \times \Sigma))$

$Pred_5$) $s.\langle(t,5p)(t',a)\rangle \in Traces(Q) \Rightarrow t+4 \leq t'$

$\forall i = 1,5$, let S_i denote the specification on TM_{FS} such that $(\forall Q \in TM_{FS} \ (\neg S_i(Q) \Leftrightarrow \exists(s,\alpha,\aleph) \in Q \ such \ that \ \neg Pred_i(s,\alpha,\aleph))$. Let $S = (S_1 \ \wedge \ S_2 \ \wedge \ S_3 \ \wedge \ S_4 \ \wedge \ S_5)$.

Now, by Theorem 5.5.2, each of S_1, S_2, S_3, S_4, and S_5 is a continuous specification. Furthermore, by inspection, $STOP$ sat S_1, S_2, S_3, S_5, and $(5p \rightarrow STOP)$ sat S_4.

Hence, by Theorem 5.5.1, to show that a given recursive process $fix(\hat{C})$ sat S, we need only to show for $i = 1,5$ and $\forall Q \in TM_{FS}$, $(Q$ sat $S_i) \Rightarrow (\hat{C}(Q)$ sat $S_i)$.

A solution in TM_{FS}

$$\begin{aligned} V_{TM_{FS}} \quad = \quad 5p \rightarrow \quad WAIT\ 4-\delta; \ &((choc \rightarrow V_{TM_{FS}} \\ &\Box \\ &toff \rightarrow V_{TM_{FS}}) \\ &\Box \\ &(tea \rightarrow V_{TM_{FS}} \\ &\sqcap \\ &coff \rightarrow V_{TM_{FS}})) \end{aligned}$$

$V_{TM_{FS}} \ = \ fix(\hat{C})$, where $\forall Q \in TM_{FS}$,

$$\begin{aligned} \hat{C}(Q) \quad = \quad 5p \rightarrow \quad WAIT\ 4-\delta; \ &((choc \rightarrow WAIT\ \delta; Q \\ &\Box \\ &toff \rightarrow WAIT\ \delta; Q) \\ &\Box \\ &(tea \rightarrow WAIT\ \delta; Q \\ &\sqcap \\ &coff \rightarrow WAIT\ \delta; Q)) \end{aligned}$$

Suppose $Q \in TM_{FS}$, then $\hat{C}(Q) \; = \; A_1 \cup A_2 \cup A_3 \cup A_4$, where

$$\begin{aligned}
A_1 &= \{(\langle\rangle, 0, \aleph) \mid \Sigma(\aleph) \cap 5p = \emptyset\} \\
A_2 &= \{(\langle(t, 5p)\rangle, t + 4 - \delta, \aleph) \mid \\
&\qquad \{choc, toff\} \cap \Sigma(\aleph \cap [(t + 4 - \delta), \infty) \times \Sigma) = \emptyset \wedge \\
&\qquad \{tea, coff\} \not\subseteq \Sigma(\aleph \cap [(t + 4 - \delta, \infty) \times \Sigma)\} \\
A_3 &= \{(\langle(t, 5p)(t', a)\rangle, t' + 2\delta + \alpha, \aleph_1 \cup (\aleph_2 + t' + 2\delta + \alpha)) \mid \\
&\qquad a \in \{choc, toff, tea, coff\} \;\wedge\; t' \geq t + 4 \wedge \\
&\qquad end(\aleph_1) \leq t' + 2\delta \;\wedge\; \aleph_1 \cap ([0, t) \times \{5p\}) = \emptyset \wedge \\
&\qquad \{choc, toff\} \cap \Sigma(\aleph_1 \cap [(t + 4 - \delta), t') \times \Sigma) = \emptyset \wedge \\
&\qquad \{tea, coff\} \not\subseteq \Sigma(\aleph_1 \cap [(t + 4 - \delta, t') \times \Sigma) \wedge \\
&\qquad (\langle\rangle, \alpha, \aleph_2) \in Q\} \\
A_4 &= \{(\langle(t, 5p)(t', a)\rangle.w + (t' + 2\delta), t' + 2\delta + \alpha, \aleph_1 \cup (\aleph_2 + t' + 2\delta)) \mid \\
&\qquad a \in \{choc, toff, tea, coff\} \;\wedge\; t' \geq t + 4 \wedge \\
&\qquad end(\aleph_1) \leq t' + 2\delta \;\wedge\; \aleph_1 \cap ([0, t) \times \{5p\}) = \emptyset \wedge \\
&\qquad \{choc, toff\} \cap \Sigma(\aleph_1 \cap [(t + 4 - \delta), t') \times \Sigma) = \emptyset \wedge \\
&\qquad \{tea, coff\} \not\subseteq \Sigma(\aleph_1 \cap [(t + 4 - \delta, t') \times \Sigma) \wedge \\
&\qquad (w, \alpha, \aleph_2) \in Q \;\wedge\; w \neq \langle\rangle\}
\end{aligned}$$

We consider only S_1; the other cases are similar (but more tedious).

Suppose Q <u>sat</u> S_1.

Clearly, $(\langle\rangle, 0, \emptyset) \in \hat{C}(Q)$.

Suppose $(s.\langle(t, a)\rangle, \alpha, \emptyset) \in \hat{C}(Q)$.

$s.\langle(t, a)\rangle \in Traces(A_1 \cup A_2) \Rightarrow \alpha \leq t + 4$

$s.\langle(t, a)\rangle \in Traces(A_3) \Rightarrow \alpha = \beta + t + 2\delta$, where $(\langle\rangle, \beta) \in Stab(Q)$
$\Rightarrow \beta = 0$, since Q <u>sat</u> S_1
$\Rightarrow \alpha = t + 2\delta < t + 4 \quad (\delta < 1)$

$s.\langle(t, a)\rangle \in Traces(A_4) \Rightarrow$
$s.\langle(t, a)\rangle = \langle(t_1, 5p)(t_2, b)\rangle.(\langle w.\langle(t - (t_2 + 2\delta), a)\rangle + (t_2 + 2\delta + \beta))$
where $(w.\langle(t - (t_2 + 2\delta), a)\rangle, \beta) \in Stab(Q)$ and $\alpha = t_2 + 2\delta + \beta$

$\Rightarrow \beta \leq t - (t_2 + 2\delta) + 4$ since Q <u>sat</u> S_1
$\Rightarrow \alpha \leq t + 4 \quad \blacksquare$

5.7 Laws of Timed CSP

$$
\begin{aligned}
P\Box P &= P\\
P\Box Q &= Q\Box P\\
P\Box(Q\Box R) &= (P\Box Q)\Box R\\
P\Box(Q\sqcap R) &= (P\Box Q)\sqcap(P\Box R)\\
P\Box STOP &= P\\
(a\to(P\sqcap Q)) &= (a\to P)\sqcap(a\to Q)\\
(a\to P)\Box(a\to Q) &= (a\to P)\sqcap(a\to Q)\\
P\sqcap P &= P\\
P\sqcap Q &= Q\sqcap P\\
P\sqcap(Q\sqcap R) &= (P\sqcap Q)\sqcap R\\
P\parallel Q &= Q\parallel P\\
P\parallel(Q\parallel R) &= (P\parallel Q)\parallel R\\
P\parallel(Q\sqcap R) &= (P\parallel Q)\sqcap(P\parallel R)\\
(a\to P)\parallel(b\to Q) &= STOP && if\ a\neq b\\
&= (a\to(P\parallel Q)) && if\ a=b\\
P\,|||\,Q &= Q\,|||\,P\\
(P\,|||\,Q)\,|||\,R &= P\,|||\,(Q\,|||\,R)\\
P\,|||\,(Q\sqcap R) &= (P\,|||\,Q)\sqcap(P\,|||\,R)\\
P;(Q;R) &= (P;Q);R\\
STOP\,|||\,Q &= Q\\
SKIP;Q &= Q\\
STOP;Q &= STOP\\
P;(Q\sqcap R) &= (P;Q)\sqcap(Q;R)\\
(P\sqcap Q);R &= (P;R)\sqcap(Q;R)\\
(a\to P);Q &= (a\to(P;Q)) && if\ a\neq\surd\\
(P\setminus X)\setminus Y &= (P\setminus Y)\setminus X\\
(P\setminus X)\setminus X &= P\setminus X\\
(P\sqcap Q)\setminus a &= (P\setminus a)\sqcap(Q\setminus a)
\end{aligned}
$$

$$
\begin{aligned}
WAIT\ 0 &= SKIP\\
WAIT\ t_1;\ WAIT\ t_2 &= WAIT\ (t_1+t_2)\\
(WAIT\ t_1 \parallel WAIT\ t_2) &= WAIT\ max\{t_1,t_2\}\\
(WAIT\ t_1\,|||\,WAIT\ t_2);P &= WAIT\ min\{t_1,t_2\};P
\end{aligned}
$$

$$
\begin{aligned}
(WAIT\ t\Box a\to P)\setminus a &= WAIT\ \delta;P\setminus a && t>0\\
((WAIT\ t\Box a\to SKIP);P)\setminus a &= WAIT\ \delta;P\setminus a && t>0\\
((WAIT\ t\Box a\to STOP);P)\setminus a &= WAIT\ \delta;STOP && t>0
\end{aligned}
$$

Note that the above laws hold in all the timed models with the exception of the last three which do not hold in TM_T, TM_S, and TM^*_{FS}.

6 Conclusions and comparisons

In this paper, we have presented a variety of timed and untimed CSP models which deal with the full complexity of distributed computing. All our models have been based on complete ultra-metric spaces. Using this common structure and our notion of stability, we have created a hierarchy from these models with well-understood links between models.

We have seen that using time as the basis of a metric space (i.e., two processes are t-alike if they are indistinguishable up to time t) allows one to be freed from the constraints of complete partial orders without losing generality or abstractness. In particular, we were able to deal effectively for the first time with infinite hiding, infinite alphabet transformations, unbounded nondeterminism, and the subtle relationship between deadlock, divergence, and the "possibility" of divergence.

The ideas behind our comprehensive model (the Timed Failures-Stability Model) are conceptually straightforward: a process is modelled by the records of experiments that an observer can carry out on it (communications accepted and communications refused). The fact that refusals must be recorded all the way through a trace is a consequence of the way timed processes interact: in some sense they can perform more delicate experiments on each other than untimed processes. Refusals only after traces are no longer properly compositional. Other authors ([J,1982],[Bo,1986]) have previously remarked on this and suggested or introduced similar solutions (based on partial orders rather than metric spaces). Phillips [Ph,1986] has studied the corresponding untimed congruence.

We have taken a view of the untimed models as containing incomplete information with respect to the "real world" of the timed models. If we take the position that time is to be modelled so as to greatly affect the interaction of processes, we *must* then accept that certain behaviours simply cannot be properly analysed without timed information. For example, in section 3.1.5, we saw that we can not necessarily establish such desirable properties as $(P \setminus a) \setminus b = (P \setminus b) \setminus a$ in our untimed models. Hence, our untimed models serve as a pessimistic guide to the analysis of timed behaviour.

We believe that our Timed Failures Model (without the complexity of stability) represents perhaps the simplest equivalence which is a full and natural congruence with respect to all the usual CSP operators. This congruence does *not* exist in the untimed case, since there consideration of divergence is necessary if one is to consider failures. However, the inclusion of stability has been crucial to our goal of developing a *uniform* theory of real-time distributed computing. Aside from the well-known arguments for wishing to distinguish a deadlocked process from a diverging one, the inclusion of stability has allowed us to develop our hierarchy of models.

In particular, since the liveness properties predicted by the Timed Failures-Stability Model for a given process can be inferred from the time of stability on, we were able to exploit this link by using reasoning in the simpler (Untimed Failures)-(Timed Stability) Model. Such reasoning can be used to infer total correctness properties for processes that do not depend on the details of timed interaction to achieve 'untimed' correctness. Indeed, it is shown in [Re,1988] how the design of quite complicated timed processes can be started in the simple Traces Model and then moved gradually up the hierarchy to the Timed Failures-Stability Model, where at each step the specification and verification techniques of the relevant model are appropriate to the complexity of the design decision.

The fact that our models are complete metric spaces and all recursions are contraction mappings make them natural vehicles for correctness proofs using the form of recursion induction described in [Ros,1982] and [RR,1986]. (A predicate that represents a non-empty closed subset and which is preserved by a recursion must contain the unique fixed point.) The introduction of stability seems to enhance the range of useful predicates which represent closed sets, since it (to a limited extent) allows us to look into the future.

The semantics we gave for timed CSP is by no means the only possible one that is reasonable, for any such semantics must make specific timing assumptions about the language and its implementation. We assumed that all events take exactly δ, while in practice each event a might take its own duration $\delta(a)$ or even a time chosen nondeterministically from some interval. In the last case our new-found ability to cope with unbounded nondeterminism would be essential. We also assumed that none of the operators except recursion consumed any time by running (i.e, there was never any setting-up or "overhead" time). Also both the parallel operators we gave were *true* parallel operators, in that the time taken by the two operands was not summed: one might well need time-sliced pseudo-parallel operators in practice. In a particular application one will always have to decide on the "right" timed semantics, but we believe that our basic timed model TM_{FS} can be adapted to meet the demands of most such applications.

Comparisons

To our knowledge, the only other attempt at a uniform theory such as the one presented here is that of [OH,1983], where several untimed CSP models were given a unified framework. In [OH,1983], the concept of a specification-oriented semantics was developed. Under this approach, the meaning of "observation of process behaviour" was defined and simple algebraic structures on such observations were introduced. Specifications were then defined as certain subsets of those observations which reflected the given algebraic structure. Finally, a process was identified with the strongest specification which it satisfied, and the set of all such strongest specifications was shown to form a complete partial order under inverse set inclusion. A denotational semantics was constructed in the usual manner with each syntactic constructor of the language mapped onto a continuous operator on specifications.

The work in [OH,1983] served as motivation for the development of our more ambitious timed theory. Although we have taken a more direct, set-theoretic approach to build our hierarchy, it would be interesting to explore the possibility of a "specification-oriented" approach to our timed models, where the appropriate observation of process behaviour would simply reflect our timed traces, timed stability values, and timed refusals. This is a topic for further research.

The reader should compare the models presented here to timed models for concurrency based on complete partial orders in [J,1982], [KSRGAK,1985], [Bo,1986], [Z,1986], and [BG,1987]. While each of these models deals effectively with certain aspects of timed behaviour, none meets our objectives as outlined in section 2.2. Our work was of course accomplished with benefit of hindsight from the work of [J,1982]. However, the other timed models listed above were developed independent and concurrent to our timed models. The Timed Stability Model was presented in [RR,1986] and the Timed Failure-Stability Model was presented in [RR,1987].

The pioneering work on the development of semantic models for timed versions of CSP was carried out by Geraint Jones in [J,1982]. This work demonstrated the basic compatibility of timed CSP with the algebraic properties of the untimed language. However, partially due to predating a solution to the proper treatment of divergence in the untimed models [BR,1985], there were several unresolved difficulties in the timed model of [J,1982]. In particular, the hiding operator was quite complicated and failed to be continuous over a continuous time domain or in general to distribute over nondeterministic choice.

In [KSRGA,1985], the authors presented a timed semantic model for a subset of the CSP of [H,1978]. Their work, although complementary, is largely independent of the aims of this paper. It was based on integer time and did not attempt to model nondeterminism nor to distinguish between divergence and deadlock.

In [Z,1986], the author, apparently unaware of Jones' work, independently gave a timed trace semantics for CSP. The semantics was again based on integer time and did not deal with hiding or nondeterminism.

The work most closely related to our models, is that of A. Boucher and R. Gerth in ([Bo,1986] and [BG,1987]). Their work shared many of our goals. The complete-partial-order timed failures model of [BG,1987] apparently is similar in may aspects to our Timed Failures Model. In particular, both they and we rediscovered (Jones' had suggested this approach in [J,1982]) the crucial necessity for recording refusals throughout a timed trace. The major difference between their work and ours is that we have developed our Timed Failures Model in the context of a hierarchical structure of untimed and timed models based on complete metric spaces. Also, whereas they chose not to distinguish between deadlock and divergence in timed processes, we have sought do make that distinction in the Timed Failures-Stability Model and to relate the models in a manner that allows an option on this and other issues with well-understood consequences.

None of the complete partial order models are able to deal with infinite hiding, infinite alphabet transformations, unbounded nondeterminism, nor a distinction between the "possibility of divergence" and divergence.

As noted in section 2.1, in any model where the partial order is based on nondeterminism or definedness ($P \sqsubseteq Q$ iff Q is more deterministic than P), the least fixed point of $\mu P.P$ (operationally, a simply diverging process) is the most nondeterministic process. One is thus forced to identify the diverging process with one that can do anything (including diverge). Of course, also a process that can do anything cannot be realistic in the timed sense of section 2.1.

Several authors, for example ([N,1979],[Ros,1982],[BZ,1982],[GR,1983], [Rou,1985]), have considered untimed models of concurrency as metric spaces. Their metrics (as was the case with ours for the untimed models) have generally been based on equivalence up to a certain number of steps or communications. As noted in section 2.3, the fact that hiding deletes communications means that a model with a metric of *visible* actions will have a discontinuous hiding operator. Also, recursions are not defined unless they represent contraction maps: something which is by no means automatic, especially when hiding is involved. Such problems have led most researchers to build models of concurrency based on a complete partial order structure, Others have chosen to retain hidden actions in their models (often synchronisation trees [N,1979]); this avoids the above problems, but leads to models which are insufficiently abstract for many purposes (indeed, semantics of this type are often termed operational). Hence, for untimed models of concurrency, the options have usually been either a complete partial order with an overly pessimistic view of divergence and a restriction to finite hiding, or a complete metric space with a non-continuous hiding operator or a loss of abstractness. Perhaps, the primary contribution of our work in this paper has been to show how for timed concurrency, one is freed from this dilemma.

Future work

Our first goal is a greater understanding of our present models. There are several research staff and graduate students at Oxford devoted to this task at present. We are (1) developing a temporal logic for our timed models and exploring an operational semantics based on timed Petri-nets, (2) adding communication to the basic timed models, establishing the properties and proof rules for piping, and doing specification and verification of protocols, (3) providing a set of inference rules for general correctness proofs based on the semantic models (see the paper by J. Davies and S. Schneider, this volume), and (4) continuing to explore our hierarchical approach to design specifications for "real-world" applications, e.g., the specification and verification of aircraft engine control software.

Our long-range goal is to complete the above work towards a unified theory of concurrency by adding real-time probability to the models in our hierarchy. Such models would allow a proper universal measure of fairness, e.g., "within 3.75 milliseconds, there is a 93.7% chance that the process will respond." We would then have as our comprehensive timed model, a model in which safety, liveness, and fairness could be adequately expressed. This would facilitate the comparison and unification of the many different methods presently used to reason about concurrent systems, and promote a far deeper understanding of concurrency in general.

We believe that our hierarchy can be extended to include probability in a natural manner. A timed probability model will be a continuous theory, and as such, will involve basic measure theoretic techniques. These techniques are more likely to be elegantly expressible in a topological context than one based on partial orders.

Finally, we plan at some stage to move the entire hierarchy up a level by considering models based upon infinite traces and infinite refusals such as the (untimed) one developed by Roscoe (see his joint paper with Barrett in this volume). These models will allow us to consider full unbounded nondeterminism, to recover the 'right' laws in the untimed models in our hierarchy, and to formulate temporal operators such as *eventually* with more precision. Again, the goal will not be to develop better models, only new models in which greater distinction is available at the price of greater complexity and in which the links between all the models are understood.

7 Acknowledgements

The work reported in this paper was done while the author was a somewhat geriatric graduate student at Oxford University pursuing a second doctorate. The inspiration for the work came from Tony Hoare. Bill Roscoe served as the thesis advisor, and the author gratefully acknowledges the free use of his keen intuition about the nature of distributed computing. In particular, Dr. Roscoe made the crucial suggestion to consider a "stability" approach to divergence.

The author would also like to state clearly that the approach to timing presented in this paper as well as the development of the Timed Stability Model and the Timed Failures-Stability Model were **joint** work with Dr. Roscoe. The conception and development of a hierarchy of untimed and timed models and all proofs of propositions about these models are due to the author, with the expressed acknowledgement of Dr. Roscoe's role as thesis advisor.

Finally, the author acknowledges many helpful comments by Stephen Blamey, Michael Goldsmith, Geraint Jones, and Steve Schneider.

8 References

[Bo,1986] A. Boucher, *A time-based model for occam,* Oxford University D.Phil. thesis 1986.

[BG,1987] A. Boucher and R. Gerth, *A timed failures model for extended communicating sequential processes,* ICALP'87,Springer LNCS.

[B,1983] S.D. Brookes, *A model for communicating sequential processes,* Oxford University D.Phil. thesis 1983.

[Br,1982] M. Broy, *Fixed point theory for communication and concurrency,* TC2 Working Conference on Formal Description of Programming Concepts II, Garmisch, 1982.

[BHR,1984] S.D. Brookes, C.A.R. Hoare and A.W. Roscoe, *A theory of communicating sequential processes,* JACM 31 (1984), 560-599.

[BR,1985] S.D. Brookes and A.W. Roscoe, *An improved failures model for communicating processes,* Proceedings of the Pittsburgh Seminar on Concurrency, Springer LNCS 197 (1985).

[BZ,1982] J.W. de Bakker and J.I. Zucker, *Processes and the denotational semantics of concurrency*, Information and Control 54 (1982), 70-120.

[E,1977] R. Engelking, *General Topology*, Polish Scientific Publishers (1977).

[GR,1983] W.G. Golson and W.C. Rounds, *Connections between two theories of concurrency: metric spaces and synchronisation trees*, Information and Control 57 (1983), 102-124.

[H,1978] C.A.R. Hoare, *Communicating sequential processes*, CACM 21 (1978), 666-677.

[H,1980] C.A.R. Hoare, *A model for communicating sequential processes*, On the construction of programs CUP (1980), 229-248.

[H,1985] C.A.R. Hoare, *Communicating Sequential Processes*, Prentice-Hall International, 1985.

[HBR,1981] C.A.R. Hoare, S.D. Brookes and A.W. Roscoe, *A theory of communicating sequential processes*, Oxford University Computing Laboratory technical monograph PRG-16 (1981).

[J,1982] G. Jones, *A timed model for communicating processes*, Oxford University D.Phil thesis, 1982.

[KSRGAK,1985] R. Koymans, R.K. Shyamasundar, W.P. de Roever, R. Gerth and S. Arun-Kumar, *Compositional semantics for real-time distributed computing*, Faculteit der Wiskunde en Natuurwetenschappen, Katholieke Universiteit, Nijmegen, technical report 68, 1985.

[M,1980] R. Milner, *A calculus of communicating systems*, Springer LNCS 92 (1980).

[M,1983] R. Milner, *Calculi for synchrony and asynchrony*, Theoretical Computer Science 25 (1983), 267-310.

[N,1979] M. Nivat, *Infinite words, infinite trees, infinite computations*, Foundations of Computer Science III (Math. Centre Tracts 109, 1979), 3-52.

[Occ,1984] *The occam programming manual*, (Inmos Ltd.) Prentice-Hall (1984).

[OH,1983] E.R. Olderog and C.A.R. Hoare, *Specification-oriented semantics for communicating processes*, Springer LNCS 154 (1983), 561-572. (Also, Acta Informatica 23 (1986), 9-66.)

[Ph,1986] I. Phillips, *Refusal testing*, Proceedings of ICALP'86, Springer LNCS 226 (1986), 304-313.

[Re,1988] G.M. Reed, *A uniform mathematical theory for real-time distributed computing*, Oxford University D.Phil. thesis 1988; Technical Monograph (Oxford University Computing Laboratory), to appear.

[RR,1986] G.M. Reed and A.W. Roscoe, *A timed model for communicating sequential processes*, Proceedings of ICALP'86, Springer LNCS 226 (1986), 314-323; Theoretical Computer Science 58 (1988) 249-261 .

[RR,1987] G.M. Reed and A.W. Roscoe, *Metric spaces as models for real-time concurrency*, Proceedings of the Third Workshop on the Mathematical Foundations of Programming Language Semantics (April,1987), LNCS 298 (1988).

[Ros,1982] A.W. Roscoe, *A mathematical theory of communicating processes*, Oxford University D.Phil. thesis 1982.

[Rou,1985] W.C. Rounds, *Applications of topology to the semantics of communicating processes*, Proceedings of the Pittsburgh Seminar on Concurrency, Springer LNCS 197 (1985).

[Z,1986] A.E. Zwarico, *A formal model of real-time computing*, University of Pennsylvania technical report (1986).

Factorizing Proofs in Timed CSP

Jim Davies and Steve Schneider[1]

Oxford University Computing Laboratory
Programming Research Group
11 Keble Road
Oxford OX1 3QD

{jdavies,sas}@uk.ac.ox.prg

Abstract. A simple notion of specification is introduced, and a complete set of inference rules given, for reasoning about real-time processes. The notation of Timed Communicating Sequential Processes is employed, and the strongest possible specification of a process is discussed. A proof of correctness of a simple protocol is given to illustrate the method of verification.

1 Introduction

Timed CSP is an extension of Communicating Sequential Processes [H85] which includes timing information. It can be used to model time-dependent properties of concurrent systems. An algebraic notation is employed in the definition of processes, capturing the behaviour of a system in a clear and intuitive manner. A uniform hierarchy of semantic models for this notation is presented in [Re88]. Each semantic model identifies a process with a set of possible behaviours: by reasoning about these sets, we may establish properties of the corresponding processes.

In untimed CSP we have a number of algebraic laws that preserve the semantics of a process. These laws allow us to rewrite a process definition to facilitate such reasoning; if necessary, we may eliminate the abstraction and parallel operators. This is not possible in Timed CSP. The semantics of the timed models are necessarily complicated, but we may use the semantic equations to derive a number of useful laws relating processes to predicates on behaviour.

These laws are central in the application of Timed CSP to the design and analysis of complex systems. We can capture the requirements of the specification using the notation of the semantic model, and formalise our intended solution in the process

[1]supported by ESPRIT project 3096

algebra. This solution allows us to move towards an implementation, should this be our aim. In any case, we are obliged to show that our proposed solution meets our requirements; we must verify it.

We consider a verification of a Timed CSP process to be a demonstration that all its possible behaviours meet a proposed specification, expressed as a predicate on a typical element of its semantics. In this case, we say that the process *satisfies* the specification. A specification in TM_{FS}, the most expressive model, can often be written as a conjunction of constraints in the simpler models; the process can then be shown to satisfy each of these independently.

Even within the simpler models, TM_F, TM_S and TM_T, the construction of such a proof directly from the semantics may be difficult and laborious. If we are to reason about complex time-critical distributed systems, we require a method of translating a proof obligation on a process into proof obligations on its syntactic subcomponents. This method will employ a number of rules grounded in the semantic mappings introduced in [RR86], [RR87] and [Re88].

In this paper, we present the notion of behavioural specifications: correctness conditions on the possible behaviours of a process. We then give a complete set of inference rules for translating such a specification on a compound process into requirements upon its subprocesses. The soundness of each rule can be established from the semantic equations for the relevant operators; example proofs are included as an appendix. To illustrate the use of these rules we present a verification of a simple stop-and-wait protocol in the Timed Failures model, TM_F.

2 Notation

In this section, we present the notation of Timed CSP, the process algebra and the semantic models, as defined in [Re88]. We then explain our concept of specification and introduce the additional notation required for this paper.

2.1 Timed CSP

Timed CSP is a simple extension of CSP [H85]. The process algebra, *TCSP*, is given in Backus-Naur form below:

$$\begin{aligned} P \;::=\; & \bot \mid STOP \mid SKIP \mid WAIT\ t \mid a \rightarrow P \mid \\ & P \mathbin{\Box} P \mid P \sqcap P \mid P \parallel P \mid P\ {}_A\!\parallel_B P \mid P \mathbin{|||} P \mid \\ & P\,;P \mid P \setminus A \mid f^{-1}(P) \mid f(P) \mid \mu X \bullet F(X) \end{aligned}$$

These operators are given interpretations in a hierarchy of semantic models, as detailed in [Re88]. These models allow us to write process specifications: a predicate on the semantics of a process corresponds to a requirement on its possible behaviours.

The semantic model TM_F consists of sets of pairs $(s, \aleph)$ satisfying the seven healthiness conditions given in [Re88]. We refer to a pair $(s, \aleph)$ as a timed failure. The semantic function $\mathcal{F}_T$ is defined on elements of *TCSP*, mapping them to failure sets in TM_F.

The first component of a timed failure represents a possible timed trace of the process: a sequence of timed observable events. The second component, $\aleph$, represents a finite union of refusal tokens, each refusal token being the product of a half-open finite time interval and a subset of the set of all events, Σ. This component denotes the $(time, event)$ pairs that may be refused if the process performs the trace s.

2.2 Specification

We consider a specification to be a predicate on a typical behaviour of a process: an arbitrary element of its semantics. If this predicate holds of all possible behaviours of a process, we say that the process satisfies the specification. We define the satisfaction operator **sat** for a process P and a specification $S(s, \aleph)$:

$$P \ \mathbf{sat} \ S(s,\aleph) \ \ \widehat{=} \ \ \forall (s,\aleph) \in \mathcal{F}_T[\![P]\!] \bullet S(s,\aleph)$$

From this definition, we can establish a number of simple inference rules:

$$\frac{}{P \ \mathbf{sat} \ true} \qquad \frac{\begin{array}{l} P \ \mathbf{sat} \ S(s,\aleph) \\ P \ \mathbf{sat} \ T(s,\aleph) \end{array}}{P \ \mathbf{sat} \ S(s,\aleph) \wedge T(s,\aleph)} \qquad \frac{\begin{array}{l} P \ \mathbf{sat} \ S(s,\aleph) \\ S(s,\aleph) \Rightarrow T(s,\aleph) \end{array}}{P \ \mathbf{sat} \ T(s,\aleph)}$$

Using the **sat** operator, we can capture any requirement that corresponds to a condition upon *all* of the possible behaviours of a process. The resulting predicate upon the *TCSP* process we call a behavioural specification. In [Re88], Reed defines specifications as predicates on the semantic set of a process, we define predicates on a typical element of that set. Behavioural specifications form a subset of Reed's specifications.

Reed's specifications permit a more detailed analysis of the process representation; ours are more suited to the capture of general requirements upon a process. For example, the predicate

$$(\langle (1,a),(2,b)\rangle, \emptyset) \ \in \ \mathcal{F}_T[\![P]\!]$$

cannot be written as a behavioural specification. It states that $(\langle (1,a),(2,b)\rangle, \emptyset)$ is a possible behaviour of P, and to decide upon its truth we need to examine the whole of the semantic set.

We are interested in the correctness of processes. Behavioural specifications reflect this: they insist that every possible behaviour is acceptable. To state that a process *may* participate in a certain event at a certain time, or refuse a certain event at another, without further information, is of little use. We are interested in what can be guaranteed about a process behaviour.

2.3 Notation

For convenience, we define a number of operators on timed failures, timed traces and timed refusals [2]. We define two functions on traces:

$$\begin{aligned} last(s^\frown\langle(t,a)\rangle) &\mathrel{\hat=} a \\ tstrip(\langle\rangle) &\mathrel{\hat=} \langle\rangle \\ tstrip(\langle(t,a)\rangle^\frown s) &\mathrel{\hat=} \langle a\rangle^\frown tstrip(s) \end{aligned}$$

The first returns the last event in the trace, the second merely strips the time information from the trace.

We define the *before*, *after*, and *during* operators on refusals:

$$\begin{aligned} \aleph \upharpoonright t &\mathrel{\hat=} \aleph \cap ([0,t) \times \Sigma) \\ \aleph \mathbin{\rceil} t &\mathrel{\hat=} \aleph \cap ([t,\infty) \times \Sigma) \\ \aleph \uparrow [t_1, t_2) &\mathrel{\hat=} \aleph \cap ([t_1,t_2) \times \Sigma) \end{aligned}$$

Recalling that Σ denotes the set of all events, we see that these restrict a refusal set to events that may be refused before, during, and after the specified times.

We define a subtraction operator on traces and refusals, translating through time:

$$\begin{aligned} \langle\rangle - t &\mathrel{\hat=} \langle\rangle \\ (\langle(t_1,a)\rangle^\frown s) - t &\mathrel{\hat=} \begin{cases} s - t & \text{if } t_1 < t \\ \langle(t_1 - t, a)\rangle^\frown(s - t) & \text{otherwise} \end{cases} \\ \aleph - t &\mathrel{\hat=} \{(t_1 - t, a) \mid (t_1, a) \in \aleph \wedge t_1 \geqslant t\} \end{aligned}$$

We define an operator σ on traces and refusals, yielding the set of events that occur in each. For convenience, we extend the definition of σ to cover failures and processes; in the latter case, the result is the set of events in which the process may participate. Observe that this operator differs from the *alphabet* concept used in earlier versions of CSP.

$$\begin{aligned} \sigma(s) &\mathrel{\hat=} \{a \in \Sigma \mid \exists t \bullet \langle(t,a)\rangle \text{ in } s\} \\ \sigma(\aleph) &\mathrel{\hat=} \{a \in \Sigma \mid \exists t \bullet (t,a) \in \aleph\} \\ \sigma(s,\aleph) &\mathrel{\hat=} \sigma(s) \cup \sigma(\aleph) \\ \sigma(P) &\mathrel{\hat=} \bigcup\{\sigma(s) \mid s \in traces(P)\} \end{aligned}$$

Similarly, we extend the definition of *end* in [Re88]:

$$end(s,\aleph) \mathrel{\hat=} max\{end(s), end(\aleph)\}$$

Finally, for use with the hiding operator, we define a predicate on failures, indexed by a set of events A:

$$\widehat{A}(s,\aleph) \mathrel{\hat=} ([0, end(s,\aleph)) \times A) \subseteq \aleph$$

This predicate holds exactly when the failure $(s,\aleph)$ is *activated* on set A.

[2] From now on, we will omit the prefix 'timed' as all subsequent specifications will be drawn from TM_F

3 Abstraction and Concurrency

As an introduction to our method of verifying processes, we consider two operators central to the language of Timed CSP: the hiding and parallel operators.

3.1 Hiding

In applying Timed CSP to complex systems, we use the hiding operator to abstract away from internal behaviour. To prove our description correct, we may need to reason about this behaviour. Hiding a set of events A from the environment of a process P restricts the set of possible behaviours to those in which P is forced to perform events from A as soon as they become available: the *A-activated* behaviours.

The events in A are no longer observable from the environment, and we may not mention them in reasoning about $P \setminus A$. Instead, we identify the A-activated behaviours of the process, establishing results that may involve events from A. These results may then be used to derive a specification that is independent of events from A. This specification is then satisfied by $P \setminus A$.

In the untimed version of CSP, we can use algebraic laws to eliminate the hiding operator from a process description: these laws preserve the equivalent set of behaviours. It would be possible to derive similar laws for Timed CSP, but their complexity would render them unusable: consider the identity below, which corresponds to the simplest non-trivial case of hiding over deterministic choice.

$$(a \rightarrow STOP \,\Box\, b \rightarrow SKIP) \setminus \{a\} \;\equiv\; ((SKIP \,\Box\, b \rightarrow SKIP)\,;\, SKIP) \,\|\, b \rightarrow SKIP$$

Our approach offers a simple, systematic solution to the problem of hiding.

We defined the $\widehat{\ }$ operator in the previous section: $\widehat{A}(s, \aleph)$ holds precisely when $(s, \aleph)$ is an A-activated failure. The following inference rule illustrates the relationship between the failures of P and those of $P \setminus A$:

$$\frac{P \text{ sat } (\widehat{A}(s, \aleph) \wedge \sigma(\aleph') \subseteq A) \Rightarrow S(s \setminus A, \aleph - \aleph')}{P \setminus A \text{ sat } S(s, \aleph)}$$

This follows from the semantic equation for the hiding operator given in [Re88], and transforms a proof obligation on $P \setminus A$ into one on P.

3.2 Parallelism

Timed CSP has three parallel operators: alphabeticised parallel, synchronised parallel and interleaving. The latter operators can be viewed as particular instances of the first. In this paper, we will illustrate the use of the most general form of parallel operator.

The alphabeticised parallel operator places a restriction on the events communicable by each argument: in the parallel combination $P\ {}_X\|_Y\ Q$, process P may perform only those events in set X. Similarly, Q is restricted to those in Y. The two processes must co-operate on events common to both sets.

As in the case of hiding, it would be impractical to eliminate alphabeticised parallelism using algebraic laws. As an illustration, consider the identity below, which holds for untimed CSP.

$$a \rightarrow P \mid\mid\mid b \rightarrow Q \quad \equiv \quad a \rightarrow (P \mid\mid\mid b \rightarrow Q) \,\Box\, b \rightarrow ((a \rightarrow P) \mid\mid\mid Q)$$

This is no longer true for Timed CSP processes because of the delay δ introduced by the prefix operator. This can arise whenever a process contains a form of parallelism which is not completely synchronised.

This means that, except for the simple case of completely synchronised parallelism, we cannot transform a process in a semantics-preserving fashion and alter the degree of parallelism present. However, our inability to do this need not detract from the applicability of the formalism; time-critical systems with communication delays have a minimum degree of parallelism. We can derive rules to allow us to establish properties of such systems.

As an introduction to the operator, consider the special case that is synchronised parallelism. The following inference rule can be derived for this operator:

$$\frac{\begin{array}{l} P_1 \ \textbf{sat}\ S_1(s, \aleph) \\ P_2 \ \textbf{sat}\ S_2(s, \aleph) \\ S_1(s, \aleph_1) \wedge S_2(s, \aleph_2) \Rightarrow S(s, \aleph_1 \cup \aleph_2) \end{array}}{P_1 \parallel P_2 \ \textbf{sat}\ S(s, \aleph)}$$

To establish that a parallel combination meets a given specification S, it is sufficient to find two specifications, one for each component, that yield S for a combination of behaviours. More precisely, a typical failure of $P_1 \parallel P_2$ must satisfy:

- any trace of $P_1 \parallel P_2$ is a trace of each component.
- any refusal set of $P_1 \parallel P_2$ will be the union of two refusal sets: one from each of the component processes.

The parallel combination refuses to participate in an event e whenever either or both of its components refuses e.

The rule for the alphabeticised parallel operator is necessarily more complicated:

$$\frac{\begin{array}{l} P_1 \ \textbf{sat}\ S_1(s, \aleph) \\ P_2 \ \textbf{sat}\ S_2(s, \aleph) \\ (\sigma(s_1, \aleph_1) \subseteq X \wedge \sigma(s_2, \aleph_2) \subseteq Y \wedge \sigma(\aleph_3) \subseteq \Sigma - (X \cup Y) \wedge S_1(s_1, \aleph_1) \\ \qquad \wedge\ S_2(s_2, \aleph_2) \wedge s_3 \in s_1\ {}_X\|_Y\ s_2) \Rightarrow S(s_3, \aleph_1 \cup \aleph_2 \cup \aleph_3) \end{array}}{P_1\ {}_X\|_Y\ P_2 \ \textbf{sat}\ S(s, \aleph)}$$

As before, we must find two specifications, one for each component, that yield S for a combination of behaviours. This time, a failure of the parallel combination must satisfy:

- any trace of $P_1 \;{}_X\|_Y\; P_2$ must be the parallel combination of a trace from each component.
- any refusal set of $P_1 \;{}_X\|_Y\; P_2$ must be the union of three refusal sets: one from each component, and an arbitrary refusal set whose alphabet lies outside $X \cup Y$.

Recall that the parallel operator on traces produces a set of traces: sequences of events drawn from $X \cup Y$, whose restriction to the sets X and Y produces the first and second arguments of the operator, respectively.

These conditions lead to the third antecedent of the rule, which allows us to transform a predicate on the failures of a parallel combination into requirements on the corresponding failures of the component processes. Together with the hiding rule, this is sufficient to treat the example of the next section.

4 A Simple Protocol

A protocol is a distributed algorithm for facilitating the communication of messages between processes. CSP is particularly suitable for the specification of protocols; the enhancements introduced in Timed CSP allow us to address the timing considerations that are often necessary for the correctness of the protocol. Using Timed CSP, we can describe and analyse processes that include *timeouts*, *interrupts* and *time-critical synchronisation*.

In this section, we consider the specification of a simple 'stop-and-wait' protocol, similar to the one described in [PS88]. This consists of two processes, P and Q, communicating across two wires: W_1 and W_2. Together, they control the flow of data between two external processes. This may be represented pictorially as follows:

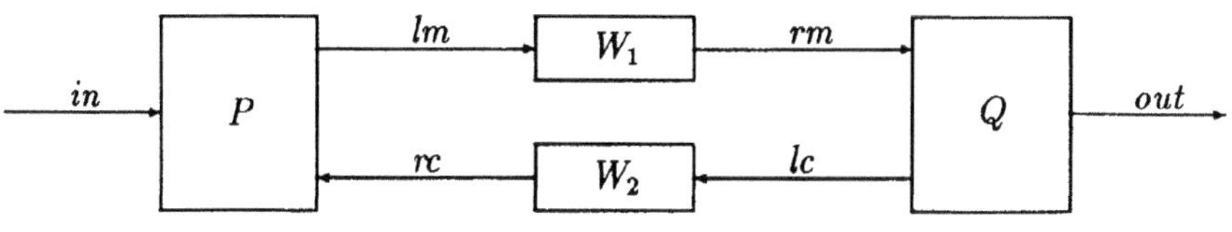

In general, protocols allow for unreliable channels, by duplicating data or requiring acknowledgements: such behaviour is easily modelled in Timed CSP. However, our purpose is to illustrate the use of the inference rules; we need not concern ourselves with these complications. Our protocol addresses only dataflow considerations, and we assume that the wires W_1 and W_2 are *reliable*: for every input, there is a corresponding output.

4.1 Specifications

There are many requirements that we could place upon the protocol, but we will consider just one: that if a message is input, then output is ready within two seconds. Formally, we wish our protocol *PROT* to meet the following timed failures specification:

$$SPEC(s, \aleph) \;\widehat{=}\; last(s) = in \Rightarrow out \notin \sigma(\aleph \upharpoonright (end(s) + 2))$$

We give conditions on the components of the protocol, and verify that they are sufficient to ensure that the protocol exhibits this behaviour.

The sending process P should meet the following specification: it should perform the three events in, lm, rc in strict rotation; after performing an event, it should be prepared to perform the next within a certain time; initially, it should be ready to receive an *in*put. We capture these requirements in the timed failures specification $SPEC_P$:

$$\begin{aligned} SPEC_P(s, \aleph) \;\widehat{=}\;& tstrip(s) \leqslant \langle in, lm, rc\rangle^* \wedge \\ & last(s) = in \Rightarrow lm \notin \sigma(\aleph \upharpoonright (end(s) + 2\delta)) \wedge \\ & last(s) = lm \Rightarrow rc \notin \sigma(\aleph \upharpoonright (end(s) + 2\delta)) \wedge \\ & last(s) = rc \Rightarrow in \notin \sigma(\aleph \upharpoonright (end(s) + 2\delta)) \wedge \\ & s = \langle\rangle \Rightarrow in \notin \sigma(\aleph) \end{aligned}$$

After accepting and transmitting a message, the sending process must await confirmation from the receiving process before accepting another. The receiving process will send a confirmation signal once the previous message has been output. Initially, the system is empty. Hence we wish the receiving process Q to satisfy $SPEC_Q$:

$$\begin{aligned} SPEC_Q(s, \aleph) \;\widehat{=}\;& tstrip(s) \leqslant \langle rm, out, lc\rangle^* \wedge \\ & last(s) = rm \Rightarrow out \notin \sigma(\aleph \upharpoonright (end(s) + 2\delta)) \wedge \\ & last(s) = out \Rightarrow lc \notin \sigma(\aleph \upharpoonright (end(s) + 2\delta)) \wedge \\ & last(s) = lc \Rightarrow rm \notin \sigma(\aleph \upharpoonright (end(s) + 2\delta)) \wedge \\ & s = \langle\rangle \Rightarrow rm \notin \sigma(\aleph) \end{aligned}$$

The wires W_1 and W_2 have a propagation delay of 1 second, and will not be required to transmit more than one message at a time. However, each must be ready to accept another input almost immediately after output. They satisfy the specifications $SPEC_{W_1}$ and $SPEC_{W_2}$ respectively, where

$$\begin{aligned} SPEC_{W_1}(s, \aleph) \;\widehat{=}\;& tstrip(s) \leqslant \langle lm, rm\rangle^* \wedge \\ & last(s) = lm \Rightarrow rm \notin \sigma(\aleph \upharpoonright (end(s) + 1)) \wedge \\ & last(s) = rm \Rightarrow lm \notin \sigma(\aleph \upharpoonright (end(s) + 2\delta)) \wedge \\ & s = \langle\rangle \Rightarrow lm \notin \sigma(\aleph) \end{aligned}$$

$$
\begin{aligned}
SPEC_{W_2}(s, \aleph) \;\widehat{=}\; & tstrip(s) \leqslant \langle lc, rc \rangle^* \wedge \\
& last(s) = lc \Rightarrow rc \notin \sigma(\aleph \upharpoonright (end(s) + 1)) \wedge \\
& last(s) = rc \Rightarrow lc \notin \sigma(\aleph \upharpoonright (end(s) + 2\delta)) \wedge \\
& s = \langle\rangle \Rightarrow lc \notin \sigma(\aleph)
\end{aligned}
$$

The protocol is a combination of the sending process, the receiving process, and the wires. We combine these in *TCSP* by way of the alphabeticised parallel operator, and hide the internal detail. If we define the sets

$$
\begin{aligned}
X &\;\widehat{=}\; \{in, lm, rc\} \\
Y &\;\widehat{=}\; \{out, rm, lc\} \\
C &\;\widehat{=}\; \{lc, rc\} \\
M &\;\widehat{=}\; \{lm, rm\} \\
A &\;\widehat{=}\; M \cup C
\end{aligned}
$$

then the protocol may be defined:

$$
PROT \;\widehat{=}\; ((P \;{}_X\|_Y\; Q) \;{}_{X \cup Y}\|_{M \cup C}\; (W_1 \;{}_M\|_C\; W_2)) \setminus A
$$

4.2 Verification

Having formalised our requirements, we can now use the inference rules given in section 3 to demonstrate that the protocol *PROT* will meet the specification *SPEC*. We wish to establish that:

$$
PROT \quad \textbf{sat} \quad SPEC(s, \aleph)
$$

The definition of *PROT* involves the hiding operator at the outermost level, so we must first apply the hiding rule. This reduces the proof requirement to:

$$
\begin{aligned}
(P \;{}_X\|_Y\; Q) \;{}_{X \cup Y}\|_{M \cup C}\; (W_1 \;{}_M\|_C\; W_2) \quad \textbf{sat} \quad & \sigma(\aleph') \subseteq A \wedge ([0, end(s, \aleph)) \times A) \subseteq \aleph \\
& \Rightarrow SPEC(s \setminus A, \aleph - \aleph')
\end{aligned}
$$

This is a proof requirement on a parallel combination, so we apply the rule for the parallel operator. We have then to find specifications S_1 and S_2 such that:

$$
\begin{array}{l}
P \;{}_X\|_Y\; Q \quad \textbf{sat} \quad S_1(s, \aleph) \\
W_1 \;{}_M\|_C\; W_2 \quad \textbf{sat} \quad S_2(s, \aleph) \\
\left.\begin{array}{l}
\sigma(s_1, \aleph_1) \subseteq (X \cup Y) \wedge \sigma(s_2, \aleph_2) \subseteq (M \cup C) \\
\sigma(\aleph_3) \subseteq \Sigma - (X \cup Y \cup M \cup C) \\
S_1(s_1, \aleph_1) \wedge S_2(s_2, \aleph_2) \wedge s_3 \in s_1 \;{}_{X \cup Y}\|_{M \cup C}\; s_2 \\
\aleph = \aleph_1 \cup \aleph_2 \cup \aleph_3 \\
\sigma(\aleph') \subseteq A \wedge ([0, end(s_3, \aleph)) \times A) \subseteq \aleph
\end{array}\right\} \Rightarrow SPEC(s_3 \setminus A, \aleph - \aleph')
\end{array}
$$

Before we continue, we note that the specification *SPEC* is independent of the hidden set of events A, for consider the definition:

$$
SPEC \;\widehat{=}\; last(s) = in \Rightarrow out \notin \sigma(\aleph \upharpoonright (end(s) + 2))
$$

Formally, we can show that

$$SPEC(s, \aleph \upharpoonright (\Sigma - A)) \;\Rightarrow\; SPEC(s, \aleph)$$

This concurs with our intuition: the correctness of the protocol may be dependent upon hidden interactions, but our formal description of the service provided (the specification *SPEC*) should abstract away from internal detail.

Taking this in conjunction with the alphabet conditions upon the failure sets, we may reduce the third proof obligation to

$$\left.\begin{array}{l} \sigma(s_1, \aleph_1) \subseteq (X \cup Y) \wedge \sigma(s_2, \aleph_2) \subseteq (M \cup C) \\ \sigma(\aleph_3) \subseteq \Sigma - (X \cup Y) \\ S_1(s_1, \aleph_1) \wedge S_2(s_2, \aleph_2) \wedge s_3 \in s_1 \;{}_{X \cup Y}\|_{M \cup C}\; s_2 \\ ([0, end(s_3, \aleph_1 \cup \aleph_2 \cup \aleph_3)) \times A) \subseteq \aleph_1 \cup \aleph_2 \end{array}\right\} \Rightarrow SPEC(s_3 \setminus A, \aleph_1)$$

To identify S_1 we apply the parallel rule once again. We are then required to find S_4 and S_5 such that:

$$\begin{array}{lll} P & \textbf{sat} & S_4(s, \aleph) \\ Q & \textbf{sat} & S_5(s, \aleph) \end{array}$$

$$\left.\begin{array}{l} \sigma(s_4, \aleph_4) \subseteq X \wedge \sigma(s_5, \aleph_5) \subseteq Y \\ \sigma(\aleph_6) \subseteq \Sigma - (X \cup Y) \\ S_4(s_4, \aleph_4) \wedge S_5(s_5, \aleph_5) \wedge s_6 \in s_4 \;{}_X\|_Y\; s_5 \end{array}\right\} \Rightarrow S_1(s_6, \aleph_4 \cup \aleph_5 \cup \aleph_6)$$

We already have specifications for the components P and Q. Substituting these for S_4 and S_5, and using the alphabet conditions upon the traces and refusals, we can reduce this proof obligation to:

$$\left.\begin{array}{l} SPEC_P(s \upharpoonright X, \aleph \upharpoonright X) \\ SPEC_Q(s \upharpoonright Y, \aleph \upharpoonright Y) \\ \sigma(s) \subseteq (X \cup Y) \end{array}\right\} \Rightarrow S_1(s, \aleph)$$

This yields a suitable instantiation for S_1: the antecedent of the above expression. In a similar fashion, we arrive at the following instantiation for S_2:

$$\begin{array}{l} SPEC_{W_1}(s \upharpoonright M, \aleph \upharpoonright M) \wedge \\ SPEC_{W_2}(s \upharpoonright C, \aleph \upharpoonright C) \wedge \\ \sigma(s) \subseteq (M \cup C) \end{array}$$

Our proof requirement can then be written as follows:

$$\left.\begin{array}{l} \sigma(s_1, \aleph_1) \subseteq (X \cup Y) \wedge \sigma(s_2, \aleph_2) \subseteq (M \cup C) \\ \sigma(\aleph_3) \subseteq \Sigma - (X \cup Y) \\ SPEC_P(s_1 \upharpoonright X, \aleph_1 \upharpoonright X) \wedge SPEC_Q(s_1 \upharpoonright Y, \aleph_1 \upharpoonright Y) \\ SPEC_{W_1}(s_2 \upharpoonright M, \aleph_2 \upharpoonright M) \wedge SPEC_{W_2}(s_2 \upharpoonright C, \aleph_2 \upharpoonright C) \\ ([0, end(s_3, \aleph_1 \cup \aleph_2 \cup \aleph_3)) \times A) \subseteq \aleph_1 \cup \aleph_2 \\ s_3 \in s_1 \;{}_{X \cup Y}\|_{M \cup C}\; s_2 \end{array}\right\} \Rightarrow SPEC(s_3 \setminus A, \aleph_1)$$

The alphabet conditions in S_1 and S_2 are subsumed in the first two conditions above.

We have reduced the proof obligation to a predicate on traces and refusal sets: the verification may be completed using simple properties of sets and sequences: assuming the conjuncts in the above antecedent, we are trying to establish that

$$last(s_3 \setminus A) = in \Rightarrow out \notin \sigma(\aleph_1 \uparrow (end(s_3 \setminus A) + 2))$$

From $SPEC_P$, $SPEC_Q$, $SPEC_{W_1}$, $SPEC_{W_2}$, and the properties of sequences, we can deduce that

$$s_3 \leqslant \langle in, lm, rm, out, lc, rc\rangle^*$$

We then proceed by case analysis on the identity of the last event in s_3, given that $last(s_3 \setminus A) = in$, there are three possibilities.

Case: $last(s_3) = in$

By $SPEC_P$,	$lm \notin \sigma((\aleph_1 \upharpoonright X) \uparrow (end(s_1 \upharpoonright X) + 2\delta))$
In this **case**	$end(s_1) = end(s_1 \upharpoonright X)$
and we know that	$lm \notin Y$
Hence	$lm \notin \sigma(\aleph_1 \uparrow (end(s_3) + 2\delta))$
Similarly, as	$s_3 \in s_1 \; {}_{X \cup Y}\|_{M \cup C} \; s_2$,
$SPEC_{W_1}$ implies that	$lm \notin \sigma(\aleph_2 \uparrow (end(s_3) + 2\delta))$
Hence	$lm \notin \sigma((\aleph_1 \cup \aleph_2 \cup \aleph_3) \uparrow (end(s_3) + 2\delta))$
However,	$([0, end(s, \aleph_1 \cup \aleph_2 \cup \aleph_3)) \times A) \subseteq (\aleph_1 \cup \aleph_2 \cup \aleph_3)$
and	$lm \in A$
So	$end(\aleph_1 \cup \aleph_2 \cup \aleph_3) \leqslant end(s_3) + 2\delta$
But $\delta \ll 1$, so	$(\aleph_1 \cup \aleph_2 \cup \aleph_3) \uparrow (end(s_3) + 2)) = \{\}$
We conclude that	$out \notin \sigma((\aleph_1 \cup \aleph_2 \cup \aleph_3) \uparrow end(s_3 \setminus A + 2))$

Case: $last(s_3) = lm$

We establish that $end(s_3) \leqslant end(s_3 \setminus A) + 2\delta$: that the lm event occurred within time 2δ of the last input.

Assume otherwise:	$end(s_3) > end(s_3 \setminus A) + 2\delta$
If we let t be the time	$(end(s_3 \setminus A) + end(s_3) + 2\delta)/2$
Then we know that	$last(s_3 \upharpoonright t) = in$
By the previous **case**	$lm \notin \sigma(((\aleph_1 \cup \aleph_2 \cup \aleph_3) \upharpoonright t) \uparrow (end(s_3 \upharpoonright t) + 2\delta))$
From our assumptions	$([0, end(s_3, \aleph_1 \cup \aleph_2 \cup \aleph_3)) \times A) \subseteq \aleph_1 \cup \aleph_2$
And	$end(s_3 \upharpoonright t) + 2\delta = end(s_3 \setminus A) + 2\delta < t$
Hence	$lm \in \sigma(((\aleph_1 \cup \aleph_2 \cup \aleph_3) \upharpoonright t) \uparrow (end(s_3 \upharpoonright t) + 2\delta))$

Forcing a contradiction.

We can show, with a similar argument to the first **case**, in which the event rm replaces lm, that $end(\aleph_1 \cup \aleph_2 \cup \aleph_3) \leqslant end(s_3) + 1$. From above, $end(s_3) \leqslant end(s_3 \setminus A) + 2\delta$: the result follows.

Case: $last(s_3) = rm$

By a similar argument, we can establish that the event rm must occur no later than $1 + 2\delta$ after the last input. We then appeal to the specification of Q, and the result follows immediately. □

The treatment of hiding in Timed CSP is central to the construction of the above proof; the hidden events lm and rm must occur as soon as possible. Our method of proof allowed us to include these events in our reasoning, by eliminating the hiding operator from our proof obligation.

4.3 Other Requirements

Only at the final stage of the proof did we identify the protocol requirement $SPEC$. To establish that another property holds of the above protocol, it would not be necessary to perform the whole proof again. We have characterised the behaviour of the protocol in terms of the known properties of its components. To prove that the protocol satisfies an arbitrary specification S, we have only to show that the following predicate is true:

$$\left.\begin{array}{l} \sigma(s_1, \aleph_1) \subseteq (X \cup Y) \wedge \sigma(s_2, \aleph_2) \subseteq (M \cup C) \\ \sigma(\aleph_3) \subseteq \Sigma - (X \cup Y) \\ SPEC_P(s_1 \upharpoonright X, \aleph_1 \upharpoonright X) \wedge SPEC_Q(s_1 \upharpoonright Y, \aleph_1 \upharpoonright Y) \\ SPEC_{W_1}(s_2 \upharpoonright M, \aleph_2 \upharpoonright M) \wedge SPEC_{W_2}(s_2 \upharpoonright C, \aleph_2 \upharpoonright C) \\ \aleph = \aleph_1 \cup \aleph_2 \cup \aleph_3 \wedge s_3 \in s_1 \;{}_{X \cup Y}\|_{M \cup C}\; s_2 \\ \sigma(\aleph') \subseteq A \wedge ([0, end(s_3, \aleph)) \times A) \subseteq \aleph \end{array}\right\} \Rightarrow S(s_3 \setminus A, \aleph - \aleph')$$

For a particular specification S, we will be able to discard most of the conditions in the antecedent: the residual proof requirement is often easy to discharge.

5 Recursion and Delay

The inference rules presented in section 3 were sufficient for the example proof above. If we wish to provide implementations for the components mentioned in the previous section, we will require other $TCSP$ operators; to verify these implementations, we will require other inference rules.

5.1 Prefixing

The simplest $TCSP$ process is deadlock, or $STOP$. It cannot engage in any event, so any trace must be empty. It may refuse any event at any time, so there are no restrictions upon refusal set $\aleph$:

$$\frac{}{STOP \ \mathbf{sat} \ (s = \langle\rangle)}$$

This process will be useful in showing that certain specifications are *satisfiable*: that there is a process that will satisfy them.

More interesting processes will be able to perform events: for these, we will require the *prefix* operator. In Timed CSP, this operator introduces a delay, corresponding to the time taken to recover from participation:

$$\frac{\begin{array}{l} P \ \mathbf{sat} \ T(s, \aleph) \\ \left.\begin{array}{l} s = \langle\rangle \wedge a \notin \sigma(\aleph) \\ \vee \\ s = \langle(t, a)\rangle ^\frown s' \wedge a \notin \sigma(\aleph \upharpoonright t) \wedge T(s' - t - \delta, \aleph - t - \delta) \end{array}\right\} \Rightarrow S(s, \aleph) \end{array}}{(a \rightarrow P) \ \mathbf{sat} \ S(s, \aleph)}$$

Any behaviour of the process $a \rightarrow P$ must involve the non-refusal of event a until it has been performed. If event a occurs at time t, the subsequent behaviour will be that of process P, but starting at time $t + \delta$ instead of time 0. If process P meets the specification $T(s, \aleph)$, then these subsequent behaviours will be described by the predicate $T(s' - t - \delta, \aleph - t - \delta)$.

5.2 Recursion

Almost any application of *TCSP* will involve repetitive behaviour: to model this, we can use the recursion operator μ. If F is a function defined on *TCSP* processes, we define the function:

$$\begin{aligned} C_{\widehat{F}} &: TM_F \rightarrow TM_F \\ C_{\widehat{F}}(X) &\mathrel{\widehat{=}} \mathcal{F}_T[\![WAIT\,\delta\,; F(X)]\!] \end{aligned}$$

The process $\mu\, X \bullet F(X)$ behaves as the fixed point of $C_{\widehat{F}}$ in the model TM_F:

$$\mu\, X \bullet F(X) \equiv F(WAIT\,\delta\,; \mu\, X \bullet F(X))$$

The recursion induction theorem introduced by Roscoe in [Ro82], developed by Reed in [Re88], provides the basis for an inference rule for recursively-defined processes:

$$\frac{\begin{array}{l} \forall X : TCSP \bullet X \ \mathbf{sat} \ S(s, \aleph) \Rightarrow F(WAIT\,\delta\,; X) \ \mathbf{sat} \ S(s, \aleph) \\ \exists P : TCSP \bullet P \ \mathbf{sat} \ S(s, \aleph) \end{array}}{\mu\, X \bullet F(X) \ \mathbf{sat} \ S(s, \aleph)}$$

The topological result underlying this rule requires that the predicate " **sat** $S(s, \aleph)$ " be both continuous and satisfiable on *TCSP* processes. It is a consequence of the definition of the **sat** operator that all such predicates are continuous; this leaves the rule with only one side condition: included as the second antecedent above. We also require that $C_{\widehat{F}}$ is a contraction mapping on TM_F, and that the specification $S(s, \aleph)$ is preserved by each recursive call. The first of these follows from the continuity of all basic *TCSP* operators, the second becomes the first antecedent of the rule.

It is possible that the specification $S(s, \aleph)$ may only be satisfiable by a recursive process. In this case, the side condition cannot be established without a separate inductive proof. By extending the contraction mapping that corresponds to F, we can produce a rule that does not have this problem:

$$\frac{\forall X \bullet X \textbf{ sat } S(s,\aleph) \Rightarrow F(\mathit{WAIT}\,\delta\,;X) \textbf{ sat } S(s,\aleph)}{\mu X \bullet F(X) \textbf{ sat } S(s,\aleph)}$$

This follows from the same topological result as the previous rule, given a simple extension to the semantic function $\mathcal{F}_T$, as detailed in appendix B. We have eliminated the second antecedent. The first antecedent is stronger: we may no longer assume that the semantics of X satisfies the axioms of TM_F: we have lost the implicit assumption that X is a *TCSP* process.

The set of inference rules in this paper is independent of the axioms of the model TM_F, so each rule may be applied to arbitrary sets of failures: they can therefore be used to establish the new antecedent. Further, the fact that all of our specifications are behavioural means that this rule is no weaker than the recursion rule in TM_F.

5.3 Delay

Finally, we will need to reason about the behaviours of processes involving delays. We may derive a simple rule from the inference rules for the sequential composition and delay operators:

$$\frac{\begin{array}{l} P \textbf{ sat } T(s,\aleph) \\ \left.\begin{array}{l} s = \langle\rangle \wedge end(\aleph) \leqslant t \\ \vee \\ begin(s) \geqslant t \wedge T(s-t, \aleph - t) \end{array}\right\} \Rightarrow S(s,\aleph) \end{array}}{\mathit{WAIT}\ t\,; P \textbf{ sat } S(s,\aleph)}$$

The inclusion of arbitrary refusals $\aleph'$ before time t reflects the fact that $\mathit{WAIT}\ t\,; P$ may refuse any event before time t.

Whenever we apply the recursion rule, we will be left with a proof obligation on $\mathit{WAIT}\,\delta; X$, given that X satisfies a certain specification. In this case, an alternative form of the above rule will be more useful:

$$\frac{P \textbf{ sat } S(s,\aleph)}{\mathit{WAIT}\ t\,; P \textbf{ sat } \begin{array}{l} s = \langle\rangle \wedge end(\aleph) \leqslant t \\ \vee \\ begin(s) \geqslant t \wedge S(s-t, \aleph - t) \end{array}}$$

No event may occur before time t, and the subsequent behaviours are simply the failures of process P translated through time t.

We have now presented all the rules required to verify a simple implementation of the protocol specified in section 4.

6 Implementing the Protocol

In section 4 we used Timed CSP to establish the correctness of a simple protocol: this result was dependent upon the correct behaviour of each component of the protocol. We now propose *TCSP* implementations of the components, and use the inference rules given in section 5 to demonstrate that they meet the appropriate specifications.

6.1 Implementation

The protocol consists of two components, transmitter P and receiver Q, communicating across two wires W_1 and W_2. The transmitter process should accept an input on channel *in*, and be prepared to transmit it along W_1, via channel *lm*. After this transmission has occurred, P waits for a confirmation event from wire W_2, on channel *rc*, before repeating this behaviour. Our intuition suggests the following as an implementation:

$$P \mathrel{\widehat{=}} \mu X \bullet in \rightarrow lm \rightarrow rc \rightarrow X$$

We have yet to establish that this implements our requirements: that it meets the formal specification $SPEC_P$.

A similar set of conditions applies to the receiving process Q. It should be prepared to receive a signal from wire W_1, on channel *rm*, before offering output on channel *out*. It should then send a confirmation signal along wire W_2, on channel *lc*, before returning to its initial state. Our proposed solution:

$$Q \mathrel{\widehat{=}} \mu Y \bullet rm \rightarrow out \rightarrow lc \rightarrow Y$$

Again, we will have to verify that this is an implementation of the specification $SPEC_Q$.

We could also model wires W_1 and W_2 in the *TCSP* process algebra. Consider wire W_1: the propagation delay, the delay between input on channel *lm* and availability of output on channel *rm*, should be no more than one second. There will be a very small ($O(\delta)$) recovery time after output has occurred. In the context of [S88], it behaves as a stable one-place timed buffer:

$$\begin{aligned} W_1 &\mathrel{\widehat{=}} \mu X \bullet lm \rightarrow WAIT\,(1-\delta)\,;\, rm \rightarrow X \\ W_2 &\mathrel{\widehat{=}} \mu Y \bullet lc \rightarrow WAIT\,(1-\delta)\,;\, rc \rightarrow Y \end{aligned}$$

Note that the explicit delay between the occurrence of lm and the availability of rm is shortened by δ to allow for the delay introduced by the prefix operator: the time taken to recover from performing an event. Although we would not wish to implement wires in this fashion, the *TCSP* description could be used to produce a software simulation of their behaviour.

6.2 Verification

We wish to show that the transmitting process P meets the specification placed upon it:

$$\mu X \bullet in \rightarrow lm \rightarrow rc \rightarrow X \quad \textbf{sat} \quad SPEC_P(s, \aleph)$$

This is a recursive process; the second recursion rule requires us to find a specification $S(s, \aleph)$ such that:

$$\begin{aligned} X \ \textbf{sat} \ S(s,\aleph) &\Rightarrow in \rightarrow lm \rightarrow rc \rightarrow (WAIT\ \delta\,; X) \ \textbf{sat} \ S(s,\aleph) \\ S(s,\aleph) &\Rightarrow SPEC_P(s,\aleph) \end{aligned}$$

Our strategy for finding such a specification would be to consider S to be $SPEC_P \wedge S'$, strengthening S' until the conjunction, which must still be satisfiable, is preserved by the recursive call. In this example, the specification $SPEC_P$ is strong enough to be preserved by the recursion, and no other conditions are required. We instantiate S with $SPEC_P$. We have then to show that:

$$X \ \textbf{sat} \ SPEC_P(s,\aleph) \ \Rightarrow \ in \rightarrow lm \rightarrow rc \rightarrow (WAIT\ \delta\,; X) \ \textbf{sat} \ SPEC_P(s,\aleph)$$

Assume that X **sat** $SPEC_P(s, \aleph)$. We wish to establish that:

$$in \rightarrow lm \rightarrow rc \rightarrow (WAIT\ \delta\,; X) \quad \textbf{sat} \quad SPEC_P(s,\aleph)$$

Applying the prefix rule three times transforms this proof obligation to the following requirement: we must find a specification $U(s, \aleph)$ such that:

$$WAIT\ \delta\,; X \ \textbf{sat} \ U(s,\aleph)$$

$$\left.\begin{array}{l} s = \langle\rangle \wedge in \notin \sigma(\aleph) \\ \vee \\ s = \langle(t, in)\rangle^\frown s' \wedge in \notin \sigma(\aleph \upharpoonright t_1) \wedge \\ \quad s' - t_1 - \delta = \langle\rangle \wedge lm \notin \sigma(\aleph - t_1 - \delta) \\ \quad \vee \\ \quad s' - t_1 - \delta = \langle(t_2, lm)\rangle^\frown s'' \wedge lm \notin \sigma(\aleph - t_1 - \delta \upharpoonright t_2) \wedge \\ \qquad s'' - t_2 - \delta = \langle\rangle \wedge rc \notin \sigma(\aleph - t_1 - t_2 - 2\delta) \\ \qquad \vee \\ \qquad s'' - t_2 - \delta = \langle(t_3, rc)\rangle^\frown s''' \wedge \\ \qquad\quad rc \notin \sigma(\aleph - t_1 - t_2 - 2\delta \upharpoonright t_3) \wedge \\ \qquad\quad U(s''' - t_1 - t_2 - t_3 - 3\delta, \aleph - t_1 - t_2 - t_3 - 3\delta) \end{array}\right\} \Rightarrow SPEC_P(s,\aleph)$$

With a suitable choice of τ_1, τ_2, τ_3, this can be transformed to:

$$WAIT\ \delta\,;\, X \ \textbf{sat}\ U(s, \aleph)$$

$$\left.\begin{array}{l}
s = \langle\rangle \wedge in \notin \sigma(\aleph) \\
\vee \\
s = \langle(\tau_1, in)\rangle \wedge \in \notin \sigma(\aleph \upharpoonright \tau_1) \wedge lm \notin \sigma(\aleph \upharpoonleft \tau_1 + \delta) \\
\vee \\
s = \langle(\tau_1, in), (\tau_2, lm)\rangle \wedge in \notin \sigma(\aleph \upharpoonright \tau_1) \\
\qquad\qquad\qquad\qquad\quad \wedge\ lm \notin \sigma(\aleph \uparrow [\tau_1 + \delta, \tau_2)) \\
\qquad\qquad\qquad\qquad\quad \wedge\ rc \notin \sigma(\aleph \upharpoonleft \tau_2 + \delta) \\
\vee \\
s = \langle(\tau_1, in), (\tau_2, lm), (\tau_3, rc)\rangle^\frown u \wedge in \notin \sigma(\aleph \upharpoonright \tau_1) \\
\qquad\qquad\qquad\qquad\qquad\qquad\qquad \wedge\ lm \notin \sigma(\aleph \uparrow [\tau_1 + \delta, \tau_2) \\
\qquad\qquad\qquad\qquad\qquad\qquad\qquad \wedge\ rc \notin \sigma(\aleph \uparrow [\tau_2 + \delta, \tau_3)) \\
\qquad\qquad\qquad\qquad\qquad\qquad\qquad \wedge\ U(u - \tau_3 - \delta, \aleph - \tau_3 - \delta)
\end{array}\right\} \Rightarrow SPEC_P(s, \aleph)$$

Applying the second form of the delay rule, we can instantiate U as follows:

$$U(s, \aleph) \ \equiv\ SPEC_P(s - \delta, \aleph - \delta) \wedge begin(s) \geqslant \delta$$

Having discharged the first proof obligation, the proof can be completed with a simple case analysis on trace s. This becomes clear when we recall the form of specification $SPEC_P$:

$$\begin{array}{rcl}
SPEC_P & \hat{=} & tstrip(s) \leqslant \langle in, lm, rc\rangle^* \wedge \\
 & & last(s) = in \Rightarrow lm \notin \sigma(\aleph \upharpoonleft (end(s) + 2\delta)) \wedge \\
 & & last(s) = lm \Rightarrow rc \notin \sigma(\aleph \upharpoonleft (end(s) + 2\delta)) \wedge \\
 & & last(s) = rc \Rightarrow in \notin \sigma(\aleph \upharpoonleft (end(s) + 2\delta)) \wedge \\
 & & s = \langle\rangle \Rightarrow in \notin \sigma(\aleph)
\end{array}$$

The only non-trivial case corresponds to $s = \langle(\tau_1, in), (\tau_2, lm), (\tau_3, rc)\rangle^\frown u$. Here we require two arguments, one for each of the cases: $u = \langle\rangle$, $u \neq \langle\rangle$. Expanding the specification $SPEC_P$ makes the solution obvious.

This completes the verification of our transmitter process P. It will not be necessary to perform a similar proof for the receiver Q; we can exploit the symmetry present in our descriptions.

6.3 Renaming

The operator f in *TCSP* allows us to relabel the events performed by a process. In the case of injective functions, this allows us to re-use a process description. By renaming events, we can transform processes while retaining their structure. The relationships between different events are maintained: given that a particular result holds for all the behaviours of a process, we can infer a corresponding result about the behaviours of the image of that process under such a transformation:

$$\frac{\begin{array}{l} P \textbf{ sat } S_1(s,\aleph) \\ S_1(s,\aleph) \Rightarrow S(f(s),f(\aleph)) \end{array}}{f(P) \textbf{ sat } S(s,\aleph)}$$

For example, we can use the result of the previous section to show that Q **sat** $SPEC_Q$, by defining injective function f such that:

$$\begin{aligned} f(in) &\mathrel{\widehat{=}} rm \\ f(lm) &\mathrel{\widehat{=}} out \\ f(rc) &\mathrel{\widehat{=}} lc \end{aligned}$$

We then observe that:

$$\begin{aligned} SPEC_P(s,\aleph) &= SPEC_Q(f(s),f(\aleph)) \\ Q &= f(P) \end{aligned}$$

The inference rule allows us to conclude that:

$$Q \quad \textbf{sat} \quad SPEC_Q(s,\aleph)$$

Which completes our verification of the protocol.

This method of re-using implementation/specification pairs helps to eliminate redundant verifications: by observing and exploiting symmetry, we can re-use process components and their specifications.

7 Completing the Picture

The laws presented above, together with the others in the appendix, are complete with respect to the semantics: any specification provable from the semantics is provable using these laws. This becomes clear when we consider the *strongest specification* of a process.

7.1 Strongest Specifications

The identification of a process with the strongest specification that it can satisfy has been discussed before. It provides an alternative method for eliminating the process algebra from our proof obligations. The inference rules presented in this paper are more flexible in this: our specification may reflect only one of the properties of the system. Using our intuition, we need consider only the relevant properties of each component: those necessary to establish that the system meets the specification. As an example, consider the law:

$$\frac{\begin{array}{l} P_1 \ \mathbf{sat}\ S_1(s,\aleph) \\ P_2 \ \mathbf{sat}\ S_2(s,\aleph) \\ S_1(s,\aleph_1) \wedge S_2(s,\aleph_2) \Rightarrow S(s,\aleph_1 \cup \aleph_2) \end{array}}{P_1 \parallel P_2 \ \mathbf{sat}\ S(s,\aleph)}$$

For $P_1 \parallel P_2$ to meet specification S, we require that P_1 and P_2 meet specifications S_1 and S_2 respectively. These need only be strong enough to fulfil the third antecedent of the rule.

If we lack this intuition, we can use the strongest specifications of P_1 and P_2 as instantiations for S_1 and S_2. If suitable instantiations exist, they can be no stronger than these: any property of a process is a logical consequence of its strongest specification. We write $\mathcal{SS}[\![P]\!]$ to denote the strongest specification of process P. For example, the strongest specification of deadlock is given by:

$$\mathcal{SS}[\![STOP]\!](s,\aleph) \quad \equiv \quad s = \langle\rangle$$

This is all we can possibly know about the behaviours of $STOP$, we can draw no conclusions about the refusal set: $STOP$ may refuse any event at any time.

For a compound process, the strongest specification is defined in terms of the strongest specifications of its proper syntactic subcomponents:

$$\begin{aligned} \mathcal{SS}[\![a \rightarrow P]\!](s,\aleph) \quad \equiv \quad & s = \langle\rangle \wedge a \notin \sigma(\aleph) \\ & \vee \\ & \exists\, s', t \bullet (s = \langle(t,a)\rangle^\frown s') \wedge a \notin \sigma(\aleph \restriction t) \wedge \\ & \qquad \mathcal{SS}[\![P]\!](s', \aleph - (t+\delta)) \end{aligned}$$

These definitions are equivalent to the semantic equations for the model TM_F. The equivalence

$$\mathcal{SS}[\![P]\!](s,\aleph) \quad \equiv \quad (s,\aleph) \in \mathcal{F}_T[\![P]\!]$$

can be established by structural induction upon process P.

Strongest specifications may be used to reduce the proof requirement on a compound process to a predicate on traces and refusals, similar to the one at the end of 4.2. The inference rules given in this paper may provide a much simpler predicate; we can discard unnecessary information. But strongest specifications provide a more *mechanical* method; there are no choices to be made, even in the case of recursion.

$$\begin{aligned} \mathcal{SS}[\![\mu X \bullet F(X)]\!](s,\aleph) \quad \equiv \quad & \bigwedge_{i \in \mathbb{N}} (end(s,\aleph) < \delta i \Rightarrow \mathcal{SS}[\![\widehat{F}^i(STOP)]\!](s,\aleph)) \\ & \text{where } \widehat{F}(X) = WAIT\ \delta\ ; F(X) \end{aligned}$$

We consider the recursive process $\mu X \bullet F(X)$ to be the limit of the finite approximations $\widehat{F}^i(STOP)$. A given behaviour of the recursive process must be a behaviour of all the finite approximations involving a sufficient number of recursions. If the behaviour in question is described by the failure $(s, \aleph)$, then all of the approximations $\widehat{F}^i(STOP)$, where $i > end(s, \aleph)/\delta$, must also exhibit that behaviour.

Hence the strongest specification of $\mu X \bullet F(X)$ can be written as the conjunction of the strongest specifications of its finite approximations, guarded by an applicability condition $end(s, \aleph) < \delta i$. We are spared the task of finding a sufficient specification that will be preserved by each recursive call.

Strongest specifications provide a complete description of the possible behaviours of a process. To decide whether a component is adequate for use in a given situation, we can use the inference rules in this paper to confirm that it meets the requirements. If a component is to be re-used in different systems, then it should be supplied with its strongest specification. The comprehensive nature of strongest specifications also allows us to demonstrate that the inference rules presented in this paper are *complete* with respect to the semantics.

7.2 Completeness

The inference rules presented in this paper are easily seen to be *sound*; example proofs are presented in appendix B. If we can use the rules to show that process P satisfies a specification $S(s, \aleph)$ then predicate $S(s, \aleph)$ must hold for all behaviours of P: it must be true of all the elements of the set of failures corresponding to P in the semantic model TM_F.

These rules also form a *complete* set. If a predicate $S(s, \aleph)$ holds for all behaviours of P, then we can use the rules to establish that P **sat** $S(s, \aleph)$. We can demonstrate this by showing that the rules preserve strongest specifications: they yield the strongest specification of a compound process in terms of the strongest specifications of its components. For example, consider the case of the parallel operator.

Suppose that the parallel combination $P_1 \parallel P_2$ meets the specification $S(s, \aleph)$. In our proof, we would employ the following inference rule:

$$\frac{\begin{array}{l} P_1 \ \mathbf{sat} \ S_1(s, \aleph) \\ P_2 \ \mathbf{sat} \ S_2(s, \aleph) \\ S_1(s, \aleph_1) \wedge S_2(s, \aleph_2) \Rightarrow S(s, \aleph_1 \cup \aleph_2) \end{array}}{P_1 \parallel P_2 \ \mathbf{sat} \ S(s, \aleph)}$$

This requires that we exhibit specifications S_1 and S_2 for which the three antecedents of the rule hold. The first two antecedents insist that these are no stronger than the corresponding strongest specifications, so if the third is also to hold, it must hold with the following instantiation:

$$\mathcal{SS}[\![P_1]\!](s, \aleph_1) \wedge \mathcal{SS}[\![P_2]\!](s, \aleph_2) \quad \Rightarrow \quad S(s, \aleph_1 \cup \aleph_2)$$

However, as $S(s, \aleph)$ is true of all behaviours of $P_1 \parallel P_2$, it can be no stronger than the strongest specification of that process, i.e.

$$\mathcal{SS}[\![P_1 \parallel P_2]\!](s, \aleph) \quad \Rightarrow \quad S(s, \aleph)$$

But the strongest specification is given by:

$$\mathcal{SS}[\![P_1 \parallel P_2]\!](s, \aleph) \equiv \exists \aleph_1, \aleph_2 \bullet \mathcal{SS}[\![P_1]\!](s, \aleph_1) \wedge \mathcal{SS}[\![P_2]\!](s, \aleph_2) \wedge \aleph = \aleph_1 \cup \aleph_2$$

So the inference rule is sufficient to establish that $P_1 \parallel P_2$ **sat** $S(s, \aleph)$. The same is true for the other operators, and we have shown that the equivalences that define our strongest specifications are no weaker than the semantic equations: we lose no information. Hence our inference rules form a complete set with respect to the semantics.

8 Stability

In this paper, we have been working within the Timed Failures model of TCSP, TM_F. Timed CSP identifies a further aspect of a process's behaviour: the *stability* value corresponding to each (trace, refusal) pair. In [Re88], this is defined to be the earliest time at which it can be guaranteed that the process can make no further internal progress. This notion has been refined by Blamey in [Bl89]. Here, he associates with each (trace,refusal) pair an "instability" set rather than a stability value: the set of times at which the process might not be stable.

One advantage of this approach is that it allows us to extend the work in this paper to models which include stability. Using instability sets, we can express the behaviour of a compound process in terms of the behaviours of its components. This is not possible in the original stability models, TM_S and TM_{FS}: to see why, consider the processes defined below.

$$\begin{aligned} P_1 &\mathrel{\hat{=}} a \rightarrow STOP \\ P_2 &\mathrel{\hat{=}} a \rightarrow WAIT\,1\,; STOP \end{aligned}$$

The stabilities associated with this process are given by:

$$\begin{aligned} \mathcal{S}_T[\![P_1]\!] &\equiv \{(\langle\rangle, 0)\} \cup \{(\langle(t, a)\rangle, t + \delta) \mid t \geqslant 0\} \\ \mathcal{S}_T[\![P_2]\!] &\equiv \{(\langle\rangle, 0)\} \cup \{(\langle(t, a)\rangle, t + 1 + \delta) \mid t \geqslant 0\} \end{aligned}$$

Now consider the behaviours of the process $P_1 \parallel\!\mid P_2$, given the semantic equation for the interleaving operator:

$$\begin{aligned} \mathcal{S}_T[\![P_1 \parallel\!\mid P_2]\!] \mathrel{\hat{=}}\; & SUP\{(s, max\{\alpha_1, \alpha_2\}) \mid \\ & \exists (s_1, \alpha_1) \in \mathcal{S}_T[\![P_1]\!], (s_2, \alpha_2) \in \mathcal{S}_T[\![P_2]\!] \bullet s \in Tmerge(s_1, s_2)\} \end{aligned}$$

The compound process can engage in a single a event, from each of its components, and give rise to a stability value that cannot be inferred from the properties of a typical behaviour of either process acting independently. The trace $\langle(0, a)\rangle$ has a stability value of $1 + \delta$: this can only be deduced by considering *all* of the stability values associated with that trace.

However, if we identify instability sets rather than stability values, no such difficulties arise. The properties of a typical instability set of a compound process behaviour can be deduced from the properties of arbitrary behaviours of the component processes. As with the timed failures model, we can restrict our attention to a typical element of the semantics. We can thus formulate a set of inference rules for reasoning about specifications involving stability conditions. As an example, we can derive the following rule for the nondeterministic choice operator:

$$\frac{\begin{array}{l} P_1 \ \mathbf{sat} \ S_1(s,\gamma,\aleph) \\ P_2 \ \mathbf{sat} \ S_2(s,\gamma,\aleph) \\ (S_1(s,\gamma,\aleph) \vee S_2(s,\gamma,\aleph)) \Rightarrow S(s,\gamma,\aleph) \end{array}}{P_1 \sqcap P_2 \ \mathbf{sat} \ S(s,\gamma,\aleph)}$$

In the above specifications γ represents an arbitrary instability value, and the **sat** operator is extended in the obvious way. The rule illustrates that an instability value of $P_1 \sqcap P_2$ must be an instability value for one of the components P_1, P_2. The converse is also true; this is not the case in TM_{FS}, in which an arbitrary behaviour requires more information. Similar results are obtained for the other *TCSP* operators.

9 Conclusions

In this paper, we have shown how we can factor out the complexity inherent in reasoning about timed distributed systems. We introduced behavioural specifications, capturing correctness conditions as simple predicates on a typical element of the semantics. We have given inference rules, derived from the semantic mappings, for reasoning about these specifications. These rules allow us to reduce proof obligations on a composite Timed CSP process to requirements on the syntactic subcomponents.

The lack of sufficient algebraic laws means that we cannot construct a proof system for Timed CSP similar to the one developed in [Br83], but we can produce a complete set of inference rules for proofs of correctness. Further, we have presented the rules in such a form as to make their application completely mechanical: an automated proof assistant could be developed similar to the one employed in [D87].

As an illustration of the use of the rules, we have presented a verification of a simple flow control protocol, whose definition involved both abstraction and concurrency. The correctness of this example depends upon the subtle treatment of hiding in Timed CSP: any hidden events are forced to occur as soon as they become available. An implementation of the protocol was proposed and verified; this required a useful result about the properties of recursive processes.

We have exhibited strongest specifications for Timed CSP processes and used these to verify that our rules form a complete set with respect to the semantics. Our intention is to work towards a specification-oriented semantics for Timed CSP, similar to the one described in [OH83], using the enhanced timed failures-stability model and the hierarchy of lower models. This will allow us to work towards a powerful specification and development methodology for real-time concurrent systems.

Acknowledgements

The authors would like to thank Bill Roscoe and Mike Reed for their advice and encouragement; Stephen Blamey, Steve Brookes and Michael Goldsmith for their suggestions; Alice King-Farlow and Elizabeth Schneider for inspiration. We are also indebted to our colleagues in the Programming Research Group for their friendship. This work was supported by grants from SERC and BP.

References

[Bl89] S.R. Blamey, *TCSP Processes as Predicates,* (to appear) Oxford 1989.

[Br83] S.D. Brookes, *A Model for Communicating Sequential Processes,* Oxford University D.Phil thesis 1983.

[D87] J.W. Davies, *Assisted Proofs for Communicating Sequential Processes,* Oxford University M.Sc. thesis 1987.

[H85] C.A.R. Hoare, *Communicating Sequential Processes,* Prentice-Hall International 1985.

[OH83] E.R. Olderog and C.A.R. Hoare, *Specification-oriented Semantics for Communicating Processes* Springer LNCS **154** 1983, **561-572**. (Also, Acta Informatica 23 1986, 9-66.).

[PS88] K. Paliwoda and J.W. Sanders, *The Sliding-Window Protocol in CSP,* Oxford University Programming Research Group Technical Monograph 1988, **66**.

[Re88] G.M. Reed, *A Uniform Mathematical Theory for Real-time Distributed Computing,* Oxford University D.Phil thesis 1988.

[RR86] G.M. Reed and A.W. Roscoe, *A Timed Model for Communicating Sequential Processes* Proceedings of ICALP'86, Springer LNCS **226** (1986), **314-323**; Theoretical Computer Science **58** 1988, **249-261**.

[RR87] G.M. Reed and A.W. Roscoe, *Metric Spaces as Models for Real-time Concurrency* Proceedings of the Third Workshop on the Mathematical Foundations of Programming Language Semantics, LNCS **298** 1987, **331-343**.

[Ro82] A.W. Roscoe *A Mathematical Theory of Communicating Processes* Oxford University D.Phil thesis 1982.

[S88] S.A. Schneider *Communication in Timed Distributed Computing* Oxford University M.Sc. thesis 1988.

A Inference Rules

In this appendix, we present a complete set of inference rules for behavioural specifications. A rule is presented for each TCSP operator.

Rule *STOP*

$$\frac{}{STOP \text{ sat } (s = \langle\rangle)}$$

Rule $\perp$

$$\frac{}{\perp \text{ sat } (s = \langle\rangle)}$$

Rule *SKIP*

$$\frac{}{\begin{array}{ll} SKIP \text{ sat } & (s = \langle\rangle \wedge \checkmark \notin \sigma(\aleph)) \\ & \vee \\ & (s = \langle(t, \checkmark)\rangle \wedge \checkmark \notin \sigma(\aleph \restriction t) \wedge t \geqslant 0) \end{array}}$$

Rule *WAIT t*

$$\frac{}{\begin{array}{ll} WAIT\ t \text{ sat } & s = \langle\rangle \wedge \checkmark \notin \sigma(\aleph \upharpoonleft t) \\ & \vee \\ & s = \langle(t', \checkmark)\rangle \wedge t' \geqslant t \wedge \checkmark \notin \sigma(\aleph \uparrow [t, t')) \end{array}}$$

The following rules apply to compound processes. When a process variable is present, it is more convenient to match proof obligations to consequents: the form in which the rules are presented makes this possible.

Rule $a \to P$

$$\frac{\begin{array}{l} P \text{ sat } T(s, \aleph) \\ \left.\begin{array}{l} s = \langle\rangle \wedge a \notin \sigma(\aleph) \\ \vee \\ s = \langle(t, a)\rangle^\frown s' \wedge a \notin \sigma(\aleph \restriction t) \wedge T(s' - t - \delta, (\aleph - t - \delta) \upharpoonleft 0) \end{array}\right\} \Rightarrow S(s, \aleph) \end{array}}{(a \to P) \text{ sat } S(s, \aleph)}$$

Rule $P_1 \Box P_2$

$$\frac{\begin{array}{l} P_1 \textbf{ sat } S_1(s,\aleph) \\ P_2 \textbf{ sat } S_2(s,\aleph) \\ \left.\begin{array}{l}(S_1(s,\aleph) \vee S_2(s,\aleph) \\ \wedge \\ S_1(\langle\rangle, \aleph \restriction begin(s)) \wedge S_2(\langle\rangle, \aleph \restriction begin(s))\end{array}\right\} \Rightarrow S(s,\aleph) \end{array}}{P_1 \Box P_2 \textbf{ sat } S(s,\aleph)}$$

Rule $a : A \rightarrow P_a$

$$\frac{\begin{array}{l} \forall a \in A \bullet P_a \textbf{ sat } S_a(s,\aleph) \\ \left.\begin{array}{l}(\aleph \cap ([0, begin(s)) \times A) = \emptyset \\ \wedge \\ \forall a \in A \bullet (s = \langle(t,a)\rangle ^\frown s') \Rightarrow S_a(s' - t - \delta, (\aleph - t - \delta) \uparrow 0))\end{array}\right\} \Rightarrow S(s,\aleph) \end{array}}{a : A \rightarrow P_a \textbf{ sat } S(s,\aleph)}$$

Rule $P_1 \sqcap P_2$

$$\frac{\begin{array}{l} P_1 \textbf{ sat } S_1(s,\aleph) \\ P_2 \textbf{ sat } S_2(s,\aleph) \\ S_1(s,\aleph) \vee S_2(s,\aleph) \Rightarrow S(s,\aleph) \end{array}}{P_1 \sqcap P_2 \textbf{ sat } S(s,\aleph)}$$

Rule $P_1 \parallel P_2$

$$\frac{\begin{array}{l} P_1 \textbf{ sat } S_1(s,\aleph) \\ P_2 \textbf{ sat } S_2(s,\aleph) \\ S_1(s,\aleph_1) \wedge S_2(s,\aleph_2) \Rightarrow S(s,\aleph_1 \cup \aleph_2) \end{array}}{P_1 \parallel P_2 \textbf{ sat } S(s,\aleph)}$$

Rule $P_1 \ {}_X\!\parallel_Y P_2$

$$\frac{\begin{array}{l} P_1 \textbf{ sat } S_1(s,\aleph) \\ P_2 \textbf{ sat } S_2(s,\aleph) \\ \left.\begin{array}{l}(\sigma(s_1,\aleph_1) \subseteq X \wedge \sigma(s_2,\aleph_2) \subseteq Y \\ \wedge \\ \wedge\ \sigma(\aleph_3) \subseteq \Sigma - (X \cup Y) \\ \wedge \\ S_1(s_1,\aleph_1) \wedge S_2(s_2,\aleph_2) \wedge s_3 \in s_1 \ {}_X\!\parallel_Y s_2)\end{array}\right\} \Rightarrow S(s_3, \aleph_1 \cup \aleph_2 \cup \aleph_3) \end{array}}{P_1 \ {}_X\!\parallel_Y P_2 \textbf{ sat } S(s,\aleph)}$$

Rule $P_1 \mathbin{|||} P_2$

$$\frac{\begin{array}{l} P_1 \ \mathbf{sat}\ S_1(s,\aleph) \\ P_2 \ \mathbf{sat}\ S_2(s,\aleph) \\ (s \in \mathit{Tmerge}(u,v) \wedge S_1(u,\aleph) \wedge S_2(v,\aleph)) \Rightarrow S(s,\aleph) \end{array}}{P_1 \mathbin{|||} P_2 \ \mathbf{sat}\ S(s,\aleph)}$$

Rule $P_1 \,;\, P_2$

$$\frac{\begin{array}{l} P_1 \ \mathbf{sat}\ S_1(s,\aleph) \\ P_2 \ \mathbf{sat}\ S_2(s,\aleph) \\ (\checkmark \notin \sigma(s) \wedge \forall I \in \mathit{TINT} \bullet S_1(s, \aleph \cup (I \times \{\checkmark\}))) \Rightarrow S(s,\aleph) \\ \left.\begin{array}{l} s \cong s_1{}^\frown(s_2 + t) \wedge \checkmark \notin \sigma(s_1) \wedge \mathit{end}(\aleph_1) \leqslant t \\ \wedge \\ S_1(s_1{}^\frown\langle(t,\checkmark)\rangle, \aleph_1 \cup ([0,t) \times \{\checkmark\})) \wedge S_2(s_2,\aleph_2) \end{array}\right\} \Rightarrow S(s, \aleph_1 \cup (\aleph_2 + t)) \end{array}}{(P_1 \,;\, P_2) \ \mathbf{sat}\ S(s,\aleph)}$$

Rule $P \setminus A$

$$\frac{P \ \mathbf{sat}\ \hat{A}(s,\aleph) \wedge \sigma(\aleph') \subseteq A \Rightarrow S(s \setminus A, \aleph - \aleph')}{P \setminus A \ \mathbf{sat}\ S(s,\aleph)}$$

Rule $f^{-1}(P)$

$$\frac{\begin{array}{l} P \ \mathbf{sat}\ S_1(s,\aleph) \\ S_1(f(s), f(\aleph)) \Rightarrow S(s,\aleph) \end{array}}{f^{-1}(P) \ \mathbf{sat}\ S(s,\aleph)}$$

Rule $f(P)$

$$\frac{\begin{array}{l} P \ \mathbf{sat}\ S_1(s,\aleph) \\ S_1(s,\aleph) \Rightarrow S(f(s), f(\aleph)) \end{array}}{f(P) \ \mathbf{sat}\ S(s,\aleph)}$$

Rule $\mu X \bullet F(X)$

$$\frac{\forall X \bullet X \ \mathbf{sat}\ S(s,\aleph) \Rightarrow F(\mathit{WAIT}\,\delta \,;\, X) \ \mathbf{sat}\ S(s,\aleph)}{\mu X \bullet F(X) \ \mathbf{sat}\ S(s,\aleph)}$$

B Example Proofs

In this appendix we present a proof of soundness for the prefixing rule. We then extend the semantic function $\mathcal{F}_T$ to permit a proof of the second recursion rule given in section 5. We verify that the proposed extension is consistent with the original formulation, and provide a simple proof of the hiding rule.

B.1 Prefixing

Rule

$$\frac{\begin{array}{l} P \ \mathbf{sat}\ T(s,\aleph) \\ \left.\begin{array}{l} s = \langle\rangle \wedge a \notin \sigma(\aleph) \\ \vee \\ s = \langle (t,a)\rangle^\frown s' \wedge a \notin \sigma(\aleph \upharpoonright t) \wedge T(s' - t - \delta, (\aleph - t - \delta) \uparrow 0) \end{array}\right\} \Rightarrow S(s,\aleph) \end{array}}{(a \rightarrow P)\ \mathbf{sat}\ S(s,\aleph)}$$

Semantics

$$\begin{array}{rcl} \mathcal{F}_T[\![a \rightarrow P]\!] & \triangleq & \{(\langle\rangle,\aleph) \mid a \notin \sigma(\aleph)\} \\ & & \cup \\ & & \{(\langle (t,a)\rangle^\frown(s + (t+\delta)), \aleph_1 \cup \aleph_2 \cup (\aleph_3 + (t+\delta))) \mid \\ & & \quad t \geqslant 0 \wedge (I(\aleph_1) \subseteq [0,t) \wedge a \notin \sigma(\aleph_1)) \\ & & \quad \wedge\ I(\aleph_2) \subseteq [t, t+\delta) \wedge (s,\aleph_3) \in \mathcal{F}_T[\![P]\!]\} \end{array}$$

Proof

$$P\ \mathbf{sat}\ T(s,\aleph)$$

$$\begin{array}{rcl} (s,\aleph) \in \mathcal{F}_T[\![a \rightarrow P]\!] & \Rightarrow & s = \langle\rangle \wedge a \notin \sigma(\aleph) \\ & & \vee \\ & & \exists \aleph_1, \aleph_2, \aleph_3, s' \bullet s = \langle (t,a)\rangle^\frown(s' + t + \delta) \\ & & \quad \wedge\ \aleph = \aleph_1 \cup \aleph_2 \cup (\aleph_3 + t + \delta) \wedge t \geqslant 0 \\ & & \quad \wedge\ I(\aleph_1) \subseteq [0,t) \wedge a \notin \sigma(\aleph_1) \\ & & \quad \wedge\ I(\aleph_2) \subseteq [t, t+\delta) \wedge (s',\aleph_3) \in \mathcal{F}_T[\![P]\!] \end{array}$$

$$\begin{array}{rrcl} \vdash & \forall (s,\aleph) \in \mathcal{F}_T[\![a \rightarrow P]\!] & \bullet & s = \langle\rangle \wedge a \notin \sigma(\aleph) \\ & & & \vee \\ & & & s = \langle (t,a)\rangle^\frown(s' + t + \delta) \wedge t' \geqslant 0 \wedge a \notin \sigma(\aleph \upharpoonright t) \\ & & & \qquad \wedge\ (s', (\aleph - (t+\delta)) \uparrow 0) \in \mathcal{F}_T[\![P]\!] \end{array}$$

$$\begin{array}{rrl} \vdash & a \rightarrow P\ \mathbf{sat} & s = \langle\rangle \wedge a \notin \sigma(\aleph) \\ & & \vee \\ & & s = \langle (t,a)\rangle^\frown(s' + t + \delta) \wedge t' \geqslant 0 \wedge a \notin \sigma(\aleph \upharpoonright t) \\ & & \qquad \wedge\ T(s', (\aleph - (t+\delta)) \uparrow 0) \end{array}$$

The inference rule for prefixing follows immediately, by a simple property of the **sat** operator (see the third inference rule given in section 2). We conclude that the rule rests soundly upon the semantics.

B.2 The Semantic Function $\overline{\mathcal{F}}_T$

As mentioned in section 5, we obtain a more powerful rule for reasoning about the behaviour of recursive processes if we extend the semantic function $\mathcal{F}_T$. First, we must define the type of failure sets, $\overline{TF}$:

$$\overline{TF} \;\widehat{=}\; \mathbf{P}(T\Sigma^*_{\leqslant} \times RSET)$$

where $T\Sigma^*_{\leqslant}$ and $RSET$ are as defined in [Re88]. We then extend the syntax of Timed CSP:

$$TCSP^+ \;::=\; TCSP \mid X_E$$

where E ranges over the whole of $\overline{TF}$. Finally, we extend the semantic function $\mathcal{F}_T$ in the following fashion:

$$\overline{\mathcal{F}}_T[\![X_E]\!] \;\widehat{=}\; E$$
$$\overline{\mathcal{F}}_T[\![P \setminus A]\!] \;\widehat{=}\; \{(s \setminus A, \aleph - \aleph') \mid (s,\aleph) \in \overline{\mathcal{F}}_T[\![P]\!] \wedge \hat{A}(s,\aleph) \wedge \sigma(\aleph') \subseteq A\}$$

The remaining clauses are entirely similar to the defining equations for $\mathcal{F}_T$ given in [Re88]. To show that the new semantic function is an extension of $\mathcal{F}_T$ we must demonstrate that the two functions agree on the intersection of their domains: $TCSP$. A simple structural induction will suffice: the only non-trivial case is that of the hiding operator. In this case, recalling the relevant semantic equations

$$\mathcal{F}_T[\![P \setminus A]\!] \;\widehat{=}\; \{(s \setminus A, \aleph) \mid (s, \aleph \cup ([0, end(s,\aleph)) \times A)) \in \mathcal{F}_T[\![P]\!]\}$$
$$\overline{\mathcal{F}}_T[\![P \setminus A]\!] \;\widehat{=}\; \{(s \setminus A, \aleph - \aleph') \mid (s,\aleph) \in \overline{\mathcal{F}}_T[\![P]\!] \wedge \hat{A}(s,\aleph) \wedge \sigma(\aleph') \subseteq A\}$$

and the definition

$$\hat{A}(s,\aleph) \;\widehat{=}\; ([0, end(s,\aleph)) \times A) \subseteq \aleph$$

we proceed as follows:

Assume that $\mathcal{F}_T[\![P]\!] = \overline{\mathcal{F}}_T[\![P]\!]$ and that P is a process.

$$(s, \aleph) \in \mathcal{F}_T[\![P \setminus A]\!]$$

$$\vdash \quad \exists\, s_1, \aleph_1, \aleph_2 \bullet s = s_1 \setminus A \wedge (s_1, \aleph \cup ([0, end(s_1, \aleph)) \times A)) \in \mathcal{F}_T[\![P]\!] \\ \wedge\, \aleph_1 = \aleph \cup ([0, end(s_1, \aleph)) \times A) \wedge \aleph = \aleph_1 - \aleph_2 \wedge \sigma(\aleph_2) \subseteq A$$

$$\vdash \quad \exists\, s_1, \aleph_1, \aleph_2 \bullet s = s_1 \setminus A \wedge \aleph = \aleph_1 - \aleph_2 \wedge (s_1, \aleph_1) \in \overline{\mathcal{F}}_T[\![P]\!] \\ \wedge\, [0, end(s_1, \aleph_1)) \times A \subseteq \aleph_1 \wedge \sigma(\aleph_2) \subseteq A$$

since $end(s_1, \aleph) = end(s_1, \aleph_1)$

$$\vdash \quad \exists\, s_1, \aleph_1, \aleph_2 \bullet s = s_1 \setminus A \wedge \aleph = \aleph_1 - \aleph_2 \wedge (s_1 \setminus A, \aleph_1 - \aleph_2) \in \overline{\mathcal{F}}_T[\![P \setminus A]\!]$$

$$\vdash \quad (s, \aleph) \in \overline{\mathcal{F}}_T[\![P \setminus A]\!]$$

Conversely,

$$(s, \aleph) \in \overline{\mathcal{F}}_T[\![P \setminus A]\!]$$

$$\vdash \quad \exists\, s_1, \aleph_1, \aleph_2 \bullet s = s_1 \setminus A \wedge \aleph = \aleph_1 - \aleph_2 \wedge (s_1, \aleph_1) \in \overline{\mathcal{F}}_T[\![P]\!] \\ \wedge\, [0, end(s_1, \aleph_1)) \times A \subseteq \aleph_1 \wedge \sigma(\aleph_2) \subseteq A$$

$$\vdash \quad \exists\, s_1, \aleph_1, \aleph_2 \bullet s = s_1 \setminus A \wedge (s_1, \aleph \cup ([0, end(s_1, \aleph_1)) \times A)) \in \mathcal{F}_T[\![P]\!] \\ \wedge\, \aleph_1 = \aleph \cup ([0, end(s_1, \aleph_1)) \times A) \wedge \aleph = \aleph_1 - \aleph_2 \wedge \sigma(\aleph_2) \subseteq A$$

by Axiom 6 of TM_F

$$\vdash \quad \exists\, s_1 \bullet s = s_1 \setminus A \wedge (s_1, \aleph \cup ([0, end(s_1, \aleph)) \times A)) \in \mathcal{F}_T[\![P]\!]$$

by Axiom 6 again, since $end(s_1, \aleph) \leqslant end(s_1, \aleph_1)$

$$\vdash \quad (s, \aleph) \in \mathcal{F}_T[\![P \setminus A]\!]$$

□

B.3 Hiding

Having verified that $\overline{\mathcal{F}}_T$ is an extension of $\mathcal{F}_T$, we can easily establish the soundness of the rule for the hiding operator:

$$\frac{P \textbf{ sat } \widehat{A}(s, \aleph) \wedge \sigma(\aleph') \subseteq A \Rightarrow S(s \setminus A, \aleph - \aleph')}{P \setminus A \textbf{ sat } S(s, \aleph)}$$

Given the semantic equation

$$\overline{\mathcal{F}}_T[\![P \setminus A]\!] \;\; \widehat{=} \;\; \{(s \setminus A, \aleph - \aleph') \mid (s, \aleph) \in \overline{\mathcal{F}}_T[\![P]\!] \wedge \widehat{A}(s, \aleph) \wedge \sigma(\aleph') \subseteq A\}$$

we proceed as follows:

$$P \textbf{ sat } \widehat{A}(s, \aleph) \wedge \sigma(\aleph') \subseteq A \Rightarrow S(s \setminus A, \aleph - \aleph')$$

$$\begin{aligned} (s, \aleph) \in \overline{\mathcal{F}}_T[\![P \setminus A]\!] \;\; &\Rightarrow \;\; \exists\, s_1, \aleph_1, \aleph_2 \bullet s = s_1 \setminus A \wedge \aleph = \aleph_1 - \aleph_2 \wedge \widehat{A}(s_1, \aleph_1) \\ &\qquad \wedge (s_1, \aleph_1) \in \overline{\mathcal{F}}_T[\![P]\!] \wedge \sigma(\aleph_2) \subseteq A \end{aligned}$$

$$\begin{aligned} \vdash \;\; \forall (s, \aleph) \in \overline{\mathcal{F}}_T[\![P \setminus A]\!] \;\; &\bullet \;\; \exists\, s_1, \aleph_1, \aleph_2 \bullet s = s_1 \setminus A \wedge \aleph = \aleph_1 - \aleph_2 \\ &\qquad \wedge S(s_1 \setminus A, \aleph_1 - \aleph_2) \end{aligned}$$

$$\vdash \qquad P \setminus A \textbf{ sat } S(s, \aleph)$$

B.4 Recursion

Finally, we establish the result that provides the motivation for the extension to the semantics: the second inference rule for recursion:

$$\frac{\forall X \bullet X \textbf{ sat } S(s, \aleph) \Rightarrow F(\mathit{WAIT}\,\delta\,; X) \textbf{ sat } S(s, \aleph)}{\mu X \bullet F(X) \textbf{ sat } S(s, \aleph)}$$

We begin by extending the topology on TM_F defined in [Re88] to $\overline{TF}$ in the obvious way: Reed's proof that all of the basic *TCSP* operators are non-expanding is independent of the axioms. That all basic $TCSP^+$ operators are non-expanding follows immediately.

If F is a function on $TCSP^+$ composed of basic operators, there is a corresponding function C_F defined on $\overline{TF}$ by:

$$C_F(E) \;\; \widehat{=} \;\; \overline{\mathcal{F}}_T[\![F(X_E)]\!]$$

From the above result, it follows that F is non-expanding, and that if any of the components of F are contracting, then so is F. The function $\mathit{WAIT}\,\delta\,; X$ is always contracting; if we define

$$\widehat{F}(X) \;\; \widehat{=} \;\; F(\mathit{WAIT}\,\delta\,; X)$$

then, for any F, the function $\widehat{F}$ will be contracting; the corresponding mapping on $\overline{TF}$, $C_{\widehat{F}}$, will be a contraction mapping on a metric space: it will have a unique fixed point. This fixed point is the semantics of $\mu X \bullet F(X)$.

If we consider the sequence $\{E_n\}$, where

$$E_n \;\; \widehat{=} \;\; C_{\widehat{F}}^n(\emptyset)$$

we observe that

$$\lim_{n \to \infty} (E_n) \;=\; \overline{\mathcal{F}}_T[\![\mu X \bullet F(X)]\!]$$

The antecedent of the recursion rule

$$\forall X \bullet X \textbf{ sat } S(s, \aleph) \;\Rightarrow\; \widehat{F}(X) \textbf{ sat } S(s, \aleph)$$

allows us to conclude that

$$\forall X, n \bullet X \textbf{ sat } S(s, \aleph) \;\Rightarrow\; \widehat{F}^n(X) \textbf{ sat } S(s, \aleph)$$

However, it is easy to show that $\forall S \bullet X$ **sat** $S(s, \aleph)$, and so

$$\forall n \;\bullet\; \widehat{F}^n(X) \textbf{ sat } S(s, \aleph)$$

and it can be shown that all predicates of the form **sat** $S(s, \aleph)$ correspond to closed predicates in $\overline{TF}$: if such a predicate holds of all the elements of a sequence, it must hold of the limit. Hence

$$\mu X \bullet F(X) \quad \textbf{sat} \quad S(s, \aleph)$$

Hence the recursion rule is sound with respect to the new semantics. □

Unbounded Nondeterminism in CSP

by A.W. Roscoe[1] and Geoff Barrett[2]

Oxford University Computing Laboratory
8-11 Keble Road
Oxford OX1 3QD

ABSTRACT. *We extend the failures/divergences model for CSP to include a component of infinite traces. This allows us to give a denotational semantics for a version of CSP including general nondeterministic choice and infinite hiding. Unfortunately the model is an incomplete partial order, so it is by no means obvious that the necessary fixed points exist. We have two proofs of this result, one via a congruence theorem with operational semantics and one via a careful analysis of operators' behaviour on a subset of the model.*

0. Introduction

As is well known to the theoretical community, it is generally far easier to model finite nondeterminism (where a process can only choose between finitely many options at any one time) than unbounded nondeterminism (where no such restriction applies). The difficulties encountered with unbounded nondeterminism have hitherto forced us to restrict the language and semantics of CSP to avoid it: the most obvious restrictions being our inability to define the hiding operator $P\backslash B$ when B is infinite and the absence of an infinite nondeterminism operator $\sqcap S$ for arbitrary nonempty sets S of processes.

In an earlier paper [R2] one of us showed how many of the restrictions on unbounded nondeterminism could be lifted by separating the nondeterminism order from the order used for finding fixed points. Unfortunately the structure of the model used there (failures and divergences using only finite traces) means that the semantics given by that model to unboundedly nondeterministic operators is not sufficiently discriminating. That model can successfully model a process which will, on its first step, nondeterministically choose any integer, but cannot tell between a process which can communicate any finite number of a's and one which may also choose to communicate an infinite number. The main purpose of this paper is to develop a

[1]A.W. Roscoe gratefully acknowledges support from ONR grant N00014-87-G-0242.

[2]Geoff Barrett's current address: inmos Ltd., 1000 Aztec West, Almondsbury, Bristol BS12 4SQ

more refined model which can make this sort of distinction. This is done by adding a component of infinite traces so that any CSP process is represented by $\langle F, D, I\rangle$ where F is its set of failures (still with finite traces), D is its set of (finite) divergence traces and I is the set of infinite traces it can communicate.

Unfortunately the obvious orders on this new model fail to be complete, though they do have greatest lower bounds for arbitrary nonempty sets, which means that the standard iterative technique will produce the least fixed point of monotone f provided there is any x with $f(x) \leq x$. The first reaction to this failure was to look for a new order coarser than the obvious one which was complete (for this was precisely what had been done in the paper mentioned above for the $\langle F, D\rangle$ model without the finite subsets axiom). However one can prove that no order which gives the right semantics can be complete. Specifically we find an ω-sequence of CSP-definable processes whose semantic values are provably ordered in any sensible order but which can have no least upper bound.

If recursions are well defined we must therefore find some special property of CSP-definable functions which leads them to have fixed points. We have found two methods for proving the existence of these fixed points. The first was to define an operational semantics for the language and to prove simultaneously that the fixed points exist and that the denotational semantics is congruent to the operational one via the natural abstraction map. Barrett's contribution to this paper was to find a much shorter proof[3] which is also more satisfactory in some ways because it rests entirely within the model itself, rather than going outside to operational semantics.

The rest of the paper is structured as follows. In the next section we develop the new model, discover its partial order properties, and show how to define the CSP operators over it. The second section gives Barrett's proof that all CSP definable functions have fixed points. Because the semantics contains a number of features which are difficult or unusual, it is even more important than usual to have evidence that they are 'right'. For example many of the operators definable turn out to be non-continuous (though monotone) and require iteration past ω to reach their fixed points – and on first inspection it is not obvious whether the meaning of a recursion $\mu P.F(P)$ should be the ωth iterate $\bigsqcup\{F^n(\bot) \mid n \in \mathbb{N}\}$ or the least fixed point (the latter is the right answer). Therefore in the final section we outline the proof of the congruence theorem which was formerly used to prove the existence of fixed points, but now stands in its own right.

The way this congruence theorem can be used to establish the existence of fixed points is explained in an earlier presentation of the new model [R3], which also gives the proofs of various results on operational semantics that have been omitted from this article due to lack of space.

[3]The new proof was discovered in May 1989, shortly after the Tulane workshop.

1. Adding infinite traces to the failures model

The failures/divergences model, developed in [R1,B,BHR,BR], has become the standard abstract model for CSP. CSP is based on atomic, handshaken communications drawn from an alphabet Σ, which may be finite or infinite. The model describes every process as a pair $\langle F, D\rangle$, where $F \subseteq \Sigma^* \times \mathcal{P}(\Sigma)$ is the (nonempty) set of a process' *failures* ($(s, X) \in F$ is failure of the process if it can perform the trace s and then refuse to accept any communication from the set X) and $D \subseteq \Sigma^*$ is the set of its *divergences* (traces on which the process can loop – perform an infinite sequence of internal actions). The usual version of this model – often called $\mathcal{N}$ – is defined by a number of axioms, (1)-(5) below plus an axiom of bounded nondeterminism

$$\forall Y \subseteq^{\mathrm{fin}} X.(s, Y) \in F \Rightarrow (s, X) \in F\,.$$

This axiom is necessary to make the nondeterminism partial order

$$\langle F, D\rangle \sqsubseteq \langle F', D'\rangle \Leftrightarrow F \supseteq F' \wedge D \supseteq D'$$

complete. However, in [R2], a stronger order (in that it orders less things) was developed, which gives exactly the same fixed point theory but which no longer requires this axiom for completeness. This new 'definedness' order is defined

$$\begin{aligned} P \leq Q \;\Leftrightarrow\; & \mathcal{D}[\![Q]\!] \subseteq \mathcal{D}[\![P]\!] \wedge \\ & s \notin \mathcal{D}[\![P]\!] \Rightarrow \mathcal{R}[\![P]\!]s = \mathcal{R}[\![Q]\!]s \wedge \\ & \mu(\mathcal{D}[\![P]\!]) \subseteq \mathit{traces}(Q) \end{aligned}$$

where μT denotes the minimal elements of a set T of finite traces and $\mathcal{R}[\![P]\!]s$ denotes $\{X \mid (s, X) \in \mathcal{F}[\![P]\!]\}$. $P \leq Q$ means that Q has less divergences than P, but that all of P's non-divergent behaviour is copied exactly in Q. This extended model was termed $\mathcal{N}'$.

Our new model will have the same structure as $\mathcal{N}'$ except that it will have an extra component representing infinite traces. Thus a process P will be a triple $\langle F, D, I\rangle$, where $F \subseteq \Sigma^* \times \mathcal{P}(\Sigma)$, $D \subseteq \Sigma^*$ and $I \subseteq \Sigma^\omega$. F should be nonempty and the eight axioms must be satisfied. The first seven are tabulated below.

$$\begin{array}{lrcl}
(1) & (st, \emptyset) \in F & \Rightarrow & (s, \emptyset) \in F \\
(2) & (t, X) \in F \wedge Y \subseteq X & \Rightarrow & (t, Y) \in F \\
(3) & (t, X) \in F \wedge \forall a \in Y.(t\langle a\rangle, \emptyset) \notin F & \Rightarrow & (t, X \cup Y) \in F \\
(4) & s \in D & \Rightarrow & st \in D \\
(5) & s \in D & \Rightarrow & (st, X) \in F \\
(6) & su \in I & \Rightarrow & (s, \emptyset) \in F \\
(7) & s \in D & \Rightarrow & su \in I
\end{array}$$

Here, and in the rest of this paper, $a, b, \ldots$ range over Σ, $X, Y, A, B, \ldots$ over $\mathcal{P}(\Sigma)$, s, t, v, w over the finite traces Σ^* and u over infinite traces Σ^ω.

Axioms (6) and (7) are both new but straightforward because they are simple extensions to axioms (1) and (4) respectively. One more axiom is required, which can be thought of as an infinite trace analogue to axiom (3). The latter says that anything which, on one step, cannot be refused, must be a possible communication. The new axiom will say that when one, from the finite convergent behaviour, can show that there must be infinite traces, then there are enough of them.

One can often prove from the failures of a nondivergent process that some infinite trace is possible because one can formulate a strategy for *forcing* one. The most simple-minded form of strategy is that based on a single infinite trace u. If $(s, \{a\}) \notin F$ for all $s\langle a\rangle < u$ then it is intuitively clear that a user single mindedly striving for the infinite trace u must be successful. However there are more subtle versions of this. Consider a process whose failure-set is

$$F_0 = \{(s, X) \mid s \in \{a, b\}^* \wedge \{a, b\} \not\subseteq X\}\,.$$

Imagine always offering this process the set $\{a, b\}$: it is never refused, so we can guarantee that an infinite trace must arise. However we have no finer control over exactly which infinite trace it is, though on further reflection we can observe that, since every finite sequence s of a's and bs is possible there must be an infinite trace su extending every such s. The necessity of some axiom reflecting the forcing of infinite traces is demonstrated by the definition of the hiding operator below. Studying this will reveal that if a process P with the above failures did not have an infinite trace, then $P\backslash\{a, b\}$ would not have any failures, divergences or infinite traces!

The final axiom proved quite hard to find – several quite plausible versions turned out to be incorrect. There are a number of equivalent (in the presence of the other 7) versions of axiom (8). Several are given in [R3] and [Blam]. Perhaps the nicest formulation is the following, which was derived from the first author's earlier version by Stephen Blamey. Here T ranges over finite prefix closed sets of finite traces and $\overline{T} = \{u \in \Sigma^\omega \mid \forall t < u.t \in T\}$.

$$(8)\quad (s, \emptyset) \in F \Rightarrow \exists T.(\forall t \in T.(st, \{a \mid t\langle a\rangle \notin T\}) \in F \wedge \{su \mid u \in \overline{T}\} \subseteq I)$$

This can be interpreted as saying that one can never tell on a single interaction that a process is not deterministic, unless it diverges. T represents one of the deterministic forms the process might take after the trace s. It is worth noting that this axiom in no way restricts the sets of failures that are possible – if $\langle F, D\rangle$ is any element of $\mathcal{N}'$ and $T = \{s \mid (s, \emptyset) \in F\}$ then $\langle F, D, \overline{T}\rangle$ is always in $\mathcal{U}$. Furthermore, if $\langle F, D, I\rangle \in \mathcal{U}$ and $I \subseteq I' \subseteq \overline{T}$, then $\langle F, D, I'\rangle \in \mathcal{U}$.

The reader might like to check that the elements of $\mathcal{U}$ with failure set F_0 as defined above are precisely $\langle F_0, \emptyset, I\rangle$ where I is a set of nonempty infinite traces such that every element of $\{a, b\}^*$ is a prefix of some element of I. This follows in part from the fact that, if P is a process with the given failure set and T is a prefix closed set of finite traces satisfying the conditions of axiom (8) (relative to any s), $\overline{T}$ must have an infinite trace as it is easy to prove that it has arbitrarily long finite traces. Some possible I's are $\{a, b\}^\omega$, and $\{su \mid s \in \{a, b\}^*\}$ for any fixed $u \in \{a, b\}^\omega$.

The notation of $\mathcal{N}'$ is easily extended to cover the new model. If $P = \langle F, D, I\rangle$ then we write $\mathcal{F}[\![P]\!] = F$, $\mathcal{D}[\![P]\!] = D$ and $\mathcal{I}[\![O]\!] = I$. The set $\{s \mid (s, \emptyset) \in F\}$ of finite traces of P will be denoted *traces*(P); while *Traces*(P) = *traces*$(P) \cup I$ will denote the set of its finite and infinite traces.

The nondeterminism order $\sqsubseteq$ extends trivially to the new model. If $P = \langle F, D, I\rangle$ and $P' = \langle F', D', I'\rangle$ are any two triples we say

$$P \sqsubseteq P' \equiv F \supseteq F' \wedge D \supseteq D' \wedge I \supseteq I' .$$

If S is any nonempty set of processes we define its nondeterministic composition $\sqcap S$ to be $\langle F, D, I\rangle$, where

$$\begin{aligned} F &= \bigcup\{F' \mid \langle F', D', I'\rangle \in S\} \\ D &= \bigcup\{D' \mid \langle F', D', I'\rangle \in S\} \\ I &= \bigcup\{I' \mid \langle F', D', I'\rangle \in S\} . \end{aligned}$$

This is just the process which can exhibit any behaviour of any element of S. It is straightforward to verify that $\sqcap S$ is always in $\mathcal{U}$.

It is clear that $P \sqsubseteq \sqcap S$ whenever $P \in S$. We say that a process $\langle F, D, I\rangle$ is *deterministic* if it satisfies $D = \emptyset$ and

$$(s, X) \in F \Leftrightarrow X \cap \{a \mid (s\langle a\rangle, \emptyset) \in F\} . \qquad (*)$$

Thus it never has the choice of accepting or rejecting an event. The infinite traces of a deterministic process $P = \langle F, D, I\rangle$ are completely determined by the failures – $I = \overline{traces(P)}$. To see this, observe that the only T satisfying the first conclusion of axiom (8) for any s is $\{t \mid st \in traces(P)\}$. Actually, a deterministic process P is completely determined by *traces*(P). The deterministic processes are precisely the greatest elements of $\mathcal{U}$ under $\sqsubseteq$ (just as over the model without infinite traces) – for it is easy to see that they *are* maximal and also to show that each process has a deterministic process above it. It is not true that for every P

$$P = \bigsqcap\{Q \mid Q \text{ is deterministic } \wedge P \sqsubseteq Q\}$$

because the right hand side always has $D = \emptyset$. However it is possible to extend the class of deterministic processes in such a way that this becomes true – say that a process $P = \langle F, D, I\rangle$ is *pre-deterministic* if whenever $s \notin D$ equation $(*)$ above holds. In other words, a process is predeterministic if it is deterministic until it diverges. We write $\mathcal{P}$ for the set of all predeterministic processes. A predeterministic process is completely determined by its sets of traces and divergences.

Ordered by $\sqsubseteq$, $\mathcal{P}$ forms a complete partial order whose least element is the immediately diverging processes and whose greatest elements are the deterministic ones. $\mathcal{P}$ plays a very important role in the next section when we come to discuss fixed points of CSP operators over $\mathcal{U}$.

We get the following fundamental result. (Indeed, before the discovery of the form of axiom (8) quoted above, this took its place.)

Lemma 1.1 For all $P \in \mathcal{U}$,

$$P = \bigsqcap imp(P), \quad \text{where} \quad imp(P) = \{Q \in \mathcal{P} \mid P \sqsubseteq Q\}.$$

Proof. The fact that $imp(P)$ is nonempty is a trivial consequence of axiom (8), since the T produced by the empty trace $\langle\rangle$ yields an implementation. Thus $\bigsqcap imp(P)$ is well-defined, and trivially $\bigsqcap imp(P) \sqsupseteq P$. Thus the Lemma will be proved if we can demonstrate the existence of an element of $imp(P)$ containing each behaviour (failure, divergence, infinite trace) of P. For failures it is guaranteed by axiom (8) as follows; suppose the failure is (s, X) and for any trace t, T_t is a set of traces generated by axiom (8) relative to t. Let the set S of finite traces be

$$\{t \mid t \le s\} \cup \{t\langle a\rangle v \mid t < s \wedge t\langle a\rangle \in traces(P) \wedge t\langle a\rangle \not\le s \wedge v \in T_{t\langle a\rangle}\}$$

$$\cup\{s\langle a\rangle v \mid a \notin X \wedge s\langle a\rangle \in traces(P) \wedge v \in T_{s\langle a\rangle}\}.$$

It is easy to show that the deterministic process corresponding to S is an element of $imp(P)$ with the failure (s, X). Similarly, given a divergence s, take the sets $S_1 = \{t \mid s \le s\}$ and

$$S_2 = \{t \mid t \le s \vee s \le t\} \cup \{t\langle a\rangle v \mid t < s \wedge t\langle a\rangle \in traces(P) \wedge t\langle a\rangle \not\le s \wedge v \in T_{t\langle a\rangle}\}.$$

The unique predeterministic process with divergence set S_1 and trace set S_2 is the required element of $imp(P)$.

It only remains to consider infinite traces u, where we define S to be

$$\{t \mid t < u\} \cup \{t\langle a\rangle v \mid t < u \wedge t\langle a\rangle \in traces(P) \wedge t\langle a\rangle \not< u \wedge v \in T_{t\langle a\rangle}\}.$$

The only difficulty in proving that the deterministic process determined by this trace set is in $imp(P)$ is in proving $\overline{S} \subseteq I$. This is a consequence of the fact that if $u' \in \overline{S}$ then *either* $u' = u$ *or*, letting $s\langle a\rangle$ be chosen (necessarily uniquely) such that $s\langle a\rangle < u' \wedge s < u \wedge s\langle a\rangle \not< u$, we know that $u'' \in T_{s\langle a\rangle}$, where $u' = s\langle a\rangle u''$. ∎

Since divergence (and hence undefinedness) always appears after a finite length of trace, there is no obvious way of extending the idea of definedness to infinite traces. We therefore extend $\le$ in the same way as above: the order on the infinite traces being by reverse inclusion.

$$\begin{aligned} P \le Q \Leftrightarrow\ & \mathcal{D}[\![Q]\!] \subseteq \mathcal{D}[\![P]\!] \wedge \\ & s \notin \mathcal{D}[\![P]\!] \Rightarrow \mathcal{R}[\![P]\!]s = \mathcal{R}[\![Q]\!]s \wedge \\ & \mu(\mathcal{D}[\![P]\!]) \subseteq traces(Q) \wedge \\ & \mathcal{I}[\![P]\!] \supseteq \mathcal{I}[\![Q]\!] \end{aligned}$$

In general we have $P \le Q \Rightarrow P \sqsubseteq Q$ but not the reverse; however it is interesting to note that if P and Q are two predeterministic processes then $P \le Q$ if and only if $P \sqsubseteq Q$. Obviously all deterministic processes are maximal under $\le$, but there are other maximal elements as well – however they do not seem to be as useful or tangible a class as the deterministic processes and so we do not discuss them further here.

We will return to examine some more properties of the partial orders shortly.

All the usual operators may be defined over $\mathcal{U}$. As one would expect, in most cases the finite parts of these definitions are exactly the same as before (with the notable exception of hiding). They are given in full below.

STOP and *SKIP* are defined

$$STOP = \langle\{(\langle\rangle, X) \mid X \subseteq \Sigma\}, \emptyset, \emptyset\rangle$$

$$SKIP = \langle\{(\langle\rangle, X) \mid \surd \notin X\} \cup \{(\langle\surd\rangle, X) \mid X \subseteq \Sigma\}, \emptyset, \emptyset\rangle .$$

Let $P = \langle F, D, I\rangle$, $P' = \langle F', D', I'\rangle$ and, for $b \in B$, $P_b = \langle F_b, D_b, I_b\rangle$ be processes. Then

$$\begin{aligned}
\mathcal{D}[\![a \to P]\!] &= \{\langle a\rangle s \mid s \in D\} \\
\mathcal{I}[\![a \to P]\!] &= \{\langle a\rangle u \mid u \in I\} \\
\mathcal{F}[\![a \to P]\!] &= \{(\langle\rangle, X) \mid a \notin X\} \cup \{(\langle a\rangle s, X) \mid (s, X) \in F\}
\end{aligned}$$

$$\begin{aligned}
\mathcal{D}[\![x : B \to P_x]\!] &= \{\langle b\rangle s \mid b \in B \wedge s \in D_b\} \\
\mathcal{I}[\![x : B \to P_x]\!] &= \{\langle b\rangle u \mid b \in B \wedge u \in I_b\} \\
\mathcal{F}[\![x : B \to P_x]\!] &= \{(\langle\rangle, X) \mid B \cap X = \emptyset\} \cup \{(\langle b\rangle s, X) \mid b \in B \wedge (s, X) \in F_b\}
\end{aligned}$$

$$\begin{aligned}
\mathcal{D}[\![P \sqcap P']\!] &= D \cup D' \\
\mathcal{I}[\![P \sqcap P']\!] &= I \cup I' \\
\mathcal{F}[\![P \sqcap P']\!] &= F \cup F'
\end{aligned}$$

$$\begin{aligned}
\mathcal{D}[\![P \Box P']\!] &= D \cup D' \\
\mathcal{I}[\![P \Box P']\!] &= I \cup I' \\
\mathcal{F}[\![P \Box P']\!] &= \{(\langle\rangle, X) \mid (\langle\rangle, X) \in F \cap F'\} \cup \\
&\quad \{(s, X) \mid s \neq \langle\rangle \wedge (s, X) \in F \cup F'\} \cup \\
&\quad \{(s, X) \mid s \in \mathcal{D}[\![P \Box P']\!]\}
\end{aligned}$$

$$\begin{aligned}
\mathcal{D}[\![P\ {}_B\|_C\ P']\!] &= \{st \mid s \in (B \cup C)^* \wedge s \upharpoonright B \in D \wedge s \upharpoonright C \in traces(P')\} \\
&\quad \cup \{st \mid s \in (B \cup C)^* \wedge s \upharpoonright B \in traces(P) \wedge s \upharpoonright C \in D'\} \\
\mathcal{I}[\![P\ {}_B\|_C\ P']\!] &= \{u \in (B \cup C)^\omega \mid u \upharpoonright B \in Traces(P) \wedge u \upharpoonright C \in Traces(P')\} \\
&\quad \cup \{su \mid s \in \mathcal{D}[\![P\ {}_B\|_C\ P']\!]\} \\
\mathcal{F}[\![P\ {}_B\|_C\ P']\!] &= \{(s, (X \cap B) \cup (Y \cap C) \cup Z) \mid s \in (B \cup C)^* \wedge (s \upharpoonright B, X) \in F \wedge \\
&\quad (s \upharpoonright C, Y) \in F' \wedge Z \cap (B \cup C) = \emptyset\} \\
&\quad \cup \{(s, X) \mid s \in \mathcal{D}[\![P\ {}_B\|_C\ P']\!]\}
\end{aligned}$$

$$\begin{aligned}
\mathcal{D}[\![P \,|||\, P']\!] &= \bigcup\{merge\langle s, t\rangle \mid s \in D \wedge t \in traces(P')\} \\
&\quad \cup \bigcup\{merge\langle s, t\rangle \mid s \in D' \wedge t \in traces(P)\} \\
\mathcal{I}[\![P \,|||\, P']\!] &= \bigcup\{merge\langle s, t\rangle \mid s \in Traces(P) \wedge t \in Traces(P') \wedge s \text{ or } t \text{ infinite}\} \\
\mathcal{F}[\![P \,|||\, P']\!] &= \{(s, X) \mid \exists t, t'.\, s \in merge\langle t, t'\rangle \wedge (t, X) \in F \wedge (t', X) \in F'\} \\
&\quad \cup \{(s, X) \mid s \in \mathcal{D}[\![P \,|||\, P']\!]\}
\end{aligned}$$

$$
\begin{aligned}
\mathcal{D}[\![P;P']\!] &= \{st \mid s \in D \wedge s \text{ tick-free}\} \\
&\quad \cup \{st \mid s\langle\surd\rangle \in \mathit{traces}(P) \wedge t \in D' \wedge s \text{ tick-free}\} \\
\mathcal{I}[\![P;P']\!] &= \{u \mid u \in I \wedge u \text{ tick-free}\} \\
&\quad \cup \{su \mid s\langle\surd\rangle \in \mathit{traces}(P) \wedge u \in I' \wedge s \text{ tick-free}\} \\
&\quad \cup \{su \mid s \in \mathcal{D}[\![P;P']\!]\} \\
\mathcal{F}[\![P;P']\!] &= \{(s,X) \mid (s, X \cup \{\surd\}) \in F \wedge s \text{ tick-free}\} \\
&\quad \cup \{(st,X) \mid s\langle\surd\rangle \in \mathit{traces}(P) \wedge s \text{ tick-free} \wedge (t,X) \in F'\} \\
&\quad \cup \{(s,X) \mid s \in \mathcal{D}[\![P;P']\!]\}
\end{aligned}
$$

$$
\begin{aligned}
\mathcal{D}[\![P\backslash X]\!] &= \{(u\backslash X)t \mid u \in I \wedge u\backslash X \text{ is finite}\} \cup \{(s\backslash X)t \mid s \in D\} \\
\mathcal{I}[\![P\backslash X]\!] &= \{u\backslash X \mid u \in I \wedge u\backslash X \text{ is infinite}\} \cup \{su \mid s \in \mathcal{D}[\![P\backslash X]\!]\} \\
\mathcal{F}[\![P\backslash X]\!] &= \{(s\backslash X, Y) \mid (s, X \cup Y) \in F\} \cup \{(s,Y) \mid s \in \mathcal{D}[\![P\backslash X]\!]\}
\end{aligned}
$$

$$
\begin{aligned}
\mathcal{D}[\![f[P]]\!] &= \{(f(s))t \mid s \in D\} \\
\mathcal{I}[\![f[P]]\!] &= \{f(u) \mid u \in I\} \cup \{(f(s))u \mid s \in D\} \\
\mathcal{F}[\![f[P]]\!] &= \{(f(s),X) \mid (s, f^{-1}(X)) \in F\} \cup \{(s,X) \mid s \in \mathcal{D}[\![f[P]]\!]\}
\end{aligned}
$$

$$
\begin{aligned}
\mathcal{D}[\![f^{-1}[P]]\!] &= \{s \mid f(s) \in D\} \\
\mathcal{I}[\![f^{-1}[P]]\!] &= \{u \mid f(u) \in I\} \\
\mathcal{F}[\![f^{-1}[P]]\!] &= \{(s,X) \mid (f(s), f(X)) \in F\}
\end{aligned}
$$

We have already seen how $\sqcap S$ is defined.

The only definition that really requires comment is that of hiding. The definition of $\mathcal{D}[\![P\backslash X]\!]$ is rather simpler than in earlier models, since a divergence caused by the hiding now arises from a single infinite behaviour rather than from an infinite collection of finite ones. With this exception, failures and divergences never depend on the infinite traces of the operands.

We can easily extend the *after* operator to the new model. If $P = \langle F, D, I\rangle \in \mathcal{U}$ and $s \in \mathit{traces}(P)$ then P *after* s is defined to be $\langle F', D', I'\rangle$, where

$$
\begin{aligned}
F' &= \{(t,X) \mid (st,X) \in F\} \\
D' &= \{t \mid st \in D\} \\
I' &= \{u \mid su \in I\}\,.
\end{aligned}
$$

P *after* s is the process which behaves like P after communicating the trace s.

Theorem 1.2. All the operators are well defined (i.e., preserve the axioms) and monotonic with respect to both orders. All operators are both finitely and infinitely distributive: i.e., $F(\sqcap S) = \sqcap\{F(P) \mid P \in S\}$ for all operators F and nonempty $S \subseteq \mathcal{U}$.

We should perhaps note that no claim has been made for the continuity of the operators, which is because many of them are not continuous as a consequence of unbounded nondeterminism. The main consequence of this lack of continuity is that the fixed points of recursively defined programs need not have appeared by the ωth iteration from $\bot$ so familiar to computer scientists. However, once we can show that necessary least upper bounds exist there is no problem in defining the meaning of any recursive term to be the least fixed point of the appropriate monotone function: it is given by $f^{\alpha}(\bot)$ for sufficiently large α. Once one can do this, we can define a semantic function $\mathcal{S} : \mathbf{E} \to UEnv \to \mathcal{U}$, where $\mathbf{E}$ is the set of all CSP terms and $UEnv$ is the set of mappings from process variables to $\mathcal{U}$, in the obvious way. ∎

Properties of the partial orders

The following Lemma records some of the facts we have already noted and one or two other elementary facts about the two partial orders.

Lemma 1.3.

a) $P \sqsubseteq Q$ if, and only if, $imp(P) \supseteq imp(Q)$.

b) $P \leq Q \Rightarrow P \sqsubseteq Q$

c) $\bot = \langle \Sigma^* \times \mathcal{P}(\Sigma), \Sigma^*, \Sigma^{\omega} \rangle$ is the least element of $\mathcal{U}$ for both orders.

d) If $P \leq R$ and $P \sqsubseteq Q \sqsubseteq R$, then $P \leq Q$.

e) A process P is pre-deterministic if and only if there is a deterministic Q such that $P \leq Q$.

f) The $\sqsubseteq$-maximal elements of $\mathcal{U}$ are precisely the deterministic processes.

Proof. (a), (b) and (c) are trivial. For (d), we observe that $P \leq Q$ if and only if $P \sqsubseteq Q$ and

(i) $(s, X) \in \mathcal{F}[\![P]\!] \wedge s \notin \mathcal{D}[\![P]\!] \Rightarrow (s, X) \in \mathcal{F}[\![Q]\!]$, and

(ii) $\mu(\mathcal{D}[\![P]\!]) \subseteq traces(Q)$,

so to prove the result it will be sufficient to prove (i) and (ii). If $(s, X) \in \mathcal{F}[\![P]\!] \wedge s \notin \mathcal{D}[\![P]\!]$ then, since $P \leq R$, we know $(s, X) \in \mathcal{F}[\![R]\!]$. Hence $(s, X) \in \mathcal{F}[\![Q]\!]$ as $Q \sqsubseteq R$. Exactly the same argument applies for (ii).

Part (e) is elementary once we observe that if P is not pre-deterministic then its nondeterministic convergent behaviour must be present in any Q such that $P \leq Q$.

It is easy to show that if P and Q are both deterministic and $P \sqsubseteq Q$ then $P = Q$. It follows that if P is deterministic then $imp(P) = \{P\}$, and, by (a), that all deterministic processes are maximal. It is easy to see that, for any $P \in \mathcal{U}$, $imp(P)$ contains a deterministic process Q (since any pre-deterministic process is weaker than some deterministic one by (e)). It follows that $P \sqsubseteq Q$ and hence that no nondeterministic process can be maximal. This proves (f). ∎

We cannot hope that $\sqsubseteq$ is complete in general, for it is not complete over $\mathcal{N}'$ when Σ is infinite. Unfortunately, *neither* order is complete, even when $\Sigma = \{a, b\}$. It is easy to construct increasing $\leq$-sequences of processes, all with $F = F_0$ as defined above and $D = \emptyset$ which can have no upper bound. As a simple example, let $u_n = (\langle a\rangle^n\langle b\rangle)^\omega$ be the infinite trace which has n a's then a b cyclically. It is clear that the sets $\{su_n \mid s \in \{a,b\}^*\}$ are disjoint as n varies, and therefore that, if we set $I_n = \{su_m \mid s \in \{a,b\}^* \wedge m \geq n\}$, any upper bound for the sequence $\langle\langle F_0, \emptyset, I_n\rangle \mid n \in \mathbb{N}\rangle$ must have an empty set of infinite traces. This is impossible for $\leq$ as, since all the processes are divergence-free, any upper bound must have failure set F_0. (And we have already observed that all such elements of $\mathcal{U}$ have nonempty I.) It is also impossible for $\sqsubseteq$ since any upper bound must have an implementation Q (necessarily deterministic). Q must also be an implementation of all processes in the sequence and therefore have an infinite trace – a contradiction.

We will return to this incompleteness shortly and show that it is, to some extent at least, inevitable. Before we do this, however, it will be nice to establish a few positive properties.

Theorem 1.4

a) Any nonempty subset S of $\mathcal{U}$ has greatest lower bounds with respect to both $\leq$ and $\sqsubseteq$. In general, $\bigsqcap_{\leq} S \sqsubseteq \bigsqcap_{\sqsubseteq} S$.

b) In either order, any subset of $\mathcal{U}$ with any upper bound has a least upper bound.

c) If $\bigsqcup_{\leq} S$ is defined then so is $\bigsqcup_{\sqsubseteq} S$ and the two are equal. Furthermore $\bigsqcup_{\leq} S = P^* = \langle F^*, D^*, I^*\rangle$, where $F^* = \bigcap\{F \mid \langle F, D, I\rangle \in S\}$, $D^* = \bigcap\{D \mid \langle F, D, I\rangle \in S\}$ and $I^* = \bigcap\{I \mid \langle F, D, I\rangle \in S\}$.

d) If S is a nonempty set then $\bigsqcup_{\sqsubseteq} S$ exists if and only if $\bigcap\{imp(P) \mid P \in S\}$ is nonempty, and in that case $\bigsqcup_{\sqsubseteq} S = \bigsqcap(\bigcap\{imp(P) \mid P \in S\})$.

e) If $f : \mathcal{U} \to \mathcal{U}$ is a function which is monotone with respect to one of the orders and there is $P \in \mathcal{U}$ such that $f(P) \leq P$ (respectively $f(P) \sqsubseteq P$), then f has a least fixed point given by $f^\alpha(\perp)$ for some ordinal α.

f) If $f : \mathcal{U} \to \mathcal{U}$ is monotone with respect to both orders then any least fixed point for one order is also the least fixed point for the other.

Proof. It is easy to see that $\bigsqcap S$ is the $\sqsubseteq$-greatest lower bound of any nonempty set S. It does not work in general for the definedness order $\leq$, however, since one does not in general have $P \in S \Rightarrow \bigsqcap S \leq P$. The greatest lower bound of $S = \{\langle F_i, D_i, I_i\rangle \mid i \in \Lambda\}$ is, as was the case in $\mathcal{N}'$, constructed so that it diverges as soon as the finite behaviour of any two elements of S starts to differ. We define $\bigsqcap_{\leq} S$ to be $\langle F, D, I\rangle$, where

- $D = \bigcup\{D_i \mid i \in \Lambda\} \cup \{st \mid \exists i, j.(\exists Y.(s, Y) \in F_i \setminus F_j) \vee (\exists a.(s\langle a\rangle, \emptyset) \in F_i \setminus F_j)\}$
- $F = \bigcup\{F_i \mid i \in \Lambda\} \cup \{(s, X) \mid s \in D\}$

- $I = \bigcup\{I_i \mid i \in \Lambda\} \cup \{su \mid s \in D\}$

It is easy to show that this process is in $\mathcal{U}$ and is indeed the $\leq$ greatest lower bound of S. Trivially $\bigsqcap_{\leq} S \sqsubseteq \bigsqcap S$. This completes the proof of (a).

(b) follows because, as is fairly well known, *any* partial order which has greatest lower bounds for nonempty sets has this property. The usual argument is repeated here. If S is a set with an upper bound, then U_S, the set of upper bounds of S is nonempty and so $x = \bigsqcap U_S$ exists. Since $y \leq z$ whenever $y \in S$ and $z \in U_S$ it follows that each $y \in S$ is a lower bound for U_S. As x is the *greatest* lower bound for S it follows that $x \geq y$ for all $y \in S$ and therefore that $x \in U_S$. Plainly x is the least element of U_S and is therefore the least upper bound of S.

The first part of (c) follows trivially from the formula which is the second part. However it has an interesting separate proof. Note that, since $Q \leq P \Rightarrow Q \sqsubseteq P$, if $P = \bigsqcap_{\leq} S$ exists then it is a $\sqsubseteq$-upper bound for S and hence $Q = \bigsqcup_{\sqsubseteq} S$ exists and $Q \sqsubseteq P$. Whenever $R \in S$ we then have $R \sqsubseteq Q \sqsubseteq P$ and $R \leq P$. Lemma 1.3 (d) above then tells us that $R \leq Q$. It follows that Q is a $\leq$-upper bound for S and hence that $Q \geq P$. We then have $Q \sqsubseteq P$ and $P \sqsubseteq Q$. The result follows immediately.

For the second part, we show first that if $P' = \langle F', D', I' \rangle$ is the actual least upper bound of S then $D^* = D'$. For trivially $D' \subseteq D^*$ so let $s \in \mu D^*$ (where recall $D^* = \bigcap\{D \mid \langle F, D, I \rangle \in S\}$). Note that there must be $P = \langle F, D, I \rangle \in S$ such that $s \in \mu D$. Since $P \leq P'$ we must have $s \in traces(P')$. If $s \notin D'$ then consider $P'' = \langle F'', D'', I'' \rangle$ defined as

$$\begin{aligned} F'' &= F' \cup \{(st, X) \mid t \in \Sigma^* \wedge X \subseteq \Sigma\} \\ D'' &= D' \cup \{st \mid t \in \Sigma^*\} \\ I'' &= I' \cup \{su \mid u \in \Sigma^\omega\}\,. \end{aligned}$$

traces(P'') is prefix closed by the observation above. It is thus easy to see that P'' is a process, that $P \leq P''$ for all $P \in S$ and that $P' \not\leq P''$. It follows that P' cannot be the least upper bound on S, a contradiction. Hence $\mu D^* \subseteq D'$; it easily follows that $D^* \subseteq D'$, so the two are equal as desired.

That P^* defined in the statement of the theorem satisfies axioms (1), (2), (4), (5), (6) and (7) is trivial. We next note that trivially $F^* \supseteq F'$. Now by the above paragraph those parts of F^* and F' implied by divergence and axiom (5) are equal. Suppose that $s \notin D' = D^*$. Then there is $P = \langle F, D, I \rangle \in S$ such that $s \notin D$. Necessarily $\mathcal{R}[\![P]\!]s = \mathcal{R}[\![P']\!]s$ as $P \leq P'$. It follows that $\mathcal{R}[\![P']\!]s \supseteq \mathcal{R}[\![P^*]\!]s$ (for the latter is the intersection of a set containing $\mathcal{R}[\![P]\!]s$). Putting these facts together yields $F' \supseteq F^*$, proving that in fact $F' = F^*$. Note that this implies that P^* satisfies axiom (3).

Since we have now shown that $D^* = D'$ and $F^* = F'$, and it is trivial that $I^* \supseteq I'$ and $\langle F', D', I' \rangle \in \mathcal{U}$ it follows directly that P^* satisfies axiom (8) and is therefore in $\mathcal{U}$. The fact that it is the $\leq$-least upper bound for S is then trivial. This completes the proof of (c).

(d) follows easily from Lemma 1.3 and (b) above.

(e) is true in any partial order with property (a). By another standard argument, if f is monotonic and $x = \bigsqcap\{P \mid f(P) \leq P\}$ exists in a partial order then it is the least fixed point of f. We still have to show that the least fixed point can also be found by iterating $f^\alpha(\perp)$. The only place at which the standard cpo proof of this could go wrong is where, for limit ordinals λ, one defines $f^\lambda(\perp) = \bigsqcup\{f^\alpha(\perp) \mid \alpha \in \lambda\}$ since this least upper bound might not be defined. But it always is, since it is easy to prove by transfinite induction that all the $f^\alpha(\perp)$ are bounded above by the least fixed point x constructed above so that we can always apply (b) when constructing $f^\lambda(\perp)$.

(f) follows easily from (c) and (e). If f is monotonic with respect to both orders and has any fixed point then it follows easily from (e) that it has least fixed points $f^\alpha_\leq(\perp)$ and $f^\beta_\sqsubseteq(\perp)$ with respect to these two orders. But one can prove from (c) that if both of these exist then the value of $f^\gamma(\perp)$ is independent of whether it was defined using $\leq$ or $\sqsubseteq$ by an easy transfinite induction on γ. From this it is easily seen that both processes reach the same fixed point, and do so at the same time.

(f) can alternatively be proved by observing that, by (e), if f has a fixed point then it has a least fixed point with respect to both orders. If x and y denote the $\leq$-least and $\sqsubseteq$-least fixed points respectively, we have $x \leq y$ and hence $x \sqsubseteq y$ by Lemma 1.2. But we know $y \sqsubseteq x$ so it follows that $x = y$. ∎

We should remark now that all of the properties of the partial orders identified in Lemma 1.3 and Theorem 1.4 extend easily (some of them appropriately amended) to products of $\mathcal{U}$, i.e., $\mathcal{U}^\Lambda (= \Lambda \to \mathcal{U})$ for an arbitrary nonempty set Λ, with the order $\underline{P} \leq \underline{Q}$ (or $\underline{P} \sqsubseteq \underline{Q}$) if and only if $P_\lambda \leq Q_\lambda$ (or $P_\lambda \sqsubseteq Q_\lambda$) for all $\lambda \in \Lambda$. Some of the more useful properties of these product spaces, which are important in the consideration of mutual recursions and in the definition of the partial abstraction functions later on, are summarised below. All the proofs are either standard or straightforward extensions of what we have already seen.

Theorem 1.5.

a) $\perp^\Lambda$ is least element of $\mathcal{U}$ with respect to both orders.

b) Any nonempty subset S of $\mathcal{U}^\Lambda$ has greatest lower bounds with respect to both $\leq$ and $\sqsubseteq$. In general, $\bigsqcap_\leq S \sqsubseteq \bigsqcap_\sqsubseteq S$. In either case the greatest lower bound's λ-component is given by $\bigsqcap\{P_\lambda \mid \underline{P} \in S\}$, where $\bigsqcap$ here denotes the greatest lower bound operator over $\mathcal{U}$ in the appropriate order.

c) In either order, any subset of $\mathcal{U}$ with any upper bound has a least upper bound. In that case its λ-component is given by $\bigsqcup\{P_\lambda \mid \underline{P} \in S\}$.

d) If $\bigsqcup_\leq S$ is defined then so is $\bigsqcup_\sqsubseteq S$ and the two are equal. Furthermore $(\bigsqcup_\leq S)_\lambda = P^*_\lambda = \langle F^*_\lambda, D^*_\lambda, I^*_\lambda\rangle$, where $F^*_\lambda = \bigcap\{\mathcal{F}[\![P_\lambda]\!] \mid \underline{P} \in S\}$, $D^*_\lambda = \bigcap\{\mathcal{D}[\![P_\lambda]\!] \mid \underline{P} \in S\}$ and $I^*_\lambda = \bigcap\{\mathcal{I}[\![P_\lambda]\!] \mid \underline{P} \in S\}$.

e) If $f : \mathcal{U}^\Lambda \to \mathcal{U}^\Lambda$ is a function which is monotone with respect to one of the orders and there is $\underline{P} \in \mathcal{U}^\Lambda$ such that $f(\underline{P}) \leq \underline{P}$ (respectively $f(\underline{P}) \sqsubseteq \underline{P}$), then f has a least fixed point given by $f^\alpha(\perp^\Lambda)$ for some ordinal α.

f) If $f : \mathcal{U}^\Lambda \to \mathcal{U}^\Lambda$ is monotone with respect to both orders then any least fixed point for one order is also the least fixed point for the other. ∎

These theorems and what we have shown up to now show that $\leq$ and $\sqsubseteq$ are exceptionally well-behaved partial orders. It is interesting to note that $\sqsubseteq$ has its lower bounds given by union and $\leq$ has its upper bounds given by intersection, but that the reverse facts are not true. For example $\bigsqcap\{a \to STOP, b \to STOP\} = \bot$ or $(a \to STOP) \sqcap (b \to STOP)$ depending on which order is chosen, and $\bigsqcup\{(a \to STOP) \sqcap (b \to STOP, (a \to STOP) \sqcap (b \to SKIP)\} = a \to STOP$ under $\sqsubseteq$ which is not the intersection of the two. Indeed even in cases where S is a chain, $\bigsqcup_{\sqsubseteq} S$ might exist but not be given by component-wise intersection. If P_n is the nth process in the chain seen earlier with no upper bound, then, if we define

$$Q_n = (c \to STOP) \sqcap (d \to P_n)$$

the least upper bound of this sequence is $c \to STOP$ even though $(\langle d \rangle, \emptyset)$ is a failure of every Q_n.

The first author's first reaction on finding that the two "natural" partial orders were incomplete was to try to find another one that was but which gave the same semantics. After all, that had been one of the main reasons for the development of the $\leq$ order over $\mathcal{N}'$ since it gave exactly the same least fixed point semantics but was complete, showing that all desired fixed points actually exist. We should perhaps remark at this point that the given orders do actually compute the correct values for CSP definable recursions and that the least upper bounds required to compute them always exist. Of course the proof of these facts will be the subject of much work later, but it is worthwhile seeing some examples here.

Examples. Abbreviate by a^n the process that performs n a's and then $STOP$s. Set $P = \bigsqcap\{a^n \mid n \in \mathbb{N}\}$, so that P can perform any finite number of a's but not an infinite sequence of them. Operationally we can think of P as a process which, as its first action, takes a secret decision on exactly how many a's to perform. Now consider the recursively defined process

$$Q \;=\; (a \to Q)_{\{a\}}\|_{\{a\}}P$$

and let $F : \mathcal{U} \to \mathcal{U}$ be the function associated with the right hand side of this recursion. Since the right hand side of the highest level parallel construct initially imposes a bound on the number of a's Q can perform, it is clear that Q itself cannot perform an infinite sequence of them. On the other hand it is clear that Q can perform as large a finite number of a's as it pleases. We would therefore expect $P = Q$. However, as is easily verified, $F^\omega(\bot)$ can perform an infinite sequence of a's (it is equal to $P \sqcap R$, where $R = a \to R$). On the other hand, $F^{\omega+1}(\bot) = (a \to (P \sqcap R))_{\{a\}}\|_{\{a\}}P = P$ and $F(P) = P$, so this recursion reaches the operationally correct fixed point at $\omega + 1$. Some more examples of recursions, their fixed points and the ordinal required to reach them are summarised below. The reader might enjoy constructing a few of his own.

- If $f : \Sigma \to \Sigma$ is such that $f^n(a) \neq f^m(a)$ when $n \neq m$ then the recursion

$$Q_1 = STOP \sqcap a \to ((Q_1\ {}_\Sigma\|_\Sigma\ P) \sqcap f(Q_1))$$

(with P as above) reaches its fixed point (which is the same as that of the recursion $P' = P \sqcap a \rightarrow f[P']$ which converges in ω steps), in exactly $\omega.2$ iterations.

- Let α be an infinite ordinal and $\Sigma = \alpha$ (the set of all $\beta < \alpha$). Then the recursion

$$Q_2 = \beta : \alpha \rightarrow ((\gamma : \beta \rightarrow STOP)\ {}_{\Sigma}\|_{\Sigma}\ Q_2)\backslash\alpha$$

takes exactly α steps to converge to its fixed point $\beta : \alpha \rightarrow STOP$. Q_2 is a process that inputs any element β of α and then outputs any element of β to a copy of itself or deadlocks if $\beta = 0$. (The fact that this is the natural fixed point is an easy consequence of the fact that there is no infinite descending sequence of ordinals.)

Suppose $\preceq$ is some partial order which does all we want: namely give the same fixed point theory and make $\mathcal{U}$ complete. Clearly it must make all CSP operators monotonic and have the same minimal element $\perp$. To give the same fixed point theory it must have the property that, when C is a linearly ordered subset of $\mathcal{U}$ with respect to $\preceq$ and one of our existing orders, then a least upper bound for $\preceq$ is also a least upper bound for the other. (Note that $\sqsubseteq$ and $\leq$ are in this relationship.) It must also make $P' \prec Q$, where Q is defined as in the example above and $P' = STOP \sqcap a \rightarrow P'$. For Q is a fixed point of this recursion but is distinct from the natural fixed point (by assumption the $\prec$-least) which has the infinite sequence of a's. ($P' \prec Q$ can also be proved by looking at the recursion of Q, where P' is the ωth iterate.)

From these simple facts and assumptions we will be able to prove that $\preceq$ cannot exist: for there is a sequence of processes in $\mathcal{U}$ which are provably ordered by $\preceq$ but which can have no upper bound. Set $\Sigma = \{a, b\}$. Recall that the set F_0 of failures was defined

$$F_0 = \{(s, X) \mid s \in \{a, b\}^* \wedge \{a, b\} \not\subseteq X\}.$$

The corresponding set where a process can refuse anything at any time is

$$F_1 = \{(s, X) \mid s \in \{a, b\}^* \wedge X \subseteq \{a, b\}\}.$$

Recall that the triples $\langle F_0, \emptyset, I\rangle$ satisfying the axioms were those where I contains an extension of every finite trace. All triples $\langle F_1, \emptyset, I\rangle$ satisfy the axioms.

We will now construct some subsets of $\{a, b\}^\omega$ to go along with F_0 and F_1. If $u \in \{a, b\}^\omega$ and $n \in \mathbb{N}$, define $r_n(u)$ to be the ratio of the number of a's to the number of b's plus one in the first n elements of u. (The "plus one" is to make this always defined.) We should perhaps remark that some traces u have $\lim_{n\rightarrow\infty} r_n(u)$ existing and some do not. (In fact, there are uncountably many u's with any given limit in $[0, \infty)$.) In the first author's experience the ratios $r_n(u)$ are very useful when it comes to choosing pathological subsets of $\{a, b\}^\omega$ and similar.

For $n \in \{1, 2, 3, \ldots\}$ we define

$$I_n = \{u \in \{a, b\}^\omega \mid \exists \epsilon > 0. \exists m. \forall k \geq m. \epsilon < r_k(u) < \frac{1}{n} - \epsilon\}.$$

Thus $u \in I_n$ if and only if the ratios eventually stay within $(0, \frac{1}{n})$ and away from the boundaries of that interval. This last condition means, amongst other things, that I_n contains no sequence with limit 0 or $\frac{1}{n}$. Notice that $u \in I_n$ does not imply that $\lim_{n\to\infty} r_n(u)$ exists. The sets I_n have some interesting properties. First, the I_n all contain elements beginning with any chosen $s \in \{a,b\}^*$ (in fact, uncountably many). Also $I_{n+1} \subseteq I_n$ and $\bigcap\{I_n \mid n \in \{1,2,\ldots\}\} = \emptyset$. Perhaps the most interesting property is that, if $m \leq n$ then

$$\bigcup\{merge\langle s,t\rangle \mid s \in I_n \cup \{a,b\}^* \wedge t \in I_m \cup \{a,b\}^* \wedge s \text{ or } t \text{ is infinite}\} = I_m\,.$$

Also, the insertion or deletion of finitely many elements of a sequence u does not effect membership of any I_n since the limiting behaviour $r_n(u)$ is not affected by such manipulations. We can now define some processes

$$\begin{array}{rcll} P_n &=& \langle F_0, \emptyset, I_n\rangle & \text{for } n \in \{1,2,3,\ldots\} \\ Q_n &=& \langle F_1, \emptyset, I_n\rangle & \text{for } n \in \{1,2,3,\ldots\} \\ P_0 &=& \langle F_0, \emptyset, \{a,b\}^\omega\rangle & \\ Q_0 &=& \langle F_1, \emptyset, \{a,b\}^\omega\rangle & \\ Q_\infty &=& \langle F_1, \emptyset, \emptyset\rangle & \end{array}$$

We will prove that the P_n are a $\preceq$-increasing sequence.

Now if $f : \Sigma \to \Sigma$ is defined by $f(a) = f(b) = a$, we have $f^{-1}[P'] = Q_0$ and $f^{-1}[Q] = Q_\infty$, where P' and Q are as described at the start of this discussion. Hence $Q_0 \preceq Q_\infty$ as f^{-1} is monotonic.

Now for all n it is not too hard to see that $P_n \,|||\, Q_0 = P_0$ and $P_n \,|||\, Q_\infty = P_n$. It follows that $P_0 \preceq P_n$ for all $n \geq 1$ as $|||$ is monotonic.

Next, observe that $P_n \;{}_\Sigma\|_\Sigma\; P_m = Q_n$ if $m \leq n$. (The transition from F_0 to F_1 arises because one side of the parallel may refuse a and the other b.) It follows that $Q_m = (P_0 \;{}_\Sigma\|_\Sigma\; P_m) \preceq (P_n \;{}_\Sigma\|_\Sigma\; P_m) = Q_n$ when $m \leq n$.

The property of the I_n described above implies that $P_m \,|||\, Q_n = P_k$, where k is the lesser of n and m. Hence, when $m \leq n$, $P_m = P_n \,|||\, Q_m \preceq P_n \,|||\, Q_n = P_n$. This completes the proof that the P_n form an increasing sequence.

The fact that the P_n are $\preceq$-increasing is unsurprising, since they are increasing with respect to $\sqsubseteq$ and $\leq$. We have specified that all $\preceq$ least upper bounds of sequences increasing in both orders are also $\sqsubseteq$ least upper bounds. Since $\bigcap\{I_n \mid n \in \mathbb{N}\}$ is empty, any $\sqsubseteq$ least upper bound for this sequence has $I = \emptyset$. But there is no element of $\mathcal{U}$ with $F \subseteq F_0$ and $I = \emptyset$. It follows that this sequence has no upper bound with respect to $\preceq$. Therefore $\preceq$ cannot be complete.

We therefore have to give up all hope of a conventional fixed point theory, though note that, by Theorem 1.4, if we can show every CSP term has some fixed point, or even maps some point down in either order, then we essentially have one. The first author's proof that these fixed points exist was via a congruence theorem with operational semantics; this was both complex and, because it relied on structures outside the model, not fully satisfying. Recently the second author has discovered a much simpler proof, within the model, which is described in the next section.

2. The fixed point theorem

To show that all the CSP operators have least fixed points, we appeal to a sort of 'dominated convergence theorem', which states that if $F \sqsubseteq G$ and G has a fixed point, ϕG, then F has a least fixed point for:

$$F(\phi G) \sqsubseteq G(\phi G) = \phi G$$

so that the fixed point of G is mapped down by F and then Theorem 1.4 (e) implies the existence of a least fixed point for F.

The usefulness of this observation is that we may find a dominating G which is monotonic and preserves predeterminism so that the completeness of that subspace guarantees a fixed point for G. Indeed, we can go one step further for, suppose we can find a monotonic function $G : \mathcal{P} \to \mathcal{P}$ such that $F{\restriction}\mathcal{P} \sqsubseteq G$, then we may extend G by:

$$G^\star(P) = \bigsqcap\{G(Q) \mid Q \in imp(P)\}$$

which agrees with G on $\mathcal{P}$ since if $P \in \mathcal{P}$, then $P \in imp(P)$ and since G is monotonic, $G(P) \sqsubseteq G(Q)$ for all $Q \in imp(P)$ giving $G^\star(P) = G(P)$. Furthermore, G dominates F everywhere for:

$$\begin{aligned} F(P) &= F\left(\bigsqcap imp(P)\right) \\ &\sqsubseteq \bigsqcap\{F(Q) \mid Q \in imp(P)\} \\ &\sqsubseteq \bigsqcap\{G(Q) \mid Q \in imp(P)\} \\ &= G^\star(P) \end{aligned}$$

We know that any fixed point of G is a fixed point of $G^\star$ and that G has a fixed point, so we can now see that F has a least fixed point.

Let us see this restriction in action. We will exhibit a monotonic function with no fixed point and show that its restriction has no dominating function.

$$F(X) = \begin{cases} (a \to X)\,{}_\Sigma\|_\Sigma P, & P \not\sqsubseteq X \\ a \to X, & P \sqsubseteq X \end{cases}$$

where

$$P = \bigsqcap_{n<\omega} a^n$$

This function really is monotonic for if $X \sqsubseteq Y$ and $P \sqsubseteq X$ or $P \not\sqsubseteq Y$, then both X and Y follow the same branch of F; otherwise $P \not\sqsubseteq X$ and $P \sqsubseteq Y$ and

$$F(X) = (a \to X)\,{}_\Sigma\|_\Sigma P \sqsubseteq (a \to Y)\,{}_\Sigma\|_\Sigma P \sqsubseteq a \to Y = F(Y)$$

The chain got by applying this function to $\perp$ is:

$$\begin{array}{c} \perp \\ STOP \sqcap a\perp \\ \vdots \\ \bigsqcap_{k<n} a^k \sqcap a^n\perp \\ \vdots \\ P \\ \vdots \\ \bigsqcap_{n\leq k<\omega} a^k \\ \vdots \end{array}$$

which has no supremum since any limit would be unable to refuse $\{a\}$ at any time but would not have the infinite sequence of a's among its infinite traces. If we restrict ourselves to a model whose alphabet is just $\{a\}$, the only predeterministic processes are a^n, a^ω and $a^n\perp$ and the application of F to each of these gives:

$$\begin{array}{rcl} F(a^n) & = & a^{n+1} \\ F(a^\omega) & = & P \\ F(a^n\perp) & = & a^{n+1}\perp \sqcap \bigsqcap_{k\leq n} a^k \end{array}$$

We know that any dominating G with a fixed point must fix one of the predeterministic processes. However, there is no predeterministic process which is mapped down by F so we cannot find such a G. (Clearly F would map any fixed point of G down, because G is assumed to dominate F.)

Proceeding with the proof, note that composition is a monotonic function on the function space of a partial order. Therefore, if two CSP functions are dominated by predeterminism-preserving functions, the composition is, also. Further, the property is preserved through recursion because if $F(P,Q)$ is dominated by $G(P,Q)$ and $Q \in \mathcal{P}$, then $\mu p.F(p,Q)$ exists (by the argument given earlier) and is dominated by $\mu p.G(p,Q)$, which is predeterministic.

So, we only need to show that the restriction of each primitive CSP function to $\mathcal{P}$ is dominated by a predeterminism-preserving function. We list a set of algebraic laws which the CSP operators satisfy and which show just where the operators introduce nondeterminism and use these laws to motivate the definition of a bounding function for each CSP operator. Each of the following operators is strict and distributive in each of its arguments. If we let P abbreviate $x : B \rightarrow P_x$ and Q abbreviate $y : C \rightarrow Q_y$, we have:

$$P \Box Q = z : (B \cup C) \rightarrow R_z$$

$$\textit{where} \quad R_z = \begin{cases} P_z, & z \in B - C \\ P_z \sqcap Q_z, & z \in B \cap C \\ Q_z, & z \in C - B \end{cases}$$

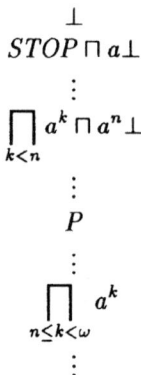

$$P_X\|_Y Q = z : D \to R_z$$

$$\begin{array}{rcl} \textit{where}\ \ D & = & (B \cap (X - Y)) \cup (B \cap C \cap X \cap Y) \cup (C \cap (Y - X)) \\ R_z & = & \left\{ \begin{array}{ll} P_{zX}\|_Y Q, & z \in B \cap (X - Y) \\ P_{zX}\|_Y Q_z, & z \in B \cap C \cap X \cap Y \\ P_X\|_Y Q_z, & z \in C \cap (Y - X) \end{array} \right. \end{array}$$

$$P \,|||\, Q = \left(x : B \to \left(P_x \,|||\, Q\right)\right) \Box \left(y : C \to \left(P \,|||\, Q_y\right)\right)$$

$$P;Q = \left\{ \begin{array}{ll} x : B \to (P_x;Q), & \surd \notin B \\ ((x : B - \{\surd\} \to (P_x;Q)) \Box Q) \sqcap Q, & \surd \in B \end{array} \right.$$

$$P \backslash X = \left\{ \begin{array}{ll} x : B \to (P_x \backslash X), & B \cap X = \emptyset \\ ((x : B - X \to (P_x \backslash X)) \Box Q) \sqcap Q, & B \cap X \neq \emptyset \end{array} \right.$$
$$\textit{where}\ \ Q = \bigsqcap \{P_x \backslash X \mid x \in B \cap X\}$$

$$f\,[P] = y : f\,(B) \to \bigsqcap \left\{ f\,[P_x] \mid x \in f^{-1}\,(y) \cap B \right\}$$

$$f^{-1}\,[P] = y : f^{-1}\,(B) \to f^{-1}\left[P_{f(y)}\right]$$

It is helpful to note that any predeterministic process P is either $\bot$ or can be written $x : B \to P_x$ for some set $B \subseteq \Sigma$ and predeterministic processes P_x. For those operators which introduce nondeterminism on $\mathcal{P}$, we aim to define new operators which make a particular nondeterministic choice. This is done by cases. We give below the equations which we expect the dominating operators to satisfy. In fact, we define the new operators (over $\mathcal{P}$ only) to be the least ones which satisfy the given equations.

C (a constant process) Let Q be some fixed predeterministic implementation of C. Dominating function: Q.

$F(P)$ **where** $F_\lambda(\underline{P}) = P_{\mu_\lambda}$ (This function is needed so that functions such as $F(p) = p$ and $F(p) = p_X\|_Y p$ can be written as compositions of primitive functions; it is a sort of syntactic glue.) Dominating function: itself.

$x : B \to P_x$ Dominating function: itself.

$\bigsqcap_{\lambda \in \Lambda} P_\lambda$ Choose $\lambda_0 \in \Lambda$. Dominating function: P_{λ_0}.

$P \Box Q$ Dominating function: $P \boxminus Q$ defined to be bi-strict (*i.e.*, $P \boxminus \bot$ and $\bot \boxminus Q$ are both $\bot$) and to satisfy:

$$(x : B \to P_x) \boxminus (y : C \to Q_y) = z : (B \cup C) \to R_z$$
$$\textit{where}\ \ R_z = \left\{ \begin{array}{ll} P_z, & z \in B \\ Q_z, & z \in C - B \end{array} \right.$$

$P_X\|_Y Q$ Dominating function: itself.

$P \mathbin{|||} Q$ Dominating function: $P \mathbin{\#} Q$ defined to be bi-strict and to satisfy:

$$P \mathbin{\#} Q = \left(x : B \to \left(P_x \mathbin{\#} Q\right)\right) \boxminus \left(y : C \to \left(P \mathbin{\#} Q_y\right)\right)$$

where $P = x : B \to P_x$ and $Q = y : C \to Q_y$.

$P;Q$ Nondeterminism is only introduced when P offers termination and other events. We choose to make it terminate immediately. Dominating function: $P \boxed{;} Q$ defined to be strict in its first argument and to satisfy:

$$(x : B \to P_x) \boxed{;} Q = \begin{cases} x : B \to \left(P_x \boxed{;} Q\right), & \surd \notin B \\ Q, & \surd \in B \end{cases}$$

$P \backslash X$ Nondeterminism arises through choices of hidden events. Let c be a choice function on X. Dominating function: $P \backslash^c X$ defined to be strict and the least operator to satisfy:

$$(x : B \to P) \backslash^c X = \begin{cases} x : B \to (P \backslash^c X), & B \cap X = \emptyset \\ P_{c(B \cap X)} \backslash^c X, & B \cap X \neq \emptyset \end{cases}$$

$f[P]$ Nondeterminism is introduced by mapping two events to the same event. Let c be a choice function on the domain of f. Dominating function: $f\{P\}$ defined to be strict and to satisfy:

$$f\{x : B \to P_x\} = x : f(B) \to f\left\{P_{c(f^{-1}(x) \cap B)}\right\}$$

$f^{-1}[P]$ Dominating function: itself.

We must now verify that the functions $\boxminus$, $\boxed{;}$, $\#$, $\backslash^c$ and $f\{\cdot\}$ exist and that they do indeed bound the CSP operators on $\mathcal{P}^{\Lambda}$. First note that if Y is complete (consistently complete) and X is a partial order, then the space of monotonic functions $X \to Y$ is also complete (consistently complete). Each of the above operators is defined to be the least fixed point of some function on $\mathcal{P}^{\Lambda} \to \mathcal{P}$, *e.g.* $\backslash^c X$ is the least fixed point of the function F where:

$$\begin{aligned} F\,G(\bot) &= \bot \\ F\,G(x : B \to P_x) &= \begin{cases} x : B \to G(P_x), & B \cap X = \emptyset \\ G\left(P_{c(B \cap X)}\right), & B \cap X \neq \emptyset \end{cases} \end{aligned}$$

That F has a least fixed point which is a monotonic function $\mathcal{P} \to \mathcal{P}$ can be seen because if G is monotonic then $F\,G$ is monotonic and if G is predeterminism-preserving then so is $F\,G$. Furthermore, if $G \sqsubseteq G'$ then $F\,G \sqsubseteq F\,G'$ so that F is a monotonic function on the complete space $\mathcal{P} \to \mathcal{P}$.

We now turn to showing that these operators dominate the CSP operators on $\mathcal{P}^{\Lambda}$. We continue with the example of hiding. First of all we note that $\backslash X$ is a fixed point of a monotonic function F' on $\mathcal{P} \to \mathcal{U}$ given by the algebraic law above. Since

$F' G(P) \sqsubseteq F G(P)$ for all $P \in \mathcal{P}$ and $G : \mathcal{P} \to \mathcal{U}$ we have that $F' \sqsubseteq F$. Now, since the monotonic functions $\mathcal{P} \to \mathcal{U}$ form a consistently complete space, we may infer the existence of a least fixed point of F' which is weaker than the least fixed point of F. All we have to do is to show that the CSP operators (restricted to $\mathcal{P}$) are indeed the least fixed points of those laws. The rest of this proof is devoted to establishing this fact. We shall refer to the algebraic laws as the fixed point equations for the operators.

We can put a sort of metric on the space as follows. We first define the nth-restriction operator, $P \downarrow n$, which gives a process which behaves like P for the first n steps and then diverges:

$$\begin{aligned} \mathcal{D}[\![P \downarrow n]\!] &= \{st \mid s \in \mathit{traces} P \wedge \#s \geq n\} \cup \mathcal{D}[\![P]\!] \\ \mathcal{F}[\![P \downarrow n]\!] &= \mathcal{F}[\![P]\!] \cup \{(s, X) \mid s \in \mathcal{D}[\![P \downarrow n]\!]\} \\ \mathcal{I}[\![P \downarrow n]\!] &= \{u \in \alpha P^{\omega} \mid \forall s < u .\, s \in \mathit{traces} P \downarrow n\} \end{aligned}$$

The 'metric' is defined by:

$$d(P, Q) = \inf \left\{2^{-n} \mid n < \omega \wedge P \downarrow n = Q \downarrow n\right\}$$

which satisfies the ultra-metric form of the triangle inequality, namely:

$$d(P, R) \leq \max(d(P, Q), d(Q, R))$$

but if $d(P, Q) = 0$, we may only deduce that the failure and divergence sets of P and Q are equal. In fact, if we define the closure of a process, $\overline{P}$, to be P with all possible infinite traces, *i.e*:

$$\begin{aligned} \mathcal{I}[\![\overline{P}]\!] &= \{u \in \alpha P^{\omega} \mid \forall s < u .\, s \in \mathit{traces} P\} \\ \mathcal{D}[\![\overline{P}]\!] &= \mathcal{D}[\![P]\!] \\ \mathcal{F}[\![\overline{P}]\!] &= \mathcal{F}[\![P]\!] \end{aligned}$$

then we notice that the distance between processes is equal just when their closures are equal:

$$d(P, Q) = 0 \Leftrightarrow \overline{P} = \overline{Q}$$

This sort of 'metric' is usually known as a *pseudo-metric*.

Note also that the closure of a process is always weaker than the process in both orderings.

Since the pseudo-metric is bounded (it is never bigger than 1), we may define a corresponding pseudo-metric on any function space $X \to \mathcal{U}$ by the usual construction:

$$d(f, g) = \sup_{x \in X} d(f(x), g(x))$$

and note that $d(f, g) = 0$ just when $\overline{f} = \overline{g}$ where $\overline{f}(x) = \overline{f(x)}$.

If we study the fixed point equations for the operators, we find that in all cases except that for hiding, the recursions are 'guarded'. That is, all the recursions are given in the form $F\,G\,P = P'$ and we can easily show that $d(F\,G, F\,G') \leq \frac{1}{2}d(G,G')$ so that if we have two operators G and G' which satisfy the equations then

$$d(G,G') = d(F\,G, F\,G') \leq \frac{1}{2}d(G,G')$$

giving $d(G,G') = 0$ and so $\overline{G} = \overline{G'}$. In all cases except hiding and forward renaming ($f[-]$) the result of applying the operators to a tuple of predeterministic processes is a closed process (as can be verified by inspection of the operator definitions in the last section – effectively this is because these are the only operators other than $\sqcap$ which can introduce unbounded nondeterminism). This means that all possible infinite traces are present and there can be no smaller fixed point.

Of these guarded recursions we are only left to dispose of forward renaming. Since any other fixed point has the same divergence and failure set, we need only consider the non-divergent infinite traces. Suppose G satisfies the equation and that u is a non-divergent infinite trace of $G(P)$. We shall construct a sequence of traces s_i such that $s_i < s_{i+1}$, $f(s_i)\,u_i = u$ and u_i is an infinite trace of $G(P\ \mathit{after}\ s_i)$. Then the existence of the traces s_i imply that P must have an infinite trace whose image under f is u, so u is an infinite of $f[P]$.

We take s_0 to be $\langle\rangle$ and $u_0 = u$. Since u is non-divergent, s_n is not a divergence of $G(P\ \mathit{after}\ s_n)$, therefore, there is a B such that $P\ \mathit{after}\ s_n = x : B \rightarrow P\ \mathit{after}\ s_n\langle x\rangle$ so that the equation which G satisfies tells us that:

$$G(P\ \mathit{after}\ s_n) = y : f(B) \rightarrow \bigsqcap\left\{G(P\ \mathit{after}\ s_n\langle x\rangle) \mid x \in f^{-1}(y) \cap B\right\}$$

Now, if b is the first element of u_n, then $b \in f(B)$ and there must be some $a \in f^{-1}(b) \cap B$ such that the tail of u_n is an infinite trace of $G(P\ \mathit{after}\ s_n\langle a\rangle)$. We take $s_{n+1} = s_n\langle a\rangle$ and u_{n+1} to be the tail of u_n. Lastly, $f(s_{n+1})\,u_{n+1} = f(s_n)\,u_n = u$ as required.

The only operator left is hiding (whose fixed point equation is not guarded). We will assume that the set X to be hidden is nonempty, the result being trivial if $X = \emptyset$ since both $\backslash\emptyset$ and $\backslash^c\emptyset$ are the identity function. The first observation to make is that the fixed point theory of $\leq$ and $\sqsubseteq$ are the same. This follows because each fixed point equation is $\leq$-monotonic and preserves $\leq$-monotonic functions; since $\perp$ is the least element for each order and $\bigsqcup_{\leq} S = \bigsqcup_{\sqsubseteq} S$ whenever the first exists, the standard iterative technique of finding the least fixed point must produce the same result. Now, if G is the least fixed point of the equation for hiding, then $G \leq \backslash X$. All we have to do is check that the minimal divergences of any fixed point are divergences of $\backslash X$ and that convergent infinite traces of a fixed point are infinites of $\backslash X$. Both parts of the proof are achieved by a construction similar to that which we used for the forward renaming operator.

If $t \in \mu\mathcal{D}[\![G(P)]\!]$ ($u \in \mathcal{I}[\![G(P)]\!]$), the idea is to construct a sequence of traces s_i such that $s_i < s_{i+1}$ and

$$s_i \backslash X t_i = t \quad (s_i \backslash X u_i = u)$$
$$t_i \in \mu\mathcal{D}[\![G(P\ \mathit{after}\ s_i)]\!] \quad (u_i \in \mathcal{I}[\![G(P\ \mathit{after}\ s_i)]\!])$$

for then P has an infinite trace, v, which is the limit of the s_i with $v\backslash X = t$ ($v\backslash X = u$) giving $t \in \mathcal{D}[\![P\backslash X]\!]$ ($u \in \mathcal{I}[\![P\backslash X]\!]$).

Choose $s_0 = \langle\rangle$ and $t_0 = t$ ($u_0 = u$). We define the $n+1$th sequences from the nth. If P *after* $s_n = \bot$ then we must have $t_n = \langle\rangle$ so we may set s_{n+1} to be any extension of s_n by an element of X (the situation cannot arise if u is a non-divergent infinite); otherwise P_n *after* $s_n = x : B \to P$ *after* $s_n \langle x \rangle$ for some B. If $B \cap X = \emptyset$, then

$$G\,(P\ \textit{after}\ s_n) = x : B \to G\,(P\ \textit{after}\ s_n\,\langle x\rangle)$$

by the fixed point equation so there must be a $b \in B$ and t_{n+1} (resp. u_{n+1}) such that $t_n = \langle b\rangle\, t_{n+1}$ (resp. $u_n = \langle b\rangle\, u_{n+1}$) and $t_{n+1} \in \mu\mathcal{D}[\![G\,(P\ \textit{after}\ s_n\,\langle b\rangle)]\!]$ (resp. $u_{n+1} \in \mathcal{I}[\![G\,(P\ \textit{after}\ s_n\,\langle b\rangle)]\!]$) so take $s_{n+1} = s_n\,\langle b\rangle$.

If $B \cap X \neq \emptyset$, then

$$G\,(P\ \textit{after}\ s_n) = ((x : B - X \to G\,(P\ \textit{after}\ s_n\,\langle x\rangle))\,\Box Q) \sqcap Q$$

where

$$Q = \bigsqcap_{x \in B \cap X} G\,(P\ \textit{after}\ s_n\,\langle x\rangle)$$

so that either the required behaviour comes from performing some action from $B - X$ immediately, in which case we employ the same construction as in the last case; or else, the behaviour comes from $G\,(P$ *after* $s_n\,\langle b\rangle)$ for some $b \in B \cap X$, in which case we define $s_{n+1} = s_n\,\langle b\rangle$ and $t_{n+1} = t_n$ ($u_{n+1} = u_n$).

The results of this section are summarised in the following theorem.

2.1 Theorem. Every CSP definable function has a least fixed point, and therefore its denotational semantics over $\mathcal{U}$ is well-defined. ∎

3. Operational semantics

In this section we present an operational semantics for CSP with unbounded nondeterminism, and summarise the main details of the congruence proof referred to in the introduction. But first we will define the abstraction functions from transition systems to $\mathcal{U}$ that will play a crucial role in the statement and proof of this theorem. The proofs of all but a very few results are omitted from this presentation – readers wishing more details should consult [R3].

Summary of notation, nomenclature and results. A *transition system* is a set of states with a binary relation $\xrightarrow{\delta}$ for each element δ of the set $\Sigma^+ = \Sigma \cup \{\tau\}$ of transitions, where τ denotes an internal transition. We should note that Σ (the set of visible actions) is an implicit parameter of almost everything we do from now on, as indeed it was in the last section.

A *morphism* [R1,R4] is a function from one transition system to another which characterises the property of indistinguishability in that no experimenter who can only see transitions (visible or invisible) should be able to tell P from $F(P)$ if F is a morphism. $F : C \to D$ is said to be a morphism if and only if:

(i) $P \xrightarrow{\delta} Q \Rightarrow F(P) \xrightarrow{\delta} F(Q)$, and

(ii) $F(P) \xrightarrow{\delta} X \Rightarrow \exists Q.P \xrightarrow{\delta} Q \wedge F(Q) = X$.

Morphisms are closely related to the idea of bisimulation but differ in that they treat internal actions in exactly the same rigid way that they treat visible ones, and that they are functions rather than relations.

The *index of nondeterminism* $i(C)$ of a transition system C is the smallest infinite regular cardinal[4] which is strictly larger than $card\{Q \mid P \xrightarrow{\delta} Q\}$ for all $P \in C$ and $\delta \in \Sigma^+$.

The functions. Given an element P of a transition system, we can construct its sets of failures, divergence and infinite traces in natural ways which are described below.

We first define two multi-step versions of the transition relation. If $P, Q \in C$ and $s = \langle x_i \mid 0 \leq i < n \rangle \in (\Sigma^+)^*$ we say $P \stackrel{s}{\longmapsto} Q$ if there exist $P_0 = P, P_1, \ldots, P_n = Q$ such that $P_k \xrightarrow{x_k} P_{k+1}$ for $k \in \{0, 1, \ldots, n-1\}$. Unlike this first version, the second ignores τs. For $s \in \Sigma^*$ we write $P \stackrel{s}{\Longrightarrow} Q$ if there exists $s' \in (\Sigma^+)^*$ such that $P \stackrel{s'}{\longmapsto} Q$ and $s' \setminus \tau = s$. The following properties of $\stackrel{s}{\Longrightarrow}$ and $\stackrel{s}{\longmapsto}$ are all obvious.

Lemma 3.1.

(a) $P \stackrel{\langle\rangle}{\Longrightarrow} P \wedge P \stackrel{\langle\rangle}{\longmapsto} P$

(b) $P \stackrel{s}{\Longrightarrow} Q \wedge Q \stackrel{t}{\Longrightarrow} R$ implies $P \stackrel{st}{\Longrightarrow} R$

(c) $P \stackrel{s}{\longmapsto} Q \wedge Q \stackrel{t}{\longmapsto} R$ implies $P \stackrel{st}{\longmapsto} R$

(d) $P \stackrel{st}{\Longrightarrow} R$ implies $\exists Q.P \stackrel{s}{\Longrightarrow} Q \wedge Q \stackrel{t}{\Longrightarrow} R$

(e) $P \stackrel{st}{\longmapsto} R$ implies $\exists Q.P \stackrel{s}{\longmapsto} Q \wedge Q \stackrel{t}{\longmapsto} R$ ∎

Suppose C is a transition system and $P \in C$. We say P can *diverge*, written $P\uparrow$, if there exist $P_0 = P, P_1, P_2, \ldots$ such that, for all $n \in \mathbb{N}$, $P_n \xrightarrow{\tau} P_{n+1}$.

$$divergences(P) = \{st \mid \exists Q.P \stackrel{s}{\Longrightarrow} Q \wedge Q\uparrow\}$$

Notice that we have said that st is a divergence trace whenever s is. This is motivated by a desire (inspired by our abstract semantics) to make all possibly divergent processes undefined. (As will be apparent from a careful reading of the proofs below, the fact that our semantic models and functions are strict with respect to divergence is sometimes of great importance.)

Say $P \in C$ is *stable* provided there is no Q such that $P \xrightarrow{\tau} Q$ (in other words, if P cannot make any internal progress). If $B \subseteq \Sigma$ we say P *ref* B if

$$\forall a \in B \cup \{\tau\}. \neg \exists Q \in C.P \xrightarrow{a} Q .$$

[4]A regular cardinal λ is one which is not the union of less than λ sets all of which are of size less than λ. There are arbitrarily large regular cardinals, since for example every *successor* cardinal is regular. The combinatorial properties which make *regular* cardinals the natural bounds for nondeterminism are well illustrated in [R3].

Thus P *ref* B implies that P is stable. We can now define

$$\mathit{failures}(P) = \{(s,B) \mid \exists Q.P \overset{s}{\Longrightarrow} Q \wedge Q \mathrel{\mathit{ref}} B\} \cup \{(s,B) \mid s \in \mathit{divergences}(P)\}\,.$$

The point of these definitions is that a process can properly refuse B only when it is in a stable state, for as long as it is performing internal actions one cannot be sure that it will not come into a state where a desired event is possible. On the other hand, when a process diverges it also refuses (in a different sense perhaps) all communications offered to it. The second part of the definition is also motivated by the desire to make a divergent process undefined.

If $u \in \Sigma^\omega$ is an infinite trace and $P \in C$, we write $P \overset{u}{\Longrightarrow}$ if there are $P = P_0, P_1, P_2, \ldots \in C$ and $x_i \in \Sigma^+$ such that $\forall k.P_k \overset{a_k}{\longrightarrow} P_{k+1}$ and $\langle a_k \mid k \in \mathbb{N} \wedge a_k \neq \tau\rangle = u$. This lets us define

$$\mathit{infinites}(P) = \{u \in \Sigma^\omega \mid P \overset{u}{\Longrightarrow}\} \cup \{su \mid s \in \mathit{divergences}(P) \wedge u \in \Sigma^\omega\}\,.$$

Similarly, if $\langle x_i \mid i \in \omega\rangle = u \in (\Sigma^+)^\omega$ we can write $P \overset{u}{\longmapsto}$ if there exist $P = P_0, P_1, P_2, \ldots$ such that, for all i, $P_i \overset{x_i}{\longrightarrow} P_{i+1}$.

Clearly it is possible to define other functions, and to vary these definitions for another definition of divergence. However the above are exactly the required maps to define the abstraction map into $\mathcal{U}$.

Definition. If C is any transition system then we define the abstraction map $\Phi : C \to \mathcal{U}$ as follows.

$$\Phi(P) = \langle \mathit{failures}(P),\ \mathit{divergences}(P), \mathit{infinites}(P)\rangle$$

We now state a theorem which establishes some basic properties of Φ.

Theorem 3.2. Φ is well defined, and furthermore

a) If $F : C \to D$ is a morphism then $\Phi(F(P)) = \Phi(P)$ for all $P \in C$.

b) If $P \in C$ and C is a sub-system of D (i.e., a subset closed under all the transition relations) then $\Phi(P)$ does not depend on whether we think of P as an element of C or of D.

c) Given any transition system C there is another one C' such that C is a subsystem of C' and $\Phi : C' \to \mathcal{U}$ is onto. ∎

It might seem a little curious that we have gone to the trouble of extending an arbitrary transition to one on which Φ is onto. The reason for this will become apperent when this result is used later.

In proving our congruence theorem later we will need not only the map $\Phi : C \to \mathcal{U}$ but also a sequence of approximations to it. We will define a map $\Phi_\alpha : C \to \mathcal{U}$ for each ordinal α. (Once again, C is here an arbitrary transition system.) It is convenient to define Φ_α in terms of a functional

$$\mathcal{G} : (C \to \mathcal{U}) \to (C \to \mathcal{U})\,.$$

If $\Psi : C \to \mathcal{U}$ and $P \in C$, we define $\mathcal{G}(\Psi)(P) = \langle F', D', I' \rangle$, where

$$\begin{array}{rcl} F' & = & \{(\langle\rangle, X) \mid P \; ref \; X\} \\ & & \cup\{(s, X) \mid \exists Q.P \xrightarrow{\tau} Q \wedge (s, X) \in \mathcal{F}[\![\Psi(Q)]\!]\} \\ & & \cup\{(\langle a\rangle s, X) \mid \exists Q.P \xrightarrow{a} Q \wedge (s, X) \in \mathcal{F}[\![\Psi(Q)]\!]\} \\ D' & = & \{s \mid \exists Q.P \xrightarrow{\tau} Q \wedge s \in \mathcal{D}[\![\Psi(Q)]\!]\} \\ & & \cup\{\langle a\rangle s \mid \exists Q.P \xrightarrow{a} Q \wedge s \in \mathcal{D}[\![\Psi(Q)]\!]\} \\ I' & = & \{u \mid \exists Q.P \xrightarrow{\tau} Q \wedge u \in \mathcal{I}[\![\Psi(Q)]\!]\} \\ & & \cup\{\langle a\rangle u \mid \exists Q.P \xrightarrow{a} Q \wedge u \in \mathcal{I}[\![\Psi(Q)]\!]\} \end{array}$$

The following Theorem establishes some useful properties of $\mathcal{G}$.

Theorem 3.3.

a) $\mathcal{G}$ is well defined and monotonic with respect to both orders.

b) Φ, as defined earlier in this section, is a fixed point of $\mathcal{G}$.

Proof. The whole of part (a) follows immediately from the fact that $\mathcal{G}$ can be re-written entirely in CSP. The operator $P \rhd Q$ used below is an abbreviation for $(P \Box Q) \sqcap Q$ (the process which can offer the choice between P and Q but which must eventually make an internal transition to become Q if no action occurs). It is a useful operator since it allows more conciseness, and has appeared before in similar circumstances in the literature, e.g. [HH].

$$\begin{array}{rcll} \mathcal{G}(\Psi)(P) & = & x : P^0 \to \sqcap\{\Psi(Q) \mid P \xrightarrow{x} Q\} & \text{if } \nexists Q.P \xrightarrow{\tau} Q \\ \mathcal{G}(\Psi)(P) & = & ((x : P^0 \to \sqcap\{\Psi(Q) \mid P \xrightarrow{x} Q\}) & \text{otherwise} \\ & & \rhd \sqcap\{\Psi(Q) \mid P \xrightarrow{\tau} Q\} & \end{array}$$

where P^0 denotes $\{a \in \Sigma \mid \exists Q.P \xrightarrow{a} Q\}$. It is easy to see that our two definitions of $\mathcal{G}$ are equivalent. Note that the overall structure of this CSP definition depends only on the transitions within C, and is therefore independent of the value of Ψ. It is this last fact which proves that $\mathcal{G}$ is monotone with respect to both orders.

Part (b) is intuitively obvious. Consider, for example, the divergence component. It follows immediately from the definition of Φ that $\mathcal{D}[\![\Phi(P)]\!] = divergences(P)$ is equal to

$$\{st \mid P \xrightarrow{\tau} Q \wedge Q \xRightarrow{s} R \wedge R\uparrow\} \cup \{\langle a\rangle st \mid P \xrightarrow{a} Q \wedge Q \xRightarrow{s} R \wedge R\uparrow\}$$

which in turn is equal to

$$\{s \mid P \xrightarrow{\tau} Q \wedge s \in divergences(Q)\} \cup \{\langle a\rangle s \mid P \xrightarrow{a} Q \wedge s \in divergences(Q)\}$$

which is $\mathcal{D}[\![\mathcal{G}(\Phi)(P)]\!]$ by definition of $\mathcal{G}$. Both the other cases are similar and depend on this one. The failures case divides into three components rather than two for obvious reasons. ∎

By Theorem 1.5 applied to the product space $\mathcal{U}^C$ $(= C \to \mathcal{U})$, it follows from the existence of one fixed point that $\mathcal{G}$ has a least fixed point which is equal to Φ_α for some α where

$$\begin{aligned} \Phi_0(P) &= \bot & \text{for all } P \in C \\ \Phi_\mu(P) &= \bigsqcup\{\Phi_\beta(P) \mid \beta \in \mu\} & \text{if } \mu \text{ is a limit ordinal} \\ \Phi_{\beta+1} &= \mathcal{G}(\Phi_\beta) \end{aligned}$$

since Φ_0 is the least element of the product space and $\Phi_\beta = \mathcal{G}^\beta(\Phi_0)$. (Since $\mathcal{G}$ is CSP-definable it also follows from the result of Section 2.) These Φ_β will play a crucial role in the main congruence theorem in the next section. This is essentially because of the next theorem, whose proof may be found in [R3].

Theorem 3.4. Φ is the least fixed point of $\mathcal{G}$. Hence there exists α such that $\Phi_\alpha = \Phi$. ∎

This result shows the equivalence of the natural operationally defined abstraction function and one which it obtained by iterating a CSP definition through the ordinals. This is exactly what we shall want to do on a much wider scale when we seek to prove the congruence theorem in the final section. It will turn out that this last result is perhaps the most important component of the proof of that theorem.

The operational semantics

This section is devoted to the definition of the operational semantics for CSP and closely related semantics over more general transition systems.

A crucial starting point for the creation of any semantics is the definition of the programming language. The definition we take is just the usual core CSP extended by unbounded nondeterminism and infinite hiding. For formal reasons we must fix *ab initio* the range of unbounded nondeterminism allowed. However this may be as large as we please. In particular, it is convenient to fix it strictly larger than the cardinality of the alphabet Σ. Thus the following language is implicitly parameterised both by the alphabet Σ of all possible communications and by the bound λ, an infinite regular cardinal on the unbounded nondeterminism.

Because the unbounded nondeterminism operator (unavoidably) and the guarded choice operator (avoidably at a price) are infinitary operators (take a potentially infinite number of process arguments) one should, for rigour, be rather careful over the definition of the syntax of this version of CSP. On the one hand we can write down the usual sort of BNF definition.

$$\begin{aligned} P ::= {} & p \mid STOP \mid SKIP \mid a \to P \mid x : B \to g(x) \mid P \Box Q \mid P \sqcap Q \mid \\ & P\, {}_B\|_C\, Q \mid P \,|||\, Q \mid P;Q \mid P \backslash B \mid f[P] \mid f^{-1}[P] \mid \mu p.P \mid \textstyle\bigsqcap S \end{aligned}$$

where g is any function from B (a subset of Σ) to processes, S ranges over nonempty sets of processes smaller than λ, f ranges over the set AT of (not necessarily finite-to-one) alphabet transformations, p over the set *Var* of process variables, etc.

When there are infinitary operators in a syntax, like those in this language, the idea of what is defined by a syntax like this one is less obvious than it usually is and should therefore be discussed briefly. If we are to have a principle of structural induction and have a way of defining the semantics of programs we cannot have a program of the form $\prod S$ or $x : B \to g(x)$ which is itself in S or in the range of g. One can, of course, regard BNF definitions like the above as fixed point equations, defining the smallest syntactic class which is closed under the various operations on the right. For a language with only finitary constructs this fixed point is reached by ω iterations (every program is "born on a finite day") but we have to go further, to cater for programs like $n : \mathbf{N} \to P_n$ where P_n is born on day n. The functional implied by the right hand side of the above BNF definition is clearly monotone (the more programs there are, the more it delivers) but since it is not operating over a set (rather over the proper Class of all syntactic objects) it is by no means obvious it even has a fixed point. Fortunately it does, and is guaranteed to reach it by λ iterations, where λ is the bound on nondeterminism and the size of Σ already mentioned. (See [BRW] for some more discussion of this question.) The principle of structural induction is then perfectly valid and corresponds to the principle of transfinite induction on the "birthday" of a term.

To simplify the operational semantics a little it is convenient, as was done in [BRW], to treat the constructs *STOP*, *SKIP* and $a \to P$ as special cases of the construct $x : B \to g(x)$: *STOP* has B empty, $a \to P$ has $B = \{a\}$ and $g(a) = P$, and $SKIP = \surd \to STOP$.

Let $\mathbf{E}$ be the set of all CSP terms defined by the above. An element of $\mathbf{E}$ may have free process variables, in which case it is said to be *open*. If it has none it is said to be *closed*; we denote the set of all closed terms by $\mathbf{P}$. Closed terms are of importance since their meaning is fully determined; there are no slots for processes waiting to be filled in.

If $P, Q \in \mathbf{E}$ and $p \in Var$ then $P[Q/p]$ denotes the term where Q has been substituted for all free occurrences of p is P. When Q is not closed (though for us it usually will be) some care will be necessary to prevent P binding any of Q's free variables.

The Plotkin-style semantics regards the set $\mathbf{P}$ of all closed CSP-terms as a transition system, since it describes the set of all actions each closed term can perform and which new terms it may then become. The clauses of this operational semantics are given in the usual "natural deduction" style below.

Below, a, b range over Σ and x, y over $\Sigma^+ = \Sigma \cup \{\tau\}$. Alphabet transformations (functions from Σ to Σ) are extended to Σ^+ by setting $f(\tau) = \tau$.

$$\frac{}{(x : B \to g(x)) \xrightarrow{b} g(b)} \quad (b \in B)$$

$$\frac{}{P \sqcap Q \xrightarrow{\tau} P} \qquad \frac{}{P \sqcap Q \xrightarrow{\tau} Q}$$

$$\frac{}{\mu p.P \xrightarrow{\tau} P[\mu p.P/p]}$$

$$\frac{P \xrightarrow{\tau} P'}{P \Box Q \xrightarrow{\tau} P' \Box Q} \qquad \frac{Q \xrightarrow{\tau} Q'}{P \Box Q \xrightarrow{\tau} P \Box Q'}$$

$$\frac{P \xrightarrow{a} P'}{P \Box Q \xrightarrow{a} P'} \qquad \frac{Q \xrightarrow{a} Q'}{P \Box Q \xrightarrow{a} Q'}$$

$$\frac{P \xrightarrow{\tau} P'}{P \,{}_B\|_C\, Q \xrightarrow{\tau} P' \,{}_B\|_C\, Q} \qquad \frac{Q \xrightarrow{\tau} Q'}{P \,{}_B\|_C\, Q \xrightarrow{\tau} P \,{}_B\|_C\, Q'}$$

$$\frac{P \xrightarrow{a} P'}{P \,{}_B\|_C\, Q \xrightarrow{a} P' \,{}_B\|_C\, Q} \quad (a \in B - C)$$

$$\frac{Q \xrightarrow{a} Q'}{P \,{}_B\|_C\, Q \xrightarrow{a} P \,{}_B\|_C\, Q'} \quad (a \in C - B)$$

$$\frac{P \xrightarrow{a} P' \quad Q \xrightarrow{a} Q'}{P \,{}_B\|_C\, Q \xrightarrow{a} P' \,{}_B\|_C\, Q'} \quad (a \in B \cap C)$$

$$\frac{P \xrightarrow{x} P'}{P \,|||\, Q \xrightarrow{x} P' \,|||\, Q} \qquad \frac{Q \xrightarrow{x} Q'}{P \,|||\, Q \xrightarrow{x} P \,|||\, Q'}$$

$$\frac{P \xrightarrow{x} P'}{P;Q \xrightarrow{x} P';Q} \quad (x \neq \surd)$$

$$\frac{\exists P'.P \xrightarrow{\surd} P'}{P;Q \xrightarrow{\tau} Q}$$

$$\frac{P \xrightarrow{x} P'}{P \backslash B \xrightarrow{x} P' \backslash B} \quad (x \notin B)$$

$$\frac{P \xrightarrow{a} P'}{P \backslash B \xrightarrow{\tau} P' \backslash B} \quad (a \in B)$$

$$\frac{P \xrightarrow{x} P'}{f[P] \xrightarrow{y} f[P']} \quad (y = f(x))$$

$$\frac{P \xrightarrow{x} P'}{f^{-1}[P] \xrightarrow{y} f^{-1}[P']} \quad (f(y) = x)$$

$$\frac{P \in S}{\sqcap S \xrightarrow{\tau} P}$$

Note at this point that the operationally natural element of $\mathcal{U}$ corresponding to each closed term P is given by $\Phi(P)$, where Φ is as defined in Section 2 and P is considered to be an element of the transition system $\mathbf{P}$ defined above. Theorem 3.2 shows that this is equal to $\Phi(F(P))$ for any morphism F. We can now state the main congruence result that we would like to prove, namely that for all closed CSP terms P, $\Phi(P) = \mathcal{S}[\![P]\!]$, where $\mathcal{S}[\![P]\!]$ denotes the value in $\mathcal{U}$ defined by the semantics defined earlier.

There are two structure clashes between the operational and denotational semantics. The first is the obvious one that one is given in terms of transition systems and the other in terms of the abstract model $\mathcal{U}$. But perhaps the more difficult one to resolve is the clash between the term rewriting style of the operational semantics and the denotational style of the other. Of course the latter means that the semantic value of each term is deduced from the semantic value of its subcomponents in a transparent way and that an abstract fixed point theory is used. In the earlier paper on the operational semantics of CSP [BRW] these two issues were resolved separately by creating an intermediate, denotational tree semantics. Unfortunately the complete metric spaces of trees used in that paper no longer exist because of the introduction here of infinite branching.

The main result of [R3] is that, for each infinite regular cardinal λ, there exists a transition system T_λ such that for all transition systems C with $i(C) \leq \lambda$, there exists a unique morphism $H_\lambda : C \to T_\lambda$. Thus T_λ is a final object in the category of transition systems with morphisms as arrows. Analogues of the contraction mapping theorem and related results hold which are useful when one uses these systems. T_λ can be used to give an intermediate denotational semantics to CSP in the style of [BRW]. However, because of the complexity of this new theory and thanks mainly to the construction of the Φ_α in the previous section we do not now need to do so.

It is useful to extend the operational space defined above to include non-closed terms with their variables instantiated by elements of an arbitrary transition system.

Definition. If C is any transition system then C^{CSP} is the system of CSP syntactic terms over C: namely the set of all substitutions by elements of C for all free variables of general terms in the language. All terms are distinct. Note that C^{CSP} contains every closed CSP term and every element of C. The transitions of each term are those of P if $P \in C$ (i.e., $P \xrightarrow{\delta} Q$ in C^{CSP} if and only if $P \xrightarrow{\delta} Q$ in C). The transitions of proper syntactic terms are determined from the operational semantic clauses above (from those of their subterms or otherwise).

The stipulation that all terms are distinct means that each possible construction of a term leads to a different element of the system. For example, in $(C^{CSP})^{CSP}$, for each $P \in C$ the terms $a \to a \to \ulcorner P \urcorner$, $a \to \ulcorner a \to P \urcorner$ and $\ulcorner a \to a \to P \urcorner$ are all different, where the syntactic quotes $\ulcorner \cdot \urcorner$ denote the boundary between the inner and outer syntactic construction. However the obvious map from $(C^{CSP})^{CSP}$ to C^{CSP} which "forgets" these boundaries is easily shown to be a morphism.

Note that Theorem 3.2 (a) tells us that the image under Φ of a closed term P is independent of whether it is considered to belong to the space $\mathbf{P}$ of closed terms or any C^{CSP}, since there is an obvious morphism embedding $\mathbf{P}$ into any C^{CSP}.

We are now in a position to begin the proof of the main theorem of this section, namely that the $\mathcal{U}$ semantics for CSP is congruent to the operational semantics. We will eventually complete the proof by performing a structural induction over C^{CSP}, but before we do that it is helpful to establish that the operational and denotational versions of all the non-recursive operators are congruent.

Theorem 3.5. The operational versions of the various CSP operators are all congruent to the denotational versions over $\mathcal{U}$. In other words, for each operator $\odot$ and each $P, Q \in C^{CSP}$,

$$\Phi(P \odot Q) = \Phi(P) \odot \Phi(Q).$$

Furthermore all the operators are well behaved with respect to the partial abstraction functions Φ_α in the sense that

$$\Phi_\alpha(P \odot Q) \leq \Phi_\alpha(P) \odot \Phi_\alpha(Q)$$

for each α. (The form of these clauses is modified suitably when the operator $\odot$ is not binary. The precise statement for each operator in turn can be found in the Lemmas below.)

Proof. This theorem is no more nor less than a convenient grouping of a large number of similar though separate results. These are stated below, grouped by operator, plus for each operator a further result which is crucial in the proof of the full congruence part of the Lemma. In each of these Lemmas it is assumed that the given term is an element of C^{CSP} of the given form; the immediate subterms being unrestricted elements of C^{CSP} (i.e., not necessarily elements of C itself).

The operators break into two classes as far as style of proof is concerned: prefixing and nondeterministic choice, which are easiest, and the rest of the operators, which require very similar though more difficult arguments – these proofs are omitted here but may be found in [R3]. As usual, recursion is a special case and will be dealt with on its own later.

Lemma 3.5.1 (i). For all terms P_x denoting functions from A into C^{CSP}, we have

$$\Phi(x : A \rightarrow P_x) = x : A \rightarrow \Phi(P_x).$$

Lemma 3.5.1 (ii). For all terms P_x denoting functions from A to C^{CSP} and all ordinals α we have

$$\Phi_\alpha(x : A \rightarrow P_x) \leq x : A \rightarrow \Phi_\alpha(P_x).$$

Lemma 3.5.2 (i). For all $S \subseteq C^{CSP}$ (of size less than our bound on nondeterminism) we have

$$\Phi(\bigsqcap S) = \bigsqcap\{\Phi(P) \mid P \in S\}.$$

Lemma 3.5.2 (ii). For all P, Q in C^{CSP} and all ordinals α we have

$$\Phi_\alpha(\bigsqcap S) \leq \bigsqcap\{\Phi_\alpha(P) \mid P \in S\}.$$

Lemma 3.5.3 (i) If $P, Q \in C^{CSP}$ and $\odot$ is any one of $\Box$, ;, ${}_X\|_Y$, $|||$, $\sqcap$ then $\Phi(P \odot Q) = \Phi(P) \odot \Phi(Q)$.

Lemma 3.5.3 (ii) If $P, Q \in C^{CSP}$, $\odot$ is any one of $\Box$, ;, ${}_X\|_Y$, $|||$, $\sqcap$ and α is any ordinal then $\Phi_\alpha(P \odot Q) \leq \Phi_\alpha(P) \odot \Phi_\alpha(Q)$.

Lemma 3.5.4 (i). If $P \in C^{CSP}$ and $\ddagger$ is any one of $f[\cdot]$, $f^{-1}[\cdot]$ and $\backslash X$ then $\Phi(\ddagger(P)) = \ddagger(\Phi(P))$.

Lemma 3.5.4 (ii). If $P \in C^{CSP}$,$\ddagger$ is any one of $f[\cdot]$, $f^{-1}[\cdot]$ and $\backslash X$ and α is any ordinal then

$$\Phi_\alpha(\ddagger(P)) \leq \ddagger(\Phi_\alpha(P)).$$

This completes Theorem 3.5. ∎

These results provide the building blocks of the proof of the main result, and are put together below. The next Theorem is the main result of this section.

Definitions. Given a CSP term P and a $\rho \in OEnv = Var \rightarrow C^{CSP}$, we can define an operational "semantic function": $\mathcal{O}[\![P]\!]\rho \in C^{CSP}$ is defined to be the result of substituting each free variable p in P by $\rho(p)$. (Note that P may have no free variables, finitely many, or infinitely many. This last possibility arises because of the two infinitary operations $\sqcap$ and $x : A \rightarrow P_x$.) Given $\rho \in OEnv$ we can define the corresponding element $\overline{\rho}$ of $UEnv = Var \rightarrow \mathcal{U}$ by

$$\overline{\rho}[\![p]\!] = \Phi(\rho(p))$$

and also, for each α, an approximation

$$\overline{\rho}^{\alpha}[\![p]\!] = \Phi_{\alpha}(\rho(p))\,.$$

In this theorem we will assume that the basic transition system C is such that $\Phi : C \rightarrow \mathcal{U}$ is onto (following Theorem 3.2 (c)). This is helpful in the proof, since it means that for each $\sigma \in UEnv$ there is a $\rho \in OEnv$ such that $\overline{\rho} = \sigma$.

Theorem 3.6. Suppose P is any CSP term. Then the following hold.

a) $\mathcal{S}[\![P]\!]\overline{\rho} = \Phi(\mathcal{O}[\![P]\!]\rho)$ for all $\rho \in OEnv$.

b) For each ordinal α and each $\rho \in OEnv$ we have $\mathcal{S}[\![P]\!]\overline{\rho}^{\alpha} \geq \Phi_{\alpha}(\mathcal{O}[\![P]\!]\rho)$.

Proof. This is by structural induction on P. Given the sequence of Lemmas above, the cases of all the non-recursive operators are trivial, parts (a) and (b) respectively following from the (i) and (ii) of the Lemmas under Theorem 3.5 above.

It only remains to consider the case of a recursively defined term $\mu p.P$, where the result is known to hold of P.

For part (a), observe that $\mathcal{S}[\![\mu p.P]\!]\overline{\rho}$ is defined to be the *least* fixed point of $F(Y) = \mathcal{S}[\![P]\!]\overline{\rho}[Y/p]$. Now set $X = \Phi(\mathcal{O}[\![\mu p.P]\!]\rho)$ and note that since the only transition of $\mathcal{O}[\![\mu p.P]\!]\rho$ is a τ-transition to $\mathcal{O}[\![P[\mu p.P/p]]\!]\rho$, we have

$$\begin{aligned} X &= \Phi(\mathcal{O}[\![P[\mu p.P/p]]\!]\rho) \\ &= \Phi(\mathcal{O}[\![P]\!]\rho[\mathcal{O}[\![\mu p.P]\!]\rho/p]) \\ &= \mathcal{S}[\![P]\!]\overline{\rho[\mathcal{O}[\![\mu p.P]\!]\rho/p]} \quad \text{by induction} \\ &= \mathcal{S}[\![P]\!]\overline{\rho}[X/p] \\ &= F(X) \end{aligned}$$

and so X is a fixed point of F and so is certainly greater than $\mathcal{S}[\![\mu p.P]\!]\overline{\rho}$.

For the reverse inequality we will prove by induction on α that $\Phi_{\alpha}(\mathcal{O}[\![\mu p.P]\!]\rho) \leq F^{\alpha}(\bot)$ for all α. This is enough since we know that for sufficiently large α the left hand side equals $\Phi(\mathcal{O}[\![\mu p.P]\!]\rho)$ and the right hand side is $\mathcal{S}[\![\mu p.P]\!]\overline{\rho}$. The cases of $\alpha = 0$ and α a limit ordinal are both trivial, the latter because both sides are defined to be the least upper bounds of the terms for smaller ordinals. So suppose it holds for β and $\alpha = \beta + 1$. Then $\Phi_{\alpha}(\mathcal{O}[\![\mu p.P]\!]\rho) = \Phi_{\beta}(\mathcal{O}[\![P[\mu p.P/p]]\!]\rho)$ because of the initial τ-transition and the definition of $\mathcal{G}$. But this by induction and monotonicity is weaker than $F^\alpha(\bot)$, as required.

It only remains to prove (c), in other words that, given ρ and α,

$$\Phi_\alpha(\mathcal{O}[\![\mu p.P]\!]\rho) \leq \mathcal{S}[\![\mu p.P]\!]\overline{\rho}^\alpha .$$

Once again we prove this by transfinite induction on α. Again the result is easy for $\alpha = 0$ since the left hand side is $\bot$ and also for the limit ordinal case since the left hand side at α is then the least upper bound of the previous left hand sides, and $\mathcal{S}[\![P]\!]$ is monotone. So suppose $\alpha = \beta + 1$ and that the result holds at β. Then

$$\begin{array}{rcll}
\Phi_{\beta+1}(\mathcal{O}[\![\mu p.P]\!]\rho) & = & \Phi_\beta(\mathcal{O}[\![P[\mu p.P/p]]\!]\rho) & \\
& = & \Phi_\beta(\mathcal{O}[\![P]\!]\rho[\mathcal{O}[\![\mu p.P]\!]\rho/p]) & \\
& \leq & \mathcal{S}[\![P]\!]\overline{\rho[\mathcal{O}[\![\mu p.P]\!]\rho/p]}^\beta & \text{by (c) of } P \\
& \leq & \mathcal{S}[\![P]\!]\overline{\rho}^\beta[\Phi_\beta(\mathcal{O}[\![\mu p.P]\!]\rho)/p] & \\
& \leq & \mathcal{S}[\![P]\!]\overline{\rho}^\beta[\mathcal{S}[\![\mu p.P]\!]\overline{\rho}^\beta/p] & \text{induction and monotonicity} \\
& = & \mathcal{S}[\![\mu p.P]\!]\overline{\rho}^\beta & \text{as recursions denote fixed points} \\
& \leq & \mathcal{S}[\![\mu p.P]\!]\overline{\rho}^{\beta+1} & \text{by monotonicity}
\end{array}$$

which proves it for $\beta + 1$. This completes the proof of Theorem 3.5. ∎

The reader may have noticed that this section has not discussed the operational semantics of processes defined by mutual recursion, whether finite or infinite. This was because the proof for single recursion was quite difficult enough, and that for the more general case adds little except complexity. Also, as is pointed out in [R3], there is a simple CSP transformation which allows one to re-cast any mutual recursion as a single recursion. It would certainly be possible to base a proof of the operational validity of mutual recursion on that.

4. Conclusions

We have seen how to construct the infinite traces model $\mathcal{U}$, how it has unusual partial order properties, and how to overcome the incompleteness of the underlying orders. We have also seen the main points of the proof that our denotational semantics are congruent to the natural operational semantics.

The two orders $\leq$ and $\sqsubseteq$ have both been used throughout the paper in various ways. This leads to the same question as was posed in [R2], namely that of which is the natural order to use when presenting the model and semantics, given that both work. Here the arguments are slightly different. On the one hand now neither order is complete (whereas only $\leq$ was over $\mathcal{N}'$). However $\leq$ does still have a nicer theory of least upper bounds than $\sqsubseteq$, for they are always given by intersection where they exist while this is not even true for directed sets for $\sqsubseteq$. On the other, $\sqsubseteq$ is simpler to define and is perhaps more intuitive, but it does not have such a claim over $\mathcal{U}$ to be the "established" order as over $\mathcal{N}$ or $\mathcal{N}'$. And also $\sqsubseteq$ played an important part in the proof of the existence of fixed points seen in Section 2. This question will be best resolved by time and experience.

On the technical side we have seen in this work that completeness and continuity are natural casualties of the introduction of unbounded nondeterminism, but that their absence does not matter unduly except in the sense that proofs become more difficult. It will be interesting to see whether similar problems arise in other formalisms.

We have also sketched the proof that our semantics is operationally valid. Perhaps the most interesting feature of the proof is the way the approximate abstraction functions Φ_α show that the least fixed point corresponds with the operationally natural one via a type of "non-destructiveness" argument.

Future work on this model will include a fuller investigation of its algebraic properties. Another issue will be the study of other unboundedly nondeterministic constructs such as fair hiding operators. We should note that it is only permissible to add a new operator (other than one derived from existing operators) to this version of CSP if its restriction to $\mathcal{P}$ can is dominated by an operator which preserves predeterminism, to allow the proof in Section 2 to carry through. It will also be interesting to see what use can be made of the infinite traces component in the *specifications* of processes. For example one could add a clause to the usual specification of a buffer which stated that the buffer never does infinitely many inputs without an output, so that anything one puts in is eventually going to come out (even in the presence of an environment which eagerly places as much as possible into the buffer at all times).

The difficulties one encounters when dealing with unbounded nondeterminism, particularly the sort which is only detectable from infinite behaviours, are certainly not restricted to the models seen in this paper. Hopefully some of the work reported here will transfer to other formalisms for concurrency. One place where valuable work could be done is in timed CSP (see [RR1, RR2, Re]). The incorporation of infinite behaviours there (were it possible) would allow more abstract and general expressions of such modalities as "eventually" which appear in some forms of temporal logic.

Acknowledgements

Stephen Blamey has put a lot of work into analysing, refining and understanding the axioms for $\mathcal{U}$. In addition this work has been assisted by conversations with a number of colleagues, notably Paul Gardiner, Michael Goldsmith, Alan Jeffrey and David Walker.

References

[B] Brookes, S.D., *A Model for Communicating Sequential Processes*, Oxford University D.Phil. thesis, 1983.

[Blam] Blamey, S.R., *The soundness and completeness of axioms for CSP processes*, submitted for publication.

[BHR] Brookes, S.D., Hoare, C.A.R., and Roscoe, A.W., *A theory of communicating sequential processes*, JACM Vol. 31, No. 3 (July 1984) 560-599.

[BR] Brookes, S.D. and Roscoe, A.W., *An improved failures model for communicating processes* in Proc. of Pittsburgh symposium on concurrency, Springer LNCS !97 (1985).

[BRW] Brookes, S.D., Roscoe A.W., and Walker, D.J., *An operational semantics for CSP*, Submitted for publication.

[H] Hoare, C.A.R., *Communicating sequential processes*, Prentice-Hall, 1985

[HH] He Jifeng and Hoare, C.A.R., *Algebraic specification and proof of properties of communicating sequential processes*, Technical monograph PRG-52, Oxford University Computing Laboratory.

[R1] Roscoe, A.W., *A mathematical theory of communicating processes*, Oxford University D.Phil. thesis, 1982.

[R2] Roscoe, A.W., *An alternative order for the failures model*, in 'Two Papers on CSP', Technical monograph PRG-67, Oxford University Computing Laboratory.

[R3] Roscoe, A.W., *Unbounded nondeterminism in CSP*, in 'Two Papers on CSP', Technical monograph PRG-67, Oxford University Computing Laboratory.

[R4] Roscoe, A.W., *Analysing infinitely branching trees*, in preparation.

[Re] Reed, G.M., *A uniform mathematical theory for real-time distributed computing*, Oxford University D.Phil. thesis, 1988.

[RR1] Reed, G.M., and Roscoe, A.W., *A timed model for communicating sequential processes*, Proceedings of ICALP'86, Springer LNCS 226 (1986), 314-323.

[RR2] Reed, G.M., and Roscoe, A.W., *Metric spaces as models for real-time concurrency*, in the proceedings of MFPLS87 Springer LNCS 298 (1988).

The Semantics of Priority and Fairness in occam

Geoff Barrett
Programming Research Group
Oxford University Computing Laboratory
11 Keble Road
Oxford
OX1 3QD

Abstract. *This paper presents an operational semantics for priority and fairness in* occam. *The semantics is based on the state transitions made by a transputer in the execution of an* occam *program. It is possible to abstract sufficiently from the transputer implementation that a clear semantics is produced but to maintain enough detail that the particular meanings intended for priority and fairness can be expressed precisely as the practical programmer would understand them.*

1 Introduction

The programming language occam is based on an early version of CSP or Communicating Sequential Processes and consequently bears a close relationship to modern CSP, sometimes referred to as TCSP or Theoretical CSP. It is not surprising, therefore, that many of the models devised for the semantics of CSP can be adapted for occam in a way similar to [8]. However, these models are not adequate to describe some of the peculiar features of occam, namely the priority constructs and the notion of fairness embodied in the parallel operator. The solution to this problem given in [8] is to specify that the implementation of priority be no less deterministic than the specification of the corresponding non-priority constructs.

Originally, the priority constructs of occam were intended to be used in a way which had no consequence except in details of the timing or efficiency of a program. Thus, a program could be proved correct using the failures/divergence or behavioural equivalence semantics without fairness and then constructs could be prioritised for the sake of efficiency. This solution might suffice were it not for the fact that the substitution in a program of non-priority constructs for priority constructs may affect its safety properties. For instance, consider the program in Figure 1 which is guaranteed to terminate after some indeterminate but finite time. The program describes two processes which run in parallel. The first process repeatedly polls the

```
CHAN interrupt:
PAR
  VAR going:
  SEQ
    going:= TRUE
    WHILE going
      PRI ALT
        interrupt ?  ANY
          going:= FALSE
        SKIP
          SKIP
  interrupt !  ANY
```

Figure 1: This program is guaranteed to terminate

interrupt channel until its environment is ready to communicate along it; the process then communicates and terminates. The second process communicates along the **interrupt** channel and terminates. The whole program terminates because the **interrupt** channel is guaranteed to become ready by the fairness of the **PAR** construct; that guard is guaranteed to be chosen at some point because of the asymmetry in the arbitration. The program is therefore equivalent to **SKIP** in some sense, whereas Roscoe's semantics cannot give a stronger specification than that the implementation should be no worse than the completely divergent process (which is really no specification at all). Furthermore, it is important to be able to treat this program correctly because it is an idiom which allows the solution of many otherwise tricky or impossible problems. It is an abstraction which describes, for example, the action of a functional language interpreter which is evaluating a non-terminating expression but must be able to respond to an interrupt.

In order to model priority in arbitrations, the notion of 'ready guards' is introduced. A channel becomes ready after an output process indicates that it is ready to output along that channel. A delay guard becomes ready at the appropriate time and a **SKIP** guard is always ready. An asymmetric arbitration may only choose the ready guard of highest priority.

The key concept in the semantics which allows the discussion of fairness and process priority as understood by one familiar with the implementation is that of assigning 'initiative' for each action either to the process or to its environment. This idea corresponds closely to the implementation situation in which the change of state of the executing machine is caused by the execution of an instruction belonging to the process or by some interference from the environment. This extends naturally to a number of concurrently executing processes so that an action which is the initiative of the concurrent composition may be assigned to one of the component processes. The notions of fairness and process priority can be adequately discussed within this framework.

2 A Horizontal Syntax for occam

We shall not use the (for our purposes) unwieldy vertical syntax of occam but, instead, a horizontal syntax which only expresses those aspects of the language in which we are interested. That is, the state of program variables is excluded so that assignment, conditionals, value passing in communication, *etc.* are absent. The syntax is 'horizontalized' by using only binary forms of the associative operators like `PAR` and `SEQ` which require a list of subcomponents in occam. The iterative `WHILE` loop is replaced by a general recursive construct; the `SKIP` guard in an arbitration is distinguished from the process `SKIP` by writing the guard as τ. Delay guards are omitted for the sake of simplicity.

In the following informal, BNF-style syntax, c is drawn from some set of channel identifiers, *CID*, and p from some set of process identifiers, *PID*.

$$
\begin{array}{rrcl}
 & c & \in & \mathit{CID} \\
 & p & \in & \mathit{PID} \\
(\mathit{GUARD}) & g & ::= & c \mid \tau \\
(\mathit{PROC}) & P & ::= & \textbf{stop} \mid \textbf{skip} \mid c \rightarrow P \mid c \,!\!\rightarrow P \mid \textbf{alt}\; G \mid g \,?\!\rightarrow P \mid P_0 \,;\, P_1 \\
 & & & \mid p \mid \mu p.\, P \mid P_0 \;\textbf{par}\; P_1 \mid P_0 \;\underleftarrow{\textbf{par}}\; P_1 \\
(\mathit{GUARDP}) & G & ::= & g \rightarrow P \mid G_0 \boxplus G_1 \mid G_0 \underleftarrow{\boxplus}\, G_1
\end{array}
$$

Informally, the process terms stand for the following processes:

stop is the canonical deadlocked process which can make no further progress.

skip is the process which has terminated successfully.

$c \rightarrow P$ is a process which indicates to its environment that it is ready to output along channel c before behaving as $c\,!\!\rightarrow P$.

$c\,!\!\rightarrow P$ is a process which has indicated to its environment that it is ready to output along c and once the output has been discharged behaves as P.

alt G is an arbitration which has not yet chosen a branch. Once the branch $g \rightarrow P$ has been chosen, the process behaves as $g\,?\!\rightarrow P$. Branches are combined by symmetric choice, $\boxplus$, or asymmetric choice, $\underleftarrow{\boxplus}$.

$g\,?\!\rightarrow P$ is a branch of an arbitration which has been chosen. Once the synchronisation suggested by the guard, g, has been discharged, the process behaves as P.

$P_0 \,;\, P_1$ is sequential composition.

$\mu p.\, P$ is the general recursive construction. A free p in the body, P, behaves just like the whole process.

P_0 **par** P_1 is the concurrent composition of P_0 and P_1.

$P_0 \;\underleftarrow{\textbf{par}}\; P_1$ is the asymmetric concurrent composition of P_0 and P_1 which favours P_0. It bears little relation to the transputer implementation of `PRI PAR` for it is associative and the scheduling of concurrent components of P_0 is fair.

2.1 Types and Well-Formedness

Each process has three sets of channel identifiers associated with it. These three sets will be called the type of a process. The type of P will be denoted $(\mathcal{I}, \mathcal{O}, \mathcal{H})$ where $\mathcal{I}$ is the set of input channels along which a process may communicate, $\mathcal{O}$ is the set of output channels and $\mathcal{H}$ is the set of channels which are reserved for communication between sub-components of P. Since no channel may be used both for input and output, *etc.*, the sets must be disjoint. For reasons discussed in Section 5, each of these sets must be finite.

Certain syntactic constructions are only valid under restrictions on the types of subprocesses, and the type of the overall construction is determined by its components. There is one process **stop** and one process **skip** of each type. The processes $c \to P$, $c\,! \!\to P$ and $g\,? \!\to P$ all have the same type as P and are only well-formed when $c \in \mathcal{O} \cup \mathcal{H}$ and $g \in \mathcal{I} \cup \mathcal{H} \cup \{\tau\}$. The channel c is allowed to be an element of $\mathcal{H}$ because the process may have been expecting to communicate with some concurrent process which has terminated and disappeared from the process state (see the operational semantics, below). The arbitration **alt** G has the same type as G. If the types of P_i are $(\mathcal{I}_i, \mathcal{O}_i, \mathcal{H}_i)$ and of the six sets only $\mathcal{I}_0$ and $\mathcal{O}_1$ or $\mathcal{I}_1$ and $\mathcal{O}_0$ may possibly intersect, then P_0 **par** P_1 and $P_0 \underleftarrow{\textbf{par}} P_1$ are well-formed and their type is given by:

$$(\mathcal{I}_0 - \mathcal{O}_1 \cup \mathcal{I}_1 - \mathcal{O}_0, \mathcal{O}_0 - \mathcal{I}_1 \cup \mathcal{O}_1 - \mathcal{I}_0, \mathcal{H}_0 \cup \mathcal{H}_1 \cup (\mathcal{I}_0 \cap \mathcal{O}_1) \cup (\mathcal{I}_1 \cap \mathcal{O}_0))$$

If both P_0 and P_1 have the same type, then $P_0\,;\,P_1$ is well-formed and also has that type. Process identifiers, p, may have any type. The type of $\mu p.\,P$ may be the type of P for any assignment to the type of p for which P is well-formed and has the same type as p. The guarded processes $g \to P$, $G_0 \boxplus G_1$ and $G_0 \underleftarrow{\boxplus} G_1$ are well-formed when $g \in (\mathcal{I} \cup \mathcal{H}) \cup \{\tau\}$ and G_0 and G_1 have the same type and then their types are the types of P, G_0 and G_0 respectively.

2.2 Proper Processes

We identify a sub-class of processes which are the possible initial states of programs. In an initial state, no channel can be ready and no arbitration can have been invoked so that the constructs $c\,! \!\to P$ and $g\,? \!\to P$ are not present. Further, no concurrent processes may have terminated and so the well-formedness condition of inputs and outputs may be strengthened to $c \in \mathcal{O}$ and $g \in \mathcal{I} \cup \{\tau\}$.

3 Ready Guards and Sequential Processes

We turn first to the semantics of sequential processes for it is here that we can see the use of 'ready guards' in the description of asymmetric arbitrations; furthermore, one must first understand the mechanism by which processes synchronise in order to be able to understand how initiative for external actions may be assigned either to the environment or to a process.

To understand the meaning of asymmetric arbitrations, consider how the two-branched arbitration $P = \textbf{alt}\,(a \to P_a) \underleftarrow{\boxplus} (b \to P_b)$ may choose the branch along which it will communicate. If P is placed in an environment which wishes to communicate along b but not along a, it must choose the second branch; so, clearly, it may not behave like **alt** $a \to P_a$. On the other hand, if the environment is offering to communicate along a, it must choose the first branch.

However, at the risk of anthropomorphising, we must consider how P senses that its environment is offering a communication. It would be unrealistic to suppose that P can sense anything about its environment other than that which has been directly notified to it for its environment may reside, at least in part, in a physically distant machine. Those channels along which the process's environment is offering communication will be known as the 'ready set' and it will be the duty of output processes to notify the corresponding input by adjoining its channel to this set.

The operational semantics is given as a labelled transition system. This gives a formalism in which to describe the execution of a program in terms of labelled transitions between states. In this system, the states are process/ready channel pairs. The ready channel component is a set of channel identifiers which contains all those channels along which either the environment is ready to output or some component process is prepared to output to another component process. It is therefore a subset of $\mathcal{I} \cup \mathcal{H}$. Let $\textit{STATE}$ be the set of states.

A labelled transition system is a collection $\left\{\xrightarrow{\delta} \subseteq \textit{STATE} \times \textit{STATE} \mid \delta \in L\right\}$ of relations between states, one relation for each label in the set L. There is one set of relations for each process type and each relation is homogeneous, *i.e.* it relates only states which represent processes of the same type. The set of labels, L, is $A \cup \widehat{A}$, where $A = \{g_j \mid g \in \textit{GUARD} \wedge j \in \textit{SUB}\}$ and $\widehat{A} = \left\{\widehat{\delta} \mid \delta \in A\right\}$; the function $\widehat{\cdot}$ is injective and its range is disjoint from A. Subscripts, $\textit{SUB}$, are used to indicate that an event was performed by the whole process on behalf of a particular concurrently executing process. They are necessary for the discussion on fairness. Just which process is considered to benefit from a transition is explained when the concurrency operators are defined. Transitions labelled with an element of A denote the execution of some output-, input- or SKIP-guard, recursion unwinding or the passing of control from a process to its sequential successor. Labels of the form $\widehat{g}_j$ are, essentially, delaying transitions. They correspond to the action of an output process notifying the corresponding guard that it is to be ready.

3.1 Operations on Subscripts

When describing the semantics of the component operators, each transition will be subscripted in order to indicate the concurrent process upon behalf of which the transition was made. For simplicity, we shall take the set $\textit{SUB}$ to be strings of 0s and 1s so that the subscript gives the index of the component in the syntax tree of the whole process. Transitions which are made on behalf of the environment are indicated by subscripting with the empty string. Concatenation of two strings is denoted by juxtaposition.

3.2 Notation

A number of abbreviations are introduced for common predicates and set expressions. The predicate $\exists P'.P \xrightarrow{x} P'$ will be rendered as $P \xrightarrow{x}$. Set adjunction $X \cup \{x\}$ is written X^x and is only valid when $x \notin X$. If the type of the process P is $(\mathcal{I}, \mathcal{O}, \mathcal{H})$, the set of external channels $\mathcal{I} \cup \mathcal{O}$ is rendered as αP, and the set of channels, $\mathcal{I} \cup \mathcal{H}$, which may have been made ready either by the environment or by a component outputting process is rendered as ιP.

3.3 The Transition System

The semantics given here differs in certain respects from those to be found in [2]. The versions which may be found there do not include the explicit states $g\ ?\!\!\rightarrow P$. This makes a certain number of contortions necessary in order to be able to produce a semantics for which there is a correctness proof of the mechanism described also in [3].

When the environment indicates its willingness to output along a channel, that fact is recorded in the state by adjoining the new channel to the set of ready channels:

$$\frac{-}{(P, X) \xrightarrow{\hat{c}} (P, X^c)}\ [c \in \mathcal{I} - X] \tag{1}$$

Other than these actions, the processes **stop** and **skip** have no transitions; the process **skip** behaves differently from **stop** in sequential and parallel composition, as will be seen later.

Before communication along an output channel may take place, the channel must be made ready:

$$\frac{-}{(c \rightarrow P, X) \xrightarrow{\hat{c}} (c\,!\!\rightarrow P, X)}\ [c \in \mathcal{O}] \tag{2}$$

and, if the channel is a blocked internal channel (see rule (21) below), the channel may also be made ready but rule (4) does not allow the output itself to be discharged:

$$\frac{-}{(c \rightarrow P, X) \xrightarrow{\hat{\tau}} (c\,!\!\rightarrow P, X^c)}\ [c \in \mathcal{H}] \tag{3}$$

(Note that so long as the environment obeys the channel separation rules, it is guaranteed that $c \notin X$, for only one process may use c as an internal or output channel.)

Once ready, output may take place at any time:

$$\frac{-}{(c\,!\!\rightarrow P, X) \xrightarrow{c} (P, X)}\ [c \in \mathcal{O}] \tag{4}$$

An arbitration with one branch chooses that branch only when its guard is ready:

$$\frac{-}{(\textbf{alt}\ g \rightarrow P, X) \xrightarrow{\tau} (g\ ?\!\!\rightarrow P, X)}\ [g \in (\mathcal{I} \cap X)^{\tau}] \tag{5}$$

If two guarded processes are combined symmetrically, the whole arbitration may choose a branch which either of the components may:

$$\frac{(\textbf{alt } G_0, X) \xrightarrow{\tau} (P, X)}{(\textbf{alt } G_0 \boxplus G_1, X) \xrightarrow{\tau} (P, X)} \tag{6}$$

$$\frac{(\textbf{alt } G_1, X) \xrightarrow{\tau} (P, X)}{(\textbf{alt } G_0 \boxplus G_1, X) \xrightarrow{\tau} (P, X)} \tag{7}$$

An asymmetric arbitration chooses one of the highest priority branches available. Locally, this means that a lower priority branch may only be chosen when there are none of higher priority. However, a higher priority branch may always be chosen:

$$\frac{(\textbf{alt } G_0, X) \xrightarrow{\tau} (P, X)}{\left(\textbf{alt } G_0 \underset{\leftarrow}{\boxplus} G_1, X\right) \xrightarrow{\tau} (P, X)} \tag{8}$$

$$\frac{\neg (\textbf{alt } G_0, X) \xrightarrow{\tau} \wedge (\textbf{alt } G_1, X) \xrightarrow{\tau} (P, X)}{\left(\textbf{alt } G_0 \underset{\leftarrow}{\boxplus} G_1, X\right) \xrightarrow{\tau} (P, X)} \tag{9}$$

A chosen branch discharges its guard and behaves as the guarded process:

$$\frac{-}{(g\, ?\!\!\rightarrow P, X) \xrightarrow{g} (P, X - \{g\})} \tag{10}$$

The sequential composition $P_0; P_1$ behaves as P_0 until $P_0 = \textbf{skip}$ and then behaves as P_1:

$$\frac{(P_0, X) \xrightarrow{\delta} (P_0', X')}{(P_0\, ; P_1, X) \xrightarrow{\delta} (P_0'\, ; P_1, X')} \tag{11}$$

$$(\textbf{skip}\, ; P_1, X) \xrightarrow{\tau} (P_1, X) \tag{12}$$

A recursive process unwinds by substituting the whole process for recursive calls:

$$(\mu p.\, P, X) \xrightarrow{\tau} (P\,[\mu p.\, P/p]\,, X) \tag{13}$$

4 Initiative and Concurrent Processes

It is at this stage that we must decide by which process particular transitions are initiated. We first note that there are some transitions whose existence may be deduced from the history of a state. *I.e.*, if we consider any sequence of transitions $\gamma_0 \xrightarrow{\delta_0} \ldots \xrightarrow{\delta_{n-1}} \gamma_n$ where $\gamma_0 = (P, \emptyset)$ and P is proper, then we may deduce:

i. if $c \in \mathcal{I}$ and there are equally many δ_i equal to c as there are equal to $\widehat{c}$, then $\gamma_n \xrightarrow{\widehat{c}}$ for the ready set only contains c when the environment has an outstanding output.

ii. If $c \in \mathcal{O}$ and there is one less δ_i equal to c as there are equal to $\widehat{c}$, then $\gamma_n \xrightarrow{c}$ since the process has an outstanding output.

In fact, this is all we can deduce. On the other hand, it is never possible to deduce that $\gamma_n \xrightarrow{\widehat{c}}$ when $c \in \mathcal{I}$ and there are more δ_i equal to $\widehat{c}$ than to c; and, similarly, $\neg\gamma_n \xrightarrow{c}$ when $c \in \mathcal{I}$ and there are equally many δ_i equal to $\widehat{c}$ as to c. Thus, it is the environment which has control over events $\widehat{c}$ for $c \in \mathcal{I}$ and c for $c \in \mathcal{O}$ and we can assign the initiative to the environment for these classes of events. All other transitions, namely those with labels in $\widehat{\mathcal{O}^\tau} \cup \mathcal{I}^\tau$, are the initiative of the process.

By the rules of well-formed phrases in the language *PROC*, it is immediate that ιP_0 and ιP_1 are disjoint if P_0 **par** P_1 is well-formed. Thus, $X \subseteq \iota(P_0 \textbf{ par } P_1)$ is partitioned by the sets $X_0 = X \cap \iota P_0$ and $X_1 = X \cap \iota P_1$ and these are the parts of the ready channel set which are privately controlled by each process.

Concurrent processes have to co-operate on channel enablements and communication along channels where the channel is in the alphabet of both processes. The resulting internal action is subscripted according to the process which initiated it, either the process which is performed the input or that which enabled the channel:

$$\frac{(P_i, X_i) \xrightarrow{c_j} (P_i', X_i')\,(i = 0,1)}{(P_0 \textbf{ par } P_1, X_0 \cup X_1) \xrightarrow{\tau_{js}} (P_0' \textbf{ par } P_1', X_0' \cup X_1')}\,[c \in \alpha P_0 \cap \alpha P_1 \cap \mathcal{I}_s] \tag{14}$$

$$\frac{(P_i, X_i) \xrightarrow{\widehat{c_j}} (P_i', X_i')\,(i = 0,1)}{(P_0 \textbf{ par } P_1, X_0 \cup X_1) \xrightarrow{\widehat{\tau_{js}}} (P_0' \textbf{ par } P_1', X_0' \cup X_1')}\,[c \in \alpha P_0 \cap \alpha P_1 \cap \mathcal{O}_s] \tag{15}$$

Interactions between one process and the environment or actions internal to either of the processes may happen independently of the other process:

$$\frac{(P_0, X_0) \xrightarrow{c} (P_0', X_0')}{(P_0 \textbf{ par } P_1, X_0 \cup X_1) \xrightarrow{c} (P_0' \textbf{ par } P_1, X_0' \cup X_1)}\,[c \in \mathcal{O}_0 - \mathcal{I}_1] \tag{16}$$

$$\frac{(P_0, X_0) \xrightarrow{\delta_j} (P_0', X_0')}{(P_0 \textbf{ par } P_1, X_0 \cup X_1) \xrightarrow{\delta_{j0}} (P_0' \textbf{ par } P_1, X_0' \cup X_1)}\,\left[\delta \in \mathcal{I}_0^\tau - \mathcal{O}_1 \cup \widehat{\mathcal{O}_0^\tau} - \widehat{\mathcal{I}_1}\right] \tag{17}$$

$$\frac{(P_1, X_1) \xrightarrow{c} (P_1', X_1')}{(P_0 \textbf{ par } P_1, X_0 \cup X_1) \xrightarrow{c} (P_0 \textbf{ par } P_1', X_0 \cup X_1')}\,[c \in \mathcal{O}_1 - \mathcal{I}_0] \tag{18}$$

$$\frac{(P_1, X_1) \xrightarrow{\delta_j} (P_1', X_1')}{(P_0 \textbf{ par } P_1, X_0 \cup X_1) \xrightarrow{\delta_{j1}} (P_0 \textbf{ par } P_1', X_0 \cup X_1')}\,\left[\delta \in \mathcal{I}_1^\tau - \mathcal{O}_0 \cup \widehat{\mathcal{O}_1^\tau} - \widehat{\mathcal{I}_0}\right] \tag{19}$$

When a process terminates, it disappears from the composition. The target of the transition is coerced to have the same type as the source giving rise to the possibility that the channel of an output process, *e.g.*, may be in the internal alphabet of a process. No further communication is possible on the channels of the terminated process because they remain in the internal alphabet of the target of the transition.

Any component process of P which has a channel in common with the terminated process must be blocked on that channel, for the terminated process is the only component which could have completed the communication. Hence, any channel which a component has in common is an element of that component's internal alphabet in the target.

$$(P\ \mathbf{par\ skip}, X) \xrightarrow{\tau_1} (P, X) \tag{20}$$

$$(\mathbf{skip\ par}\ P, X) \xrightarrow{\tau_0} (P, X) \tag{21}$$

The derivation rules for asymmetric concurrent composition are largely the same as those for the symmetric version. In asymmetric concurrent compositions, the lower priority process may only initiate actions when the higher priority process cannot. There are three agents which may initiate actions, including the environment. The only pair which may pre-empt one another in any way is the high priority and the low priority process. What the required modifications must do, in effect, is to prevent any action whose subscript terminates with 1 from taking place when an action whose subscript terminates with 0 is possible.

Let us first consider co-operative actions between the two processes. The requisite modification to rules (14) and (15) involves dividing each rule into two. One of these divisions of each is got by fixing $s = 0$ and substituting $\underleftarrow{\mathbf{par}}$ for **par**. The second is got by fixing $s = 1$, substituting $\underleftarrow{\mathbf{par}}$ for **par** and adding the negative hypothesis that (P_0, X_0) cannot initiate any actions, whence the restricted rules are:

$$\frac{\forall \delta \in \widehat{\mathcal{O}_0^\tau} \cup \mathcal{I}_0^\tau . \neg (P_0, X_0) \xrightarrow{\delta} \wedge (P_i, X_i) \xrightarrow{c_j} (P_i', X_i')(i = 0, 1)}{\left(P_0\ \underleftarrow{\mathbf{par}}\ P_1, X_0 \cup X_1\right) \xrightarrow{\tau_{j1}} \left(P_0'\ \underleftarrow{\mathbf{par}}\ P_1', X_0' \cup X_1'\right)}\ [c \in \alpha P_0 \cap \alpha P_1 \cap \mathcal{I}_1] \tag{22}$$

$$\frac{\forall \delta \in \widehat{\mathcal{O}_0^\tau} \cup \mathcal{I}_0^\tau . \neg (P_0, X_0) \xrightarrow{\delta} \wedge (P_i, X_i) \xrightarrow{\widehat{c_j}} (P_i', X_i')(i = 0, 1)}{\left(P_0\ \underleftarrow{\mathbf{par}}\ P_1, X_0 \cup X_1\right) \xrightarrow{\widehat{\tau_{j1}}} \left(P_0'\ \underleftarrow{\mathbf{par}}\ P_1', X_0' \cup X_1'\right)}\ [c \in \alpha P_0 \cap \alpha P_1 \cap \mathcal{O}_1] \tag{23}$$

Any action described by rules (16) or (18) is initiated by the environment and so both stand with only the obvious modifications. Rules (17) and (21) describe actions initiated by the higher priority process and so stand with the obvious modification. Rules (19) and (20) must be modified by adding the negative hypothesis as above.

4.1 What the Deduction System Defines

It is usual practice to state that what is intended to be defined by a deduction system such as the one presented here is the least relation which satisfies the rules. In a system where transitions are deduced positively from the existence of other transitions, the existence of this minimal relation is obvious. However, the definitions of $\underleftarrow{\boxplus}$ and $\underleftarrow{\mathbf{par}}$ rely on negative information and so there is no *a priori* minimal relation for this system.

What we intend to define is the relation in which every transition is inferred from some rule. For instance, the state $(\mathbf{alt}\ a \rightarrow P, X)$ cannot be deduced to have any transition other than $\xrightarrow{\tau}$ when $a \in X$. It would therefore be counter-intuitive to endow the state with a τ-transition when $a \notin X$, say, so that $\left(\mathbf{alt}\ a \to P \underline{\boxplus}\ b \to Q, \{b\}\right)$ cannot be deduced to select the lower priority branch.

The deduction rules can be viewed as defining a functional on the transition relations so that the validity of the deduction rules is guaranteed by $\mathcal{F}\left\{\xrightarrow{\delta} \mid \delta \in L\right\} \subseteq \left\{\xrightarrow{\delta} \mid \delta \in L\right\}$. That every transition can be deduced from some rule is the assertion of containment in the opposite direction. When the hypotheses of the rules do not rely on any negative information, the functional is monotonic and it is easy to show the existence of a minimal relation for which the rules are valid. The functional of the system defined here is not monotonic so we have to work a little harder to prove the existence of a minimal fixed point.

Notice that the rules which contain a negation in the hypothesis, namely those for the priority constructs, rely only on the transitions of proper subcomponents. This allows an inductive definition. The relations which are required can be defined to be $\xrightarrow{\delta} = \bigcup_{i<\omega} \xrightarrow{\delta}_i$ where each $\xrightarrow{\delta}_i$ is defined on processes which contain at most i priority operators. Assuming that all the relations $\xrightarrow{\delta'}_j$ are defined for $j < i$ and $\delta' \in L$, the only negative hypotheses in the rules for $\xrightarrow{\delta}_i$ are already fixed so we may substitute $\neg\exists\delta \in T.\gamma \xrightarrow{\delta}_j$ with appropriate T and $j < i$ for $\neg\exists\delta \in T.\gamma \xrightarrow{\delta}$ and define the relations $\xrightarrow{\delta}_i$ to be the minimal vector $\left\langle \xrightarrow{\delta}_i \mid \delta \in L\right\rangle$ satisfying the rules.

5 Computations

We now define which computations, *i.e.* sequences of transitions are admissible for each process.

Let $\surd$ be some event which is not in the alphabet of any process. Let α abbreviate $\alpha P^{\surd}$.

We extend the definition of the transition system with the axiom:

$$(\mathbf{skip}, X) \xrightarrow{\surd} (\mathbf{stop}, X)$$

(Indeed, we could have used this definition earlier as in [4].) We use the term $(P, X) \overset{s}{\longmapsto} (P', X')$ where $s = \langle \delta_i \mid i < n\rangle$ to denote that there exists some sequence $\langle (P_i, X_i) \mid i \leq n\rangle$ with $P_0 = P$ and $X_0 = X$ and such that $(P_i, X_i) \xrightarrow{\delta_i} (P_{i+1}, X_{i+1}) (i < n)$ where $P_n = P'$ and $X_n = X'$. Also, $(P, X) \overset{u}{\longmapsto}$ where $u = \langle \delta_i \mid i < \omega\rangle$ is used to denote the 'fair' infinite computations, *i.e.* that there exists a sequence $\langle (P_i, X_i) \mid i < \omega\rangle$ with $P_0 = P$ and $X_0 = X$ such that $(P_i, X_i) \xrightarrow{\delta_i} (P_{i+1}, X_{i+1}) (i < \omega)$ along with a fairness constraint which we shall discuss further.

The transitions in an infinite computation can be divided conveniently into those which are the initiative of the process and those which are the initiative of its environment. Because the process may be placed in a context in which the process has higher or lower priority, *i.e.* there are contexts in which an infinite, unbroken sequence of process initiatives or environment initiatives are possible, it is not possible to say anything about the infinite interleavings of process/environment events. Furthermore, we may not make any stipulation about the fairness of the environment.

Thus, the only part of an infinite computation which is of interest is the subsequence of process initiatives. If $(P, X) \stackrel{u}{\longmapsto}$, then the subsequence of process initiatives is given by $w = u \upharpoonright \left(\widehat{\mathcal{O}^{\tau}} \cup \mathcal{I}^{\tau}\right)$.

If w is finite, then u must eventually consist entirely of environment initiatives. Since an environment initiative is either the discharge of an output or an indication of willingness to communicate over an input channel, we see that w can only be finite when $\mathcal{I}$ is infinite; there cannot be an infinite sequence of outputs to discharge because only a finite number of outputs can become ready in a finite time. Thus, the sequence must eventually be all elements of $\widehat{\mathcal{I}}$. This sort of behaviour would appear to be very much like divergence; by demanding that $\mathcal{I}$ be finite, there cannot be an infinite sequence of $\widehat{\mathcal{I}}$ since that would mean some channel becoming ready more than once. Correspondingly, the sets $\mathcal{O}$ and $\mathcal{H}$ must also be finite.

So, in any infinite computation there is an infinite subsequence of process initiatives. To see the sort of fairness which a round robin scheduler can guarantee, consider first only processes which do not involve $\overleftarrow{\textbf{par}}$. Once some component process becomes able to execute an action of its own initiative, the ability is not lost until the whole process performs this action on behalf of the component. We can say that process i ($\in$ SUB) is *enabled* in state γ when $\gamma \stackrel{\delta}{\longrightarrow}$ and the subscript of δ is i. Given a round robin scheduler, we can see that once a process is enabled in some state γ_n of an infinite computation, then there must eventually be some state γ_m with $m \geq n$ such that $\gamma_m \stackrel{\delta}{\longrightarrow} \gamma_{m+1}$ is an action performed on behalf of the enabled process.

In the literature, there are two common notions of fairness. In the terminology of, *e.g.*, [7] these are, respectively, justice/fairness and weak/strong fairness. They are defined as follows:

i. an infinite computation is just/weakly fair if there is no component process which is enabled in all states after the nth (for some n) but on behalf of which the whole process never performs an action.

ii. an infinite computation is fair/strongly fair if there is no component process which is enabled in infinitely many states but on behalf of which the whole process only performs a finite number of actions.

In the case of a process which does not involve $\overleftarrow{\textbf{par}}$, these two are clearly equivalent.

In the particular case of occam, the difference between these two ideas is demonstrated by the process $P = I_0 \; \overleftarrow{\textbf{par}} \; (O_0 \; \textbf{par} \; Q)$ where:

$$
\begin{array}{ll}
I_0 = \mu p.\, \textbf{alt}\; a \rightarrow p & O_0 = \mu q.\, a \rightarrow q \\
I_1 = \textbf{alt}\; a \rightarrow I_0 & O_1 = a \rightarrow O_0 \\
I_2 = a \,?\!\rightarrow I_0 & O_2 = a \,!\!\rightarrow O_0
\end{array}
$$

The whole process may pass through a sequence of states whose process parts are:

$$\begin{aligned} P_{00} &= I_0 \underleftarrow{\mathbf{par}}\, (O_0 \;\mathbf{par}\; Q) \\ P_{10} &= I_1 \underleftarrow{\mathbf{par}}\, (O_0 \;\mathbf{par}\; Q) \\ P_{11} &= I_1 \underleftarrow{\mathbf{par}}\, (O_1 \;\mathbf{par}\; Q) \\ P_{12} &= I_1 \underleftarrow{\mathbf{par}}\, (O_2 \;\mathbf{par}\; Q) \\ P_{22} &= I_2 \underleftarrow{\mathbf{par}}\, (O_2 \;\mathbf{par}\; Q) \\ P_{00} & \end{aligned}$$

In states P_{00}, P_{12}, and P_{22}, the process Q may not initiate any action because its initiative is pre-empted by that of I_0, I_1, or I_2, respectively, and so the infinite cyclic sequence which extends the above is weakly fair but not strongly fair because Q may initiate some action in states P_{10} and P_{11}.

Now, we must make a decision about which of the sorts of fairness is correct for occam. The author prefers the strong fairness requirement for the following reason. The argument comes from compositionality. If one has given Q a fair meaning, then one would wish to use that meaning in giving $P \underleftarrow{\mathbf{par}}\, Q$ a meaning. If one cannot guarantee fairness between components of Q in this construction then one cannot use the meaning already given. One cannot guarantee fairness between the components of low priority processes unless one adopts the strong fairness guarantee.

6 Comparison with Other Work

Previous research in this area has treated the topics of priority and fairness quite separately. Indeed, existing treatments cannot easily mix the concepts. There has been much more effort devoted to the latter issue, see [7]. Very few of these accounts seem to be able to provide a compositional definition for processes such as $P = (\mu p.\, a \rightarrow p)\;\mathbf{par}\;(b \rightarrow \mathbf{skip})$. The reasoning of most expositions is that any infinite computation of this process must contain the event b for otherwise the scheduler has not been fair to the right-hand process. However, if this process were put in parallel with $Q = \mu p.\, \mathbf{alt}\; a \rightarrow p$ with the type $(\{a, b\}, \emptyset, \emptyset)$, one would expect an infinite sequence of internal actions in which the process which wishes to communicate b never advances. However, if the process P is only allowed infinite sequences which contain the event b, the process P **par** Q must eventually deadlock. The semantics given here circumvents this problem because the fairness condition only applies to events in which the environment of each component process cannot refuse to participate.

As we shall see, similar problems can arise in defining the semantics of priority. Attempts in this area are reviewed more fully below. Complicating the matter further, there are also concerns about whether a semantics gives a distributable definition of **par**. One must be careful that internal actions of processes executing on different machines are not required to pre-empt one another, for an implementation of such would necessitate more inter-processor communication than is acceptable.

The work of [1] is aimed at producing an algebraic semantics for an interrupt mechanism in which some events have priority over others. The endowment of relative priorities for each event is, formally, outside the syntax of the language and is given by a pre-order defined outside the language. The imposition of priority is expressed formally by the application of an operator, θ, to a process, this operator being implicitly parameterised by the pre-order. The meaning of the new construction is given by a complete set of laws whose normal form involves only the familiar constructs of sequencing and symmetric alternation. For example, if a has higher priority than b, then the prioritised choice between them, $\theta(a+b)$, is equal to a. This semantic equation is almost forced on the treatment because there is no consideration of the environment. To adapt this semantics for occam would, at the very least, require the parameterisation of the priority operator to be made explicit for the $\underleftarrow{\boxplus}$ construct of occam gives a more dynamic description of priority than the θ of [1]. The example equivalence would require the description of a simple construct like $(b \rightarrow Q)\, \mathbf{par} \left(\mathbf{alt}\,(a \rightarrow P_a) \underleftarrow{\boxplus} (b \rightarrow P_b)\right)$ with $a \in \mathcal{O}_Q$, to be quite subtle in order to avoid equating the process to deadlock.

The aim of [6] is similar. That work gives an operational semantics to an interrupt mechanism which has two priority levels for events. Events are given their priority by operators which enforce high or low priority for a particular event within the scope of the operator. High priority events are denoted by symbols $\underline{a}$ and complementary events synchronise to give a prioritised internal action, $\underline{\tau}$. In fact, two semantics are given. In the first, any prioritised action has the power to pre-empt any non-prioritised action. This has the unfortunate property that, *e.g.*, $a.p + \underline{b}.q$ is strongly bisimilar to $\underline{b}.q$ and leads to the equivalence losing the property of congruence, for putting $\underline{b}.q$ in an environment which will not accept $\underline{b}$ leads to deadlock whereas $a.p + \underline{b}.q$ can still make progress. The second semantics overcomes this failing by only allowing the prioritised internal action $\underline{\tau}$ to pre-empt non-prioritised actions. A bisimulation equivalence, a complete set of laws for finite (*i.e.* non-recursive) terms and proof rules for guarded recursion are then given in the style of most CCS semantics. The calculus models interrupt mechanisms in a theoretically more tractable and succinct way, perhaps, than occam but it is not clear that there is an efficient, distributed implementation. The fact that actions from quite separate component processes must be sequentialised in a fixed order implies the need for some central control. Furthermore, the calculus does not allow the implementation of a 'fair' arbitration as the asymmetric arbitration of occam does (see [9]) and hence is less expressive.

In [5], an operational semantics for every construct in occam except the priority concurrent composition is presented. A meaning is given to $\underleftarrow{\boxplus}$ which is more restrictive than the meaning given in this paper. The method employed relies on explicit information about the environment, namely a set of channels along which the environment refuses to communicate. The asymmetry of the definition is got by allowing only the highest priority guards of an arbitration which the environment cannot refuse to be selected. The approach is very similar to that employed here. However, the definition of refusal sets is flawed. As we have seen, the important concept in procuring the asymmetry is that of the channels along which the environment has indicated its willingness to communicate. This is subtly different from a refusal set for the output process $c \rightarrow P$ cannot refuse to output along c but it may not yet have indicated its readiness to communicate. The consequence is that the process $(a \to Q_a)$ **par** $(b \to Q_b)$ **par** $\left(\mathbf{alt}\,(a \to P_a) \underleftarrow{\boxplus} (b \to P_b)\right)$ is modelled as **skip** ; $(Q_a$ **par** $(b \to Q_b)$ **par** $P_a)$ which is an unreasonably strong specification of the behaviour of the asymmetric arbitration because it cannot select the alternative which it is going to execute before its environment has stabilised enough to have indicated its willingness to communicate along a and b. Furthermore since, by Camilleri's definition, the process **skip** ; $(a \to Q_a)$ refuses to communicate along a, replacing $a \to Q_a$ by the above process gives a very different result: $\mathbf{alt}\,(\tau \to (Q_b\ \mathbf{par}\ P_b\ \mathbf{par}\ a \to Q_a)) \boxplus (\tau \to (Q_a\ \mathbf{par}\ P_a\ \mathbf{par}\ b \to Q_b))$; so that **skip** is no longer an identity for sequential composition. It is also unclear whether this system could be implemented without central control.

7 Conclusions

This paper has presented an operational semantics for occam. The introduction of the concepts of 'ready guards' and 'initiative' was essential for the description of the various priority constructs available in occam. While maintaining a reasonable degree of abstraction, these ideas allow a description of priority and fairness which is closely related to the way in which programs would be implemented. In turn, this allows the semantics to get a firm, intuitive grasp of the higher level notions.

The danger of giving an operational semantics for a language with concurrency and priority is that, although the semantics appears to allow a reasonable uni-processor implementation, the concurrent operators may have no distributed implementation. In any reasonable distributed environment, total state changes must be initiated by one particular element. One must ensure that operators which are intended to have a truly concurrent implementation do not favour the initiative of one element over that of any other. It is intuitively obvious that the $\underleftarrow{\mathbf{par}}$ construction of our language cannot be distributed but that the **par** construct could, with some effort to ensure that any pair of signals sent from one processor to another cannot overtake one another. The assignment of initiative for each action was only possible because of the asymmetry between input and output in occam. Describing a symmetric language in this framework could involve describing the intended synchronisation mechanism, albeit schematically, as was done here.

Perhaps the most attractive feature of the concept of initiatives is that it allows a compositional approach to fairness and priority, a feature which is lacking from other accounts. By dividing the actions of a process into two sets, one of which belongs to the process and the other to the environment, one can give a definition in which priority and fairness requirements apply only to the set of actions belonging to the process, thus ensuring that there are no conflicts when the process is considered as part of a larger context.

References

[1] Baeten, J.C.M., Bergstra, J.A. & Klop, J.W., *Syntax and Defining Equations for an Interrupt Mechanism in Process Algebra.* Technical Report CS-R8503, Department of Computer Science, Center for Mathematics and Computer Science, Amsterdam, February 1985

[2] Barrett, G., *The Semantics and Implementation of* occam. Thesis DPhil, Oxford, September 1988

[3] Barrett, G., Goldsmith, M.H., Jones, G. & Kay, A., The Meaning and Implementation of PRI ALT in occam. in occam *and the Transputer – Research and Applications*, ed. C Askew, IOS, Amsterdam, 1988

[4] Brookes, S.D., Roscoe, A.W. & Walker, D.J., *An Operational Semantics for CSP*, forthcoming

[5] Camilleri, J., *An Operational Semantics for* occam. University of Cambridge, Computer Laboratory Technical Report 144, August 1988

[6] Cleaveland, R. & Hennessy, M., *Priorities in Process Algebras.* Report No. 2/88, Computer Science, University of Sussex, March 1988

[7] Francez, N., *Fairness.* Springer Verlag, New York, 1986

[8] Roscoe, A.W., A Denotational Semantics for occam. in *Seminar on Concurrency*, ed. S.D. Brookes, A.W. Roscoe, & G. Winskel, Lecture Notes in Computer Science, vol. 197, pp. 306-329, Springer Verlag, Berlin, 1985

[9] Roscoe, A.W., Routing messages through networks: an exercise in deadlock avoidance. in *Proc. 7th oUG Tech. Meeting*, ed. T. Muntean, Grenoble, September 1987

Inductively Defined Types in the Calculus of Constructions

Frank Pfenning

School of Computer Science
Carnegie Mellon University
5000 Forbes Avenue
Pittsburgh, Pennsylvania 15213

Internet: fp@cs.cmu.edu

Christine Paulin-Mohring

INRIA and LIENS, URA CNRS 1327
Ecole Normale Superiéure
45 Rue d'Ulm
75005 Paris, France

Internet: mohring@ens.ens.fr

Abstract

We define the notion of an *inductively defined type* in the Calculus of Constructions and show how inductively defined types can be represented by closed types. We show that all primitive recursive functionals over these inductively defined types are also representable. This generalizes work by Böhm & Berarducci on synthesis of functions on term algebras in the second-order polymorphic λ-calculus (F_2). We give several applications of this generalization, including a representation of F_2-programs in F_3, along with a definition of functions `reify`, `reflect`, and `eval` for F_2 in F_3. We also show how to define induction over inductively defined types and sketch some results that show that the extension of the Calculus of Construction by induction principles does not alter the set of functions in its computational fragment, F_ω. This is because a proof by induction can be *realized* by primitive recursion, which is already definable in F_ω.

1 Introduction

The motivation for the this paper comes from two sources: work on the extraction of programs from proofs in the Calculus of Constructions (CoC) [23, 24] and work on the implementation of LEAP [25], an explicitly polymorphic ML-like programming language (here we only consider the pure F_ω fragment of LEAP). The former emphasizes the *logical* aspects of CoC, the latter its *computational* aspects. The basic relationship is simple: an extraction process relates proofs in CoC to programs in F_ω. In other words, in F_ω we can express the computational contents of proofs in CoC. Said yet another way: programs in F_ω realize propositions in CoC.[1]

[1]For the purposes of this paper, we are ignoring the distinction between *Data*, *Prop*, and *Spec* made in [23, 24]. For practical purposes, this distinction is extremely important. Here it is more

Both on the logical and computational level, inductively defined propositions or types play a central rôle in any applications. Their logical aspect, that is, proving properties by induction, and their computational aspect, that is, defining functions by primitive recursion, are very closely related: the computational content of a proof by induction is a function definition by primitive recursion. Said another way: primitive recursion realizes induction. One of our results is that, even though induction principles are not provable in CoC, their computational content is already definable in F_ω. Thus augmenting CoC by induction principles over inductively defined types is in some sense "conservative" over its computational fragment: even though we can prove more specifications, any function which we might be able to extract from such proofs is already definable in pure F_ω—we just would not be able to show in CoC without induction that it satisfies its specification.

Closely related is work by Girard [13, 14], Fortune, Leivant & O'Donnell [12], and Leivant [17, 18] who are concerned with the relationship between higher-order logic and polymorphic λ-calculi.

Mendler [19, 20] studied inductive types in the setting of the second-order polymorphic λ-calculus and the NuPrl type theory. He adds to the system F a new scheme for defining recursive types. The system is extended with new constants for representing the type, its constructor and the primitive recursion operator. The rules of conversion of the system are also extended for each new recursive type. In our presentation the inductive types are internally represented using higher-order quantification and the only reduction rule used is β-reduction. An advantage of our approach is that types that in some sense "are already there" are not also added artificially. On the other hand, a significant drawback of our approach is the relative weakness of our notion of equality induced by this representation, even if one adds η-conversion. For example, let R be the closed term for primitive recursion over the natural numbers, defined using iteration and pairing as in Section 5. Then the equality between $R\,\beta\,h'_z\,h'_s\,(\mathtt{succ}\,n)$ and $h'_s\,(\mathtt{pair}\,n\,(R\,\beta\,h'_z\,h'_s\,n))$ is not an internal equality (as it is in Mendler's system) but is only provable using induction on n. The types given for primitive recursion in Mendler's work and in this paper are slightly different but equivalent. Work along Mendler's lines for the Calculus of Constructions is presented by Coquand and Paulin-Mohring [9] and for Martin-Löf's type theory by Dybjer [11].

On the purely computational level, we generalize Böhm & Berarducci's [4] construction of functions on term algebras in the second-order polymorphic λ-calculus (F_2) to F_ω. One of their results does not generalize in unmodified form beyond algebraic types: not every closed term of the representation type will be $\beta\eta$-convertible to the representation of a term in the inductive type. This does not appear to be computationally relevant. One can consider alternative definitions of inductive types outside F_ω (but still inside CoC) which have the same computational content as our definitions. Another alternative would be to strengthen the notion of equality. We conjecture that one can use Reynolds' condition of *parametricity* [26] to recover uniqueness of representations at least in the F_ω fragment.

---

convenient to simply use $*$ to encompass all of them. We thus use the terms "proposition" and "specification" interchangeably.

A facility to generate the definition of inductively defined types, the constructors, and the primitive recursion operator from specifications like the ones in Examples 3 to 9 has been added to the implementation of the Calculus of Constructions V4.10 developed at INRIA. Work on the efficient implementation of inductively defined types and primitive recursion over such types in F_ω is currently under way in the framework of the Ergo project at Carnegie Mellon University.

2 The Calculus of Constructions

The Calculus of Constructions (CoC) of Coquand & Huet (see [7, 6, 16, 8]) is a very powerful type theory, yet it can be formulated very concisely. It encompasses Girard's system F_ω (see [13, 14]) and the type theory of LF, the Edinburgh Logical Framework (see Harper, Honsell & Plotkin [15]) and may be considered the result of combining these two type theories (see Barendregt [2]). The formulation we present here is a very brief summary of the concrete syntax, notation, and inference system given in [8].

We use $M, N, \ldots$ for terms in general and x, y, z for variables (abstractly, though, they are de Bruijn indices [10], where the occurrences of x in $(\lambda x{:}M)\,N$ and $[x{:}M]\,N$ are binding occurrences). We have

$$M ::= x \mid (\lambda x{:}M)\,N \mid (M\,N) \mid [x{:}M]\,N \mid *$$

Following [8] we call $[x{:}M]\,N$ a *product.* $*$ is the universe of all types, but is itself not a type. *Contexts* (denoted by Γ, Δ) are products over $*$ and thus have the form $[x_1{:}M_1]\ldots[x_n{:}M_n]\,*$, all other terms will be referred to as *objects.* Contexts serve as types, but do not have types themselves. When it is clear that a term is a context, we sometimes omit the trailing $*$.

The inference system defines two judgments: $\Gamma \vdash \Delta$ means that Δ is a valid context in the valid context Γ, and $\Gamma \vdash M : P$ means that M is a well-typed term of type P in the valid context Γ. We use $P, Q, \ldots$ for *types*, that is, terms which can appear in the place of P in the judgments below. The inference system below entails that a type P will either be a context, or have the property that $\Gamma \vdash P : *$. $[N/x]Q$ is the notation for substituting N for x in Q (abstractly defined using the de Bruijn notation, and therefore avoiding the issues of name clashes).

Valid Contexts.

$$\vdash * \qquad \frac{\Gamma \vdash \Delta}{\Gamma[x{:}\Delta] \vdash *} \qquad \frac{\Gamma \vdash P : *}{\Gamma[x{:}P] \vdash *}$$

Product Formation.

$$\frac{\Gamma[x{:}P] \vdash \Delta}{\Gamma \vdash [x{:}P]\Delta} \qquad \frac{\Gamma[x{:}P] \vdash N : *}{\Gamma \vdash [x{:}P]N : *}$$

Variables, Abstraction, and Application.

$$\frac{\Gamma \vdash *}{\Gamma \vdash x : P}\,[x{:}P] \text{ in } \Gamma \qquad \frac{\Gamma[x{:}P] \vdash N : Q}{\Gamma \vdash (\lambda x{:}P)\,N : [x{:}P]\,Q} \qquad \frac{\Gamma \vdash M : [x{:}P]\,Q \qquad \Gamma \vdash N : P}{\Gamma \vdash (M\,N) : [N/x]Q}$$

We will consider β-conversion ($\cong$) in the "full" form (see [8, Page 102]) and have the following rule of type conversion:

$$\frac{\Gamma \vdash M : P \qquad \Gamma \vdash P \cong Q}{\Gamma \vdash M : Q}$$

η-conversion does not play a very important role, but we will have occasion to use it when considering the representation of inductively defined types.

The calculus shares the basic properties of the LF type theory and F_ω, such as strong normalization, decidability of type-checking, and the Church-Rosser property for well-typed terms. We will make use of the properties in the development below. We formulate the basic induction principle over normal forms of types in CoC separately as a lemma, since we will need it frequently. Its proof is immediate from the Lemmas in [8].

Lemma 1 (Normal forms of types) *Given a type R, that is, a term R such that for some Γ and N we have $\Gamma \vdash N : R$. Then the β-normal form of R has the shape $N_0\, N_1 \ldots N_p$, $*$, or $[x{:}R_0]\, R_1$. In particular, the β-normal form of R cannot be an abstraction.*

We say that a type R is *atomic* if it is in normal form and does not begin with a product, that is, is not of the form $[x{:}P]\, Q$.

We will use $P \rightarrow Q$ as an abbreviation for any $[x{:}P]\, Q$, if x does not occur free in Q. We will sometimes omit the parentheses surrounding applications in which case application is written simply as juxtaposition and associates to the left. Juxtaposition binds tighter than "$\rightarrow$", which associates to the right. Abstraction and product also associate to the right and bind less tightly than "$\rightarrow$". The equality in the metalanguage is "$=$". Definitional equality is written as "$\equiv$" and may be thought of as introducing an abbreviation at the level of the Calculus of Construction as available in its implementation at INRIA. We will use this notion of notational definition in examples without formalizing it.

3 Inductively Defined Types

Intuitively, an inductively defined type is given by a complete list of constructors for terms of the type. We reason about the type with an appropriate induction principle, and we write functions over the type using iteration, which is powerful enough to define primitive recursive functionals over elements of the type. This notion encompasses the usual notions of free term algebras with associated induction principles, but it is more general and allows the definition of types such as natural numbers, pairs, lists, ordinal notations, logical quantifiers and connectives, or programs in F_2, a significant fragment of CoC of independent interest.

Below is our concrete syntax for the definition of an inductive type. We refer to α as the *inductively defined type*, and $c_1, \ldots, c_n$ as the *constructors for α*.

$$\begin{array}{l}\textbf{indtype}\ \alpha : [z_1{:}Q_1]\ldots[z_m{:}Q_m] * \textbf{ with}\\ \quad c_1 : [x_1{:}P_{11}]\ldots[x_{k_1}{:}P_{1k_1}]\,\alpha M_{11}\ldots M_{1m}\\ \quad \vdots\\ \quad c_n : [x_1{:}P_{n1}]\ldots[x_{k_n}{:}P_{nk_n}]\,\alpha M_{n1}\ldots M_{nm}\\ \quad \textbf{end}\end{array}$$

In such an inductive definition, α may not occur in Q_j, nor in any M_{ij}. However, α may occur in P_{il}, but only *positively* (see Definition 2). Throughout the paper, we will use the names α, c_i, Q_j, P_{il}, M_{ij} when we need to refer to the components of a given inductive type definition. Annotating a P_{il}^{α} serves only as a reminder that α may be free in P_{il}, and P_{il}^{β} is the result of substituting β for α in P_{il}. We will also use throughout this paper:

$$\begin{array}{l}Q = [z_1{:}Q_1]\ldots[z_m{:}Q_m] *\\ P_i^{\alpha} = [x_1{:}P_{i1}^{\alpha}]\ldots[x_{k_i}{:}P_{ik_i}^{\alpha}]\,\alpha M_{i1}\ldots M_{im} \quad \text{for} \quad 1 \leq i \leq n\end{array}$$

Besides positivity, we make an additional assumption that greatly simplifies the presentation and holds in all examples we are aware of, but is not essential. We require that for any quantifier $[y{:}R_0^{\alpha}]\,R_1^{\alpha}$ appearing in the definition of α, either y does not occur in R_1^{α} or α does not occur in R_0^{α}. For a development without this restriction see Paulin-Mohring [24]. The additional complexity arises primarily in the definition of Φ below (Definition 11)—all theorems remain valid when appropriately modified.

We define by simultaneous induction when a variable occurs only positively and only negatively in a type R, where R is in β-normal form. Since R is a type and assumed to be in normal form the (omitted) case $R = (\lambda z{:}R_0)\,R_1$ cannot arise (see Lemma 1).

Definition 2 (Positive and negative occurrences of variables) *We define by simultaneous induction: a variable x occurs only positively in the β-normal type R if*

Case $R = x\,N_1 \ldots N_m$ *and x does not occur in $N_1, \ldots N_m$,*

Case *R is atomic and x does not occur in R,*

Case $R = [z{:}R_0]\,R_1$ *and x occurs only negatively in R_0 and only positively in R_1.*

and a variable x occurs only negatively in the β-normal type R if

Case *R is atomic and x does not occur in R,*

Case $R = [z{:}R_0]\,R_1$ *and x occurs only positively in R_0 and only negatively in R_1.*

We begin with some examples for inductively defined types. The first one is algebraic (as in [4]).

Example 3 (Natural Numbers) *This is the canonical example for an inductively defined type.*

indtype nat : $*$ **with**
  zero : nat
  succ : nat $\rightarrow$ nat
  end

Pairs and lists, the next two examples, are parameterized types which are hereditarily algebraic: once instantiated with algebraic types, the result will be algebraic. The representation of the parameterized type itself, however, is beyond the framework of [4].

Example 4 (Pairs) *Pairs are definable in this calculus. They will be used in Section 5 in order to define primitive recursion from iteration.*

indtype prod : $* \rightarrow * \rightarrow *$ **with**
  pair : $[A{:}*]\,[B{:}*]\,A \rightarrow B \rightarrow \mathsf{prod}\,A\,B$
  end

We will have occasion to use a generalized notion of pair in the metalanguage that applies to parameterized types. Given R and S of type $[z_1{:}Q_1]\ldots[z_m{:}Q_m]*$. *We define* $R \times S = [z_1{:}Q_1]\ldots[z_m{:}Q_m]\,\mathsf{prod}\,(R\,z_1\ldots z_m)\,(S\,z_1\ldots z_m)$.

Example 5 (Lists) *This is a simple example for a parameterized type that involves a non-trivial induction. As we will see later in Example 21 the representation of this parameterized type in our framework is somewhat different from the representation, for example, given by Reynolds [27].*

indtype list : $* \rightarrow *$ **with**
  nil : $[A{:}*]\,\mathsf{list}\,A$
  cons : $[A{:}*]\,A \rightarrow \mathsf{list}\,A \rightarrow \mathsf{list}\,A$
  end

Ordinal notations, the next example, are not algebraic for a different reason: the argument to one of the constructors ranges over sequences (which are naturally represented as functions).

Example 6 (Ordinal Notations) *This example is due to Coquand [6] and generalized by Huet [16, Section 10.3.5]. The limit constructor* olim *is applied to a sequence of ordinals which is represented as a function from natural numbers to ordinals.*

indtype ord : $*$ **with**
  ozero : ord
  osucc : ord $\rightarrow$ ord
  olim : $[A{:}*](A \rightarrow \mathsf{ord}) \rightarrow \mathsf{ord}$
  end

The next example is a representation of programs in the polymorphic λ-calculus (F_2). This type is clearly not hereditarily algebraic.

Example 7 (Programs in F_2) *This inductive type is noteworthy for several reasons. Its representation will lie in F_3, the third-order polymorphic λ-calculus. Moreover, one can program an evaluation function for F_2 in F_3 over this representation. For a more detailed account, see [25].*

$$
\begin{array}{l}
\textbf{indtype } \mathsf{prog} : * \to * \textbf{ with} \\
\quad \texttt{rep} : [A{:}*]\, A \to \mathsf{prog}\, A \\
\quad \texttt{lam} : [A{:}*]\,[B{:}*]\,(A \to \mathsf{prog}\, B) \to \mathsf{prog}\,(A \to B) \\
\quad \texttt{app} : [A{:}*]\,[B{:}*]\,\mathsf{prog}\,(A \to B) \to \mathsf{prog}\, A \to \mathsf{prog}\, B \\
\quad \texttt{typlam} : [A{:}* \to *]\,([B{:}*]\,\mathsf{prog}\,(A\,B)) \to \mathsf{prog}\,([B{:}*]\,(A\,B)) \\
\quad \texttt{typapp} : [A{:}* \to *]\,\mathsf{prog}\,([B{:}*](A\,B)) \to [B{:}*]\,\mathsf{prog}\,(A\,B) \\
\quad \textbf{end}
\end{array}
$$

All the examples so far lie within the F_ω fragment of CoC. The following examples deal with aspects of dependent types in CoC which can be used to define logical notions.

Example 8 (Leibniz' Equality) *Leibniz' equality and other logical connectives can be defined as inductive types. We express here that equality is the least relation which relates every element to itself.*

$$
\begin{array}{l}
\textbf{indtype } \mathsf{eq} : [A{:}*]\, A \to A \to * \textbf{ with} \\
\quad \texttt{refl} : [A{:}*]\,[x{:}A]\,\mathsf{eq}\, A\, x\, x \\
\quad \textbf{end}
\end{array}
$$

Example 9 (Existential Quantification) *We express the usual inference rule for existential quantification and (since the type is inductive) that this is the only way we can establish an existentially quantified proposition.*

$$
\begin{array}{l}
\textbf{indtype } \mathsf{exists} : [A{:}*]\,(A \to *) \to * \textbf{ with} \\
\quad \texttt{exists-intro} : [A{:}*]\,[P{:}A \to *]\,[x{:}A]\, P\, x \to \mathsf{exists}\, A\, P \\
\quad \textbf{end}
\end{array}
$$

Similar to the way we generalized prod *to* $\times$ *we can generalize dependent pairs. This will be used in the definition of induction in Section 6. Given* $R : [z_1{:}Q_1] \ldots [z_m{:}Q_m]\, *$ *and* $P : [z_1{:}Q_1] \ldots [z_m{:}Q_m]\, R\, z_1 \ldots z_m \to *$. *We define the type*

$$R \otimes P = [z_1{:}Q_1] \ldots [z_m{:}Q_m]\, \mathsf{exists}\,(R\, z_1 \ldots z_m)\,(P\, z_1 \ldots z_m)$$

Counterexample 10 (LF encoding of logical systems) *LF, the Logical Framework, is a very weak subsystem of* CoC *in which one can encode inference systems as signatures. Judgments of the inference system become types or type families, logical connectives and quantifiers and inference rules become typed constants. See Harper, Honsell & Plotkin [15] for a description of LF and Avron, Honsell & Mason [1] for LF representations of a variety of logics. These signatures resemble inductive type definitions, but upon closer inspection the analogy fails. Consider the following two problematic declarations which would be part of an inductive type definition derived from an encoding of first-order arithmetic.*

indtype $\vdash : \mathsf{o} \rightarrow *$ **with**

$\ldots$

$\supset\mathrm{I} : [A{:}\mathsf{o}]\,[B{:}\mathsf{o}]\,(\vdash A \rightarrow \vdash B) \rightarrow \vdash A \supset B$

$\forall\mathrm{I} : [A{:}\mathsf{nat} \rightarrow \mathsf{o}]\,([x{:}\mathsf{nat}]\vdash A\,x) \rightarrow \vdash \forall A$

end

In the case of $\supset$I, *the first occurrence of* $\vdash A$ *is* negative, *and therefore falls outside of our framework of inductive definitions. This is a simple example of a type that is non-empty, even though it may not have a "base case" when one tries to consider it as an inductively defined type, ignoring the negative occurrence of* $\vdash$. *In the case of* $\forall$I, *the rule may become too powerful and actually formalize a version of the* ω*-rule (and not universal introduction) when we make induction over natural numbers available at the level of LF. This failure of induction is not a defect of LF, since induction is done once and for all when the LF type theory itself is defined inductively. However, it does make it considerably more difficult to extend LF while preserving adequacy of representations of logical systems in LF.*

4 Representing Inductively Defined Types

There are two aspects of inductively defined types that we are interested in. The first one might be called the *computational aspect*, the second the *logical aspect.*

When investigating the computational aspect of an inductive type, we consider F_ω only and assume that we have a new (possibly parameterized) type constant α and new term constructors c_i. Functions over α may be defined using primitive recursion at higher type (see Definition 31). We ask if there is already a type in pure F_ω itself that can be used to represent terms built from the constructors such that the functions that are definable by primitive recursion are also definable. The answer here is "yes", though there will be a delicate point about the exact formulation of the theorem to that effect.

The logical aspect is based on the simple premise that one would like to reason inductively about inductive types. Since the various induction principles themselves are not provable in CoC, they have to be added as primitive constants. What are the properties of such an extension? We do not have a complete answer here, but at least we ascertain one pleasant property: when considering the computational content of proofs of specifications under this extension, it is *conservative*: we have new theorems (and proofs), but no new functions in F_ω.

We begin by giving a method for representing inductively defined types. An important property we would like to preserve is that an inductive type in F_ω will also be represented in F_ω. This fact is used vitally in the implementation of LEAP [25].

Now assume we are given an inductively defined type α in the notation at the beginning of Section 3. In this section we show that there is actually a closed type $\underline{\alpha}$ in CoC such that any well-typed term that can be built with the constructors of α and terms in CoC has a representation of type $\underline{\alpha}$. The converse, namely that every closed term M of type $\underline{\alpha}$ can be expressed in terms of the constructors of α is not true if one takes $\beta\eta$-conversion as the notion of term equality. We conjecture that the converse is true in models that satisfy Reynolds' condition of *parametricity* [26]. This conjecture is based on the intuition that completeness fails because $\beta\eta$-equality is too weak to identify indistinguishable terms, under some reasonable assumptions about when terms should be indistinguishable (see Mitchell and Meyer [21]). Computationally this failure of completeness is not a problem, and the logical characterization of an inductive type in terms of an induction axiom is satisfactory from the logical point of view (though, of course, also incomplete in another sense).

Of course, there may be many ways an inductively defined type could be represented in CoC. We give here a canonical construction in which the representation of an element of the inductive type is its own iteration function. This representation has some drawbacks which we will return to in Section 5, where we show how to define primitive recursion at all types over an inductively defined type.

Before launching into the description of the representation of inductive types, we need an important technical tool. In its simplest form, we define a map Φ on terms that lifts a function $F : P \to Q$ to a function $\Phi_R : R\,P \to R\,Q$ where $R : * \to *$ and R is positive in its argument (that is, $R = (\lambda x{:}*)\,R'$ and x is only positive in R').

Definition 11 (Maps Φ and Ψ) *Given S and T of type $[z_1{:}Q_1]\dots[z_m{:}Q_m]*$ and a function $F : [z_1{:}Q_1]\dots[z_m{:}Q_m]\,S\,z_1\dots z_m \to T\,z_1\dots z_m$. Furthermore, we are given a type $R = R^x$ with some free occurrences of $x{:}[z_1{:}Q_1]\dots[z_m{:}Q_m]*$. We define Φ_R for R^x with only positive occurrences of x such that for any term $N : R^S$, $\Phi_R(N) : R^T$, and simultaneously we define Ψ_R for R^x with only negative occurrences of x such that for any term $N : R^T$, $\Psi_R(N) : R^S$.*

Case *$R^x = x\,N_1\dots N_m$. Then let $\Phi_R(N) = F\,N_1\dots N_m\,N : R^T$, since x does not occur in $N_1,\dots,N_m$ by positivity.*

Case *R^x is atomic and x does not occur in R^x. Then $R^S = R^T$ and we let $\Phi_R(N) = N$.*

Case *$R^x = [z{:}R_0^x]\,R_1^x$. Then $\Phi_R(N) = (\lambda z{:}R_0^T)\,\Phi_{R_1}(N\,\Psi_{R_0}(z))$. Note that x will occur only negatively in R_0^x since it occurs only positively in R^x.*

Remember that the case $R^x = (\lambda z{:}R_0^x)\,R_1^x$ cannot arise, since R^x is a type *in normal form (see Lemma 1). Now for R^x with x only occurring only negatively, we define:*

Case *$R^x = x\,N_1\dots N_m$. This case cannot arise, since x is positive in R^x, but we assumed that x occurs only negatively in R^x.*

Case *R^x is atomic and x does not occur in R^x. Then $R^S = R^T$ and we let $\Psi_R(N) = N$.*

Case *$R^x = [z{:}R_0^x]\,R_1^x$. Then $\Psi_R(N) = (\lambda z{:}R_0^S)\,\Psi_{R_1}(N\,\Phi_{R_0}(z))$.*

The construction of Φ depends on F and its type. If we want to make the dependency explicit, we write Φ^F for the map Φ that is constructed from F.

The term constructed according to this definition will not always be correctly typed. We need an additional restriction that is satisfied in all of our examples and in particular is always satisfied for inductive type in the F_ω fragment of CoC.

Lemma 12 *In the context of Definition 11 and under the assumption that for any quantifier* $[z{:}R_0^x]\,R_1^x$ *in* R^x*, either* z *does not occur in* R_1^x *or* x *does not occur in* R_0^x*,* Φ *and* Ψ *are well-defined and* Φ *satisfies*

$$\Phi_R(N) : R^T \text{ for any } N : R^S$$

The proof is by a simple induction on the structure of R^x. The definition of Φ and Ψ with the same property can be made in full generality, but is quite complex. Details can be found in Paulin-Mohring [24, page 107].

Now we are prepared to state and prove the representation of inductive types.

Definition 13 (Representation $\underline{\alpha}$ of an inductively defined type α) *Given* α*, defined inductively as in Section 3. We will use the notation* P_{il}^{α} *for* P_{il} *and* P_{il}^{β} *for the result of substituting* β *for* α *in* P_{il} *and* P_i^{β} *for the result of substituting* β *for* α *in* P_i*. We let*

$$\underline{\alpha} = (\lambda z_1{:}Q_1)\ldots(\lambda z_m{:}Q_m)\,[\beta{:}Q]\,P_1^{\beta} \to \cdots \to P_n^{\beta} \to \beta\, z_1 \ldots z_m$$

It is easy to see that $\underline{\alpha} : Q$. The definition of the representations of the constructors c_i will make use of the function $()^+$ defined below with the property that if $N : R^{\underline{\alpha}}$ then $N^+ : R^{\beta}$.

Definition 14 (Representation $\underline{c}_i$ of constructor c_i)

$$\underline{c}_i = (\lambda x_1{:}P_{i1}^{\underline{\alpha}})\ldots(\lambda x_{k_i}{:}P_{ik_i}^{\underline{\alpha}})\,(\lambda\beta{:}Q)\,(\lambda y_1{:}P_1^{\beta})\ldots(\lambda y_n{:}P_n^{\beta})\,y_i\,x_1^+\ldots x_{k_i}^+$$

Given the property of $()^+$ stated above, it is easy to verify that $\underline{c}_i : P_i^{\underline{\alpha}}$. We now define the map $()^+$ using Φ and its properties.

Definition 15 (Map $()^+$) *Given a context* $[\beta{:}Q]\,[y_1{:}P_1^{\beta}]\ldots[y_n{:}P_n^{\beta}]$ *where all occurrences of* β *in the* P_i *are positive. In order to be able to apply* Φ *such that it coerces* $N : R^{\underline{\alpha}}$ *to* $N^+ : R^{\beta}$ *, we have to define a function* $F : [z_1{:}Q_1]\ldots[z_m{:}Q_m]\,\underline{\alpha}\,z_1\ldots z_m \to \beta\,z_1\ldots z_m$*. But* $\underline{\alpha}\,z_1\ldots z_m = [\beta{:}Q]\,P_1^{\beta} \to \cdots \to P_n^{\beta} \to \beta\,z_1\ldots z_m$ *and so we let*

$$F = (\lambda z_1{:}Q_1)\ldots(\lambda z_m{:}Q_m)\,(\lambda g{:}[\beta{:}Q]\,P_1^{\beta} \to \cdots \to P_n^{\beta} \to \beta\,z_1\ldots z_m)\,g\,\beta\,y_1\ldots y_n$$

and define N^+ *as* $\Phi_{R^{\alpha}}^{F}(N)$*.*

Definition 16 (Γ_α) *Given a type* α *defined inductively as above. Then*

$$\Gamma_\alpha = [\alpha{:}[z_1{:}Q_1]\ldots[z_m{:}Q_m]*]\,[c_1{:}P_1^{\alpha}]\ldots[c_n{:}P_n^{\alpha}]\,*$$

We also extend $\underline{()}$ *homomorphically from* α *and constructors* c_i *to any term* N *that is well-formed in a context* Δ, Γ_α*. We sometimes refer to a term in the context* Γ_α *as a* constructor term.

For the adequacy theorem it is convenient to consider η-conversion in addition to β-conversion.

Theorem 17 (Adequacy) *For any inductively defined type α and closed terms $N_1, \ldots, N_m$ such that $\Gamma_\alpha \vdash \alpha\, N_1 \ldots N_m : *$, $\underline{()}$ is a bijection between $\beta\eta$-equivalence classes of terms N such that $\Gamma_\alpha \vdash N : \alpha\, N_1 \ldots N_m$ and equivalence classes of terms M such that $\vdash M : \underline{\alpha}\, N_1 \ldots N_m$.*

Proof sketch: It is easy to verify by calculation as in [4] using Lemma 12 that $\underline{()}$ has the injection properties. The inverse map $\mathcal{F}(M) = M\, \alpha\, c_1 \ldots c_n$ applies the representation M of a term in an inductive type to the constructors of that type to yield the term that it represents. □

It is important to note that the inverse map $\mathcal{F}$ does not need to examine the structure of its argument M to determine what constructor term M represents. This means that even in an implementation where the intensional structure of functions is inaccessible (for example, when functions are compiled into machine code) we can still extract the constructor term that is represented by a function by applying it to the constructor constants.

The adequacy theorem is somewhat weaker than Böhm and Berarducci's representation theorem. This is because the mappings $\underline{()}$ and $\mathcal{F}$ do not go between $\beta\eta$-equivalence classes: as the following counterexample shows, non-convertible terms may represent the same constructor term.

Counterexample 18 (Non-uniqueness of representation under $\beta\eta$) *Consider the following inductively defined type with one constructor, where* nat *is defined as in Example 19:*

> **indtype** cex : $*$ **with**
> c : $(\mathsf{nat} \to \mathsf{nat}) \to \mathsf{cex}$
> **end**

This type would be represented as

$$\mathsf{cex} \equiv [p{:}*]\,((\mathsf{nat} \to \mathsf{nat}) \to p) \to p$$
$$\mathsf{c} \equiv (\lambda f{:}\mathsf{nat} \to \mathsf{nat})\,(\lambda p{:}*)\,(\lambda y{:}(\mathsf{nat} \to \mathsf{nat}) \to p)\, y\, f$$

The following term is not $\beta\eta$-equivalent to a term c f *for any f, even though it has type* cex*:*

$$\begin{aligned} M \;=\; & (\lambda p{:}*)\,(\lambda y{:}(\mathsf{nat} \to \mathsf{nat}) \to p) \\ & y\,((\lambda n{:}\mathsf{nat})\, n\, (p \to \mathsf{nat})\,((\lambda x{:}p)\, \mathsf{zero})\,((\lambda x{:}p \to \mathsf{nat})\, x)\,(y\,((\lambda n{:}\mathsf{nat})\, n))) \end{aligned}$$

Using the inverse mapping $\mathcal{F}$ one can calculate what constructor term is represented by M:

$$\mathcal{F}(M) = \mathsf{c}\,((\lambda n{:}\mathsf{nat})\, n\,(\mathsf{cex} \to \mathsf{nat})\,((\lambda x{:}\mathsf{cex})\,\mathsf{zero})\,((\lambda x{:}\mathsf{cex} \to \mathsf{nat})\, x)\,(\mathsf{c}\,((\lambda n{:}\mathsf{nat})\, n)))$$

One can easily see that $\underline{\mathcal{F}(M)}$ and M are not $\beta\eta$-convertible, though they both represent $\mathcal{F}(M)$.

One can recover uniqueness by using dependency: in essence, a term of a constructor type is represented as the proof that it is well-formed. Such a more complex proof term has the same computational contents as our representation (see [24] or [18]). One can also formulate a simple criterion on the types P_i of the constructors that ensures uniqueness of the representation under $\beta\eta$-conversion (see [24, page 125]). Finally, one could claim that the failure of uniqueness is due to incompleteness of $\beta\eta$-conversion in the polymorphic λ-calculus and that they really should be equivalent. We conjecture that Reynolds' condition of *parametricity* [26] can be used to justify this claim, but under parametricity even more terms might be identified than under our notion of equivalence that is induced by the function $\mathcal{F}$. For example, under parametricity, the term M in the counterexample would also be equivalent to $\mathsf{c}\,((\lambda n{:}\mathsf{nat})\,\mathtt{zero})$.

Example 19 (Natural Numbers) *Here we obtain the well-known representation of the natural numbers in the second-order polymorphic λ-calculus.*

$$\mathsf{nat} \equiv [C{:}*]\, C \to (C \to C) \to C$$

Example 20 (Pairs) *Using $\underline{()}$ we obtain:*

$$\begin{array}{lcl} \mathsf{prod} & \equiv & (\lambda A{:}*)\,(\lambda B{:}*)\,[C{:}* \to * \to *]\,([A{:}*]\,[B{:}*]\,A \to B \to C\,A\,B) \to C\,A\,B \\ \mathtt{pair} & \equiv & (\lambda A{:}*)\,(\lambda B{:}*)\,(\lambda C{:}* \to * \to *)\,(\lambda f{:}[A{:}*]\,[B{:}*]\,A \to B \to C\,A\,B)\,f\,A\,B\,x\,y \end{array}$$

This is not the encoding given, for example, by Reynolds [27] and is slightly more awkward. The standard definition can be recovered by parameterizing the whole inductive definition by A and B and then abstracting over A and B to obtain global definitions (we refer to this method as uniform parameterization*). Uniform parameterization often leads to simpler equivalent representation of inductively defined parameterized types. Here, we define in the context $A{:}*, B{:}*$ (the superscripts serve only as a reminder of the dependency):*

indtype $\mathsf{prod}^{A,B} : *$ **with**
  $\mathtt{pair}^{A,B} : A \to B \to \mathsf{prod}^{A,B}$
  end

This yields the representation

$$\begin{array}{lcl} \mathsf{prod}^{A,B} & \equiv & [C{:}*]\,(A \to B \to C) \to C \\ \mathtt{pair}^{A,B} & \equiv & (\lambda x{:}A)\,(\lambda y{:}B)\,(\lambda C{:}*)\,(\lambda f{:}A \to B \to C)\,f\,x\,y \end{array}$$

One can then abstract over A and B (discharge them from the context) to obtain the usual, now global definitions of prod *and* pair*:*

$$\begin{array}{lcl} \mathsf{prod} & \equiv & (\lambda A{:}*)\,(\lambda B{:}*)\,[C{:}*]\,(A \to B \to C) \to C \\ \mathtt{pair} & \equiv & (\lambda A{:}*)\,(\lambda B{:}*)\,(\lambda x{:}A)\,(\lambda y{:}B)\,(\lambda C{:}*)\,(\lambda f{:}A \to B \to C)\,f\,x\,y \end{array}$$

Example 21 (Lists) *The representation of lists obtained this way is also different from, though equivalent to the encoding in F_2 given in [27].*

$$\mathsf{list} \equiv (\lambda B{:}*)\,[C{:}* \to *]\,([A{:}*]\,C\,A) \to ([A{:}*]\,A \to C\,A \to C\,A) \to C\,B$$

As in Example 20, one can obtain the usual definition by uniform parameterization.

Example 22 (Ordinal Notations)

$$\mathsf{ord} \equiv [C{:}*]\,C \to (C \to C) \to ((\mathsf{nat} \to C) \to C) \to C$$

Example 23 (Programs in F_2) *This is an example where uniform parameterization is not possible, since* prog *is applied to different arguments at different occurrences in the types of the constructors in Example 7. Thus a representation of this F_2-type will lie in F_3. We conjecture that no F_2 representation is possible such that the normalization function over the representation is definable.*

$$\begin{array}{lll}
\mathsf{prog} & \equiv & (\lambda D{:}*)\,[C{:}* \to *] \\
 & & ([A{:}*]\,A \to C\,A) \qquad \textit{from}\ \mathtt{rep} \\
 & & \to ([A{:}*][B{:}*]\,(A \to C\,B) \to C\,(A \to B)) \qquad \textit{from}\ \mathtt{lam} \\
 & & \to ([A{:}*][B{:}*]\,C\,(A \to B) \to C\,A \to C\,B) \qquad \textit{from}\ \mathtt{app} \\
 & & \to ([A{:}* \to *]\,([B{:}*]C\,(A\,B)) \to C\,([B{:}*](A\,B))) \qquad \textit{from}\ \mathtt{typlam} \\
 & & \to ([A{:}* \to *]\,C\,([B{:}*](A\,B)) \to [B{:}*]C\,(A\,B)) \qquad \textit{from}\ \mathtt{typapp} \\
 & & \to C\,D
\end{array}$$

Example 24 (Leibniz' Equality) *In order to show that Example 8 actually defines Leibniz' equality, we use uniform parameterization (see Example 20) to modify the previous definition. Assume we are in the context $A{:}*, x{:}A$. We would like to define the type of elements equal to x inductively. We define*

indtype $\mathsf{eq}^{A,x} : A \to *$ **with**
$\mathtt{refl}^{A,x} : \mathsf{eq}^{A,x}\,x$
end

Our representation function yields

$$\begin{array}{rcl}
\mathsf{eq}^{A,x} & \equiv & (\lambda y{:}A)\,[C{:}A \to *]\,(C\,x \to C\,y) \\
\mathtt{refl}^{A,x} & \equiv & (\lambda C{:}A \to *)\,(\lambda z{:}C\,x)\,z
\end{array}$$

After abstracting over A and x we obtain the usual definition of Leibniz' equality in the setting of CoC *or higher-order logic.*

Example 25 (Existential Quantification) *Here, too, we apply uniform parameterization in order to expose the similarity to the usual definition of existential quantification in* CoC *or higher-order logic. In the context $A{:}*, P{:}A \to *$ we define*

indtype $\mathsf{exists}^{A,P} : *$ **with**
$\mathtt{exists\text{-}intro}^{A,P} : [x{:}A]\,(P\,x \to \mathsf{exists}^{A,P})$
end

Our representation function yields

$$\begin{array}{rcl}
\mathsf{exists}^{A,P} & \equiv & [C{:}*]\,([x{:}A]\,(P\,x \to C)) \to C \\
\mathtt{exists\text{-}intro}^{A,P} & \equiv & (\lambda x{:}A)\,(\lambda v{:}P\,x)\,(\lambda C{:}*)\,(\lambda w{:}[x{:}A]\,(P\,x \to C))\,w\,x\,v
\end{array}$$

After discharging A and P from the context, we obtain the usual definitions.

5 Computing with Inductively Defined Types

Enriching CoC by inductively defined types must go along with some method for defining recursive functions over these types. We choose iteration rather than primitive recursion since it is a simpler notion and primitive recursion is definable from iteration. For an implementation of a programming language based on an enriched F_ω one would probably need to choose primitive recursion, since its implementation through iteration is provably inefficient in some cases (see Colson [5] or Parigot [22]).

Definition 26 (Definition by iteration) *Let an α be an inductively defined data type as in Section 3. Given a $\beta : Q$ and functions $h_1{:}P_1^\beta, \ldots, h_n{:}P_n^\beta$. Then the function*

$$f : [z_1{:}Q_1] \ldots [z_m{:}Q_m]\, \alpha\, z_1 \ldots z_m \rightarrow \beta\, z_1 \ldots z_m$$

is defined by iteration over α at type β from $h_1, \ldots, h_n$ *if it satisfies the following equations:*

$$\begin{aligned} f\, M_{11} \ldots M_{1m}\, (c_1\, x_1 \ldots x_{k_1}) &= h_1\, \overline{x}_1 \ldots \overline{x}_{k_1} \\ &\vdots \\ f\, M_{n1} \ldots M_{nm}\, (c_n\, x_1 \ldots x_{k_n}) &= h_n\, \overline{x}_1 \ldots \overline{x}_{k_n} \end{aligned}$$

where $\overline{N}$ is defined below.

The idea in the definition of $\overline{N}$ is to replace occurrences of variables whose type has the form $\alpha\, N_1 \ldots N_m$ by recursive calls to f. The map Φ is already of the right form to define $\overline{()}$.

Definition 27 (Map $\overline{()}$) *For $f : [z_1{:}Q_1] \ldots [z_m{:}Q_m]\, \alpha\, z_1 \ldots z_m \rightarrow \beta\, z_1 \ldots z_m$ and $N : R^\alpha$ we define $\overline{N}$ such that $\overline{N} : R^\beta$ by $\overline{N} = \Phi_R^f(N)$.*

Given the basic representation $\underline{()}$, how can we define iteration on the representation? A basic insight is that a constructor is implemented as an iterator, thus applying the representation of a constructor term as a function will perform iteration.

Theorem 28 *Given the type β and $h_1, \ldots, h_n$, then*

$$\underline{f} \equiv (\lambda z_1{:}Q_1) \ldots (\lambda z_m{:}Q_m)\, (\lambda x{:}\underline{\alpha}\, z_1 \ldots z_m)\, x\, \beta\, h_1 \ldots h_n$$

is defined from $h_1, \ldots, h_n$ by iteration over type α at type β. Thus we have

$$\underline{f} M_{i1} \ldots M_{im}\, (\underline{c}_i\, x_1 \ldots x_{k_i}) \cong h_i\, \overline{\underline{x}}_1 \ldots \overline{\underline{x}}_{k_i}$$

where $\overline{\underline{x}}_l$ is like $\overline{x_l}$ except that it inserts recursive calls to $\underline{f}$ rather than to f, that is, $\overline{\underline{x}}_l = \Phi_{P_{il}}^{\underline{f}}(x_l)$.

Proof sketch: By simple inductions as in [4]. □

Note that we claim convertibility only for terms in the image of the $\underline{()}$ translation function, not for any term that represents $c_i\, x_1 \ldots x_{k_i}$. We conjecture that under the assumption of parametricity (for the F_ω fragment) a stronger theorem also holds: the equivalence classes of representations from Theorem 17 satisfy the equations for iteration, given the definition of $\underline{f}$ above.

Example 29 (Existential Quantification) *For pairs or dependent pairs, the schema of iteration simply allows access to the components of the pair. We show only the dependent case.*

$$f\,A\,P\,(\texttt{exists-intro}\,A\,P\,x\,w) = h_1\,A\,P\,x\,w$$

with types $f : [A{:}*]\,[P{:}A \to *]\,\mathsf{exists}\,A\,P \to \beta\,A\,P$ *and* $h_1 : [A{:}*]\,[P{:}A \to *]\,[x{:}A]\,[w{:}P\,x]\,\beta\,A\,P$. *The first projection function* `fst` *for the usual pairs is easily definable, as is the function* `dfst` *for extracting the first component of a dependent pair shown here. In terms of the notation above we have*

$$\texttt{dfst}\,A\,P\,(\texttt{exists-intro}\,A\,P\,x\,w) = x$$

$$\beta = (\lambda A{:}*)\,(\lambda P{:}A \to *)\,A$$
$$h_1 = (\lambda A{:}*)\,(\lambda P{:}A \to *)\,(\lambda x{:}A)\,(\lambda w{:}P\,x)\,x$$

Example 30 (Programs in F_2) *We now give definition of* **`reify`**, **`reflect`** *and* **`eval`** *in the form of an iteration. These definitions are in the F_3 fragment of* CoC. *The crucial function is* **`reflect`** $: [A{:}*]\,\mathsf{prog}\,A \to A$. *In terms of the above definition,* $\beta = (\lambda A{:}*)\,A$

$$
\begin{array}{lcl}
\texttt{reflect}\,A\,(\texttt{rep}\,A\,x) & = & x \\
\texttt{reflect}\,(A \to B)\,(\texttt{lam}\,A\,B\,x) & = & (\lambda y{:}A)\,\texttt{reflect}\,B\,(x\,y) \\
\texttt{reflect}\,B\,(\mathsf{app}\,A\,B\,x\,y) & = & (\texttt{reflect}\,(A \to B)\,x)\,(\texttt{reflect}\,A\,y) \\
\texttt{reflect}\,([B{:}*](A\,B))\,(\texttt{typlam}\,A\,x) & = & (\lambda B{:}*)\,\texttt{reflect}\,(A\,B)\,(x\,B) \\
\texttt{reflect}\,(A\,B)\,(\texttt{typapp}\,A\,x\,B) & = & \texttt{reflect}\,([B{:}*]\,A\,B)\,x\,B
\end{array}
$$

From this the other definitions follow easily:

$$
\begin{array}{rcl}
\texttt{reify} & : & [A{:}*]\,A \to \mathsf{prog}\,A \\
\texttt{reify} & \equiv & \texttt{rep}
\end{array}
$$

$$
\begin{array}{rcl}
\texttt{eval} & : & [A{:}*]\,\mathsf{prog}\,A \to \mathsf{prog}\,A \\
\texttt{eval} & \equiv & (\lambda A{:}*)\,(\lambda x{:}\mathsf{prog}\,A)\,\texttt{reify}\,A\,(\texttt{reflect}\,A\,x)
\end{array}
$$

In [25] we give the expanded definition of `reflect` *in F_3 using Theorem 28.*

Primitive recursion at all types is somewhat more difficult, but as shown in various places for the second-order polymorphic λ-calculus (see, for example, Reynolds [27]) it can be reduced to iteration. We briefly state only the form of primitive recursion and the type of the primitive recursive operator pr_α over an inductively defined type α. $\times$ is the generalized product from Definition 4.

Definition 31 (Definition by primitive recursion at arbitrary type) *Let an α be an inductively defined data type as in Section 3. Given a $\beta : Q$ and functions $h'_1, \ldots, h'_n$ where $h'_i : [x'_1{:}P^{\alpha\times\beta}_{i1}]\ldots[x'_{k_i}{:}P^{\alpha\times\beta}_{ik_i}]\,\beta\, M_{i1}\ldots M_{im}$. A function $f : [z_1{:}Q_1]\ldots[z_m{:}Q_m]\ \alpha\, z_1 \ldots z_m \rightarrow \beta\, z_1 \ldots z_m$ is defined by primitive recursion over α at type β from $h'_1, \ldots, h'_n$ if it satisfies the following equations:*

$$f\, M_{11} \ldots M_{1m}\, (c_1\, x_1 \ldots x_{k_1}) = h'_1\, \hat{x}_1 \ldots \hat{x}_{k_1}$$
$$\vdots$$
$$f\, M_{n1} \ldots M_{nm}\, (c_n\, x_1 \ldots x_{k_n}) = h'_n\, \hat{x}_1 \ldots \hat{x}_{k_n}$$

where $\hat{x}_l = \Phi^F_{P_{il}}(x_l)$ for $F = (\lambda z_1{:}Q_1)\ldots(\lambda z_m{:}Q_m)\,(\lambda x{:}\alpha\, z_1 \ldots z_m)\,\mathtt{pair}\, x\, (f\, z_1 \ldots z_m\, x)$ and thus $\hat{()} : R^\alpha \rightarrow R^{\alpha\times\beta}$.

Note that the occurrences of M_{ij} are not binding occurrences: they are determined by the type of the constructor c_i. In the simplest case, $\hat{x}$ is merely x (if the type of x does not involve α), or the pair of x and fx (if the type of x is α). In general, the variable pr_α which generates the definition of f given β and functions $h'_1, \ldots, h'_n$ has type

$$\begin{array}{lll}
\mathrm{pr}_\alpha & : & [\beta{:}[z_1{:}Q_1]\ldots[z_m{:}Q_m]\,*] & \text{for } \beta \\
& & ([x'_1{:}P^{\alpha\times\beta}_{11}]\ldots[x'_{k_1}{:}P^{\alpha\times\beta}_{1k_1}]\,\beta\, M_{11}\ldots M_{1m}) & \text{for } h'_1 \\
& & \vdots & \vdots \\
& & \rightarrow ([x'_1{:}P^{\alpha\times\beta}_{n1}]\ldots[x'_{k_n}{:}P^{\alpha\times\beta}_{nk_n}]\,\beta\, M_{n1}\ldots M_{nm}) & \text{for } h'_n \\
& & \rightarrow [z_1{:}Q_1]\ldots[z_m{:}Q_m]\,\alpha\, z_1\ldots z_m \rightarrow \beta\, z_1 \ldots z_m &
\end{array}$$

Example 32 (Primitive Recursion over Lists) *To illustrate the schema of primitive recursion we use lists as defined in Example 5. Given $\beta : * \rightarrow *$, primitive recursion can define a function $f : [A{:}*]\,\mathsf{list}\, A \rightarrow \beta\, A$. The schema looks like*

$$\begin{array}{rcl}
f A\, (\mathtt{nil}\, A) & = & h'_1\, A \\
f A\, (\mathsf{cons}\, A\, x\, l) & = & h'_2\, A\, x\, (\mathtt{pair}\, l\, (f A\, l))
\end{array}$$

where $h'_1 : [A{:}*]\,\beta\, A$ and $h'_2 : [A{:}*]\,[x{:}A]\,(\mathsf{prod}\,(\mathsf{list}\, A)\,(\beta\, A)) \rightarrow \beta\, A$.

As a concrete example consider the function `tl` *which takes a list and a default value and returns the tail of the list or the default value (if the list is empty). We could program this as a primitive recursion with*

$$\mathtt{tl} : [A{:}*]\,\mathsf{list}\, A \rightarrow \mathsf{list}\, A \rightarrow \mathsf{list}\, A$$

$$\mathtt{tl}\, A\, (\mathtt{nil}\, A) = (\lambda d{:}\mathsf{list}\, A)\, d$$
$$\mathtt{tl}\, A\, (\mathsf{cons}\, A\, x\, l) = (\lambda d{:}\mathsf{list}\, A)\, l$$

In the notation above we would have

$$\beta = (\lambda A{:}*)\,\mathsf{list}\, A \rightarrow \mathsf{list}\, A$$
$$h'_1 = (\lambda A{:}*)\,(\lambda d{:}A)\, d$$
$$h'_2 = (\lambda A{:}*)\,(\lambda x{:}A)\,(\lambda p{:}\mathsf{prod}\,(\mathsf{list}\, A)\,(\beta\, A))\,(\lambda d{:}\mathsf{list}\, A)\,\mathtt{fst}\,(\mathsf{list}\, A)\,(\beta\, A) p$$

6 Reasoning with Induction

One of the motivations behind inductively defined types is that we would like to reason about elements of these types using induction. In particular, we would like to extract provably correct functions from proofs. In this section we state the natural notion of induction over an inductively defined type, and show how induction relates to the notion of primitive recursive functionals.

Induction principles are not definable (that is, provable) in CoC itself, but one could *assume* such induction principles and associated reduction rules (see [8, Section 8] or [24, Section 4.4]). Such an extension of the calculus is in some sense "benign." This can be formalized as saying the computational content of a proof that used induction is already present in pure F_ω. The proof of this fact is surprisingly simple (see Theorem 35). Thus, if one is interested only in the computational content of proofs, the extension of CoC by induction over inductively defined types does not change the set of definable functions. However, with the addition of induction one will in general be able to prove many more specifications. Other conservative extension results for polymorphic λ-calculi have been obtained by Breazu-Tannen & Gallier [3].

Definition 33 (Induction principle ind_α for inductively defined α) *Let α be an inductively defined type as before. We define* ind_α*, the* induction principle over α *by*

$$\begin{array}{rl} \mathrm{ind}_\alpha : & [A{:}[z_1{:}Q_1]\ldots[z_m{:}Q_m]\,\alpha\,z_1\ldots z_m \to *] \\ & ([x'_1{:}P_{11}^{\alpha\otimes A}]\ldots[x'_{k_1}{:}P_{1k_1}^{\alpha\otimes A}]\,A\,M_{11}\ldots M_{1m}\,(c_1\,\check{x}'_1\ldots\check{x}'_{k_1})) \\ & \vdots \\ & \to ([x'_1{:}P_{n1}^{\alpha\otimes A}]\ldots[x'_{k_n}{:}P_{nk_n}^{\alpha\otimes A}]\,A\,M_{n1}\ldots M_{nm}\,(c_n\,\check{x}'_1\ldots\check{x}'_{k_n})) \\ & \to [z_1{:}Q_1]\ldots[z_m{:}Q_m]\,[x{:}\alpha\,z_1\ldots z_m]\,A\,z_1\ldots z_m\,x \end{array}$$

where $\check{x}'$ is defined below and $\alpha\otimes A$ is the type of generalized dependent pairs (see Definition 9).

In the simplest case $\check{x}'$ will simply turn out to be x' (if the type of x' does not involve α) or $\mathtt{dfst}\,\alpha\,A\,x'$, extracting the element x from the pair consisting of an x and the proof that x satisfies property A (if x' has type α).

Definition 34 (Map $\check{x}$) *Let F be the generalized first projection function (derived easily from* `dfst`*, see Example 29) on elements of dependent pair type $\alpha\otimes A$. Then*

$$F : [z_1{:}Q_1]\ldots[z_m{:}Q_m]\,\mathsf{exists}\,(\alpha\,z_1\ldots z_m)\,(A\,z_1\ldots z_m) \to \alpha\,z_1\ldots z_m$$

and for R^x and $N : R^{\alpha\otimes A}$ we define $\check{N} = \Phi_R^F(N) : R^\alpha$.

Coquand & Huet define ν, the *stripping map*, which extracts an untyped λ-term as the computational content of a proof in CoC. We use a less drastic erasure in the proof of our conservative extension result below, which maps terms in CoC into terms in F_ω. The partial erasure map $\mathcal{E}$ is defined in detail in [23, 24].

Theorem 35 (Primitive recursion realizes induction) *We use* pind_α *and* ppr_α *as abbreviation for the types of* ind_α *and* pr_α*, respectively. Then* $\mathcal{E}(\mathrm{pind}_\alpha) \cong \mathcal{E}(\mathrm{ppr}_\alpha)$.

Proof sketch: The map $\mathcal{E}$ will erase $\alpha\, z_1 \ldots z_m$ from the type of A, and all corresponding arguments to A at all occurrences of A (notation as in definition 33). The resulting term is a valid type and $\beta\eta$-equivalent to the type of pr_α (see Definition 31). The crucial observation is that $\mathcal{E}(P_{il}^{\alpha \otimes A}) = \mathcal{E}(P_{il}^{\alpha \times \mathcal{E}(A)})$. □

This theorem means that the set of functions that can be extracted from induction proofs over α can already be defined explicitly by primitive recursion at arbitrary types. This corollary generalizes one direction of results obtained by Girard [14], and Fortune, Leivant & O'Donnell [12], and Leivant [17, 18] which may be summarized as "The number-theoretic functions representable in F_n are exactly the functions provably recursive in n^{th}-order arithmetic."

Example 36 (Induction over Lists) *Here we obtain a principle of induction over the construction of lists. Since induction is a logical statement, it best to think of* [] *as universal quantification.*

$$
\begin{aligned}
\mathrm{ind}_{\mathsf{list}} \;:\; & [P{:}[A{:}*]\,\mathsf{list}\,A \to *] \\
& ([A{:}*]\,P A\,(\mathtt{nil}\,A)) \\
& \to ([A{:}*]\,[x{:}A]\,[l'{:}\mathsf{exists}\,(\mathsf{list}\,A)\,(P A)]\,P A\,(\mathsf{cons}\,A\,x\,(\mathtt{dfst}\,A\,P\,l'))) \\
& \to [A{:}*]\,[l{:}\mathsf{list}\,A]\,P A\,l
\end{aligned}
$$

The induction principle will look more familiar after we curry at the argument l' to eliminate the dependent pair and also apply uniform parameterization over the argument A. We then get:

$$
\begin{aligned}
& [P{:}\mathsf{list}\,A \to *] \\
& \quad (P\,(\mathtt{nil}\,A)) \\
& \quad \to [x{:}A]\,[l{:}\mathsf{list}\,A]\,P A\,l \to P A\,(\mathsf{cons}\,A\,x\,l) \\
& \quad \to [l{:}\mathsf{list}\,A]\,P\,l
\end{aligned}
$$

Acknowledgments

We would like to thank Thierry Coquand, Jean Gallier, Bob Harper, Peter Lee, and Dan Leivant for helpful discussions. The first author was supported in part by the Office of Naval Research under contract N00014-84-K-0415 and in part by the Defense Advanced Research Projects Agency (DOD), ARPA Order No. 5404, monitored by the Office of Naval Research under the same contract.

References

[1] Arnon Avron, Furio A. Honsell, and Ian A. Mason. Using typed lambda calculus to implement formal systems on a machine. Technical Report ECS-LFCS-87-31, Laboratory for Foundations of Computer Science, University of Edinburgh, Edinburgh, Scotland, June 1987.

[2] Henk Barendregt. The forest of lambda calculi with types. Talk given at the Workshop on Semantics of Lambda Calculus and Category Theory, Carnegie Mellon University, April 1988.

[3] Val Breazu-Tannen and Jean Gallier. Polymorphic rewriting conserves algebraic strong normalization and confluence. In G. Ausiello, M. Dezani-Ciancaglini, and S. Ronchi Della Rocca, editors, *Proceedings of the 16th International Colloquium on Automata, Languages and Programming, Stresa, Italy*, pages 137–150. Springer-Verlag LNCS 372, July 1989.

[4] Corrado Böhm and Alessandro Berarducci. Automatic synthesis of typed Λ-programs on term algebras. *Theoretical Computer Science*, 39:135–154, 1985.

[5] Loïc Colson. About primitive recursive algorithms. In G. Ausiello, M. Dezani-Ciancaglini, and S. Ronchi Della Rocca, editors, *Proceedings of the 16th International Colloquium on Automata, Languages and Programming, Stresa, Italy*, pages 194–206. Springer-Verlag LNCS 372, July 1989.

[6] Thierry Coquand. *Une Théorie des Constructions*. PhD thesis, University Paris VII, January 1985.

[7] Thierry Coquand and Gérard Huet. Constructions: A higher order proof system for mechanizing mathematics. In *EUROCAL85*. Springer-Verlag LNCS 203, 1985.

[8] Thierry Coquand and Gérard Huet. The Calculus of Constructions. *Information and Computation*, 76(2/3):95–120, February/March 1988.

[9] Thierry Coquand and Christine Paulin-Mohring. Inductively defined types. Talk presented at the *Workshop on Programming Logic*, University of Göteborg and Chalmers University of Technology, May 1989.

[10] N. G. de Bruijn. Lambda-calculus notation with nameless dummies: a tool for automatic formula manipulation with application to the Church-Rosser theorem. *Indag. Math.*, 34(5):381–392, 1972.

[11] Peter Dybjer. An inversion principle for Martin-Löf's type theory. Talk presented at the *Workshop on Programming Logic*, University of Göteborg and Chalmers University of Technology, May 1989.

[12] Steven Fortune, Daniel Leivant, and Michael O'Donnell. The expressiveness of simple and second-order type structures. *Journal of the ACM*, 30:151–185, 1983.

[13] Jean-Yves Girard. Une extension de l'interprétation de Gödel à l'analyse, et son application a l'élimination des coupures dans l'analyse et la théorie des types. In J. E. Fenstad, editor, *Proceedings of the Second Scandinavian Logic Symposium*, pages 63–92, Amsterdam, London, 1971. North-Holland Publishing Co.

[14] Jean-Yves Girard. *Interprétation fonctionelle et élimination des coupures de l'arithmétique d'ordre supérieur*. PhD thesis, Université Paris VII, 1972.

[15] Robert Harper, Furio Honsell, and Gordon Plotkin. A framework for defining logics. In *Symposium on Logic in Computer Science*, pages 194–204. IEEE, June 1987.

[16] Gérard Huet. Formal structures for computation and deduction. Lecture notes for a graduate course at Carnegie Mellon University, May 1986.

[17] Daniel Leivant. Reasoning about functional programs and complexity classes associated with type disciplines. In *Proceedings of the Twenty Fourth Annual Symposium on the Foundations of Computer Science*, pages 160–169. IEEE, 1983.

[18] Daniel Leivant. Contracting proofs to programs. In P. Odifreddi, editor, *Logic and Computer Science*. Academic Press, 1990. To appear.

[19] N. P. Mendler. First- and second-order lambda calculi with recursive types. Technical Report TR 86–764, Department of Computer Science, Cornell University, Ithaca, New York, July 1986.

[20] Paul Francis Mendler. *Inductive Definition in Type Theory.* PhD thesis, Department of Computer Science, Cornell University, September 1987.

[21] John C. Mitchell and Albert Meyer. Second-order logical relations. In Rohit Parikh, editor, *Logics of Programs*, pages 225–236. Springer-Verlag LNCS 193, June 1985.

[22] Michel Parigot. On the representation of data in lambda-calculus. Draft, 1988.

[23] Christine Paulin-Mohring. Extracting F_ω programs from proofs in the calculus of constructions. In *Sixteenth Annual Symposium on Principles of Programming Languages*, pages 89–104. ACM Press, January 1989.

[24] Christine Paulin-Mohring. *Extraction de programmes dans le Calcul des Constructions.* PhD thesis, Université Paris VII, January 1989.

[25] Frank Pfenning and Peter Lee. Metacircularity in the polymorphic lambda-calculus. *Theoretical Computer Science*, 1990. To appear. A preliminary version appeared in *TAPSOFT '89, Proceedings of the International Joint Conference on Theory and Practice in Software Development, Barcelona, Spain*, pages 345–359, Springer-Verlag LNCS 352, March 1989.

[26] John Reynolds. Types, abstraction and parametric polymorphism. In R. E. A. Mason, editor, *Information Processing 83*, pages 513–523. Elsevier Science Publishers B. V., 1983.

[27] John Reynolds. Three approaches to type structure. In Hartmut Ehrig, Christiane Floyd, Maurice Nivat, and James Thatcher, editors, *Mathematical Foundations of Software Development*, pages 97–138. Springer-Verlag LNCS 185, March 1985.

On Some Semantic Issues in the Reflective Tower

Karoline Malmkjær*
DIKU – Computer Science Department, University of Copenhagen
Universitetsparken 1, DK-2100 Copenhagen Ø, DENMARK
e-mail: karoline@diku.dk

Abstract

Introducing meta-level access in a programming language is a delicate task, surrounded by threats of vicious circularity, infinite regress and unresolveable paradoxes on the one hand and triviality on the other hand. The reflective tower is one of the attempts to structure computational reflection, that is, access from a running process to its computational state.

This paper gives the framework of a denotational semantics of a reflective tower, isolating the features which we have found to be sufficient to describe this model. The framework is then used to analyze some concrete problems in a reflective tower based on the programming language Scheme with continuations as applicable objects.

The paper first gives a brief description of the reflective tower and discusses the unstable points in the design. Then the denotational description is outlined and used to analyze these points and to clarify the consequences of the necessary design decisions.

1 Introduction

The reflective tower [Smith 82, Wand & Friedman 88, Danvy & Malmkjær 88] is a design introducing new structured programming facilities. This design still holds ambiguities, and for some facilities it is not clear whether they are desirable or not and whether they are optional or intrinsic. A description of a reflective tower in denotational semantics provides the means for identifying a number of the arbitrary choices and their consequences and it gives a clear description of why and how this conceptually infinite design can be implemented finitely.

The reflective tower is designed to structure facilities of meta-level access (sometimes called reflective or introspective facilities). The aim is to be able to access, test, modify, and reinstall the state of any program from the program itself while maintaining a proper distance from the program when it is being modified.

To fulfill the criteria of proper distance, access is only given to special procedures (called *reifiers*) which, when they are applied, run in the interpreter of the program

*Current address: Dept. of Computing and Information Sciences, Kansas State University, Manhattan, KS 66506. e-mail: karoline@ksuvax1.cis.ksu.edu

rather than in the program itself. In order to be able to write the reifiers in the same language as the rest of the program, the interpreter is assumed to be meta-circular [Reynolds 72]. Since there are no restrictions on the actions that can be performed in the reifiers (*e.g.*, applying another reifier) this leads to the picture of an infinite tower of interpreters, each interpreting the interpreter below it, until the user program.

This can then be simulated using some circular structure to represent the infinity of interpreter states — it is not necessary to represent the infinity of interpreter programs since they are identical.

With the facilities of the reflective tower it is possible – in the language itself – to define environment and control manipulating procedures such as Scheme's call-with-current-continuation (abbreviated call/cc) [Rees & Clinger 86]. Call/cc, however, may also serve as an example of the problems of the reflective tower. [Wand & Friedman 88] attempts two definitions of call/cc, and neither of them corresponds to the Scheme procedure. In order to see why, we first have to take a closer look at the access that is provided in the reflective tower and the movements in the tower that accompany the use of this access.

The next section presents the reflective tower. Section 3 sketches a denotational description of a reflective tower. This is then used in section 4 and section 5 to clarify some problems of the access to control and of the concept of scope in a multi-level setting and to outline solutions. Section 6 discusses the particular problems of reinstalling expressions.

2 The Reflective Tower

The tower is pictured as an infinite tower of levels, where each level represents a running program. The user program runs at the bottom of the tower and each level above it is an interpreter interpreting the level below itself (if it was not interpreting, there would not be any levels below).

```
.         .
.         .
+---------+
|    2    |
+---------+
|    1    |
+---------+
|    0    |
+---------+
```

The languages of the reflective towers designed so far are mainly Scheme-like expression languages.

The interpreter is a continuation-passing, tail-recursive program manipulating three state variables representing the current expression, the environment in which to evaluate the expression, and the continuation (or control) that requests the result of the evaluation. Since it is continuation-passing, the control of the interpreted program is explicitly represented in the continuation, rather than being implicitly present in the state of the interpreter.

A program can access the state variables in its interpreter by applying a reifier. A reifier is a kind of abstraction and has at least two parameters. When it is applied, its level of application ceases to run and it will not be resumed unless explicitly requested in the reifier. The body of the reifier will be evaluated at the level above with its parameters bound to the state variables – the first two to the environment and continuation of the application and the rest bound to the unevaluated arguments to which the reifier is applied.

```
.         .                                 .         .
.         .                                 .         .
+---------+                                 +---------+
|    2    |                                 |    2    |
+---------+                                 +---------+
|    1    |   (<reifier> <arguments>)       |    1    |
+---------+          ------>                +---------+
|    0    |
+---------+
```

The means to resume the computation of a level (denoted "reflection" in [Friedman & Wand 84]) are in principle quite simple. It is sufficient to call the interpreter which by design is supposed to be globally defined at each level. In Blond [Danvy & Malmkjær 88'], the dialect we will use here, the identifier `meaning` is bound to the interpreter. It takes three arguments: the three state variables (which might of course have been modified or replaced).

```
.         .                                 .         .
.         .                                 .         .
+---------+                                 +---------+
|    1    |                                 |    1    |
+---------+                                 +---------+
|    0    |        (meaning ...)            |    0    |
+---------+          ------>                +---------+
                                            |   new   |
                                            +---------+
```

Blond is a higher-order, lexically scoped, call-by-value expression language, reminiscent of Scheme. Reifiers are defined like other procedures except with the keyword `mu` instead of `lambda`.

This design increases the existing conceptual overloading of the relatively few syntactic constructions in Scheme since applying a reifier is conceptually a different action from applying a procedure or even from reducing a special form [1]. This could argue for a design with more syntactic constructions, closer to that of 2-Lisp [Smith 82].

The current Blond implementation is interactive as illustrated in the following Scheme session:

[1] Though special forms can be simulated by reifiers.

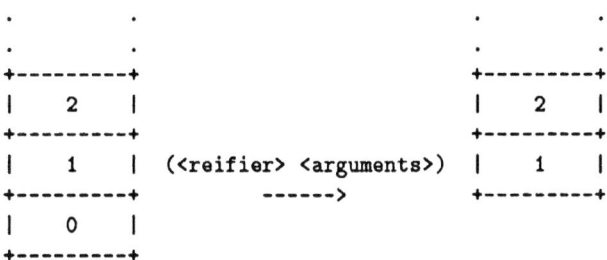

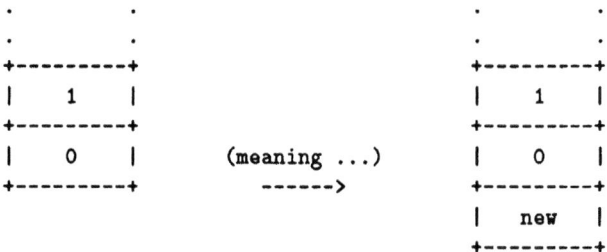

```
> (load "mublond.ss")
()
> (mublond)
0-0: "bottom-level"
0-1> ((mu (r k e)
          (meaning e r k))
      "hello")
0-1: "hello"
0-2> ((mu (r k e)
          "one up")
      "lost")
1-0: "one up"
1-1> (meaning "new-level"
              (reify-new-environment)
              (reify-new-continuation "new"))
new-0: "new-level"
new-1>
```

The Blond prompt contains two identifications: the first is the identification of the level in the tower where the user program is running and the second is a counter displaying the number of iterations in the interactive loop. The identifications of the levels are initially numbers, starting with 0 at the lowest level and growing upwards in the tower, but it is possible to use any denotable value as an identification.

In this scenario we first apply a reifier to the string `"hello"`. The reifier is the reflective version of identity: it reifies to the level above and then restores the level it came from by applying `meaning` to the newly reified state variables. The result is to return the string at the same iteration of level 0.

Then we apply a reifier which does not restore the level but simply returns the string `"one up"`. This is returned at the first iteration of the level above (level 1).

Finally we create a completely new level with `meaning` and the primitives `reify-new-environment`, that returns a representation of the initial Blond environment, and `reify-new-continuation`, that returns a representation of an initial continuation like the ones originally running at each level in the tower. `reify-new-continuation` takes a level-identification as argument because this information is closed in the continuation – it is the continuation that constitutes the level.

Since representations of continuations are applicable objects in Blond, it is also possible to resume a level simply by applying a representation of a continuation:

```
new-1> ((mu (r k e)
            (k "hello")) "hi")
new-1: "hello"
new-2>
```

This reifier resumes the level `new` by applying its continuation to the result of evaluating `"hello"`. Since the argument is evaluated before it is passed to the continuation, it is not necessary to specify an environment when a level is resumed by applying its continuation.

The representation of an environment is also an applicable object in Blond – if applied to the representation of an identifier, it returns the value the identifier denotes in the environment.

3 A Denotational Description

It is natural to describe the reflective tower in denotational semantics since the meta-theory employed in the design is very close to a continuation semantics.

A denotational description of Blond looks, for the most parts, like any denotational description of an expression language. It is a continuation semantics in order to model the access to the computational state. Furthermore, the main valuation function has an extra argument to model the infinity of interpreter states, the so-called *meta-continuation* [Wand & Friedman 88]. But except in the descriptions of **meaning** and of the definition and application of reifiers, the meta-continuation can be η-reduced. Thus the parts not connected with the reflective facilities look quite standard and will only be outlined here. A more complete denotational semantics for Blond and an account of its development can be found in [Malmkjær 88].

Abstract syntax

Blond has a standard parenthesised, Scheme-like syntax:

$\mathtt{E} \in \mathit{Expression}$, $\mathtt{C} \in \mathit{Constant}$, $\mathtt{I} \in \mathit{Identifier}$

$\mathtt{E} ::= \mathtt{C} \mid \mathtt{I} \mid \mathtt{(lambda\ (I_1 \ldots I_n)\ E)} \mid \mathtt{(E_0\ E_1 \ldots E_n)} \mid \mathtt{(if\ E_0\ E_1\ E_2)} \mid$
$\quad \mathtt{(quote\ E)} \mid \mathtt{(mu\ (I_r\ I_k\ I_n \ldots E_n)\ E)} \mid \mathtt{(meaning\ E_e\ E_r\ E_k)} \mid \ldots$

We do not give all expressions here; the language also has **define**, **let**, **letrec**, **case** and so on.

Domains

We assume standard domains of identifiers, constants, denotable values (*DenVal*), environments (*Env*) and so on. The domain of applicable values (*App*) contains reifiers and reified continuations and environments, as well as the usual primitives and lambda abstractions.

The domain of meta-continuations, whose elements represent the infinite tower, is recursive: a meta-continuation holds the environment, continuation and meta-continuation of the level above the current level in the tower.

Meta-continuations: $\tau \in \mathit{MetaCont} = \mathit{Env} \times \mathit{Cont} \times \mathit{MetaCont}$

The continuations take meta-continuations as arguments. This is to situate the level represented by the continuation in the tower. This also fits with the idea that continuations are functions at the definitional level [Strachey & Wadsworth 74]. Thus they need the continuation of the level above, and this continuation is present in the meta-continuation.

Continuations: $\kappa \in Cont = DenVal \to MetaCont \to Answer$

Correspondingly the domain of λ-abstractions also needs the meta-continuation:

λ-abstractions: $\alpha \in \lambda Abs = DenVal^{*} \to Cont \to MetaCont \to Answer$

Representation of state

To model the application of a reifier in a denotational semantics, it is necessary to be able to map elements of the domains *Expression*, *Env* and *Cont* into elements of the domain *DenVal* in order to bind them to the parameters of the reifier. This can be understood as representing them in the domain of denotable values. In order to model the application of **meaning**, it is necessary to be able to map the opposite way – to take a representation of an object to the object it represents. Let us assume that we have three pairs of such mappings

$$
\begin{array}{llll}
exp^{\wedge} : Expression \to DenVal & & exp^{\vee} : DenVal \to Expression \\
env^{\wedge} : Env \to DenVal & & env^{\vee} : DenVal \to Env \\
cont^{\wedge} : Cont \to DenVal & & cont^{\vee} : DenVal \to Cont
\end{array}
$$

The $^{\vee}$ mappings take elements of *DenVal* that do not represent anything to the bottom element of their respective codomains.

To formalize the idea that $^{\wedge}$ takes an element into a representation in *DenVal* and $^{\vee}$ takes a representation into the element it represents, we require that each pair of mappings forms a retraction pair, that is,

$$
\begin{aligned}
f^{\vee} \circ f^{\wedge} &= identity_{D} \\
f^{\wedge} \circ f^{\vee} &\sqsubseteq identity_{DenVal}
\end{aligned}
$$

for (f, D) equal to $(exp, Expression)$, (env, Env) and $(cont, Cont)$.

But even with this restriction, there are many possible mappings corresponding to many possible representations of one domain in another, also because semantically it is sufficient to consider observational equivalence.[2]

The elements of a level can be represented in the domains of the level above in two distinct ways: with a separate domain for each of the domains to be represented (a domain of represented continuations, *etc.*) or by integrating the elements into existing domains (*e.g.*, continuations as λ-abstractions, expressions as Lisp S-expressions).[3]

The representation determines the $^{\wedge}$ mappings, and for elements in the range of $^{\wedge}$ the representation also determines the $^{\vee}$ mappings. However, elements which are not (necessarily) previously reified elements can either be mapped to *error* or we can try to consider them as representations. For example for the domain λAbs it is reasonable to consider its elements as representations of continuations; this representation corresponds to considering continuations as functions at the definitional level.

[2] In an implementation, however, efficiency reasons could argue for textual identity.

[3] Note that even if, for example, continuations are represented in a separate domain, they might still be applicable objects. This simply requires an extension of the interpreter to handle more applicable objects as in our present design.

But though some representations may be more useful for special purposes, there is nothing that makes one set of representations more "reflective" than another; as long as it serves adequately as a representation, we can make a reflective tower.

Valuation functions

The main valuation function is $\mathcal{E}$, defined over expressions, since the language is an expression language. We do not show every case since it is a standard continuation semantics as can be seen from the definition of application and abstraction:

$$\mathcal{E} : \mathit{Expression} \rightarrow \mathit{Env} \rightarrow \mathit{Cont} \rightarrow \mathit{MetaCont} \rightarrow \mathit{Answer}$$

...

$$\begin{aligned}
&\mathcal{E}[\![\texttt{(E}_0\ \texttt{E}_1\ \ldots\ \texttt{E}_n\texttt{)}]\!]\,\rho\,\kappa \\
&= \mathcal{E}[\![\texttt{E}_0]\!]\,\rho\,(\lambda\, v.(\mathit{cases}\ v\ \mathit{of}\ \mathit{isConst}(c) \rightarrow\ (\mathit{error\text{-}message}\ \langle \mathit{suitable\ text}\rangle) \\
&\qquad\qquad [\!]\ \mathit{isApp}(f) \rightarrow \\
&\qquad\qquad\qquad (\mathit{cases}\ f\ \mathit{of} \ldots \\
&\qquad\qquad\qquad\qquad [\!]\ \mathit{is}\lambda\mathit{Abs}(\alpha) \rightarrow \\
&\qquad\qquad\qquad\qquad\qquad \mathcal{E}[\![\texttt{E}_1]\!]\,\rho(\lambda\, v_1.\ \ldots \mathcal{E}[\![\texttt{E}_n]\!]\,\rho(\lambda\, v_n.\alpha\,(v_1, \ldots, v_n)\,\kappa)) \\
&\qquad\qquad\qquad\qquad [\!]\ \ldots\ \mathit{end}) \\
&\qquad\qquad [\!]\ \ldots\ \mathit{end}))
\end{aligned}$$

$$\begin{aligned}
&\mathcal{E}[\![\texttt{(lambda (I}_1\ \ldots\ \texttt{I}_n\texttt{) E)}]\!]\,\rho\,\kappa \\
&= \kappa\ \mathit{inApp}(\mathit{in}\lambda\mathit{Abs}(\lambda\,(v_1, \ldots, v_n).\,\mathcal{E}[\![\texttt{E}]\!]\,([[\![\texttt{I}_1]\!] \mapsto v_1, \ldots, [\![\texttt{I}_n]\!] \mapsto v_n]\rho)))
\end{aligned}$$

...

To define the application of `meaning` we use the $^{\vee}$ mappings, but first all three arguments are evaluated in the standard way. $\mathit{exp}^{\vee}$ is then applied to the value of the first argument, and the result is to be considered as a new expression that will be evaluated by the interpreter. But the interpreter is the same meta-circular interpreter which is (supposed to be) evaluating all other expressions, so the same semantic equations can be used. Thus the meaning of evaluating the expression can be found in the semantics by using the valuation function $\mathcal{E}$. The environment and continuation passed to $\mathcal{E}$, however, have to be the ones specified by the second and third arguments. Still, that does not imply that the current environment and continuation can be discarded, since they are to be used if a reifier is applied in the expression. This is solved by "stacking" them on the meta-continuation which is then passed as an argument to $\mathcal{E}$.

$$\begin{aligned}
&\mathcal{E}[\![\texttt{(meaning E}_e\ \texttt{E}_r\ \texttt{E}_k\texttt{)}]\!]\,\rho\,\kappa\,\tau \\
&= \mathcal{E}[\![\texttt{E}_e]\!]\,\rho \\
&\quad (\lambda\, v_1\tau.\,\mathcal{E}[\![\texttt{E}_r]\!]\,\rho \\
&\qquad\quad (\lambda\, v_2\tau.\,\mathcal{E}[\![\texttt{E}_k]\!]\,\rho \\
&\qquad\qquad\quad (\lambda\, v_3\tau.\,\mathcal{E}(\mathit{exp}^{\vee} v_1)(\mathit{env}^{\vee} v_2)(\mathit{cont}^{\vee} v_3)(\rho, \kappa, \tau))\tau)\tau)\tau
\end{aligned}$$

Note that all the τ's except the innermost one can be η-reduced in the definition. This captures the intuition that the evaluation of arguments does not explicitly involve the levels above the current level in the tower.

Strictly speaking, this is of course not a denotational definition since it is not compositional (the expressions on the right hand side are not subexpressions of the expression on the left hand side) and thus only partly defined. It is not surprising that this problem arises since **meaning** is basically a three argument version of the eval function in Lisp and **eval** is precisely the reason why Lisp cannot be described compositionally.

Section 6 shows how to make a compositional description by defining *exp*^ appropriately and requiring the first argument of meaning to be a previously reified expression.

Before giving the denotations of definition and of application of reifiers, we will consider the simultaneous definition and application of reifiers. The discussion of how to separate actions between definition time and application time is deferred to section 4.

So let us consider the expression $[\![$`((mu (`$\mathtt{I}_r$ $\mathtt{I}_k$ $\mathtt{I}_1$ ... $\mathtt{I}_n$`) E) `$\mathtt{E}_1$ ... $\mathtt{E}_n$`)`$]\!]$. The body of the reifier, $[\![\mathtt{E}]\!]$, should be evaluated in an environment that is extended with the bindings of the identifiers to representations of the current environment and continuation and of $[\![\mathtt{E}_1]\!], \ldots, [\![\mathtt{E}_n]\!]$. So we map these to denotable values and apply $\mathcal{E}$ to the body of the reifier and an extended environment, just like for ordinary procedure application. In order to model that this takes place at the level above, however, we use the environment and continuation from the "top" of the meta-continuation and we use the tail of the meta-continuation as the new meta-continuation.

$$\begin{aligned}
&\mathcal{E}[\![\mathtt{((mu\ (I_r\ I_k\ I_1\ \ldots\ I_n)\ E)\ E_1\ \ldots\ E_n)}]\!]\,\rho\,\kappa\,(\rho',\kappa',\tau)\\
&= \mathcal{E}[\![\mathtt{E}]\!]\,([[\![\mathtt{I}_r]\!] \mapsto (env^\wedge\rho), [\![\mathtt{I}_k]\!] \mapsto (cont^\wedge\kappa),\\
&\qquad [\![\mathtt{I}_1]\!] \mapsto (exp^\wedge[\![\mathtt{E}_1]\!]), \ldots, [\![\mathtt{I}_n]\!] \mapsto (exp^\wedge[\![\mathtt{E}_n]\!])]\rho')\kappa'\tau
\end{aligned}$$

Summary

We have found that in order to support a reflective tower, a denotational description needs:

- No implicit state, for example, by being a continuation semantics.
- A meta-continuation, recursively defined over all arguments of the main valuation function, to represent the infinite tower.
- Conversion functions between the domains of the main valuation function and the domain of denotable values to model the representation of state.

and

- The main valuation function and the continuation must take the meta-continuation as an argument.

In order for the language being described to be a reflective tower, however, it must have constructions to push and pop the meta-continuation, e.g., **meaning** and reifiers in the case of Blond.

4 Separating Definition and Application of Reifiers

The question of in which environment to evaluate reifiers has been hidden in the denotational semantics by the fact that we have only treated direct application of reifiers.

Consider the following reflective expression:

```
(let ((x "foo"))
  (meaning '(let ((x "bar"))
              ((mu (r k) x)))
           (reify-new-environment)
           (lambda (a) "whatever")))
```

If we disregard the innermost `let` expression, the `meaning` expression is supposed to be the identity on `x`, except that the identifiers `r` and `k` become bound to the values we have specified.

If the reifier had been a λ-abstraction, the `x` in the body would of course refer to the innermost definition, that is, the body of the abstraction would be evaluated in the environment of definition. And this is also the way it works for reifiers in [Smith 82] and [Wand & Friedman 88].

But in fact, even though the μ-abstraction is declared at the same place as it is applied, it is not going to be evaluated in the context `(let ((x "bar")) ...)`. The μ-abstraction will run at the level above – and the binding of `x` to `"bar"` will be present in the environment-representation that the parameter `r` is bound to. Thus it is conceptually unsatisfactory to run at the level above with an environment of the current level.

Furthermore, it seems unsatisfactory that even if the innermost binding were not there, the outermost binding of `x` would not be visible. This is proper for a λ-abstraction, since the outermost binding belongs to the level above, but not for a μ-abstraction since this would mean that the expression `(meaning '((mu ...) ...) ...)` would change the current lexical environment. For these reasons the μ-abstractions should rather run in an environment of the level above their level of definition.

One way to obtain this is to let a μ-abstraction close the environment of the level above its level of definition. Then the domain of reifiers would be:

$$m \in \mu Abs = DenVal \rightarrow DenVal \rightarrow DenVal^{*} \rightarrow Cont \rightarrow MetaCont \rightarrow Answer$$

and the declaration of a reifier could look like this:

$$\begin{aligned}&\mathcal{E}[\![\texttt{(mu (I}_r\ \texttt{I}_k\ \texttt{I}_1\ \ldots\ \texttt{I}_n\texttt{) E)}]\!]\,\rho\,\kappa\,(\rho',\kappa',\tau')\\&= \kappa\,(inApp(in\mu Abs\,(\lambda\, v_r v_k(v_1,\ldots,v_n).\,\mathcal{E}[\![\texttt{E}]\!]\,([\![\![\texttt{I}_r]\!] \mapsto v_r, [\![\texttt{I}_k]\!] \mapsto v_k,\\&\qquad [\![\texttt{I}_1]\!] \mapsto v_1,\ldots,[\![\texttt{I}_n]\!] \mapsto v_n]\rho')))))\,(\rho',\kappa',\tau')\end{aligned}$$

Note that, even though we do not activate the level above, we do access the meta-continuation in order to find the environment ρ'.

This definition, composed with application like this:

$$\mathcal{E}[\![(\mathtt{E_0\ E_1 \ldots\ E_n})]\!]\,\rho\,\kappa$$
$$= \mathcal{E}[\![\mathtt{E_0}]\!]\,\rho\,(\lambda v(\rho',\kappa',\tau').(\mathit{cases}\ v\ \mathit{of} \ldots$$
$$[\!]\ \mathit{is}\mu\mathit{Abs}(m) \rightarrow$$
$$m\,(\mathit{env}^{\wedge}\rho)(\mathit{cont}^{\wedge}\kappa)((\mathit{exp}^{\wedge}[\![\mathtt{E_1}]\!]),\ldots,(\mathit{exp}^{\wedge}[\![\mathtt{E_n}]\!]))\kappa'\,\tau'$$
$$\ldots \mathit{end}))$$

is equivalent to the description we have given so far. This is still a "statically scoped" reifier since it extends the environment ρ'. With this definition, the reifier would know the bindings at the level above its definition, so the "syntactically outermost" definition of `x` in our example would be known in the body of the reifier. The reifier also knows the bindings at its level of application (at the time of application), but indirectly, through its second parameter.

Another possibility is to evaluate the body of the reifier in the environment of the level above the level of application of the reifier. Then the domain of reifiers would be:

$$m \in \mu\mathit{Abs} = \mathit{DenVal} \rightarrow \mathit{DenVal} \rightarrow \mathit{DenVal}^{*} \rightarrow \mathit{Env} \rightarrow \mathit{Cont} \rightarrow \mathit{MetaCont} \rightarrow \mathit{Answer}$$

And declaration could be defined like this:

$$\mathcal{E}[\![(\mathtt{mu}\ (\mathtt{I}_r\ \mathtt{I}_k\ \mathtt{I}_1\ \ldots\ \mathtt{I}_n)\ \mathtt{E})]\!]\,\rho\,\kappa$$
$$= \kappa\,(\mathit{inApp}(\mathit{in}\mu\mathit{Abs}\,(\lambda\, v_r v_k(v_1,\ldots,v_n)\rho.\ \mathcal{E}[\![\mathtt{E}]\!]\,([[\![\mathtt{I}_r]\!] \mapsto v_r, [\![\mathtt{I}_k]\!] \mapsto v_k,$$
$$[\![\mathtt{I}_1]\!] \mapsto v_1,\ldots,[\![\mathtt{I}_n]\!] \mapsto v_n]\rho))))$$

Such a definition, with a corresponding application, would give the same semantics to the example as the previous definition. This is because the reifier is declared at the place it is used.

In [Smith 82] and [Wand & Friedman 88] reifiers close their environment of definition, just like λ-abstractions. In [Wand & Friedman 88] this does not pose any conceptual problem – their language, Brown, is an experimental language and thus very small; and one of its restrictions is that all the levels share the same environment.

In Brown, the problem of environment is in fact intrinsic since reifiers are not defined directly but constructed from λ-abstractions using a procedure `make-reifier`. This procedure takes any argument and tries to turn it into a reifier; thus

```
(make-reifier (lambda (e r k) <body>))
```

evaluates to a reifier and so does

```
(let ((foo (lambda (e r k) <body>)))
   (make-reifier foo))
```

This means that a reifier in Brown does not even close its environment of definition but the environment of definition of the λ-abstraction from which it is constructed.

More generally, `make-reifier` is a symptom of seeing reifiers as a kind of λ-abstractions. This view might be misleading since the concept of a reifier seems to have no place in the λ-calculus (or its derived applications) although it can be simulated – as shown with the denotational semantics of Blond.

5 Managing Control in the Tower

In this section we consider different ways to represent continuations and reactivate reified continuations. As an example we study the definitions of Scheme's call/cc in various representations.

5.1 Semantics of Representations of Continuations

A first-class representation of a continuation can be considered in at least three ways:

- as a data structure, e.g., a control stack, that can be reinstalled by a special `continue` operator,
- as a kind of procedure that can be applied (since it is a function at the definitional level), or
- (less interesting, but safe) as basically unattainable objects, which are only continuations and can only be reactivated with `meaning`.

Following the first two methods, there are several different, but equally reasonable, possible semantics of the reactivation. The key question is what happens to the current continuation when a representation of a continuation is applied. If, for example, continuations are represented in a domain that we call *ReifCont* and whose elements are applicable and given a special treatment by the interpreter, then there are two obviously different possibilities: ignore κ (this corresponds to the so-called jumpy continuations [des Rivières 88])

$$\mathcal{E}[\![(\texttt{E}_0\ \texttt{E}_1)]\!]\,\rho\,\kappa\,\tau \;=\; \ldots\,isReifCont(k) \rightarrow \mathcal{E}[\![\texttt{E}_1]\!]\,\rho\,k\,\tau \tag{1}$$

or compose κ with τ (corresponding to pushy continuations [Danvy & Malmkjær 88])

$$\mathcal{E}[\![(\texttt{E}_0\ \texttt{E}_1)]\!]\,\rho\,\kappa\,\tau \;=\; \ldots\,isReifCont(k) \rightarrow \mathcal{E}[\![\texttt{E}_1]\!]\,\rho\,k\,(\rho_{init},\kappa,\tau)\ldots \tag{2}$$

The pushy alternative appears clumsy because of the "dummy" environment, which is pushed with the continuation. This alternative also raises another question – when is the new continuation going to be effective? With the application given above, a reification in $\texttt{E}_1$ will reify into the current level and give a representation of the continuation denoted by $\texttt{E}_0$ to the reifier.

But it is also possible to consider $\texttt{E}_1$ as an expression at the current level and only install the level represented by $\texttt{E}_0$ when (if) it can be passed the value of $\texttt{E}_1$:

$$\ldots\,isReifCont(k) \;\rightarrow\; \mathcal{E}[\![\texttt{E}_1]\!]\,\rho\,(k\,\rho\,\kappa)\,\tau\ldots \tag{3}$$

$$cont^{\wedge} = \lambda\,\kappa'.\,inApp(inReifCont(\lambda\,\rho\,\kappa\,v\,\tau.\,\kappa' v\,(\rho,\kappa,\tau)))$$

This is perhaps more natural since the expression $[\![\texttt{E}_1]\!]$ belongs with the environment ρ and the continuation κ at the current level and not at the level below (which is the one of the continuation k). So now we can avoid the dummy environment, and only the value of $[\![E_1]\!]$ is exported to the level below.

If the continuation is reified into the domain of λ-abstractions, rather than getting a special treatment, then the problem of jumpy *vs.* pushy appears in $cont^{\wedge}$ rather than in the application.

$$cont^{\wedge} = \lambda\,\kappa.inApp(in\lambda Abs(\lambda\,v\,\kappa'\tau.\,\kappa\,v\,\tau)) \quad (4)$$

$$\text{or} \quad cont^{\wedge} = \lambda\,\kappa.\,inApp(in\lambda Abs(\lambda\,v\,\kappa'\tau.\,\kappa\,v\,(\rho_{init},\kappa',\tau))) \quad (5)$$

With **continue** it appears in the denotation of this operator (both if it is a special form and if it is bound in the initial environment).

With **meaning**, however, we do not have these problems since the semantics of **meaning** specifies that the current environment and continuation should be pushed on the meta-continuation and since a **meaning** expression contains both expression and environment as well as the continuation.

5.2 Defining call/cc

We would like to define a procedure `call/cc` that, when we consider only the level where it is applied, has the same semantics as the Scheme call/cc and that does not change the state of the tower above its level of application (*i.e.*, the meta-continuation).

In Scheme, call/cc is a procedure which takes a unary procedure and applies it to a procedural representation of the current continuation. The denotation of call/cc in Scheme can be written

$inProc(\lambda\,v\,\kappa.\,(let\ v = Proc(f)\ in\ f\ inProc(\lambda\,v\,\kappa'.\,\kappa\,v)))$

The language Brown – used in [Wand & Friedman 88] – is quite close to Blond syntactically, so for simplicity [Wand & Friedman 88]'s definitions of call/cc are here given in Blond. Continuations are pushy, corresponding, for example, to definition 3. In [Wand & Friedman 88] call/cc is first defined like this

```
0-1> (define call/cc (lambda (f)
                       ((mu (r k)
                            (k ((r 'f) k))))))
0-1: call/cc
```

But, as pointed out immediately in [Wand & Friedman 88], this is only similar to the Scheme call/cc as long as we stay at one level. If we move upwards, we see that its application causes strange side-effects on the tower as, for example, in:

```
0-2> (call/cc (lambda (c) (cons (c 2) (c 3))))
0-2: 2
0-3> ((mu (r k e) e) foo)
```

```
0-2: 3
0-3> ((mu (r k e) e) bar)
0-2: (foo . bar)
0-3>
```

Here `call/cc` causes `c` to be bound to the bottom level loop, and applying `c` to 2 returns the result 2 as expected. In Scheme, that would be it – at the application of a continuation, the current continuation is discarded – Scheme has a "jumpy" behaviour. But due to its design, this is not the case in Brown. The continuation that was active at the application of `c` is saved in the meta-continuation, so when we reify, we arrive in this continuation, and not in the bottom level loop of the level above. This continuation applies `c` again, and if we reify once more, we finally get the results of our reifications. But they have been **cons**ed together and passed to the continuation **k** by the definition of `call/cc`, so we are still at the level 0. The prompt system of Blond clearly shows why: the continuation of the second iteration of the bottom level loop is reactivated again and again, first by the function and then by `call/cc`.

Our denotational model shows us the problems. With $cont^{\wedge}$ defined as in 3 `call/cc` has the denotation:

$$\begin{aligned}&inApp(in\lambda Abs(\lambda\, v\, \kappa\, (\rho', \kappa', \tau').\, (let\ v = App(\lambda Abs(\alpha))\\ &\qquad\qquad\qquad in\ \alpha\ inApp(inReifCont(\lambda\, \rho'\kappa' v\, \tau'.\, \kappa\, v\, (\rho', \kappa', \tau')))\kappa'_1 \tau')))\\ &\text{where } \kappa'_1 = \lambda\, v\, \tau.\, \kappa\, v\, (\rho'_1, \kappa', \tau')),\\ &\qquad \rho'_1 = ([[\![\mathtt{r}]\!] \mapsto (env^{\wedge}([[\![\mathtt{f}]\!] \mapsto v]\rho))\\ &\qquad\qquad [\![\mathtt{k}]\!] \mapsto inApp(inReifCont(\lambda\, \rho'\kappa' v\, \tau'.\, \kappa\, v\, (\rho', \kappa', \tau')))]\rho'),\\ &\text{and} \quad \rho \text{ is the closure environment of } \mathtt{call/cc}.\end{aligned}$$

This differs from Scheme's call/cc since the procedure α is in fact evaluated at the level above (since it is applied to the meta-continuation τ') and with a continuation that will push the level when it is applied. Furthermore, the captured continuation will also push if it is applied.

The second definition of call/cc avoids the last reactivation by changing the body of the reifier from `(k ((r 'f) k))` to `((r 'f) k)`. This puts a restriction on the use of call/cc since it assumes that the continuation is always applied in the procedure, and if continuations are applied more than once, it does not remedy the problem:

```
0-1> (define new-call/cc (lambda (f)
                           ((mu (r k)
                                ((r 'f) k)))))
0-1: new-call/cc
0-2> (new-call/cc (lambda (c) (cons (c 2) (c 3))))
0-2: 2
0-3> ((mu (r k e) e) new-foo)
0-2: 3
0-3> ((mu (r k e) e) new-bar)
1-0: (new-foo . new-bar)
1-1>
```

This scenario is identical to the previous one except in the last application of a reifier, where we now arrive at the level above, since the application of the function to the continuation is not wrapped in an application of the continuation.

In denotational semantics we now have:

$$inApp(in\lambda Abs(\lambda\, v\, \kappa\, (\rho', \kappa', \tau').\, (let\ v = \ App(\lambda Abs(\alpha)) \\ in\ \alpha\ inApp(inReifCont(\lambda\, \rho'\kappa' v\, \tau'.\, \kappa\, v\, (\rho', \kappa', \tau')))\, \kappa'\, \tau')))$$

This is obviously not better than the previous one – α is still evaluated at the wrong level and the captured continuation still has the wrong behaviour.

If we try to write a definition that corresponds to the denotation of Scheme's call/cc, it is not hard to notice the first problem: reified continuations in a reflective tower are not the same kind of objects as Scheme first-class continuations. Whereas Scheme first-class continuations discard the current continuations when they are applied, the continuations of Brown push the current continuation on the meta-continuation in order to spawn a level below the current one. They are designed to do reflection, not to escape from a context and continue another one.

This problem can be remedied either by changing the semantics of the reflective tower to provide jumpy continuations (since we also have `meaning` to reflect) or by "jumpifying" the reified continuation into an object with a behaviour similar to that of Scheme's first-class continuations.

The second problem, that α is evaluated at the wrong level, is caused by the fact that the level is restored by applying the continuation. It can be remedied by changing the semantics of reified continuations; but a more reliable method is to use `meaning` to reflect instead[4].

By designing reified continuations to be jumpy, corresponding to Scheme first-class continuations, we can write the following definition of call/cc[5].

```
(define call/cc
   (lambda (f)
      ((mu (r k)
          (meaning '(f c)
                   (extend-reified-environment '(c) (list k) r)
                   k)))))
```

The denotation of this definition is almost the same as the Scheme call/cc in a reflective tower with jumpy continuations as in 4:

$$inApp(in\lambda Abs(\lambda\, v\, \kappa\, (\rho', \kappa', \tau').\, (let\ v = \ App(\lambda Abs(\alpha)) \\ in\ \alpha\ inApp(in\lambda Abs(\lambda\, v\, \kappa' \tau.\, \kappa\, v\, \tau))\, \kappa\, (\rho'_1, \kappa', \tau'))))$$
where $\rho'_1 = ([[\![\mathtt{r}]\!] \mapsto (env^{\wedge}([[\![\mathtt{f}]\!] \mapsto v]\rho))[\![\mathtt{k}]\!] \mapsto (cont^{\wedge}\kappa)]\rho')$

If we prefer continuations to be pushy, we could use the following version instead:

---

[4]This also solves the third problem of the definition, which is that in the body of the reifier, `(k ((r 'f) k))`, it is assumed that the two occurrences of `k` represent different functions.

[5]The operator `extend-reified-environment` extends a reified environment from two equally long lists of identifiers and values.

```
(define call/cc
    (lambda (f)
        ((mu (r k)
             (meaning '(f c)
                      (extend-reified-environment '(c) (list (jumpify k)) r)
                      k)))))
(define jumpify
    (lambda (pushy-k)
        (lambda (v)
            ((mu (r k-to-be-discarded)
                 (meaning 'v r (r 'pushy-k)))))))
```

This is comparatively more complicated since the continuation has to be "jumpified" before it is passed to the function.

It is worth noticing that neither version leaves the tower above the current level unaffected. In fact any procedure that performs reification and then reflection will side-effect the meta-continuation since the application of the reifier will install its closed environment or extend the current environment and the application of `meaning` will then save this environment instead of the original one on the meta-continuation. This can be observed by deriving the meaning of the combination of reification followed by reflection (here for the "statically" scoped reifiers).

$$
\begin{aligned}
&\mathcal{E}[\![\texttt{((mu (I}_r\ \texttt{I}_k\ \texttt{I}_1\ \ldots\ \texttt{I}_n\texttt{) (meaning E}_e\ \texttt{E}_r\ \texttt{E}_k\texttt{)) E}_1\ \ldots\ \texttt{E}_n\texttt{)}]\!]\,\rho\,\kappa\,(\rho',\kappa',\tau') \Rightarrow\\
&\mathcal{E}[\![\texttt{(mu (I}_r\ \texttt{I}_k\ \texttt{I}_1\ \ldots\ \texttt{I}_n\texttt{) (meaning E}_e\ \texttt{E}_r\ \texttt{E}_k\texttt{))}]\!]\,\rho\\
&\quad(\lambda f\,(\rho',\kappa',\tau').\,(\mathit{let}\ f = \mathit{App}(\mu \mathit{Abs}(m))\\
&\qquad\qquad\qquad \mathit{in}\ m\,(\mathit{env}^\wedge\rho)(\mathit{cont}^\wedge\kappa)((\mathit{exp}^\wedge[\![\texttt{E}_1]\!]),\ldots,(\mathit{exp}^\wedge[\![\texttt{E}_n]\!]))\kappa'\tau'))\\
&\quad(\rho',\kappa,\tau) \Rightarrow\\
&(\lambda\, v_r v_k(v_1,\ldots,v_n)\kappa.\,\mathcal{E}[\![\texttt{(meaning E}_e\ \texttt{E}_r\ \texttt{E}_k\texttt{)}]\!]\,\rho''\kappa)\\
&\quad(\mathit{env}^\wedge\rho)(\mathit{cont}^\wedge\kappa)((\mathit{exp}^\wedge[\![\texttt{E}_1]\!]),\ldots,(\mathit{exp}^\wedge[\![\texttt{E}_n]\!]))\kappa'\tau' \Rightarrow\\
&\quad \text{where } \rho'' = ([[\![\texttt{I}_r]\!] \mapsto v_r, [\![\texttt{I}_k]\!] \mapsto v_k,\\
&\qquad\qquad\qquad [\![\texttt{I}_1]\!] \mapsto v_1,\ \ldots,\ [\![\texttt{I}_n]\!] \mapsto v_n]\rho')\\
&\mathcal{E}[\![\texttt{(meaning E}_e\ \texttt{E}_r\ \texttt{E}_k\texttt{)}]\!]\,\rho'''\kappa'\tau' \Rightarrow\\
&\quad \text{where } \rho''' = ([[\![\texttt{I}_r]\!] \mapsto (\mathit{env}^\wedge\rho), [\![\texttt{I}_k]\!] \mapsto (\mathit{cont}^\wedge\kappa),\\
&\qquad\qquad\qquad [\![\texttt{I}_1]\!] \mapsto (\mathit{exp}^\wedge[\![\texttt{E}_1]\!]),\ldots,[\![\texttt{I}_n]\!] \mapsto (\mathit{exp}^\wedge[\![\texttt{E}_n]\!])]\rho')\\
&\mathcal{E}[\![\texttt{E}_e]\!]\,\rho'''(\lambda\, v_1.\,\mathcal{E}[\![\texttt{E}_r]\!]\,\rho'''(\lambda\, v_2.\,\mathcal{E}[\![\texttt{E}_k]\!]\,\rho'''(\lambda\, v_3\tau.\,\mathcal{E}(\mathit{exp}^\vee v_1)(\mathit{env}^\vee v_2)(\mathit{cont}^\vee v_3)(\rho''',\kappa',\tau))))\tau'
\end{aligned}
$$

That reification followed by reflection always side-effects the tower also implies that although it is possible to write procedures `jumpify` and `pushify` converting one sort of continuations to the other, these procedures will side-effect the level above. Thus a continuation made jumpy with a user-defined procedure `jumpify` is not semantically equivalent to a jumpy continuation.[6]

6 Abstracting Expressions

One way to obtain structural compositionality of the definition of `meaning` is to let $\mathit{exp}^\wedge$ abstract expressions so that reified expressions are only accessible indirectly

---

[6] Although this is only noticed if the whole tower is considered.

through a previously supplied set of primitives. If `meaning` then only allows previously reified expressions or abstract expressions constructed from predefined constants as its first argument, the definition is compositional. This can be obtained with the definitions:

$$exp^{\wedge} \;=\; \lambda\,[\![\mathrm{E}]\!].\,inReifExpr(\lambda\,\rho\,\kappa\,\tau.\,\mathcal{E}[\![\mathrm{E}]\!]\,\rho\,\kappa\,\tau) \tag{6}$$

$$\begin{array}{l}\mathcal{E}[\![\texttt{(meaning E}_e\ \texttt{E}_r\ \texttt{E}_k\texttt{)}]\!]\,\rho\,\kappa\,\tau\\ = \mathcal{E}[\![\mathrm{E}_e]\!]\,\rho\\ \quad (\lambda\,v_1\tau.\,\mathcal{E}[\![\mathrm{E}_r]\!]\,\rho\\ \qquad (\lambda\,v_2\tau.\,\mathcal{E}[\![\mathrm{E}_k]\!]\,\rho\\ \qquad\quad (\lambda\,v_3\tau.\,(let\ v_1 = ReifExpr(e)\\ \qquad\qquad in\ e\,(env^{\vee}v_2)(cont^{\vee}v_3)(\rho,\kappa,\tau)))\tau)\tau)\tau\end{array}$$

This way there is no need for the $exp^{\vee}$ conversion, which is the source of the non-compositionality.

In order to be useful, however, we need a large set of primitives – including constants, operators, and predicates – corresponding to the syntactic algebra.

7 Comparison with Related Work

[Wand & Friedman 88] gives a Scheme-implementation of a reflective tower in a denotational semantics "style".

[Danvy & Malmkjær 88] gives equations for reification and reflections and discusses jumpy *vs.* pushy continuations.

The reflective tower has also been given a denotation as the limit of a chain of processors [Choo 87]. This contrasts with the present work, where it is rather the reflective *language* that is given a denotation.

8 Conclusion

This paper has shown how to analyze problems of the reflective tower using denotational semantics. The problems concern the semantics of continuations as first class objects in a multi-level setting and the bindings of free variables in multi-level procedures when each level has a separate environment.

The analyses of these and similar problems have strongly influenced the development of the Blond reflective tower which, for the moment, supplies both jumpy and pushy continuations as well as three kinds of reifiers [Danvy & Malmkjær 89]. Maintaining distinct but only slightly different primitives seems undesirable in the long run but is ideal at the current experimental state. Since the implementation is in Scheme, it is straightforward to include any new definition obtained in denotational semantics into the current implementation (although this might not give the most efficient result) [Malmkjær 88].

Acknowledgements

I would like to thank Olivier Danvy for our many inspiring discussions about computational reflection and Austin Melton for his careful rereading.

References

[Choo 87] Young-il Choo: *Logic from Programming Language Semantics*, Ph. D. thesis, California Institute of Technology Pasadena, California (1987)

[Danvy & Malmkjær 88] Olivier Danvy, Karoline Malmkjær: *Intensions and Extensions in a Reflective Tower*, proceedings of the 1988 ACM Conference on Lisp and Functional Programming pp 327-341, Snowbird, Utah (July 1988)

[Danvy & Malmkjær 88'] Olivier Danvy, Karoline Malmkjær: *A Blond Primer*, DIKU Rapport 88/21, DIKU, Computer Science Department, University of Copenhagen, Copenhagen, Denmark (October 1988)

[Danvy & Malmkjær 89] Olivier Danvy, Karoline Malmkjær: *Aspects of Computational Reflection in a Programming Language*, working paper, DIKU, Computer Science Department, University of Copenhagen, Copenhagen, Denmark (1989)

[des Rivières 88] Jim des Rivières: *Control-Related Meta-Level Facilities in LISP*, from *Meta-Level Architectures and Reflection*, Patti Maes & Daniele Nardi (eds.), North-Holland (1988)

[Friedman & Wand 84] Daniel P. Friedman, Mitchell Wand: *Reification: Reflection without Metaphysics*, Conference Record of the 1984 ACM Symposium on LISP and Functional Programming pp 348–355, Austin, Texas (August 1984)

[Malmkjær 88] Karoline Malmkjær: *The Reflective Tower*, student project no. 88-12-9, DIKU, Computer Science Department, University of Copenhagen, Copenhagen, Denmark (December 1988)

[Rees & Clinger 86] Jonathan Rees, William Clinger (eds): *Revised^3 Report on the Algorithmic Language Scheme*, Sigplan Notices, Vol. 21, No 12 pp 37-79 (December 1986)

[Reynolds 72] John Reynolds: *Definitional Interpreters for Higher-Order Programming Languages*, proceedings of the *25th ACM National Conference*, pp 717-740, New York (1972)

[Schmidt 86] David A. Schmidt: *Denotational Semantics: A Methodology for Language Development*, Allyn and Bacon, Inc. (1986)

[Strachey & Wadsworth 74] Christopher Strachey, C. P. Wadsworth: *Continuations: a Mathematical Semantics for Handling Full Jumps*, Technical Monograph PRG-11, Oxford University Computing Laboratory, Programming Research Group, Oxford, England (1974)

[Smith 82] Brian C. Smith: *Reflection and Semantics in a Procedural Language*, Ph. D. thesis, MIT/LCS/TR-272, MIT, Cambridge, Massachusetts (January 1982)

[Wand & Friedman 88] Mitchell Wand, Daniel P. Friedman: *The Mystery of the Tower Revealed: a Non-Reflective Description of the Reflective Tower*, Volume 1, Issue 1 of *Lisp and Symbolic Computation* (May 1988)

SEMANTIC MODELS FOR TOTAL CORRECTNESS AND FAIRNESS†

Michael G. Main and David L. Black
Department of Computer Science
University of Colorado
Boulder, CO 80309 USA
Email: main@ boulder.colorado.edu

Abstract. Assertional s-rings are introduced to provide an algebraic setting in which the finite and infinite behavior of nondeterministic programs can be expressed and reasoned about. This includes expressing the fair infinite behavior of nondeterministic iterative programs, and reasoning about termination under various fairness assumptions. We also address the question of when the reasoning techniques are semantically complete.

1. Background and Motivation

The purpose of this paper is to provide an algebraic setting for reasoning about the control structures of iterative nondeterministic programs. The algebra supports reasoning about nontermination (*i.e., total* correctness) and about *fair* nondeterministic constructions. To quote Kuich and Salomaa: "The tools from linear algebra make the proofs computational in nature and, consequently, more satisfactory from the mathematical point of view than the customary proofs." [9]

This algebraic approach to the semantics of programs also underlies dynamic logic (*e.g.* [7,8]), and our particular approach owes much to the assertional categories of Manes [13,15,14] and the use of the Boolean algebra of guards within a zerosum-free semiring [16]. The major addition of our work to this earlier research is the algebraic treatment of nontermination and fairness.

Our starting point is a familiar idea: Nondeterministic programs denote elements in an algebraic structure, which is almost a semiring. Each of the usual syntactic constructions on programs (such as IF–THEN–ELSE or WHILE–DO) has a corresponding semantic operation that is defined in the algebra. Boolean expressions, used as pre-conditions and post-conditions about programs, are elements in a Boolean algebra of "guards", which exists as a subset of the algebra. Because these guards are part of the algebra, the usual program assertions (involving a pre-condition, a program, and a post-condition) can be proved algebraically, using only the laws of the algebra. Moreover, we show algebraically that the usual Hoare rules for proving total correctness assertions are valid — and if certain conditions (given in Section 6) are valid, then the entire Hoare calculus (including rules for termination of fair nondeterministic loops) is "semantically complete".

† This research has been supported in part by National Science Foundation grant CCR-8701946. David Black has been supported by Storage Technology Corporation, Louisville, CO 80028.

This use of the Boolean algebra of guards in a semiring was proposed by Manes and Benson [16] and had some of its motivation from [10,11]. It has been further developed by Manes and Arbib [13,14,15], who use partially additive semirings which are morphism sets in a certain kind of category, called an *assertional category*. They develop a calculus for proving partial correctness assertions about programs, and this calculus was recently shown by Bloom to be sound and complete in the setting of iterative algebraic theories [2]. The semirings used by Manes and Arbib possess a unary operation * which meets the axiom $s^* = ss^* + 1$. The * is used to provide semantic operations for iterative constructions by solving iteration equations – in the spirit of Elgot [4,5] and dynamic logic. In general, these semantic operations ignore infinite iterative behavior and are not appropriate for expressing total correctness assertions (*i.e.*, assertions where termination is guaranteed).†

With this in mind, we introduce another unary operation $^\infty$, where the axiom $s^\infty = ss^\infty$ is met. Intuitively, s^∞ is the result of executing s infinitely often (in much the same way that s^* is the result of executing s finitely often). The meaning of an iterative program is still a solution to the usual iteration equation, but the solution is constructed using both the * and the $^\infty$ operations. Whereas the original approach (using only *) provided a calculus for partial correctness assertions, the new approach (with * and $^\infty$) yields a calculus for total correctness assertions.

In addition, the two unary operations can be combined in various ways that express different kinds of fair behavior within a DO–OD loop (as presented by Francez [6]).

2. S-rings and a Concrete Example

2.1 S-rings

In the previous section, we said that nondeterministic programs will denote elements in a certain kind of algebra. This algebra is *almost* a semiring: The "almost" occurs because our primary example violates the semiring law $s0=0$. But apart from this violation, the structure is a semiring. We refer to this algebraic structure as an *s-ring* (think of a semiring with something missing). Formally, an s-ring is a set S with two binary operations (+ and $\cdot$) and two distinct constants (0 and 1) such that:

1. $(S, +, 0)$ is a commutative monoid.

2. $(S, \cdot, 1)$ is a monoid.

3. Multiplication ($\cdot$) distributes over addition (+) on both sides.

† Total correctness assertions have been studied by Manes and Arbib in one assertional category (the category of partial functions — suitable for deterministic programs), but it's unclear whether their technique extends to other assertional categories.

4. For all $s \in S$: $0s = 0$.

A *zerosum-free s-ring* also has the axiom:

5. For all $s, t \in S$: $s+t = 0$ iff $s = 0 = t$.

We will use the typical semiring notation with s-rings, writing st instead of $s \cdot t$, and assuming that multiplication has precedence over addition in expressions like $s+tu$. If s is an element of an s-ring and i is a natural number, then s^i denotes the multiplication of i copies of s (with s^0 defined as 1).

2.2 The S-ring of Strict Relations

A simple example can clarify the representation of nondeterministic programs by elements of an s-ring. For this example, consider a setting where each program computes in some fixed "state space" D. Thus, each execution of a program starts in some state $d \in D$ and, if it terminates, will finish in some state $e \in D$. There may also be nonterminating executions, which start in some state, but fail to terminate.

A nondeterministic program denotes a binary relation on $D_\perp$, where $D_\perp$ is the set D plus a new element $\perp$ which represents the "result" of a program that is in an unending loop. Intuitively, the relation denoted by a program is the state-transition for the program: if s is a relation denoting a program, and $(d,e) \in s$, then the corresponding program is capable of mapping an initial state d to a final state e. If $e = \perp$, then the program has a nonterminating execution starting in state d. We also require each program's relation to map $\perp$ to $\perp$ and nowhere else (*i.e.*, $(\perp, e) \in s$ iff $e = \perp$). This means that if the input to s comes from a nonterminating program, then s cannot fix this. A relation with this behavior for $\perp$ is called *strict*.

Now we focus on the algebraic structure of the set of all strict binary relations on $D_\perp$. We call this set of relations A and note that it forms an s-ring. Addition in A is union of relations, so $(d,e) \in s+t$ iff $(d,e) \in s$ or $(d,e) \in t$. Multiplication is composition of relations, so $(d,e) \in st$ iff there exists some $c \in D_\perp$ with $(d,c) \in s$ and $(c,e) \in t$. The multiplicative identity (1) is the identity relation $\{(d,d) \mid d \in D_\perp\}$, and the zero (0) is the smallest strict relation $\{(\perp,\perp)\}$. These s-ring operations correspond to operations on programs. If s and t represent programs, then st is the composite program ("first do s, then do t"). The union relation $s+t$ is a program which can behave like either s or t ("a nondeterministic choice between s and t").

Certain relations in A do not correspond to programs, but they have another important interpretation. These are the relations which are subrelations of 1. Such a relation, called a *guard,* has two choices for each $d \in D$: either d is related to d (and nothing else) or d is related to nothing. Each such guard also relates $\perp$ to $\perp$, since we are dealing with strict relations. A predicate p on D can be interpreted as the guard $\{(d,d) \mid p(d) \text{ or } d = \perp\}$. For a guard p and a state d, we say that d *satisfies* p provided that $(d,d) \in p$. Each guard p has a complement guard $\bar{p}$, such that $p + \bar{p} = 1$ and $p\bar{p} = 0 = \bar{p}p$. A state from D satisfies exactly one of p and $\bar{p}$.

2.3 Conditional Programs and Iteration

Let b be a guard, and let s and t be relations. The relation for a conditional program **IF** b **THEN** s **ELSE** t is expressed in the s-ring of relations as $bs + \bar{b}t$.

To express the relation for an iterative program **WHILE** b **DO** s, we use two unary operations on relations. For any $u \in A$, define u^* to be the transitive and reflexive closure of u. And define u^∞ to be $\{(d, \perp) \mid d$ lies on an infinite u-path$\}$. An *infinite u-path* is a countably infinite sequence $d_0, d_1, \cdots$ such that for all $i \geq 0$, $(d_i, d_{i+1}) \in u$. With these operations, the meaning of **WHILE** b **DO** s is $(bs)^\infty + (bs)^*\bar{b}$. The left term provides the nonterminating behavior of the loop, and the right term provides the terminating behavior.

The $*$ and $^\infty$ can also be used to express various kinds of *fair* iteration. For example, consider the program

$$\textbf{WHILE } b \textbf{ DO } (s_1 \,\square\, s_2),$$

where $\square$ indicates a nondeterministic choice. In the usual semantics, this **WHILE**-loop may have infinite computations which do not execute each of the choices (s_1 and s_2) infinitely often. If we let $t = bs_1 + bs_2$, then the usual meaning of the loop is $t^\infty + t^*\bar{b}$.

But this kind of program has also been studied using various *fairness assumptions* (see [1,6]). The simplest fairness assumption is to forbid infinite computations that don't choose both branches infinitely often. In our algebra, the meaning of the loop with this fairness assumption is:

$$(t^*bs_1t^*bs_2)^\infty + t^*\bar{b},$$

where t is defined as in the previous paragraph. Intuitively, the second term represents the finite behaviors and the first term represents fair infinite behaviors. Later we will show how to express other kinds of fairness.

In a moment, we will introduce *assertional s-rings*, which are s-rings with two additional operations $*$ and $^\infty$. But first we explain how assertions about programs are given in the s-ring of strict relations.

2.4 Assertion Semantics and $ps\bar{q} = 0$

In assertion semantics, reasoning about programs occurs in terms of a pre-condition p and a post-condition q. These conditions are guards (*i.e.*, predicates on states), and in order for a program s to be totally correct with respect to the conditions, the following must hold: whenever the program s starts in a state which satisfies p, then it will end in a state which satisfies q. The notation $[p]\, s\, [q]$ is an assertion that s is totally correct for pre-condition p and post-condition q.

In the s-ring of strict relations, correctness of a program s with respect to pre-condition p and post-condition q is simply $ps\bar{q}=0$. The intuitive translation of this equality says that certain mappings are forbidden by s: Specifically, it is not possible to start in a state $d \in D$ such that $p(d)$ holds, $(d,e) \in s$, and $q(e)$ fails to hold. Also, it is not possible to start in a state $d \in D$ such that $p(d)$ holds and $(d,\perp) \in s$, since this would imply $(d,\perp) \in ps\bar{q} \neq 0$. Thus, $ps\bar{q}=0$ expresses *total* correctness: s cannot fail to terminate when it's started in a state that satisfies p.

Manes [13,15,14] used the equation $ps\bar{q}=0$ in the more general setting of *assertional categories*. However, in these papers, the equation expressed partial correctness of nondeterministic programs (termination was not guaranteed).

3. Program Denotations in Assertional S-rings

The previous section illustrated the idea that nondeterministic programs denote elements in an s-ring of relations, and Boolean expressions denote guards in this s-ring. We now generalize this idea as follows: nondeterministic programs denote elements in a certain kind of s-ring called an *assertional s-ring*, defined in this section. Boolean expressions denote a certain kind of element, which we will call a guard, and correctness assertions about programs are proved by showing algebraic identities of the form $ps\bar{q}=0$. We begin by defining the meaning of a guard in an arbitrary s-ring, and observing some of the properties of guards.

3.1. The Boolean Algebra of Guards

In the s-ring of relations, we defined special relations called guards, which correspond to "state predicates". A similar notion is available in any s-ring, as defined here:

Definition: Let S be an s-ring. A *guard* of S is an element p such that for some $\bar{p} \in S$: $p+\bar{p}=1$ and $p\bar{p}=0=\bar{p}p$. The element $\bar{p}$ is called the complement of p. The set of all guards of S is denoted $GUARD_S$.

Manes and Benson [16] showed that the set of guards in any zerosum-free semiring forms a Boolean algebra, with the partial order $s \leq t$ iff there exists u with $s+u=t$. This order, called the *sum-order*, is not always a partial order on a semiring (anti-symmetry can fail). But for a zerosum-free semiring, it is a Boolean-algebra order on the guards and 0 is the minimum element in the semiring.

Manes and Benson's results also hold for zerosum-free s-rings: the guards form a Boolean algebra under the sum-order, and 0 is the minimal element in the s-ring. In the Boolean algebra of guards, the minimum guard is 0, the maximum guard is 1, the "meet" of p and q is pq and the "join" of p and q is $p+\bar{p}q$. If W is a set of guards and the join (or least upper-bound) of W exists, then we denote this by $\bigvee W$ or sometimes $\bigvee_{p \in W} p$.

3.2. Assertional S-rings

Now we can define assertional s-rings, which we will use as semantic models of nondeterministic programming languages.

Definition: An *assertional s-ring* is a zerosum-free s-ring S, with two unary operations ($*$ and ∞) which meet these axioms: For all $s, t \in S$ and $W \subseteq GUARD_S$:

1. **Closure Axiom:** $s^* = ss^* + 1$.
2. **Iteration Axiom:** $(\forall\ i.\ rs^i t = 0)$ implies $rs^*t = 0$.
3. **Infinity Axiom:** $s^\infty = ss^\infty = s^\infty t$.
4. **Continuity Axiom:** When $\bigvee W$ exists, then $(\forall\ p \in W.ps = 0)$ implies $(\bigvee W)s = 0$.

We write the unary operations using postfix notation, and these have higher precedence than the s-ring operations in expressions. For example ss^∞ is $s\,(s^\infty)$. The motivation for the two new operations and their axioms comes from the corresponding operations in the s-ring of strict relations over $D_\perp$ (see Section 2.3). The intuition behind the axiom $ss^\infty = s^\infty t$ is that s^∞ represents infinite behaviors; therefore anything which follows s^∞ will never be reached.

3.3. Program Constructs and Correctness Assertions

Assertional s-rings provide semantic models for nondeterministic programming languages. In general, nondeterministic programs denote elements in an assertional s-ring, and Boolean expressions denote guards in the same s-ring.

Within any assertional s-ring, the usual program constructs, such as IF–THEN–ELSE and WHILE–DO can be represented algebraically, as discussed in Section 2. Also, the identity $ps\bar{q} = 0$ is important when p and q are guards and s denotes a program, since this corresponds to the correctness assertion $[p]\,s\,[q]$. Because of this correspondence, we will use the $[p]\,s\,[q]$ notation to express the equality $ps\bar{q} = 0$. This can be either total or partial correctness, depending on the particular s-ring.

The remainder of the paper gives algebraic demonstrations of rules for showing correctness assertions for various different forms of the program s. When s has the form of one of the usual program constructions (such as $bs_1 + \bar{b}s_2$ for the program IF b THEN s_1 ELSE s_2), then the demonstrated rules will be the usual Hoare rules for total correctness assertions. This includes rules for fair iterative constructions.

In effect: every assertional s-ring comes with the "semantic operations for programming" and the "Hoare calculus of programs" as standard equipment.

4. Some Basic Rules

This section provides basic rules for proving $[p]s[q]$, where s has one of the forms s_1s_2, s_1+s_2, $(s_1)^*$, or $(s_1)^\infty$. We also present rules that correspond to the Consequence Rule and the Disjunction Rule of the Hoare calculus. Some of the proofs are based on results in [15, Section 3.3] — although the proofs need changing to avoid using $s0=0$.

Throughout this section and the rest of the paper, we will assume a fixed assertional s-ring, using the letters s, t (sometimes with subscripts) for arbitrary elements. We'll use b, p, q, r (sometimes with subscripts) for arbitrary guards.

4.1. Composition

This section gives a rule for proving a correctness assertion of the form $[p]st[q]$.
Composition Rule: $[p]s[q]$ and $[q]t[r]$ imply $[p]st[r]$.
Proof: We are given $ps\bar{q}=0=qt\bar{r}$, and we must show $pst\bar{r}=0$. This is done here:

$$pst\bar{r}=ps(q+\bar{q})t\bar{r}=psqt\bar{r}+ps\bar{q}t\bar{r}=ps0+0t\bar{r}=ps0+0=ps0+ps\bar{q}=ps(0+\bar{q})=ps\bar{q}=0$$

Note the bit of extra work because we cannot immediately conclude that $ps0=0$. □

In general, this rule is not semantically complete – meaning that even when $[p]st[r]$ is true, there might not be any guard q such that $[p]s[q]$ and $[q]t[r]$. We'll address this more in Section 6.

4.2. Addition

This section gives a rule for proving a correctness assertion of the form $[p]s+t[q]$. This rule is semantically complete, meaning that it is sufficient for any assertion of this form.
Addition Rule: $[p]\ s+t\ [q]$ iff $[p]s[q]$ and $[p]t[q]$.
Proof:

$[p]\ s+t\ [q]$

$\Leftrightarrow\ ps\bar{q}+pt\bar{q}=0$

$\Leftrightarrow\ ps\bar{q}=0=pt\bar{q}$

$\Leftrightarrow\ [p]s[q]$ and $[p]t[q]$

The second equivalence is valid from the zerosum-free law. □

4.3. The Consequence Rule and the Disjunction Rule

In the usual Hoare calculus, a valid program assertion remains valid when the pre-condition is weakened or the post-condition is strengthened. This section gives the corresponding rule for assertional s-rings. In this section, we use Boolean algebra terminology on guards, so for example "q_1 implies q_2" means $q_1 q_2 = q_1$ (which is also equivalent to the program assertion $[q_1]\, 1\, [q_2]$).

Consequence Rule: Suppose p_2 implies p_1, and q_1 implies q_2, and $[p_1]\, s\, [q_1]$. Then $[p_2]\, s\, [q_2]$.

Proof: The two assumptions about implications are equivalent to the program assertions $[p_2]\, 1\, [p_2]$ and $[q_1]\, 1\, [q_2]$. From two applications of the composition rule, $[p_2]\, 1\, [p_1]$ and $[p_1]\, s\, [q_1]$ and $[q_1]\, 1\, [q_2]$ imply $[p_2]\, 1s1\, [q_2]$. This is just the result we need, since $1s1 = s$. □

Another rule in the Hoare calculus is the Disjunction Rule, which follows immediately from our Continuity Axiom:

Disjunction Rule: Let W be a set of guards such that $\vee W$ exists, and suppose that for all $p \in W$, $[p]\, s\, [q]$. Then $[\vee W]\, s\, [q]$. □

4.4. The * Operation

Here is the rule for *, which uses the Consequence Rule:

Iteration Rule: Suppose there exists a guard p such that q implies p, and $[p]\, s\, [p]$, and p implies r. Then $[q]\, s^*\, [r]$.

Proof: An induction on i shows that $[p]\, s^i\, [p]$ is valid for all i, and by the Iteration Axiom this implies $[p]\, s^*\, [p]$. Using the Consequence Rule (with "q implies p" and "p implies r"), yields the needed result: $[q]\, s^*\, [r]$. □

4.5. The $^\infty$ Operation

The correctness rule for $[p]\, s^\infty\, [q]$ is notable because it is independent of the post-condition q. This matches our intuition that s^∞ consists of the behaviors resulting from executing s infinitely often – so that the post-condition is never reached! Thus, $[p]\, s^\infty\, [q]$ really means that the pre-condition p is sufficient to guarantee there will be no "infinite paths" in s^∞. Thus, instead of proving assertions $[p]\, s^\infty\, [q]$ with arbitrary post-conditions, we'll generally only prove them with the post-condition 1 (true). Here's the formal justification:

Theorem. $[p]\, s^\infty\, [q]$ iff $[p]\, s^\infty\, [1]$.

Proof: The equivalence follows from this derivation:

$$[p]s^\infty[q] \Leftrightarrow ps^\infty\bar{q}=0 \Leftrightarrow ps^\infty=0 \Leftrightarrow ps^\infty 0=0 \Leftrightarrow [p]s^\infty[1]$$

The second and third equivalences follow from $s^\infty t = s^\infty$, which is part of the Infinity Axiom. □

Notation. Because the post-condition doesn't matter, we will sometimes write TERMINATE(p, t) as notation for $[p]t^\infty[q]$. □

Now we'll give the general rule for proving an assertion TERMINATE(p, t):

Infinity Rule: Let $(I, <)$ be a well-founded set and for each $i \in I$ let p_i be a guard such that:

$$[p_i]\, t\, [\bigvee_{j<i} p_j].$$

Also suppose that p implies $\bigvee_{i \in I} p_i$. Then TERMINATE(p, t).

Proof: To begin, we show that for any $i \in I$, $[p_i]t^\infty[1]$. The proof is by well-founded induction on I. For this induction, let $i \in I$, and suppose (for the induction hypothesis) that whenever $j<i$ then $[p_j]t^\infty[1]$. From the Disjunction Rule, this implies $[\bigvee_{j<i} p_j]t^\infty[1]$. This is combined with the given assertion $[p_i]t[\bigvee_{j<i} p_j]$ to yield $[p_i]tt^\infty[1]$. But $tt^\infty = t^\infty$ (by the Infinity Axiom), so this last assertion is just $[p_i]t^\infty[1]$, and this completes the induction.

Finally, since we have shown $[p_i]t^\infty[1]$ for all $i \in I$, the Disjunction Rule implies $[\bigvee_{i \in I} p_i]t^\infty[1]$, or equivalently TERMINATE$(\bigvee_{i \in I} p_i, t)$. Since p implies $\bigvee_{i \in I} p_i$, the needed result then follows from the Consequence Rule. □

A combination of the Composition Rule and the Infinity Rule gives us the ∞-Composition Rule, will be useful later on:

∞-Composition Rule: Suppose $t_1, t_2, ..., t_n$ are any elements in the s-ring. Let $(I, <)$ be a well-founded set and for each $i \in I$ let p_i be a guard such that:

$$\forall m\ (1 \le m \le n).\ [p_i]\, t_m\, [p_i].$$

$$\exists m\ (1 \le m \le n).\ [p_i]\, t_m\, [\bigvee_{j<i} p_j].$$

Also suppose that p implies $\bigvee_{i \in I} p_i$. Then TERMINATE$(p,\ t_1 t_2 \cdots t_n)$.

Proof: Let $t = t_1 t_2 \cdots t_n$. As in the previous proof, we can show (by well-founded induction on I) that for any $i \in I$, $[p_i]t^\infty[1]$. From this and the Disjunction Rule, it follows that $[\bigvee_{i \in I} p_i]t^\infty[1]$. Since p implies $\bigvee_{i \in I} p_i$, the needed result then follows from the Consequence Rule. □

5. Hoare Calculus in Assertional S-rings

The previous section gave rules for proving correctness assertions about programs constructed with the multiplication, addition, * and $^\infty$ operations in any assertional s-ring. In this section, these four operations are used to define compound operations in an assertional s-ring, corresponding to typical operations on programs (such as IF–THEN–ELSE and WHILE–DO). The results of Section 4 are then used to prove that the usual rules of the Hoare calculus are valid in any s-ring. This includes rules about certain fair iteration constructions.

5.1. IF–THEN–ELSE

Suppose that b is a guard denoting a Boolean expression B, and s,t are elements denoting some syntactic programs S and T. Then the syntactic program IF B THEN S ELSE T is denoted by the element $bs+\bar{b}t$.

Notation. For any guard b and any elements s,t, we use the mnemonic notation IF b THEN s ELSE t for $bs+\bar{b}t$. □

The usual Hoare calculus rule for IF–THEN–ELSE provides a necessary and sufficient condition for proving assertions about these programs:

IF **Rule:** $[p]$ IF b THEN s ELSE t $[q]$ iff $[pb]\,s\,[q]$ and $[p\bar{b}]\,t\,[q]$.

Proof:

$[p]$ IF b THEN s ELSE t $[q]$ $\Leftrightarrow$ $[p]\; bs+\bar{b}t\;[q]$

$\Leftrightarrow$ $[p]\,bs\,[q]$ and $[p]\,\bar{b}t\,[q]$

$\Leftrightarrow$ $pbs\bar{q}=0=p\bar{b}t\bar{q}$

$\Leftrightarrow$ $[pb]\,s\,[q]$ and $[p\bar{b}]\,t\,[q]$

Note that the second equivalence follows from the Addition Rule. □

5.2. Iteration

Suppose that b is a guard denoting a Boolean expression B, and s is an element denoting some syntactic program S. Then the syntactic program WHILE B DO S is denoted by the element $(bs)^\infty+(bs)^*\bar{b}$. Intuitively, $(bs)^\infty$ gives the infinite behavior of the WHILE-loop, and $(bs)^*\bar{b}$ gives the finite behaviors.

Notation. For any guard b and any element s, we use the mnemonic notation WHILE b DO s for $(bs)^{\infty} + (bs)^{*}\bar{b}$. □

The usual Hoare calculus rule for WHILE–DO is valid for proving assertions about these programs:

WHILE **Rule:** Suppose the following two conditions hold (where "**or**" is join in the Boolean algebra of guards):

PARTIAL CORRECTNESS: There exists a guard p (the "loop invariant") such that

$$q \text{ implies } p \quad \text{and} \quad [pb]\, s\, [p] \quad \text{and} \quad p \text{ implies } (b \textbf{ or } r).$$

TERMINATION: There exists a well-founded set $(I, <)$ and a guard p_i for each $i \in I$ such that:

(i) For every $i \in I$: $[p_i b]\, s\, [\bigvee_{j<i} p_j]$, and

(ii) q implies $\bigvee_{i \in I} p_i$.

Then $[q]$ WHILE b DO s $[r]$.

Proof: Assume the two conditions hold. We must prove $[q]$ WHILE b DO s $[r]$, or equivalently both

(1) TERMINATE(q, bs), and

(2) $[q]\,(bs)^{*}\bar{b}\,[r]$.

Assertion (1) will follow from the TERMINATION condition, and assertion (2) from PARTIAL CORRECTNESS.

For (1), we can rewrite part (i) of TERMINATION as this:

$$\forall i \in I. [p_i]\, bs\, [\bigvee_{j<i} p_j].$$

This rewriting makes TERMINATION equivalent to the hypothesis of the Infinity Rule (taking bs as t in that rule). Therefore, the Infinity Rule implies TERMINATE$(\bigvee_{i \in I} p_i,\, bs)$. This result, together with "q implies $\bigvee_{i \in I} p_i$", (and the Consequence Rule) implies (1).

For (2), we can rewrite PARTIAL CORRECTNESS as:

$$q \text{ implies } p \quad \text{and} \quad [p]\, bs\, [p] \quad \text{and} \quad p \text{ implies } (b \textbf{ or } r).$$

This is now in the form to apply the Iteration Rule, which implies $[q]\,(bs)^{*}\,[b \textbf{ or } r]$, or equivalently $0 = q\,(bs)^{*}\,\overline{(b \textbf{ or } r)} = q\,(bs)^{*}\,\bar{b}\,\bar{r}$. Since this last expression is 0, we have $[q]\,(bs)^{*}\bar{b}\,[r]$, as required. □

5.3. Fair Iteration

In Section 2.3 we discussed the program WHILE b DO $(s_1 \Box s_2)$. Using a simple fairness assumption which forbids an infinite computation from eventually ignoring one of the directions, this program has the algebraic meaning:

$$(t^*bs_1t^*bs_2)^\infty + t^*\bar{b}$$

where t is defined as $bs_1 + bs_2$.

In proving correctness assertions about such a program, the second term ($t^*\bar{b}$, which represents terminating behavior) can be handled using the PARTIAL CORRECTNESS condition of the WHILE-Rule. The first term – representing the fair infinite behavior – can be handled with the following rule, which is motivated by the "Unconditional Fair Termination" Rule of [6].

Fair-WHILE Termination Rule: Let $(I, <)$ be a well-founded set and for each $i \in I$ let p_i be a guard such that:

(i) For all k $(1 \le k \le 2)$: $[p_i b]\, s_k\, [p_i]$, and

(ii) For some k $(1 \le k \le 2)$: $[p_i b]\, s_k\, [\bigvee_{j<i} p_j]$.

Also suppose that p implies $\bigvee_{i \in I} p_i$.

Then TERMINATE(p, $t^*bs_1t^*bs_2$), where $t = (bs_1 + bs_2)$.

Proof: We will show that the hypothesis of the ∞-Composition Rule (Section 4.5) is met, by breaking $(t^*bs_1t^*bs_2)$ into the four pieces $t_1 = t^*$, $t_2 = bs_1$, $t_3 = t^*$, and $t_4 = bs_2$.

For the first part of the ∞-Composition hypothesis, we must show that for all $i \in I$:

$$\forall m\ (1 \le m \le 4).\ [p_i]\, t_m\, [p_i].$$

For $m=2$ and $m=4$, this is just the statement (i). For $m=1$ and $m=3$, we have $t_m = t^*$, so we must show $[p_i]\, t^*\, [p_i]$ (for all $i \in I$). Toward this goal, let i be some element of I and note that by the Iteration Axiom it is sufficient to show that $[p_i]\, t^h\, [p_i]$ for all natural numbers h. We prove this by induction on h: For the base case ($[p_i]\, t^0\, [p_i]$):

$$p_i t^0 \bar{p}_i = p_i 1 \bar{p}_i = p_i \bar{p}_i = 0.$$

For the induction step, assume $[p_i]\, t^h\, [p_i]$ for some h. Also note that $[p_i]\, t\, [p_i]$ follows from the Addition Rule since:

$$t = bs_1 + bs_2 \text{ (by definition)},$$
$$[p_i]\, bs_1\, [p_i] \text{ (by (i)), and}$$
$$[p_i]\, bs_2\, [p_i] \text{ (by (i)).}$$

Combining $[p_i]\, t^h\, [p_i]$ and $[p_i]\, t\, [p_i]$ with the Composition Rule yields $[p_i]\, t^{h+1}\, [p_i]$, which completes the induction.

For the second part of the ∞-Composition hypothesis, we must show that for all $i \in I$:

$$\exists m\ (1 \le m \le 4).\ [p_i]\, t_m\, [\bigvee_{j<i} p_j].$$

But for any $i \in I$, this follows immediately from (ii) — in fact we know that it must be valid for either $m=2$ or $m=4$. □

5.4. Weak and Strong Fair Iteration

Guarded iteration, introduced by Dijksta [3], has the form:

DO

$b_1 \to s_1$

□

$b_2 \to s_2$

OD

This is a loop where each iteration will execute one of the s_i where the corresponding guard (b_i) is true. When both the guards become false, the loop terminates. In general, the number of directions could be more than two, but for clarity we'll only handle the two-direction case.

To define the meaning of the loop, it will be useful to use the abbreviations BODY for $(b_1 s_1 + b_2 s_2)$, and b for $b_1 + b_2$. Informally, BODY is one iteration of the body of the loop, and b is the condition for continuing the loop. With these definitions, the meaning of the guarded iteration will always have the following form:

$$(\cdots)^\infty + \text{BODY}^*\bar{b},$$

The $(\cdots)^\infty$ portion indicates the infinite computations, and $\text{BODY}^*\bar{b}$ indicates the terminating computations. The $(\cdots)$ will vary, depending on the particular fairness assumptions that we choose.

In general, proving a program assertion about a DO–OD program requires partial correctness ($[p]\,\text{BODY}^*\bar{b}\,[q]$) and TERMINATE($p$, $\cdots$). Partial correctness can be handled in the same way as Section 5.2, so we won't deal with that here. Thus, this section is concerned with proving termination assertions of the form TERMINATE(p, $\cdots$) for various forms of $(\cdots)$ which arise from different fairness assumptions about guarded iteration.

No Fairness Assumption: With no fairness assumption, the algebraic meaning of the infinite part of the DO–OD loop is just BODY^∞. Here's the rule for proving its termination:

DO–OD Termination Rule: Let $(I, <)$ be a well-founded set and for each $i \in I$ let p_i be a guard such that:

For all k $(1 \le k \le 2)$: $[p_i b_k]\, s_k\, [\bigvee_{j<i} p_j]$.

Also suppose that p implies $\bigvee_{i \in I} p_i$. Then TERMINATE(p, BODY).

Proof: Consider any $i \in I$. Since $[p_i b_k] s_k [\bigvee_{j<i} p_j]$ holds for both $k=1$ and $k=2$, we also have $[p_i] b_1 s_1 + b_2 s_2 [\bigvee_{j<i} p_j]$, which is equivalent to:

$$[p_i] \text{ BODY } [\bigvee_{j<i} p_j].$$

But this is just the hypothesis of the Infinity Rule of Section 4.5; therefore TERMINATE($\bigvee_{i \in I} p_i$, BODY). Since p implies $\bigvee_{i \in I} p_i$, the needed result then follows from the Consequence Rule. □

Weak Fairness Assumption: The assumption of *weak fairness* forbids an infinite path from ignoring a direction when the corresponding guard is always true at the choice point. More precisely: for each direction k, an infinite path must take direction k infinitely often — unless b_k is false infinitely often when the choice occurs. Algebraically, the weakly-fair infinite paths are expressed as:

$$(\text{BODY}^* \ (b_1 s_1 + \bar{b}_1 b_2) \ \text{BODY}^* \ (b_2 s_2 + \bar{b}_2 b_1))^\infty.$$

Intuitively, a weakly-fair infinite path must execute direction 1 infinitely often, unless guard b_1 fails infinitely often – hence the subterm $(b_1 s_1 + \bar{b}_1 b_2)$. And similarly for direction 2.

Here's the rule for proving termination of a DO–OD loop with the weak fairness assumption. The rule is based on the Weakly-Fair Termination Rule of [6].

Weakly-Fair Termination Rule: Let $(I, <)$ be a well-founded set and for each $i \in I$ let p_i be a guard such that:

(i) For all k $(1 \le k \le 2)$: $[p_i b_k] s_k [p_i]$, and

(ii) For some k $(1 \le k \le 2)$: $[p_i b_k] s_k [\bigvee_{j<i} p_j]$, and $p_i \bar{b}_k$ implies $\bigvee_{j<i} p_j$.

Also suppose that p implies $\bigvee_{i \in I} p_i$.

Then TERMINATE(p, BODY* $(b_1 s_1 + \bar{b}_1 b_2)$ BODY* $(b_2 s_2 + \bar{b}_2 b_1)$).

Proof: We can show that the hypothesis of the $^\infty$-Composition Rule (Section 4.5) is met by breaking (BODY* $(b s_1 + \bar{b}_1 b_2)$ BODY* $(b s_2 + \bar{b}_2 b_1)$) into the four pieces

$$t_1 = \text{BODY}^*,$$
$$t_2 = b_1 s_1 + \bar{b}_1 b_2,$$
$$t_3 = \text{BODY}^*,$$
$$t_4 = b_2 s_2 + \bar{b}_2 b_1.$$

The remainder of the proof is similar to the previous Fair-WHILE Termination Rule. □

Strong Fairness Assumption: An alternate fairness assumption for a guarded iteration is called *strong fairness*. Under this assumption, an infinite path must execute each branch infinitely often, provided that the branch's guard is true infinitely often. In other words, an infinite path may eventually ignore a branch, provided that the branch's guard is eventually always false. Algebraically, the strongly-fair infinite paths can be expressed as:

$$(\text{BODY}^* \ (b_1 s_1 + (\bar{b}_1 b_2 s_2)^\infty) \ \text{BODY}^* \ (b_2 s_2 + (\bar{b}_2 b_1 s_1)^\infty))^\infty.$$

Intuitively, this expression says that a strongly-fair infinite path may execute a finite number of iterations (BODY^*), but eventually it must either execute direction 1 ($b_1 s_1$) or go into a loop where guard 1 remains false ($(\bar{b}_1 b_2 s_2)^\infty$). And similarly for direction 2.

Program assertions about strongly-fair guarded iteration can be proved with the following rule, based on the Strongly-Fair Termination Rule of [6].

Strongly-Fair Termination Rule: Let $(I, <)$ be a well-founded set and for each $i \in I$ let p_i be a guard such that:

(i) For all k $(1 \le k \le 2)$: $[p_i b_k]\, s_k\, [p_i]$.

(ii) Either $[p_i b_1]\, s_1\, [\bigvee_{j<i} p_j]$, and $\text{TERMINATE}(p_i, \ \bar{b}_1 b_2 s_2)$,

or $[p_i b_2]\, s_2\, [\bigvee_{j<i} p_j]$, and $\text{TERMINATE}(p_i, \ \bar{b}_2 b_1 s_1)$.

Also suppose that p implies $\bigvee_{i \in I} p_i$.

Then $\text{TERMINATE}(p, \ \text{BODY}^* \ (b_1 s_1 + (\bar{b}_1 b_2 s_2)^\infty) \ \text{BODY}^* \ (b s_2 + (\bar{b}_2 b_1 s_1)^\infty))$.

Proof: We can show that the hypothesis of the ∞-Composition Rule (Section 4.5) is met, by breaking $\text{BODY}^* \ (b s_1 + (\bar{b}_1 b_2 s_2)^\infty) \ \text{BODY}^* \ (b s_2 + (\bar{b}_2 b_1 s_1)^\infty)$ into the four pieces

$$t_1 = \text{BODY}^*,$$
$$t_2 = b_1 s_1 + (\bar{b}_1 b_2 s_2)^\infty,$$
$$t_3 = \text{BODY}^*,$$
$$t_4 = b_2 s_2 + (\bar{b}_2 b_1 s_1)^\infty.$$

The remainder of the proof is similar to the previous Fair-WHILE Termination Rule. □

6. Semantic Completeness

We have given rules which are sufficient for proving program assertions for programs of the forms:

st	WHILE b DO s
$s+t$	WHILE b DO $s_1 \Box s_2$ (Fair)
s^*	DO–OD loop
s^∞	Weakly Fair DO–OD loop
IF b THEN s ELSE t	Strongly Fair DO–OD loop

The rules all have a similar form: If some algebraic condition holds, then some program assertion holds — where the program assertion is about a program of one of the above forms. Such a rule is *semantically complete* if the algebraic condition is also necessary for the program assertion to hold.

Some of our rules are semantically complete. For example, the Addition Rule states that $[p]\, s+t\, [q]$ if *and only if* $[p]\, s\, [q]$ and $[p]\, t\, [q]$. But in general, the rules are not semantically complete. For example, it is quite easy to define an assertional s-ring with an element s such that $[1]\,ss\,[1]$ is valid, but there is no guard q with both $[1]\,s\,[q]$ and $[q]\,s\,[1]$.

Ideally, we would like our rules to be semantically complete, since this guarantees that the rules are as strong as possible. This section gives a condition for all the rules in this paper to be semantically complete.

Completeness Theorem. Suppose that

(i) The Composition Rule is semantically complete, so that $[p]\,st\,[r]$ if *and only if* there exists some q with $[p]\,s\,[q]$ and $[q]\,t\,[r]$.

(ii) The Infinity Rule is semantically complete, so that TERMINATE(p, s) if *and only if* the hypothesis of the Infinity Rule is met.

(iii) The Boolean algebra of guards is a complete Boolean algebra, so that for any set W of guards, $\bigvee W$ exists.

Then the Iteration Rule and each of the rules of Sections 4 and 5 are also semantically complete.

Proof: The proof consists of a sequence of lemmas, showing that the three conditions imply that each of the indicated rules is semantically complete. These lemmas are given in the remainder of this section. □

It is not hard to show that the hypotheses of the Completeness Theorem hold for the the s-ring of strict relations; hence all the indicated rules are semantically complete when programs are represented by strict relations. In particular, the fairness rules are complete — and the demonstration of that completeness comes from completeness of simpler rules. This may be easier than the usual direct proof in [6].

For the remainder of this section, we will assume the hypotheses of the Completeness Theorem, and we'll use this assumption to prove the "sequence of lemmas" mentioned in the Completeness Theorem's proof. Prior to this sequence of lemmas, we will prove a few results about useful guards called *kernels* and *domains*, which are guaranteed to exist by the hypotheses of the Completeness Theorem.

6.1 Kernels and Domains

The third hypothesis of the Completeness Theorem is that the Boolean algebra of guards is complete. Intuitively, this assures us that there are "enough" guards around. In particular, it allows us to define guards called kernels and domains, as follows.

Definition. Let s be an element in an s-ring S where the Boolean algebra of guards (denoted by $GUARDS_S$) is complete. The *kernel* of s (denoted by $\text{KER}(s)$) and the *domain* of s (denoted by $\text{DOM}(s)$) are defined by:

$$\text{KER}(s) = \bigvee\{p \in GUARDS_S \mid ps = 0\} \quad \text{and} \quad \text{DOM}(s) = \bigwedge\{p \in GUARDS_S \mid ps = s\}$$

□

In the s-ring of strict relations, a state d satisfies $\text{DOM}(s)$ provided that $(d, e) \in s$ for some e. For each element s in an s-ring, $\text{KER}(s)$ is always the complement of $\text{DOM}(s)$. This property and other properties of kernels are summarized below. The theorems make use of the first and third hypotheses of the Completeness Theorem.

Theorem 6.1.0 $\text{KER}(s) = \overline{\text{DOM}(s)}$.

Proof: By the infinite DeMorgan Law, $\overline{\text{DOM}(s)} = \bigvee\{\bar{p} \in GUARDS_S \mid ps = s\}$. But $ps = s$ iff $\bar{p}s = 0$ in any s-ring. Therefore, $\overline{\text{DOM}(s)} = \bigvee\{\bar{p} \in GUARDS_S \mid \bar{p}s = 0\}$, which is just $\text{KER}(s)$. □

Theorem 6.1.1. $\text{KER}(s)s = 0$ and $\text{DOM}(s)s = s$.

Proof: The first equality follows immediately from the Continuity Axiom and the definition of kernels. The second equality follows from:

$$s = 1s = (\text{KER}(s) + \text{DOM}(s))s = \text{KER}(s)s + \text{DOM}(s)s = 0 + \text{DOM}(s)s = \text{DOM}(s)s.$$

□

Theorem 6.1.2. $st = 0$ if and only if $s\,\text{DOM}(t) = 0$.

Proof: Assume $s\,\text{DOM}(t) = 0$. Then by 6.1.1 we have $st = s\,\text{DOM}(t)t = 0t = 0$. On the other hand, assume $st = 0$. Therefore $[1]\,st\,[0]$, and by the first hypothesis of the Completeness Theorem, there exists some guard q such that $[1]\,s\,[q]$ and $[q]\,t\,[0]$. For this value of q the two assertions can be rewritten as:

$$(1)\ s\bar{q} = 0, \text{ and } (2)\ qt = 0.$$

From (2), we have that q is below $\text{KER}(t)$ in the Boolean algebra of guards, or equivalently $\bar{q}$ is above $\text{DOM}(t)$. But recall that the order on the guards is the summation order, so there exists some p such that $\bar{q} = p + \text{DOM}(t)$. Finally, since $s\bar{q} = 0$, and the s-ring is zerosum-free, this implies that $s\,\text{DOM}(t)$ is also 0. □

Theorem 6.1.3. $\text{KER}(st) = \text{KER}(s\,\text{DOM}(t))$.

Proof:

$$\text{KER}(st) = \bigvee\{p \mid pst = 0\} = \bigvee\{p \mid ps\,\text{DOM}(t) = 0\} = \text{KER}(s\,\text{DOM}(t)).$$

The second equality is from 6.1.2. □

Theorem 6.1.4. $\text{KER}(s+t) = \text{KER}(s)\text{KER}(t)$.

Proof: Let $A = \{p \mid ps = 0\}$ and $B = \{p \mid pt = 0\}$. Then

$$\text{KER}(s+t) = \bigvee\{p \mid p(s+t)=0\} = \bigvee(A \cap B) = (\bigvee A)(\bigvee B) = \text{KER}(s)\text{KER}(t).$$

The third equality is not valid for arbitrary A and B, but in this case it is true since A and B are downward closed. □

Theorem 6.1.5. $[\text{KER}(s^*t)]\, s\, [\text{KER}(s^*t)]$.

Proof: From 6.1.1 we have $\text{KER}(s^*t)s^*t = 0$. Since $s^* = ss^* + 1$, this implies $\text{KER}(s^*t)ss^*t = 0$. From 6.1.2, this implies $\text{KER}(s^*t)s\,\text{DOM}(s^*t) = 0$. This last equality is just $[\text{KER}(s^*t)]\, s\, [\text{KER}(s^*t)]$. □

6.2 Completeness of the Iteration Rule

We can now use kernels to demonstrate the completeness of the Iteration Rule:

Lemma 6.2. Assume the hypotheses of the Completeness Theorem, and that $[q]\, s^*\, [r]$. Then there exists a guard p such that q implies p, and $[p]\, s\, [p]$, and p implies r.

Proof: Define $p = \text{KER}(s^*\bar{r})$. We will show that this choice of p meets the three requirements.

Step 1: Show that q implies p.

We must show that q is below p in the Boolean algebra order. From $[q]\, s^*\, [r]$ it follows that $qs^*\bar{r} = 0$, and therefore q is below $\bigvee\{q \mid qs^*\bar{r} = 0\} = \text{KER}(s^*\bar{r}) = p$.

Step 2: Show that p implies r.

From 6.1.1 we have $ps^*\bar{r} = \text{KER}(s^*\bar{r})s^*\bar{r} = 0$. Since $s^* = ss^* + 1$, this implies $p(ss^*+1)\bar{r} = 0$, and from the zerosum-free axiom we must have $p\bar{r} = 0$. This occurs only if p implies r in the Boolean algebra, which is the result we need.

Step 3: Show that $[p]\, s\, [p]$.

We must show $ps\bar{p} = 0$, which is done here:

$ps\bar{p} = \text{KER}(s^*\bar{r})s\bar{p}$	(Definition of p)
$= \text{KER}(ss^*\bar{r} + \bar{r})\, s\bar{p}$	($s^* = ss^* + 1$)
$= \text{KER}(\bar{r})\, \text{KER}(ss^*\bar{r})\, s\bar{p}$	(6.1.4)
$= \text{KER}(\bar{r})\, \text{KER}(s\,\text{DOM}(s^*\bar{r}))\, s\bar{p}$	(6.1.3)
$= \text{KER}(\bar{r})\, \text{KER}(s\bar{p})\, s\bar{p}$	(Definition of p)
$= \text{KER}(\bar{r})\, 0$	(6.1.1)
$= 0$	(Guard times 0 is 0)

□

6.3 Completeness of the WHILE Rule

The semantic completeness of the **WHILE** Rule follows from the semantic completeness of the Iteration Rule and the Infinity Rule.

Lemma 6.3. Assume the hypotheses of the Completeness Theorem, and that $[q]$ WHILE b DO $s\ [r]$. Then the following two conditions hold (where "**or**" is join in the Boolean algebra of guards):

PARTIAL CORRECTNESS: There exists a guard p (the "loop invariant") such that:

$$q \text{ implies } p \quad \text{and} \quad [pb]\, s\, [p] \quad \text{and} \quad p \text{ implies } (b \textbf{ or } r).$$

TERMINATION: There exists a well-founded set $(I, <)$ and a guard p_i for each $i \in I$ such that:

(i) For every $i \in I$: $[p_i b]\, s\, [\bigvee_{j<i} p_j]$, and

(ii) q implies $\bigvee_{i \in I} p_i$.

Proof:

$$[q] \text{ WHILE } b \text{ DO } s\ [r] \;=>\; [q]\ (bs)^{\infty} + (bs)^{*}\bar{b}\ [r] \qquad \text{(Definition of WHILE)}$$

$$=> \; [q]\,(bs)^{\infty}\,[r] \text{ and } [q]\,(bs)^{*}\bar{b}\,[r] \qquad \text{(Addition Rule)}$$

$$=> \; [q]\,(bs)^{\infty}\,[r] \text{ and } [q]\,(bs)^{*}\ [b \textbf{ or } r]$$

By hypothesis (ii) of the Completeness Theorem, the first assertion in the last line implies TERMINATION. By Lemma 6.2, the second assertion in the last line implies PARTIAL CORRECTNESS. □

6.4 Completeness of the Fair-WHILE Termination Rule

Semantic completeness of the Fair-**WHILE** Termination Rule follows from the semantic completeness of the Infinity Rule. The proof is omitted, since it is just a simplification of the proof of Lemma 6.6.

Lemma 6.4. Assume the hypotheses of the Completeness Theorem, and that TERMINATE(p, $t^{*}bs_1t^{*}bs_2$), where $t = (bs_1 + bs_2)$. Then there exists a well-founded set $(I, <)$ and a guard p_i for each $i \in I$ such that p implies $\bigvee_{i \in I} p_i$ and for each $i \in I$:

(i) For all k $(1 \le k \le 2)$: $[p_i b]\, s_k\, [p_i]$, and

(ii) For some k $(1 \le k \le 2)$: $[p_i b]\, s_k\, [\bigvee_{j<i} p_j]$.

□

6.5 Completeness of the DO–OD Termination Rule

In this section, we show that the DO–OD Termination Rule is semantically complete for proving termination assertions of the form TERMINATE(p, BODY), where BODY is defined as $b_1 s_1 + b_2 s_2$. Throughout the rest of Section 6, we will always take BODY to be defined this way.

Lemma 6.5. Assume the hypotheses of the Completeness Theorem, and that TERMINATE(p, BODY). Then there exists a well-founded set $(I, <)$ and a guard p_i for each $i \in I$ such that p implies $\bigvee_{i \in I} p_i$ and for each $i \in I$:

For all k $(1 \le k \le 2)$: $[p_i b_k]\, s_k\, [\bigvee_{j<i} p_j]$.

Proof: By hypothesis (ii) of the Completeness Theorem, the assertion TERMINATE(p, BODY) implies the existence of a well-founded set $(I, <)$ and a guard p_i for each $i \in I$ such that p implies $\bigvee_{i \in I} p_i$ and for each $i \in I$:

$$[p_i]\, \text{BODY}\, [\bigvee_{j<i} p_j].$$

Since BODY$= b_1 s_1 + b_2 s_2$, the addition rule states that the assertion on the previous line is equivalent to the two assertions:

$$[p_i]\, b_1 s_1\, [\bigvee_{j<i} p_j] \quad \text{and} \quad [p_i]\, b_2 s_2\, [\bigvee_{j<i} p_j].$$

And these are equivalent to the two required assertions:

$$[p_i b_1]\, s_1\, [\bigvee_{j<i} p_j] \quad \text{and} \quad [p_i b_2]\, s_2\, [\bigvee_{j<i} p_j].$$

□

6.6 Completeness of the Weakly-Fair Termination Rule

This section shows the semantic completeness of the Weakly-Fair Termination Rule for proving termination assertions of the form

$$\text{TERMINATE}(p,\ \text{BODY}^*\ (b_1 s_1 + \bar{b}_1 b_2)\ \text{BODY}^*\ (b_2 s_2 + \bar{b}_2 b_1)).$$

Recall that this is the form of termination assertion that is needed to prove termination of a DO–OD loop under the weak fairness assumption.

Lemma 6.6. Assume the hypotheses of the Completeness Theorem, and that TERMINATE(p, BODY* $(b_1 s_1 + \bar{b}_1 b_2)$ BODY* $(b_2 s_2 + \bar{b}_2 b_1)$). Then there exists a well-founded set $(I, <)$ and a guard p_i for each $i \in I$ such that p implies $\bigvee_{i \in I} p_i$ and for each $i \in I$:

(i) For all k $(1 \le k \le 2)$: $[p_i b_k]\, s_k\, [p_i]$, and

(ii) For some k $(1 \le k \le 2)$: $[p_i b_k]\, s_k\, [\bigvee_{j<i} p_j]$, and $p_i \bar{b}_k$ implies $\bigvee_{j<i} p_j$.

Proof: By hypothesis (ii) of the Completeness Theorem, the assertion TERMINATE(p, BODY* $(b_1s_1+\bar{b}_1b_2)$ BODY* $(b_2s_2+\bar{b}_2b_1)$) implies the existence of a well-founded set $(H,<)$ and a guard p_i for each $i \in H$ such that p implies $\bigvee_{i \in H} p_i$ and for each $i \in H$:

$$[p_i]\ \text{BODY*}\ (b_1s_1+\bar{b}_1b_2)\ \text{BODY*}\ (b_2s_2+\bar{b}_2b_1)\,[\bigvee_{j<i} p_j]\,.$$

We will use this well-founded set H to construct another well-founded set I which meets the requirements of the lemma.

Let $I = H \times \{1,2\}$ be the well-founded set with an ordering defined by $(j,m)<(i,n)$ if and only if $j<i$ or ($j=i$ and m=1 and n=2). For each $(i,n) \in I$, we define a guard p_{in} as follows:

$$p_{i1} = \text{KER}(\text{BODY*}t_2\ \overline{\bigvee_{j<i} p_j})$$

$$p_{i2} = \text{KER}(\text{BODY*}t_1\ \bar{p}_{i1}),$$

where $t_1 = b_1s_1+\bar{b}_1b_2$, and similarly for $t_2=b_2s_2+\bar{b}_2b_1$. It remains to show that these definitions of the set I and the guards p_{in} meet the requirements of the lemma. We do this in three steps, listed below.

Step 1: Show that p implies $\bigvee_{(i,n)\in I} p_{in}$.

We know that p implies $\bigvee_{i \in H} p_i$. Therefore, it is sufficient to show that for each $i \in H$, p_i implies p_{i2}. Here is the proof for an arbitrary $i \in H$:

$[p_i]\ \text{BODY*}t_1\text{BODY*}t_2\,[\bigvee_{j<i} p_j]$	Given about p_i
=> $p_i\text{BODY*}t_1\text{BODY*}t_2\ \overline{\bigvee_{j<i} p_j} = 0$	Definition of Assertion
=> $p_i\text{BODY*}t_1\text{DOM}(\text{BODY*}t_2\ \overline{\bigvee_{j<i} p_j}) = 0$	(6.1.2)
=> $p_i\text{BODY*}t_1\bar{p}_{i1} = 0$	Definition of p_{i1}
=> $p_i\text{DOM}(\text{BODY*}t_1\bar{p}_{i1}) = 0$	(6.1.2)
=> $p_i\bar{p}_{i2} = 0$	Definition of p_{i2}
=> p_i implies p_{i2}	Meaning of "implies"

Step 2: Show that for all $(i,n) \in I$: there exists k such that $[p_{in}b_k]\,s_k\,[\bigvee_{(j,m)<(i,n)} p_{jm}]$.

We show this only for n=2, since the case of n=1 is similar. For n=2, we can always choose the value of k to be 1, since then:

$p_{i2}\text{BODY}^*t_1\bar{p}_{i1} = 0$ (6.1.1)

$=> p_{i2}1t_1\bar{p}_{i1} = 0$ ($\text{BODY}^* = \text{BODY}(\text{BODY}^*) + 1$, and zerosum-free)

$=> p_{i2}b_1s_1\bar{p}_{i1} = 0$ ($t_1 = b_1s_1 + \bar{b}_1b_2$, and zerosum-free)

$=> [p_{i2}b_1]\, s_1\, [p_{i1}]$ Definition of Assertion

$=> [p_{i2}b_1]\, s_1\, [\bigvee_{(j,m)<(i,n)} p_{jm}]$ Consequence Rule

Step 3: Show that for all $(i,n) \in I$: and for k=1,2: $[p_{in}b_k]\, s_k\, [p_{in}]$.

We show this only for n=2, since the case of n=1 is similar. For both k=1 and k=2, we have:

$[\text{KER}(\text{BODY}^*t_1\bar{p}_{i1})]\ \text{BODY}\ [\text{KER}(\text{BODY}^*t_1\bar{p}_{i1})]$ (6.1.5)

$=> [p_{i2}]\ \text{BODY}\ [p_{i2}]$ Definition of p_{i2}

$=> [p_{i2}]\, b_k s_k\, [p_{i2}]$ ($b_k s_k$ is one term of BODY)

$=> [p_{i2}b_k]\, s_k\, [p_{i2}]$ Associativity

These three parts complete the proof. □

6.7 Completeness of the Strongly-Fair Termination Rule

This section discusses the semantic completeness of the Strongly-Fair Termination Rule, for proving termination assertions of the form

$$\text{TERMINATE}(p,\ \text{BODY}^*\ (b_1s_1 + (\bar{b}_1b_2s_2)^\infty)\ \text{BODY}^*\ (b_2s_2 + (\bar{b}_2b_1s_1)^\infty)).$$

Recall that this is the form of termination assertion that is needed to prove termination of a DO–OD loop under the strong fairness assumption.

Lemma 6.7. Assume the hypotheses of the Completeness Theorem, and that $\text{TERMINATE}(p,\ \text{BODY}^*\ (b_1s_1 + (\bar{b}_1b_2s_2)^\infty)\ \text{BODY}^*\ (b_2s_2 + (\bar{b}_2b_1s_1)^\infty))$. Then there exists a well-founded set $(I, <)$ and a guard p_i for each $i \in I$ such that p implies $\bigvee_{i \in I} p_i$ and for each $i \in I$:

(i) For all k ($1 \le k \le 2$): $[p_ib_k]\, s_k\, [p_i]$.

(ii) Either $[p_ib_1]\, s_1\, [\bigvee_{j<i} p_j]$, and $\text{TERMINATE}(p_i,\ \bar{b}_1b_2s_2)$,

or $[p_ib_2]\, s_2\, [\bigvee_{j<i} p_j]$, and $\text{TERMINATE}(p_i,\ \bar{b}_2b_1s_1)$.

Proof: The proof is similar to that of Lemma 6.6, and will be included in a longer version of the paper. □

Acknowledgement

We would like to thank David Benson for comments on a preliminary version of this paper, and we would like to thank a referee for a careful reading. Thanks also to Ernie Manes, whose comments in [12] prompted a simplification of the axioms of an assertional s-ring.

Bibliography

[1] K.R. Apt and G.D. Plotkin. Countable nondeterminism and random assignment, Internal Report CSR-98-82, Department of Computer Science, University of Edinburgh, Scotland, (January 1982). Revised version appears in *JACM 33 (1986).*

[2] S. Bloom and Z. Esik. Floyd-Hoare Logic in Iteration Theories, Technical Report #8801, Stevens Institute of Technology, Electrical Engineering and Computer Science Department (January 1988).

[3] E.W. Dijkstra. Guarded commands, nondeterminacy and formal derivation of programs, *CACM* 18 (1975), 453-457.

[4] C.C. Elgot. Monadic computations and iterative algebraic theories, in: *Logic Colloquium 73* (H.E. Rose and J.C. Sheperdson, Eds.), (North-Holland Publishing Co., 1975), 175-230.

[5] C.C. Elgot. Matricial Theories, *J. Algebra* 42 (1976), 391-421.

[6] N. Francez. *Fairness*, Springer-Verlag, New York, 1986.

[7] D. Harel. First-order dynamic logic, LNCS 68 (Springer-Verlag, 1979).

[8] D. Harel and V.R. Pratt. Nondeterminism in Logic of Programs, in: *Proceedings of the 5th ACM Symposium on Principles of Programming Languages* (1978), 203-213.

[9] W. Kuich and A. Salomaa. *Semirings, Automata, Languages*, Springer-Verlag, 1985.

[10] M.G. Main and D.B. Benson. Functional behavior of nondeterministic programs, in:

Foundations of Computation Theory, LNCS 158, (Springer-Verlag, 1983), 290-301.

[11] M.G. Main and D.B. Benson. Functional behavior of nondeterministic and concurrent programs, *Information and Control* 62 (1984), 144-189.

[12] E.G. Manes. Boolean Theories, preprint.

[13] E.G. Manes. Assertion semantics in a control category, *Theoretical Computer Science*, to appear.

[14] E.G. Manes. Assertional Categories, in: *Proceedings of the Third Workshop on Mathematical Semantics of Programming Languages*, Lecture Notes in Computer Science 298, Springer-Verlag (1988),85-120.

[15] E.G. Manes and M.A. Arbib. *Algebraic Approaches to Program Semantics*, Springer-Verlag, New York, 1986.

[16] E.G. Manes and D.B. Benson. The inverse semigroup of a sum-ordered semiring, *Semigroup Forum* 31 (1985), 129-152.

Equationally Fully Abstract Models of PCF

Allen Stoughton

Computer Science and Artificial Intelligence
School of Cognitive and Computing Sciences
University of Sussex
Falmer, Brighton BN1 9QH, England

1 Introduction

In Plotkin's applied typed lambda calculus PCF [Plo] it is natural to consider one term *operationally less defined* than another iff whenever the first term converges to a constant in a ground context, then the second term converges to the same constant in that context. Two terms are considered *operationally equivalent* iff each is less defined than the other, i.e., they have the same behaviour in all ground contexts. Terms are thus equivalent when they are interchangeable in complete programs. See [Mey] and [Sto] for detailed discussions of these concepts.

Over a decade ago, Robin Milner showed the existence of a unique order-extensional model of PCF that is *inequationally fully abstract* in the sense that one term is operationally less defined than another exactly when the meaning of the first is less than that of the second in the model [Mil]. (Models that consist of functions are called extensional; when in addition these functions are ordered pointwise, the models are called order-extensional.) Milner constructed this model using term model techniques, and considerable effort has been expended in attempts to synthesize his model in a more natural or semantic way; see [BerCurLév] for a survey of this work.

In practice, term equivalence is probably of greater interest than term ordering, and this suggests that one consider models that are *equationally fully abstract* in the sense that two terms are operationally equivalent exactly when they are mapped to the same semantic value. Milner's inequationally fully abstract model is clearly equationally fully abstract, and it is natural to ask whether there exist equationally fully abstract models that are not inequationally fully abstract. The purpose of this paper is to answer this question in the affirmative, and to begin the study of the category **E** of extensional, equationally fully abstract models and structure-preserving functions.

The paper's main results are as follows:

(i) **E** is a pre-ordering with arbitrary products and coproducts and whose initial and terminal objects are not isomorphic.

(ii) All objects of **E** are strongly algebraic (SFP) and all isolated elements of these models are definable by terms.

(iii) There is a morphism from an object $\mathcal{A}$ to an object $\mathcal{B}$ of **E** iff $\mathcal{B}$ relates at least as many pairs of terms as does $\mathcal{A}$ (i.e., if the meaning of M is less than that of N in $\mathcal{A}$, then the meaning of M is less than that of N in $\mathcal{B}$).

(iv) Objects of **E** that relate the same pairs of terms are isomorphic.

(v) The initial object of **E** is also initial in the category of (not necessarily extensional) equationally fully abstract models.

(vi) The terminal object of **E** is order-extensional and inequationally fully abstract, i.e., is Milner's original model.

2 Preliminaries

The reader is assumed to be familiar with such standard domain-theoretic concepts as complete partial orders (cpo's), continuous functions, and ω-algebraic, strongly algebraic and consistently complete cpo's.

A function $f: P \to Q$ over posets is an *order-embedding* iff for all $p_1, p_2 \in P$, $p_1 \sqsubseteq p_2$ iff $f\,p_1 \sqsubseteq f\,p_2$.

In the sequel we will make essential use of Berry's category of dI-domains and stable functions, the definitions of which we now review. A *dI-domain* P is an ω-algebraic, consistently complete cpo such that

(i) $x \sqcap (y \sqcup z) = (x \sqcap y) \sqcup (x \sqcap z)$, for all $x, y, z \in P$ such that $\{y, z\}$ is consistent; and

(ii) for all isolated $p \in P$, $\{p' \in P \mid p' \sqsubseteq p\}$ is finite.

A function $f: P \to Q$ between dI-domains is *stable* iff it is continuous and for all $p \in P$ and $q \in Q$ such that $q \sqsubseteq f\,p$, there exists a least $p' \in P$ such that $p' \sqsubseteq p$ and $q \sqsubseteq f\,p'$. Given dI-domains P and Q, the poset $P \to_s Q$ consists of the set of all stable functions from P to Q, with the *stable ordering*: $f \sqsubseteq g$ iff

(i) $f\,p \sqsubseteq g\,p$, for all $p \in P$; and

(ii) for all $p, p' \in P$ and $q \in Q$, if $q \sqsubseteq f\,p$, $p' \sqsubseteq p$ and $q \sqsubseteq g\,p'$ then $q \sqsubseteq f\,p'$.

In [Ber], it is shown that the collection of dI-domains is closed under $\to_s$, and that the category of dI-domains and stable functions, ordered with the stable ordering, is a cpo-enriched cartesian closed category.

To see that the stable ordering is finer than the pointwise ordering, define functions $f, g: N_\perp \to N_\perp$ by

$$f\,x = \begin{cases} \perp & \text{if } x = \perp, \text{ and} \\ 0 & \text{otherwise,} \end{cases}$$

and $g\,x = 0$. Then f is less than g in the pointwise ordering, but not in the stable ordering.

3 Fully Abstract Models of Programming Languages

In this section, we recall—very briefly—the definitions and results from [Sto] that will be required in the sequel. A gentle introduction to this material can be found in this reference.

The reader is assumed to be familiar with many-sorted signatures Σ over sets of sorts S, as well as algebras over such signatures. Signatures are assumed to contain distinguished constants Ω_s at each sort s, which intuitively stand for divergence. We use uppercase script letters ($\mathcal{A}$, $\mathcal{B}$, etc.) to denote algebras and the corresponding italic letters (A, B, etc.) to stand for their carriers. We write $\mathcal{T}_\Sigma$ (or just $\mathcal{T}$) for the initial (term) algebra, so that T_s is the set of terms of sort s. Given an algebra $\mathcal{A}$ and a term t of sort s, $[\![t]\!]_\mathcal{A}$ (or just $[\![t]\!]$) is the meaning of t in A_s, i.e., the image of t under the unique homomorphism from $\mathcal{T}$ to $\mathcal{A}$. An algebra is *reachable* iff all of its elements are denotable (definable) by terms. A pre-ordering over an algebra is *substitutive* iff it is respected by all of the operations of that algebra. Substitutive equivalence relations are called *congruences*, as usual. A pre-ordering over an algebra in which the Ω constants are least elements at all sorts is referred to as *Ω-least*. The congruence over $\mathcal{T}$ that is induced by an algebra $\mathcal{A}$ is called $\approx_\mathcal{A}$: two terms are congruent when they are mapped to the same element of A. When we say that $c[v_1, \ldots, v_n]$ is a *derived operator* of type $s_1 \times \cdots \times s_n \to s'$, this means that c is a context of sort s' over context variables v_i of sort s_i. We write $c_\mathcal{A}$ for the corresponding *derived operation* over an algebra $\mathcal{A}$.

Familiarity with *ordered algebras*, i.e., algebras whose carriers are S-indexed families of posets with least elements denoted by the Ω constants, and whose operations are monotone functions, is also assumed. Such an algebra is called *complete* when its carrier is a cpo and operations are continuous, and a homomorphism over complete ordered algebras is called *continuous* when it is continuous on the underlying cpo's. Two complete ordered algebras are *order-isomorphic* iff there exists a continuous homomorphism from one to the other that is a surjective order-embedding on the underlying cpo's. In any full subcategory of the category of complete ordered algebras and continuous homomorphisms, objects are isomorphic exactly when they are order-isomorphic. We write $\mathcal{OT}_\Sigma$ (or just $\mathcal{OT}$) for the initial ordered algebra, which consists of $\mathcal{T}$ with the "Ω-match" ordering: one term is less than another when the second can be formed by replacing occurrences of Ω in the first by terms. A complete ordered algebra is called *inductively reachable* iff all of its elements can be reached by the following transfinite process: start with the denotable elements, and close under lub's of directed sets. Complete ordered algebras whose carriers are ω-algebraic and whose isolated elements are all denotable are thus inductively reachable, but the converse is false. The Ω-least substitutive pre-ordering over $\mathcal{T}$ that is induced by an ordered algebra $\mathcal{A}$ is called $\preceq_{\mathcal{A}}$: one term is less than another when the meaning of the first is less than that of the second in A.

If $P \subseteq S$, $\mathcal{A}$ is an algebra and R is a pre-ordering over $A|P$ then R^c, the *contextualization* of R, is the relation over A defined by: $a\, R^c_s\, a'$ iff $c\langle a\rangle\, R_p\, c\langle a'\rangle$, for all derived operators $c[v]$ of type $s \rightarrow p$, $p \in P$.

Lemma 3.1 *If $P \subseteq S$, $\mathcal{A}$ is a reachable algebra and R is a pre-ordering (respectively, equivalence relation) over $A|P$ then R^c is the greatest substitutive pre-ordering (respectively, congruence) over $\mathcal{A}$ whose restriction to P is included in R.*

Proof. See lemmas 2.2.25 and 2.2.29 of [Sto]. □

Let $\approx$ be a congruence over $\mathcal{T}$ and $\mathcal{A}$ be an algebra. Then $\mathcal{A}$ is *$\approx$-equationally correct* iff $\approx_{\mathcal{A}} \subseteq \approx$, and *$\approx$-equationally fully abstract* iff $\approx_{\mathcal{A}} = \approx$.

Let $\preceq$ be an Ω-least substitutive pre-ordering over $\mathcal{T}$ and $\mathcal{A}$ be an ordered algebra. Then $\mathcal{A}$ is *$\preceq$-inequationally correct* iff $\preceq_{\mathcal{A}} \subseteq \preceq$, and *$\preceq$-inequationally fully abstract* iff $\preceq_{\mathcal{A}} = \preceq$.

Let $\approx$ be a congruence over $\mathcal{T}$ and $\mathcal{A}$ be an algebra. Then $\mathcal{A}$ is *$\approx$-contextually correct* iff for all derived operators $c_1[v_1,\ldots,v_n]$ and $c_2[v_1,\ldots,v_n]$ of type $s_1 \times \cdots \times s_n \rightarrow s'$,

$$\text{if } c_{1\mathcal{A}} = c_{2\mathcal{A}} \text{ then for all } t_i \in T_{s_i}, 1 \leq i \leq n,\ c_1\langle t_1,\ldots,t_n\rangle \approx_{s'} c_2\langle t_1,\ldots,t_n\rangle,$$

and $\mathcal{A}$ is *$\approx$-contextually fully abstract* iff for all derived operators $c_1[v_1,\ldots,v_n]$ and $c_2[v_1,\ldots,v_n]$ of type $s_1 \times \cdots \times s_n \rightarrow s'$,

$$c_{1\mathcal{A}} = c_{2\mathcal{A}} \text{ iff for all } t_i \in T_{s_i}, 1 \leq i \leq n,\ c_1\langle t_1,\ldots,t_n\rangle \approx_{s'} c_2\langle t_1,\ldots,t_n\rangle.$$

Theorem 3.2 *Suppose $\mathcal{A}$ is an inductively reachable complete ordered algebra and $\approx$ is a congruence over $\mathcal{T}$. Then $\mathcal{A}$ is $\approx$-fully abstract iff $\mathcal{A}$ is $\approx$-contextually fully abstract.*

Proof. See theorem 5.3.1 of [Sto]. □

A *family of least fixed point constraints* Φ is an S-indexed family of sets such that for all $s \in S$, $\Phi_s \subseteq T_s \times \mathcal{P}T_s$, and for all $\langle t, T'\rangle \in \Phi_s$, T' is a directed set in OT_s. We write $t \equiv \bigsqcup T'$ instead of $\langle t, T'\rangle$ for elements of Φ_s.

A family of least fixed point constraints Φ is *closed* iff for all $\sigma \in \Sigma$ of type $s_1 \times \cdots \times s_n \rightarrow s'$, if $t_i \equiv \bigsqcup T'_i \in \Phi_{s_i}$, $1 \leq i \leq n$, and T'' is a cofinal subset of $\sigma(T'_1 \times \cdots \times T'_n)$ then $\sigma\langle t_1,\ldots,t_n\rangle \equiv \bigsqcup T'' \in \Phi_{s'}$. We write $\overline{\Phi}$ for the *closure* of Φ, i.e., the least closed family of least fixed point constraints containing Φ.

A complete ordered algebra $\mathcal{A}$ *satisfies* Φ iff for all $t \equiv \bigsqcup T' \in \Phi_s$, $s \in S$, $[\![t]\!] = \bigsqcup\{\, [\![t']\!] \mid t' \in T'\,\}$. An Ω-least substitutive pre-ordering $\preceq$ over $\mathcal{T}$ *satisfies* Φ iff for all $t \equiv \bigsqcup T' \in \Phi_s$, $s \in S$, t is a lub of T' in $\langle T_s, \preceq_s\rangle$.

Lemma 3.3 *Let Φ be a family of least fixed point constraints and $\mathcal{A}$ be a complete ordered algebra. If $\mathcal{A}$ satisfies Φ, then $\mathcal{A}$ satisfies $\overline{\Phi}$.*

Proof. See lemma 3.2.7 of [Sto]. □

Lemma 3.4 *Let $\mathcal{A}$ be a complete ordered algebra that satisfies Φ, and $P \subseteq S$. Define a pre-ordering $\preceq$ over $T|P$ by: $t_1 \preceq_p t_2$ iff $[\![t_1]\!] \sqsubseteq_p [\![t_2]\!]$. Then $\preceq^c$ is an Ω-least substitutive pre-ordering over $\mathcal{T}$ that satisfies $\overline{\Phi}$.*

Proof. See the proofs of lemma 4.1.1 and theorem 7.1.1 of [Sto]. □

Theorem 3.5 *Suppose Φ is a closed family of least fixed point constraints and $\preceq$ is an Ω-least substitutive pre-ordering over $\mathcal{T}$ that satisfies Φ. There exists an inductively reachable, $\preceq$-inequationally fully abstract, complete ordered algebra $\mathcal{I}(\preceq, \Phi)$ satisfying Φ, such that if $\mathcal{A}$ is a complete ordered algebra satisfying Φ with the property that $\preceq \subseteq \preceq_{\mathcal{A}}$, then there is a unique continuous homomorphism $h: \mathcal{I}(\preceq, \Phi) \to \mathcal{A}$.*

Proof. See theorem 5.1.3 and corollary 5.1.6 of [Sto]. □

4 Syntax and Semantics of PCF

In this section, we collect together the various definitions and theorems about the syntax and semantics of PCF that we require in the sequel. For technical simplicity, we have chosen to work with a combinatory logic version of PCF with a single ground type ι, whose intended interpretation is the natural numbers. From the viewpoint of the conditional operations, zero is interpreted as false and non-zero as true.

We begin by defining the syntax of PCF, i.e., its signature. The sorts of this signature consists of PCF's types. The set of *sorts* S is least such that

(i) $\iota \in S$, and

(ii) $s_1 \to s_2 \in S$ if $s_1 \in S$ and $s_2 \in S$.

Define ι^n, for $n \in \omega$, by: $\iota^0 = \iota$ and $\iota^{n+1} = \iota \to \iota^n$. The *signature* Σ over S has the following operators:

(i) Ω_s of type s,

(ii) K_{s_1,s_2} of type $(s_1 \to s_2 \to s_1)$,

(iii) S_{s_1,s_2,s_3} of type $((s_1 \to s_2 \to s_3) \to (s_1 \to s_2) \to s_1 \to s_3)$,

(iv) Y_s of type $((s \to s) \to s)$,

(v) n of type ι, for $n \in \omega$,

(vi) *Succ* and *Pred* of type $(\iota \to \iota)$,

(vii) If_s of type $(\iota \to s \to s \to s)$, and

(viii) $\cdot_{s_1,s_2}$ of type $(s_1 \to s_2) \times s_1 \to s_2$,

where the compound sorts are parenthesized in order to avoid confusion. Thus $\cdot$ (application) is a binary operator, and all of the other operators are nullary. In keeping with standard practice, we usually abbreviate $M \cdot N$ to $M\,N$, and let application associate to the left.

Next, we define several combinators that will be required below. We confuse use and mention for these combinators: given a combinator C, we also write C for its denotation in any model that may be at hand.

For $s \in S$, we write I_s for the term $S_{s,s\to s,s}\,K_{s,s\to s}\,K_{s,s}$ of sort $s \to s$. I will be the identity operation in all models. For $s \in S$, define approximations Y_s^n to Y_s of sort $(s \to s) \to s$ by

$$Y_s^0 = \Omega_{(s\to s)\to s}, \qquad Y_s^{n+1} = S_{s\to s,s,s}\,I_{s\to s}\,Y_s^n,$$

so that Y_s^n is an ω-chain in $OT_{(s\to s)\to s}$. For all $n \in \omega$ and $s \in S$, define syntactic projections Ψ_s^n of sort $s \to s$ by

$$\Psi_\iota^n = Y_{\iota\to\iota}^n\,F, \qquad \Psi_{s_1\to s_2}^n = \lambda x.\,\lambda y.\,(\Psi_{s_2}^n(x(\Psi_{s_1}^n\,y))),$$

where F of sort $(\iota \to \iota) \to \iota \to \iota$ is

$$\lambda x.\,\lambda y.\,(\mathit{If}\,y\,(\mathit{Succ}(x(\mathit{Pred}\,y)))\,0).$$

Expanding the abstractions, one can see that the Ψ_s^n form an ω-chain in $OT_{s\to s}$. Let the equality test Eq of sort $\iota \to \iota \to \iota$ be

$$Y(\lambda z.\,\lambda x.\,\lambda y.(\mathit{If}\,x\,(\mathit{If}\,y\,(z(\mathit{Pred}\,x)(\mathit{Pred}\,y))\,0)\,(\mathit{If}\,y\,0\,1))).$$

Eq yields 1 for true and 0 for false. Define glb operators Inf_s of sort $s \to s \to s$ by

$$\begin{aligned} \mathit{Inf}_\iota &= \lambda x.\,\lambda y.\,(\mathit{If}\,(\mathit{Eq}\,x\,y)\,x\,\Omega), \\ \mathit{Inf}_{s_1\to s_2} &= \lambda x.\,\lambda y.\,\lambda z.\,(\mathit{Inf}_{s_2}(x\,z)(y\,z)). \end{aligned}$$

For $n \in \omega$, define operators And_n of sort ι^n by: $\mathit{And}_0 = 1$ and

$$\mathit{And}_{n+1} = \lambda x.\,\lambda y_1.\,\ldots\,\lambda y_n.\,(\mathit{If}\,x\,(\mathit{And}_n\,y_1\,\cdots\,y_n)\,0).$$

Define step operators St_n of sort $\iota \to \iota$, for $n \in \omega$, by

$$\mathit{St}_n = \lambda x.\,(\mathit{If}\,(\mathit{Eq}\,n\,x)\,1\,\Omega).$$

St_n yields true (1) if its argument is n, and diverges otherwise. Define alternative identify operators I'_s of sort $s \to s$ by

$$I'_\iota = Y_{\iota\to\iota}\,F, \qquad I'_{s_1\to s_2} = \lambda x.\,\lambda y.\,(I'_{s_2}(x(I'_{s_1}\,y))).$$

I' will be identical to I in some models.

A *model* $\mathcal{A}$ of PCF is a complete ordered algebra such that the following conditions hold:

(i) A_ι is the flat cpo

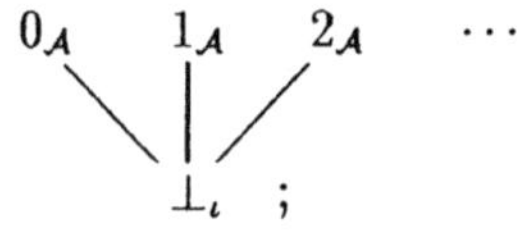

;

(ii) For all $s_1, s_2 \in S$, $a_1 \in A_{s_1}$ and $a_2 \in A_{s_2}$, $K_{s_1,s_2}\, a_1\, a_2 = a_1$;

(iii) For all $s_1, s_2, s_3 \in S$, $a_1 \in A_{s_1 \to s_2 \to s_3}$, $a_2 \in A_{s_1 \to s_2}$ and $a_3 \in A_{s_1}$, $S_{s_1,s_2,s_3}\, a_1\, a_2\, a_3 = a_1\, a_3\,(a_2\, a_3)$;

(iv) For all $s \in S$, $Y_s = \bigsqcup_{n \in \omega} Y_s^n$;

(v) For all $a \in A_\iota$, $Succ\, a$ is equal to $\bot_\iota$, if $a = \bot_\iota$, and is equal to $a+1$, if $a \in \omega$;

(vi) For all $a \in A_\iota$, $Pred\, a$ is equal to $\bot_\iota$, if $a = \bot_\iota$, is equal to 0, if $a = 0$, and is equal to $a - 1$, if $a \in N - \{0\}$.

(vii) For all $s \in S$, $a_1 \in A_\iota$, and $a_2, a_3 \in A_s$, $If_s\, a_1\, a_2\, a_3$ is equal to $\bot_s$, if $a_1 = \bot_\iota$, is equal to a_2, if $a_1 \in N - \{0\}$, and is equal to a_3, if $a_1 = 0$.

A model $\mathcal{A}$ is *extensional* iff for all $a_1, a_2 \in A_{s_1 \to s_2}$, if $a_1\, a = a_2\, a$, for all $a \in A_{s_1}$, then $a_1 = a_2$, and *order-extensional* iff for all $a_1, a_2 \in A_{s_1 \to s_2}$, if $a_1\, a \sqsubseteq_{s_2} a_2\, a$, for all $a \in A_{s_1}$, then $a_1 \sqsubseteq_{s_1 \to s_2} a_2$. Finally, *morphisms* between models are simply continuous homomorphisms between the complete ordered algebras.

Application is left-strict in all models $\mathcal{A}$ since $\bot_{s_1 \to s_2} \sqsubseteq_{s_1 \to s_2} K_{s_2,s_1}\, \bot_{s_2}$, and thus $\bot_{s_1 \to s_2}\, a \sqsubseteq_{s_2} K_{s_2,s_1}\, \bot_{s_2}\, a = \bot_{s_2}$, for all $a \in A_{s_1}$.

The following theorem introduces the *stable function model*, which features prominently below [Ber][BerCurLév].

Theorem 4.1 (Berry) *There is a unique model $\mathcal{A}$ constructed from the category of dI-domains and stable functions in the natural way, i.e., such that $A_\iota = N_\bot$, $A_{s_1 \to s_2} = A_{s_1} \underset{s}{\to} A_{s_2}$, $a_1 \cdot a_2 = a_1\, a_2$, and $n_{\mathcal{A}} = n$. $\mathcal{A}$ is extensional but not order-extensional.*

The following theorem is proved by making use of an operational semantics for PCF; see theorem 3.1 of [Plo].

Theorem 4.2 (Plotkin) *For all models $\mathcal{A}$ and $\mathcal{B}$ and terms M of sort ι, $[\![M]\!]_{\mathcal{A}} = [\![M]\!]_{\mathcal{B}}$.*

This theorem allows us to define the meaning $[\![M]\!] \in N_\bot$ of a term M of sort ι to be $[\![M]\!]_{\mathcal{A}}$, for an arbitrary model $\mathcal{A}$.

We now define notions of program ordering and equivalence for PCF. Define a pre-ordering $\precsim$ over $T|\{\iota\}$ by: $M \precsim_\iota N$ iff $[\![M]\!] \sqsubseteq [\![N]\!]$, and let $\approx$ be the equivalence relation over $T|\{\iota\}$ induced by $\precsim$. Then, $\precsim^c$ is an Ω-least substitutive pre-ordering over $\mathcal{T}$, $\approx^c$ is a congruence over $\mathcal{T}$, and $\precsim^c$ induces $\approx^c$.

Specializing the notions of the previous section, we say that a model is

(i) *inequationally correct* iff it is $\precsim^c$-*inequationally correct*;

(ii) *inequationally fully abstract* iff it is $\precsim^c$-*inequationally fully abstract*;

(iii) *equationally correct* iff it is $\approx^c$-*equationally correct*;

(iv) *equationally fully abstract* iff it is $\approx^c$-*equationally fully abstract*;

(v) *contextually correct* iff it is $\approx^c$-*contextually correct*; and

(vi) *contextually fully abstract* iff it is $\approx^c$-*contextually fully abstract.*

It is not hard to show that all models are inequationally, equationally and contextually correct. Clearly inequational full abstraction implies equational full abstraction, but the converse, as we shall see, is false. The stable function model is not even equationally fully abstract [Ber][BerCurLév].

Finally, we recall Milner's important result concerning the order-extensional nature of $\sqsubseteq^c$ and the extensional nature of $\approx^c$; see lemma 4.1.11 of [Cur].

Theorem 4.3 (Milner) (i) $\sqsubseteq^c_\iota = \sqsubseteq_\iota$ *and* $\approx^c_\iota = \approx_\iota$.

(ii) *For all* $M_1, M_2 \in T_{s_1 \to s_2}$, $M_1 \sqsubseteq^c_{s_1 \to s_2} M_2$ *iff for all* $N \in T_{s_1}$, $M_1\,N \sqsubseteq^c_{s_2} M_2\,N$.

(iii) *For all* $M_1, M_2 \in T_{s_1 \to s_2}$, $M_1 \approx^c_{s_1 \to s_2} M_2$ *iff for all* $N \in T_{s_1}$, $M_1\,N \approx^c_{s_2} M_2\,N$.

From theorem 4.3 (i), we know that for all terms M of sort ι, either $M \approx^c_\iota \Omega$ or $M \approx^c_\iota n$, for some $n \in \omega$.

5 Equationally Fully Abstract Models

This section consists of the paper's main results, concerning the category **E** of extensional, equationally fully abstract models and their morphisms. To begin with, we introduce our main technical device. Let $\mathbf{\Phi}$ be the family of least fixed point constraints such that

$$\Phi_{(s\to s)\to s} = \{Y_s \equiv \bigsqcup \{\, Y_s^n \mid n \in \omega \,\}\},$$

for all $s \in S$, and $\Phi_s = \emptyset$, whenever s does not have the form $(s' \to s') \to s'$. A *least fixed point ordering* $\preceq$ is an Ω-least substitutive pre-ordering over $\mathcal{T}$ that induces $\approx^c$ and satisfies $\overline{\Phi}$. We write **L** for the set of all least fixed point orderings, ordered by inclusion.

By lemma 3.3, all models satisfy $\overline{\Phi}$. Lemma 3.4 allows us to conclude that $\sqsubseteq^c$ is an element of **L**.

Lemma 5.1 *For all least fixed point orderings* $\preceq$ *and terms* M, N *of sort* ι, $M \preceq_\iota N$ *iff either* $M \approx^c_\iota \Omega$ *or* $M \approx^c_\iota N$. *Thus, all least fixed point orderings agree at sort* ι.

Proof. Suppose that $\preceq\, \in \mathbf{L}$ and $m \preceq_\iota n$, for $m, n \in \omega$. Define a term M of sort $\iota \to \iota$ which yields n when applied to m and m when applied to n. Then $n \approx^c_\iota M\,m \preceq_\iota M\,n \approx^c_\iota m$, showing that $n \preceq_\iota m$, and thus $m \approx^c_\iota n$. The rest follows easily. □

Lemma 5.2 *Let* $\mathcal{A}$ *be an extensional model and* $P = \{\iota, \iota \to \iota\}$. *Define a pre-ordering* $\preceq$ *over* $T|P$ *by:* $M \preceq_p N$ *iff* $[\![M]\!] \sqsubseteq_p [\![N]\!]$. *Then* $\preceq^c$ *is a least fixed point ordering.*

Proof. By lemma 3.4, all that remains to be shown is that $\preceq^c \cap \succeq^c = \approx^c$. Clearly, $\preceq^c \subseteq \sqsubseteq^c$, and thus $\preceq^c \cap \succeq^c \subseteq \approx^c$. For the opposite inclusion, suppose that $M_1 \approx^c_s M_2$, and let $c[v]$ be a derived operator of type $s \to (\iota \to \iota)$. We must show that $[\![c\langle M_1\rangle]\!] = [\![c\langle M_2\rangle]\!]$, and since $\mathcal{A}$ is extensional and all elements of A_ι are denotable, it suffices to show that $[\![c\langle M_1\rangle\, N]\!] = [\![c\langle M_2\rangle\, N]\!]$, for all terms N of sort ι. But this follows from the assumption that $M_1 \approx^c_s M_2$. □

Theorem 5.3 **L** *is a nontrivial complete lattice whose greatest element is* $\sqsubseteq^c$.

Proof. We have already observed that $\sqsubseteq^c \in \mathbf{L}$. To see that **L** is nontrivial, let $\mathcal{A}$ be the stable function model, and define $\preceq$ as in the statement of lemma 5.2. The lemma then allows us to conclude that $\preceq^c \in \mathbf{L}$. To see that $\sqsubseteq^c$ and $\preceq^c$ are distinct, define terms M, N of sort $\iota \to \iota$ by $M = \lambda x.(\mathit{If}\, x\, 0\, 0)$ and $N = \lambda x. 0$. Then $M \sqsubseteq^c_{\iota\to\iota} N$ by theorem 4.3 (ii), but $M \not\preceq^c_{\iota\to\iota} N$ since M is not less than N in $A_{\iota\to\iota}$.

Showing that **L** is closed under arbitrary nonempty intersections is straightforward, and it remains to show that $\sqsubseteq^c$ is the greatest element of **L**. Suppose that $\preceq \in \mathbf{L}$, $M \preceq_s N$ and let $c[v]$ be a derived operator of type $s \to \iota$. Then $c\langle M\rangle \preceq_\iota c\langle N\rangle$, and thus $c\langle M\rangle \sqsubseteq_\iota c\langle N\rangle$ by lemma 5.1. But then $M \sqsubseteq^c_s N$, as required. □

We write $\preceq_0$ for the least element of **L**.

Theorem 5.4 *For each least fixed point ordering* $\preceq$*, there is an inductively reachable,* $\preceq$*-inequationally fully abstract model* $\mathcal{M}(\preceq)$*, such that for all models* $\mathcal{A}$ *with the property that* $\preceq \subseteq \preceq_{\mathcal{A}}$*, there is a unique morphism from* $\mathcal{M}(\preceq)$ *to* $\mathcal{A}$*. In particular,* $\mathcal{M}(\preceq)$ *is initial in the category of* $\preceq$*-inequationally fully abstract models and their morphisms.*

Proof. By theorem 3.5, we know all that is necessary about $\mathcal{M}(\preceq)$ except conditions (i)–(iii) and (v)–(vii) of the definition of model. Condition (i) holds since the denotable elements are ordered properly (lemma 5.1) and $\mathcal{M}(\preceq)$ is inductively reachable. The remaining conditions can be expressed by sets of equations (pairs of derived operators), and these equations hold in $\mathcal{M}(\preceq)$ since it is contextually fully abstract (theorem 3.2) and all models are contextually correct. □

$\mathcal{M}(\preceq)$ is uniquely specified, up to order-isomorphism.

A model $\mathcal{A}$ is *syntactically strongly algebraic* (or *syntactically SFP*) iff the following conditions hold:

(i) $\Psi^n_s(\Psi^n_s\, a) = \Psi^n_s\, a$, for all $a \in A_s$, $n \in \omega$ and $s \in S$;

(ii) $a = \bigsqcup_{n\in\omega}(\Psi^n_s\, a)$, for all $a \in A_s$ and $s \in S$; and

(iii) $\{\, \Psi^n_s\, a \mid a \in A_s \,\}$ is finite, for all $n \in \omega$ and $s \in S$.

The carrier of any syntactically SFP model is clearly SFP. Furthermore, if such a model is inductively reachable then $\Psi^n_s\, a$ is isolated and thus denotable, for all $a \in A_s$ and $n \in \omega$.

Lemma 5.5 (Milner) (i) *Extensional models are syntactically SFP.*

(ii) *For all models $\mathcal{A}$ and $a_1, a_2 \in A_\iota$, $Inf_\iota\, a_1\, a_2$ is the glb of a_1 and a_2. If $\mathcal{A}$ is order-extensional, then for all $s \in S$ and $a_1, a_2 \in A_s$, $Inf_s\, a_1\, a_2$ is the glb of a_1 and a_2.*

(iii) *The carriers of order-extensional models are Scott domains, i.e., consistently complete, ω-algebraic cpo's.*

Proof. (i) and (ii) are straightforward inductions on S. For (iii), each A_s is ω-algebraic, by part (i). For consistent completeness, it suffices to show that each consistent pair a_1, a_2 of isolated elements of A_s has a lub. Let $n \in \omega$ be such that $\Psi^n\, a_i = a_i$, for $i = 1, 2$, and $X = \{\, \Psi^n\, a \mid a \sqsupseteq_s \{a_1, a_2\}\,\}$. Then X is nonempty and finite, and thus has a glb z, by part (ii). But z is easily seen to be the lub of a_1 and a_2. □

Lemma 5.5 tells us, in particular, that the stable function model is syntactically SFP.

Theorem 5.6 *Inductively reachable, equationally fully abstract models are syntactically SFP and extensional. If, in addition, a model is inequationally fully abstract, then it is order-extensional.*

Proof. Let $\mathcal{A}$ be such a model. Condition (i) of the definition of syntactic strong algebraicity holds, since $\mathcal{A}$ is contextually fully abstract and (i) holds in, e.g., the stable function model, which is contextually correct. Expanding the identifier abstractions, one can see that $I'_s \equiv \bigsqcup\{\, \Psi^n_s \mid n \in \omega\,\} \in \overline{\Phi}_{s \to s}$, for all $s \in S$. Since $\mathcal{A}$ is equationally fully abstract, we have that $I'_s = I_s$, and thus that $I'_s\, a = a$, for all $a \in A_s$ and $s \in S$. Thus condition (ii) holds.

For condition (iii), we prove by induction on $\mathcal{A}$ that for all $a \in A_s$, $s \in S$, and $n \in \omega$, there is a term M of sort s such that $\Psi^n_s\, a = [\![\Psi^n_s\, M]\!]$. This is obvious for denotable elements. Suppose that it is true for the elements of a directed set D. Then

$$\Psi^n_s \bigsqcup D = \bigsqcup\{\, \Psi^n_s\, d \mid d \in D\,\} = \bigsqcup\{\, [\![\Psi^n_s\, N]\!] \mid N \in T'\,\},$$

for a set of terms T'. But $\{\, [\![\Psi^n_s\, N]\!] \mid N \in T'\,\}$ is finite (since it is finite in, e.g., the stable function model) and thus contains its own lub, which is some $[\![\Psi^n_s\, N]\!]$, thus completing the induction. Then, $\{\, \Psi^n_s\, a \mid a \in A_s\,\}$ is equal to $\{\, [\![\Psi^n_s\, M]\!] \mid M \in T\,\}$, and thus is finite by the above reasoning.

For the extensionality of $\mathcal{A}$, suppose that $a_1, a_2 \in A_{s_1 \to s_2}$ and $a_1\, a' = a_2\, a'$, for all $a' \in A_{s_1}$. To show that $a_1 = a_2$, it suffices to show that $\Psi^n_{s_1 \to s_2}\, a_1 = \Psi^n_{s_1 \to s_2}\, a_2$, for all $n \in \omega$. From the above induction, we know that $\Psi^n_{s_1 \to s_2}\, a_1$ and $\Psi^n_{s_1 \to s_2}\, a_2$ are denotable. Furthermore, for all denotable $a' \in A_{s_1}$,

$$(\Psi^n_{s_1 \to s_2}\, a_1)\, a' = \Psi^n_{s_2}(a_1(\Psi^n_{s_1}\, a')) = \Psi^n_{s_2}(a_2(\Psi^n_{s_1}\, a')) = (\Psi^n_{s_1 \to s_2}\, a_2)\, a'.$$

Thus, by the obvious semantic restatement of theorem 4.3 (iii), $\Psi^n_{s_1 \to s_2}\, a_1 = \Psi^n_{s_1 \to s_2}\, a_2$, as required.

Order-extensionality under the additional hypothesis that $\mathcal{A}$ is inequationally fully abstract follows similarly, using theorem 4.3 (ii). □

Theorem 5.7 (Milner/Berry) *Extensional, equationally fully abstract models are syntactically SFP and inductively reachable.*

Proof. Adapted from theorem 3.6.18 of [Ber]. Let $\mathcal{A}$ be such a model, which is syntactically SFP by lemma 5.5. Clearly, all elements of A_ι are denotable. Suppose that $s = s_1 \to \cdots \to s_n \to \iota$, for $n \geq 1$, is such that all isolated elements of each A_{s_i} are denotable. Suppose, toward a contradiction, that there is a non-denotable isolated element a of A_s. Let $n \in \omega$ be such that $\Psi^n\, a = a$. Define a pre-ordering $\leq$ over A_s by: $a \leq a'$ iff $a\, a_1 \cdots a_n \sqsubseteq_\iota a'\, a_1 \cdots a_n$, for all $a_i \in A_{s_i}$. Let X be the set of all denotable elements of $\{\, \Psi^n\, a' \mid a' \in A_s \,\}$, $X^+ = \{\, x \in X \mid a \leq x \,\}$ and $X^- = X - X^+$.

Let $\delta_1, \ldots, \delta_p$ be the elements of X^-; here $p \geq 1$, since $\bot \in X^-$. Then, for all $1 \leq i \leq p$, there exist isolated $w^i_j \in A_{s_j}$ and $z_i \in \omega$ such that $a\, w^i_1 \cdots w^i_n = z_i$ and $\delta_i\, w^i_1 \cdots w^i_n \neq z_i$. Let W^i_j be terms denoting the w^i_j, and let Q of sort $s \to \iota$ be

$$\lambda x.\, (And_p(St_{z_1}\, (x\, W^1_1 \cdots W^1_n)) \cdots (St_{z_p}\, (x\, W^p_1 \cdots W^p_n))).$$

There are now two cases to consider:

(X^+ is nonempty) Suppose that x_1 and x_2 are elements of X^+ that are denoted by terms X_1 and X_2, respectively. Let $X_3 = \Psi^n(Inf\, X_1\, X_2)$ and x_3 be the meaning of X_3. A bit of work then shows that x_3 is a $\leq$-lower bound of $x_1 = \Psi^n\, x_1$ and $x_2 = \Psi^n\, x_2$, and that $a = \Psi^n\, a \leq x_3$, i.e., $x_3 \in X^+$. Thus we can conclude that there is a $\leq$-least element γ of X^+. There exist isolated $u_i \in A_{s_i}$ and $v \in \omega$ such that $a\, u_1 \cdots u_n = \bot$ and $\gamma\, u_1 \cdots u_n = v$. Let U_i be terms denoting the u_i, and define terms M_1 and M_2 of sort $s \to \iota$ by

$$M_1 = \lambda x.\, (Q(\Psi^n\, x)),$$
$$M_2 = \lambda x.\, (And_2(Q(\Psi^n\, x))(St_v\, (\Psi^n\, x\, U_1 \cdots U_n))).$$

Then the meaning of M_1 applied to a is 1, whereas the meaning of M_2 applied to a is $\bot$. On the other hand, we can use theorem 4.3 (iii) to show that $M_1 \approx^c_{s \to \iota} M_2$. But this contradicts the equational full abstraction of $\mathcal{A}$.

(X^+ is empty) Similar to the nonempty case, with M_1 defined as before and $M_2 = \Omega$. □

Since all objects of $\mathbf{E}$ are inductively reachable (theorem 5.7), it follows that $\mathbf{E}$ is a pre-ordering.

Lemma 5.8 *If $\mathcal{A}$ is an equationally fully abstract model then $\preceq_{\mathcal{A}}$ is a least fixed point ordering.*

Proof. Immediate from lemma 3.3. □

Theorem 5.9 *If $\mathcal{A}$ is an extensional, equationally fully abstract model then it is order-isomorphic to $\mathcal{M}(\preceq_{\mathcal{A}})$.*

Proof. Let $\mathcal{B} = \mathcal{M}(\preceq_{\mathcal{A}})$ and i be the unique continuous homomorphism from $\mathcal{B}$ to $\mathcal{A}$. By theorem 5.7, $\mathcal{A}$ is inductively reachable, and thus it suffices to show that i is an order-embedding. Suppose that $i_s\, b_1 \sqsubseteq_s i_s\, b_2$. Then, for all $n \in \omega$,

$$i_s(\Psi^n\, b_1) = \Psi^n(i_s\, b_1) \sqsubseteq_s \Psi^n(i_s\, b_2) = i_s(\Psi^n\, b_2).$$

But $\Psi^n\, b_1$ and $\Psi^n\, b_2$ are denotable, and thus $\Psi^n\, b_1 \sqsubseteq_s \Psi^n\, b_2$. Thus $b_1 \sqsubseteq_s b_2$, since $\mathcal{B}$ is syntactically SFP. □

Proposition 5.10 *Suppose $\mathcal{A}$ and $\mathcal{B}$ are extensional, equationally fully abstract models. If $\preceq_{\mathcal{A}} \subseteq \preceq_{\mathcal{B}}$, then there is a unique morphism from $\mathcal{A}$ to $\mathcal{B}$. If there is a morphism from $\mathcal{A}$ to $\mathcal{B}$, then $\preceq_{\mathcal{A}} \subseteq \preceq_{\mathcal{B}}$.*

Proof. The first part follows from theorems 5.9 and 5.4, and the second part is obvious. □

Corollary 5.11 **E** *and* **L** *are equivalent categories.*

Proof. Immediate from theorems 5.4, 5.6 and 5.9 and proposition 5.10. □

From the above results, we know that $\mathcal{M}(\preceq_0)$ and $\mathcal{M}(\precsim^c)$ are the initial and terminal objects, respectively, of **E**. It is easy to see that $\mathcal{M}(\preceq_0)$ is also initial in the category of (not necessarily extensional) equationally fully abstract models and their morphisms. $\mathcal{M}(\precsim^c)$ is the only object of **E** that is order-extensional, since models that are order-extensional, SFP and whose isolated elements are all denotable are easily seen to be inequationally fully abstract. Another fact about $\mathcal{M}(\precsim^c)$ is that its carrier is consistently complete; it is unknown whether there are other objects of **E** with consistently complete carriers. Another obvious open question is whether $\mathcal{M}(\preceq_0)$ and $\mathcal{M}(\precsim^c)$ are the only objects of **E**.

Acknowledgments

Conversations with Albert Meyer and Gordon Plotkin stimulated my attempts to show that equational and inequational full abstraction were distinct for PCF.

References

[Ber] G. Berry. *Modèles complètement adéquats et stables des lambda-calculs typés*. Thèse de Doctorat d'Etat, Université Paris VII, 1979.

[BerCurLév] G. Berry, P.-L. Curien and J.-J. Lévy. Full abstraction for sequential languages: the state of the art. In M. Nivat and J. Reynolds (editors), *Algebraic Methods in Semantics*, Cambridge University Press, 1985.

[Cur] P.-L. Curien. *Categorical combinators, sequential algorithms and functional programming*. Research Notes in Theoretical Computer Science, Pitman/Wiley, 1986.

[Mey] A. Meyer and S. Cosmadakis. Semantical paradigms: notes for an invited lecture. *Proc. 3rd LICS*, 1988.

[Mil] R. Milner. Fully abstract models of typed λ-calculi. *Theoretical Computer Science* 4, 1977.

[Plo] G. Plotkin. LCF considered as a programming language. *Theoretical Computer Science* 5, 1977.

[Sto] A. Stoughton. *Fully Abstract Models of Programming Languages*. Research Notes in Theoretical Computer Science, Pitman/Wiley, 1988.

Generalization of Final Algebra Semantics by Relativization

Lawrence S. Moss[1] Satish R. Thatte[2]

Abstract

We consider the semantics of algebraic specifications viewed as evolving rather than static entities. This leads to a relativization of final algebra semantics with respect to the space of evolutionary possibilities open to a given specification. The evolutionary space itself, which we call a *language*, has significant semantic content. The unit of application for our semantics is such a language of specifications. We formalize relevant notions of languages and language models and derive conditions to establish the existence of a final object in the category of models for a language. The results can be used to broaden the applicability of final algebra semantics and to fine tune the balance between the desired strength of semantically valid assertions and the range of evolutionary possibilities to be considered.

1 Introduction

Abstract data types are normally viewed as black boxes whose users are to be concerned only with their observable behavior; their internal mechanisms are deliberately hidden or "encapsulated." Final algebra semantics ([Wan]; [Kam]) is the formalization of this viewpoint for algebraic specifications of such types. The main technical drawback of final algebra semantics (as opposed to semantics based on initiality) is its limited applicability — the existence of a final algebra is dependent upon some form of sufficient completeness (the idea was introduced by Guttag-Horning [GH]; Wand [Wan] calls it Λ-fullness). When a specification does not possess a final algebra, it may nevertheless intuitively imply certain assertions from the black box viewpoint which do not hold in the initial algebra. The motivation for generalizing final algebra semantics is the desire to capture such observable properties for incomplete specifications.

In this paper we present a proposal for such a generalization. Our main idea is to *relativize* the semantics with respect to the evolutionary possibilities of the

[1]Mathematical Sciences Department, IBM T. J. Watson Research Center, Yorktown Heights, New York 10598.

[2]Department of Mathematics and Computer Science, Clarkson University, Potsdam, NY 13676.

specification. The space of possibilities has something of the flavor of a *Kripke model*, but instead of considering spaces of possible interpretations, we consider spaces of possible extensions of the theory — we call such spaces *languages*. Technically, our generalization amounts to changing the unit of application of the semantics from individual specifications to languages of specifications, and considering final models of the latter. The main insight here is that observational semantics is not absolute; it is subject to "engineering" to fit the needs of the application one has in mind. In our approach to the semantics of languages, for instance, assertions which hold in the meaning assigned to a particular specification are required to hold in the meanings assigned to all its extensions. This is because our languages are meant to model reasoning in the context of incremental development of specifications. As a result of this requirement, the size of the language has a strong impact on the semantics of individual specifications — one is forced to seek a balance between the desired strength of semantically valid assertions and the range of evolutionary possibilities to be considered. As an extreme case, we show in Section 4 that there is a very natural language which has a *final* model that is identical to its *initial* model. The observable semantics of *any* specification in the context of this language is therefore the same as its antithesis, the initial semantics.

Our approach can therefore be seen as a *twofold* generalization. Besides broadening the applicability of observable semantics, it also forces the semanticist to confront the issue of the *range of validity* of the semantics over possible modifications to the specification as an explicit engineering decision.

To motivate the need for applying observable semantics to specifications to which traditional final algebra semantics cannot be applied, consider the main example given by Wand [Wan], a specification for Nat-indexed arrays of Nat (natural numbers). We assume that basic types Nat and Bool are specified in some standard way, and we also assume that the signature contains an equality test eq defined on Nat. Moreover, we assume that Nat contains a value U for "undefined." The values of the original types Nat and Bool are taken to be observable. Nat + Bool is, in Wand's terminology, the *base* specification. The extension is obtained by adding a new type Array, together with three new symbols:

$$\mathsf{empty} :\rightarrow \mathsf{Array}$$
$$\mathsf{alt} : \mathsf{Nat} \times \mathsf{Nat} \times \mathsf{Array} \rightarrow \mathsf{Array}$$
$$\mathsf{val} : \mathsf{Nat} \times \mathsf{Array} \rightarrow \mathsf{Nat}$$

subject to the following two equations:

$$\mathsf{val}(x, \mathsf{alt}(y, z, a)) = \textbf{if } \mathsf{eq}(x, y) \textbf{ then } z \textbf{ else } \mathsf{val}(x, a)$$
$$\mathsf{val}(x, \mathsf{empty}) = \mathsf{U}$$

The function alt corresponds to an update. Intuitively, only the latest modification of an array with alt for a given index can affect the outcome for val for that index. Earlier values at that index can be forgotten, and distinctions arising solely out of differences among such obsolete values need not be maintained. The fact that these distinctions are preserved in the initial algebra is the motivation for the observability oriented final algebra approach.

The intended interpretation of the extension is the set of functions from finite subsets of Nat to Nat, with the obvious interpretations of the functions. It turns out that this interpretation is the *final* object of the category of reachable algebras for the extension whose reduct to the base is initial. In other words, the category of algebras where Nat + Bool has its initial interpretation, and where every element of sort Array is the interpretation of some term in the language.

This is the reason for using final algebra semantics. To see that a final algebra exists, we need only check sufficient completeness. This means that every ground term of sort Nat or Bool in the new signature rewrites using the new equations to one of the original terms of its sort. The verification is an easy induction on terms.

Now suppose we drop the second equation from the specification. Perhaps the reduced specification reflects an earlier stage in development before the possibility of anomalous expressions like val(4, empty) had been considered. The same reasoning that led us to prefer a final model before also applies here. The fact that only the latest value at any index needs to be remembered in an array representation can be deduced from the reduced specification because it holds in all its completions. There are even techniques for making this deduction mechanically [KMb]. Thus, the natural model is again similar to arrays of natural numbers. However, this specification is not sufficiently complete because the value of val(4, empty) is left unspecified — it may be equated to any natural number. This specification therefore has no final algebra. Reasoning about incomplete specifications such as the one above is a practical necessity since specifications (like programs) are actually developed incrementally and one can never be certain that a specification has reached its final form.

In Section 2, we introduce our new notions of languages and their models. Section 3 gives existence theorems for the final models of languages, and Section 4 gives examples of interesting languages for which final models exist.

2 Languages and their Models

The phenomenon we wish to capture semantically is an evolving algebraic specification. A language represents a certain "evolutionary space". Of course, no mathematical structure can usefully capture the full range of changes real specifications undergo. For instance, one may wish to change the *signature* of a specification in arbitrary ways during evolution. It is hard to see how to capture intended semantic relationships between theories with unrelated signatures. Our definition of a language is intended to capture relatively "smooth" changes that amount to refinements which do not negate anything existing beforehand. The main motivation is to use the semantics of languages to justify reasoning about incomplete specifications in such a way that the results remain valid for all possible evolutionary futures of the specification.

The reason we consider languages as the objects for semantic interpretation, rather than merely fixing a universe of evolutionary possibilities for the interpretation of individual specifications, is that any such universe itself has semantic content, and no specific universe is likely to be appropriate for all purposes. Other authors [KMb] thinking along similar lines have considered what amounts to a *fixed* language, only to find later [KMa] that a slightly different language captured their intentions better.

Before defining languages formally, we recall some standard ideas and notation in algebraic semantics. Most of our technical machinery is based on the work of the ADJ group. An excellent tutorial introduction to this material can be found in [ADJ]. We assume familiarity with the basic notions therein.

Given an S-sorted signature Σ, we use T_Σ to denote both the initial (free) Σ-algebra, and the (many-sorted) set of all Σ-terms. Given a Σ-algebra A, the unique homomorphism from T_Σ to A which simply evaluates terms according to their interpretation in A will also be denoted by A. A **theory** E consists of a signature Σ_E and a set (ambiguously called) E of Σ_E-equations (with variables). One of the basic results of algebraic semantics is that there is an initial algebra in the category of Σ_E-algebras which satisfy E – we shall refer to this as the **initial algebra** for (the theory) E, denoted by I_E. This algebra semantically identifies ground terms exactly when their equality is provable from the equations E using simple equational deduction. We use $=_E$ to denote semantic equality in I_E.

Formally, a language is a set of equational theories in which evolutionary possibilities are captured by a preorder relation[3]. The BASE of a language is a theory which describes the observable values for *all* theories in the language. Observable values can be thought of as tokens in an abstract sense. Given that a language captures the evolution of a single specification, the nature of available tokens can be taken to be common to all theories in a language, which makes it natural to use a common base theory for the entire language. The semantics of the BASE is taken to coincide with its initial algebra. In other words, we assume that the base is a complete axiomatization of the algebra of observable values. Looser assumptions are possible, but this version is technically simple and also realistic in most cases of practical interest. All elements of all carriers in I_{BASE} are considered observable, and distinctions between different observable values are also observable. For this reason, we call the set of sorts of Σ_{BASE} the *observable sorts*. We assume that *at least one* of the carriers of I_{BASE} is not a singleton — otherwise, observable distinctions are impossible.

Definition A **language** $\mathcal{L}$ is triple $\langle \mathcal{L}, \leq, \text{BASE}\rangle$, where $\mathcal{L}$ (ambiguously) denotes a set of theories, and $\leq$ is a (pre)order relation called the **extension relation** of $\mathcal{L}$. The BASE is a theory that defines observable values. It is not necessarily a member of $\mathcal{L}$. The following well-formedness conditions are imposed on all theories in $\mathcal{L}$:

(1) $\Sigma_E \supseteq \Sigma_{\text{BASE}}$

(2) $I_E|\Sigma_{\text{BASE}} \cong I_{\text{BASE}}$

(3) $E_2 \geq E_1 \Rightarrow \Sigma_{E_2} \supseteq \Sigma_{E_1}$.

(4) $E_2 \geq E_1 \Rightarrow I_{E_2}|\Sigma_{E_1}$ is a quotient of I_{E_1}.

[3]This is a partial order if theories with isomorphic initial algebras are identified.

Although the BASE need not belong to the language as a theory, condition (1) requires each theory in a language to be an extension of the BASE. Condition (2) asserts that a theory must neither imply new identifications nor new distinctions in the values created by the BASE. This a *consistency* condition motivated by the intuition that it would be visibly absurd to allow a model of some theory in a language to destroy the observable distinctions made by its BASE. A theory which satisfies this condition is said to *preserve the* BASE. Condition (3) on the order $\leq$ states that as we move up in the order, we are permitted larger signatures The notation $A|\Sigma$ in condition (4) denotes the reduct of the algebra A to the signature Σ (that is, A considered as a Σ-algebra). Of course, this is defined only when the signature of A includes Σ. Condition (4) is slightly weaker (more flexible) than requiring that $E_2 \supseteq E_1$ — it makes the extension relation a preorder instead of a partial order. Our definition views specifications as axiomatizations of their deductive consequences and permits rearrangement of equations during extension for clarity or technical convenience.

Traditional final algebra semantics applied to a *single* theory E requires an additional *sufficient completeness* condition on E:

$$\forall t \in T_{\Sigma_E} \text{ of observable sort, } \exists v \in T_{\text{BASE}} \text{ such that } t =_E v.$$

This ensures that the carriers of I_E and I_{BASE} are isomorphic for observable sorts, and leads to the following theorem:

Theorem 1 ([Wan]) *For every sufficiently complete theory $E \in \mathcal{L}$, the category consisting of* BASE*-preserving, reachable models of E has a final object.*

We now turn to the definition of the main semantic concept of this paper, the notion of a *language model.* A model for $\mathcal{L}$, which we shall call an **$\mathcal{L}$-algebra**, assigns a BASE-preserving Σ_E-algebra for each theory $E \in \mathcal{L}$. Each such algebra must satisfy the corresponding theory and must be minimal (reachable, contain no "junk"). Beyond these obvious requirements, the (extension) *structure* of $\mathcal{L}$ is reflected in a corresponding *coherence* property in its model. It seems intuitively clear that when a specification is extended, the semantic equalities implied by the original equations are meant to be preserved and extended. This is the basis for the expectation that the results of reasoning about equality assertions in an incomplete specification should hold in its possible extensions. Technically, this amounts to saying that there should be a homomorphism from the model for a theory to the models for its extensions (appropriately reduced if the extension has a bigger signature). This idea is naturally formalized in category theoretic terms by defining a language model to be a functor.

Given a language $\langle \mathcal{L}, \leq, \text{BASE} \rangle$, let $\mathbf{Mod}_{\mathcal{L}}$ be the class of all BASE-preserving minimal Σ_E-algebras for all $E \in \mathcal{L}$. $\mathbf{Mod}_{\mathcal{L}}$ can be viewed as a (large) category in the following way. Consider two algebras A and B in $\mathbf{Mod}_{\mathcal{L}}$ with signatures Σ_A and Σ_B, where $\Sigma_B \supseteq \Sigma_A$. Let a morphism from A to B be a Σ_A-morphism from A to $B|\Sigma_A$. Because A and B are minimal, such a morphism is unique when it exists. In this way, $\mathbf{Mod}_{\mathcal{L}}$ is a preorder category.

Definition Let $\mathcal{L} = \langle \mathcal{L}, \leq, \text{BASE} \rangle$ be a language. An **$\mathcal{L}$-algebra** is a *functor*

$\mathcal{A} : \mathcal{L} \to \mathbf{Mod}_{\mathcal{L}}$ such that for any $E \in \mathcal{L}$, $\mathcal{A}(E)$ is a minimal Σ_E-algebra such that $\mathcal{A}(E) \models E$.

$A \models E$ means that the equations of E hold in the algebra A, i.e., A satisfies E. An $\mathcal{L}$-algebra is an indexed collection $\mathcal{A} = \{\mathcal{A}(E) : E \in \mathcal{L}\}$ of minimal algebras such that if $E_1 \leq E_2$ in $\mathcal{L}$, then there is a morphism $\mathcal{A}_{E_1,E_2} : \mathcal{A}(E_1) \to \mathcal{A}(E_2)|\Sigma_{E_1}$. Obviously, if $E_1 \leq E_2 \leq E_3$, then $\mathcal{A}_{E_1,E_3} = \mathcal{A}_{E_2,E_3} \circ \mathcal{A}_{E_1,E_2}$, since the morphisms are closed under composition and the category is a preorder. The crucial consequence of the existence of these morphisms is that semantic equality within $\mathcal{A}(E_1)$ is preserved in $\mathcal{A}(E_2)$ whenever $E_1 \leq E_2$, thus reflecting the structure of extensions as we wanted.

The class of all models of a language forms a category in a natural way — it is a *functor category* in which a morphism between two $\mathcal{L}$-algebras $\mathcal{A}$ and $\mathcal{B}$ is just a *natural transformation.* Informally, we will often speak of an $\mathcal{L}$-algebra $\mathcal{A}$ as a collection of interpretations of the theories in $\mathcal{L}$, together with the appropriate homomorphisms, and of morphisms between $\mathcal{L}$-algebras as collections of homomorphisms. It is easy to see that an initial object exists in the category of $\mathcal{L}$-algebras for any language $\mathcal{L}$ — it simply maps each $E \in \mathcal{L}$ to its initial interpretation I_E. *Observable* semantics corresponds to the *final* object in this category, and although final models exist for many interesting and useful languages, their existence must be *proved* individually. In the next section we derive some results which help in establishing the existence of final models for many languages.

3 On the Existence of Final $\mathcal{L}$-algebras

In this section, we look at the existence question for final $\mathcal{L}$-algebras in two ways. First, we give a *characterization* of the conditions under which final $\mathcal{L}$-algebras exist, using the analogy between finality and *full abstraction* [Mil]. Second, we describe *sufficient conditions* for the existence of final $\mathcal{L}$-algebras which are much easier to verify in practice than the characterizing conditions.

Throughout the following, we fix a signature Γ as the universal signature from which all signatures in all languages considered in the following are drawn. Γ is a countable signature which contains a countable number of function symbols for each arity and coarity. One can think of Γ as a universe o' typed identifiers, and its use ensures that our languages will be proper sets of theories, which is the only realistic situation.

Fully abstract models are defined using the ideas of context and separability. To adapt these ideas for algebraic specifications, we need to introduce a few syntactic notions. Consider Σ-terms as trees labeled with a function symbol from Σ at each node. The symbols at leaves may also be variables. A path p is a string of positive integers. The empty string is denoted by Λ. We assume familiarity with what is meant by the subterm reached by a path p in a Σ-term t; we use t/p as a notation for this term. The expression $t[p \leftarrow w]$ denotes the term obtained by replacing t/p at p by w. A **context** in Σ is a pair (c, q) such that $c \in T_\Sigma$ and q is a path in c. A context in which $q = \Lambda$ is said to be *empty*. Following traditional notation, a context will be denoted by $C[\cdot]$, and the term obtained by inserting a term t in $C[\cdot]$ will be written as $C[t]$. If $C[\cdot]$ is the pair (c, q), then $C[t]$ is the term $c[q \leftarrow t]$.

Throughout the following, we shall be concerned with an arbitrary language

$$\mathcal{L} \quad = \quad \langle \mathcal{L}, \leq, \text{BASE} \rangle \ .$$

Restrictions on $\mathcal{L}$ will be introduced explicitly, when needed.

Definition Let E be a theory in $\mathcal{L}$. Two ground terms t and u are said to be **separable** in E iff there is an extension $E' \geq E$ in $\mathcal{L}$ and a context $C[\cdot]$ such that $C[t] =_{E'} v$ and $C[u] =_{E'} w$ for some distinct $v, w \in T_{\text{BASE}}$ such that $v \not=_{\text{BASE}} w$. We write $t \approx_E u$ to mean that t and u are **inseparable** (that is, not separable) in E.

Obviously, two terms are inseparable if and only if there is no observable way to distinguish between them. This suggests that semantic equality in the final $\mathcal{L}$-algebra should coincide with inseparability. Of course, semantic equality must be a congruence relation. The inseparability relation on terms is clearly reflexive and symmetric. It is also preserved under application of function symbols. However, inseparability is not necessarily transitive. It turns out that a final $\mathcal{L}$-algebra exists iff separability is transitive. For a clear statement of this result, it is convenient to introduce notions of congruences and quotients of $\mathcal{L}$-algebras.

Definition Let $\mathcal{A}$ and $\mathcal{B}$ be $\mathcal{L}$-algebras. $\mathcal{B}$ is a **quotient** of $\mathcal{A}$ if there is a morphism from $\mathcal{A}$ to $\mathcal{B}$. Recall that $\mathcal{L}$-algebras are functors. A morphism is therefore just a natural transformation, and any natural transformation from an $\mathcal{L}$-algebra $\mathcal{A}$ to another $\mathcal{L}$-algebra is called an **$\mathcal{L}$-congruence**. Informally, an $\mathcal{L}$-congruence is a collection $\{\equiv_E : E \in \mathcal{L}\}$ such that for all $E \in \mathcal{L}$, $\equiv_E$ is a Σ_E congruence on $\mathcal{A}(E)$, and if $E \leq E'$, then $\equiv_{E'}$ refines $\equiv_E$.

Theorem 2 *Suppose that for every theory E in $\mathcal{L}$, $\approx_E$ is transitive. Then the collection $\{\approx_E : E \in \mathcal{L}\}$ is an $\mathcal{L}$-congruence, and the final $\mathcal{L}$-algebra is the quotient of the initial $\mathcal{L}$-algebra by it. Conversely, if the final $\mathcal{L}$-algebra exists, then $\approx_E$ is transitive for each $E \in \mathcal{L}$.*

Proof Suppose every $\approx_E$ is transitive. The first assertion reduces to showing that for any two Σ_E-terms t and u and any $\mathcal{L}$-algebra $\mathcal{A}$, $\mathcal{A}(E)(t) = \mathcal{A}(E)(u)$ implies $t \approx_E u$. To see this, suppose towards a contradiction that the terms were separable via a context $C[\cdot]$ in an extension E' of E; i.e., $C[t] =_{E'} v$ and $C[u] =_{E'} w$, where $v, w \in T_{\text{BASE}}$ and $v \not=_{\text{BASE}} w$. Since $\mathcal{A}(E)(t) = \mathcal{A}(E)(u)$ and $\mathcal{A}(E')$ is a surjective image of $\mathcal{A}(E)$ for Σ_E, we must have $\mathcal{A}(E')(t) = \mathcal{A}(E')(u)$. Now

$$\mathcal{A}(E')(C[t]) = \mathcal{A}(E')(v) = \mathcal{A}(E')(w) = \mathcal{A}(E')(C[u]) \ .$$

Therefore $\mathcal{A}$ does not preserve the BASE and cannot be an $\mathcal{L}$-algebra.

Going the other way, we need only show that if the final $\mathcal{L}$-algebra F exists, then for all E and for any two Σ_E-terms t and u, $t \approx_E u$ implies $\mathcal{A}(E)(t) = \mathcal{A}(E)(u)$ for some $\mathcal{L}$-algebra $\mathcal{A}$. Consider the $\mathcal{L}$-algebra which assigns initial interpretations to all theories not extending E, and assigns the quotient of the initial interpretation by the least congruence including $t = u$ to E and all theories extending E. The maps between the algebras are the natural ones. This $\mathcal{L}$-algebra preserves BASE by the very assumption of inseparability. ⊣

Although this characterization is appealing in being constructive and related to full abstraction, it is often not adequate since it is hard to prove properties of inseparability for many interesting languages. We therefore give sufficient conditions for the existence of final $\mathcal{L}$-algebras which are almost always easier to verify. The notion of a *maximal* theory conditions derived below.

Definition A theory M in a language $\mathcal{L}$ is **maximal** if $E \geq M$ implies $M \geq E$ for each $E \in \mathcal{L}$.

Examples If $\mathcal{L}$ is the language of all (finite and infinite) theories with a *fixed* common signature, then every theory in $\mathcal{L}$ has a maximal extension. On the other hand, if we consider the interesting sublanguage of $\mathcal{L}$ consisting only of the *finite* theories, then there are no maximal $\mathcal{L}$-theories. These two languages will be important in this and the next section.

Before stating the results, we need to recall the properties of an important subcategory of $\mathbf{Mod}_{\mathcal{L}}$. Let $\mathbf{Norm}(E)$ denote the full subcategory of $\mathbf{Mod}_{\mathcal{L}}$ which contains only Σ_E-algebras that satisfy E. This is the category of interpretations considered in traditional final algebra semantics. $\mathbf{Norm}(E)$ can obviously be thought of as the class of all BASE-preserving quotients of I_E, and as such it carries a natural algebraic structure that is slightly weaker than the complete lattice structure carried by all quotients of a free Σ-algebra. $\mathbf{Norm}(E)$ is a complete lower semilattice. That is, the intersection of any family of BASE-preserving congruences on I_E is also a BASE-preserving congruence. Moreover, although unions of BASE-preserving congruences are not in general BASE-preserving (unless E has a final algebra), unions of such congruences corresponding to (finite and infinite) *directed* sets of quotients do preserve BASE. This can be seen by a simple *compactness* argument – any counterexample must prove the equivalence of two observable values in a finite number of equational steps based on a finite number of congruences, but the corresponding quotients have an BASE-preserving upper bound. Directed sets of quotients therefore have least upper bounds.

The close connection between maximal theories and final $\mathcal{L}$-algebras is stated in the following two results, of which the first is an obvious consequence of maximality.

Proposition 3 *If the final $\mathcal{L}$-algebra $\mathcal{F}$ exists for a language $\mathcal{L}$, then for each maximal theory M of $\mathcal{L}$, $\mathcal{F}(M)$ is the final object of* $\mathbf{Norm}(M)$.

Theorem 4 *Suppose that $\mathcal{L}$ has the following two properties:*

(1) Every E in $\mathcal{L}$ has a maximal extension M in $\mathcal{L}$.

(2) For every maximal M in $\mathcal{L}$, there is a final object F_M in **Norm**(M).

Then there is a final $\mathcal{L}$-algebra.

Proof We define an $\mathcal{L}$-algebra $\mathcal{F}$ by specifying its value on each $E \in \mathcal{L}$. Fix E, and let

$$\mathcal{F}(E) \quad = \quad \bigwedge \{F_M : M \geq E \text{ maximal}\}$$

That is, for each E, we take the meet of all the final interpretations of maximal extensions of E, look at their induced congruences, and take the intersection. The meet is well defined by the two conditions in the theorem and is a model for E. The collection $\mathcal{F}$ of algebras of the form $\mathcal{F}(E)$ obviously constitutes an $\mathcal{L}$-algebra, since $E' \geq E$ implies that $\mathcal{F}(E')$ is formed by the meet of a smaller set of algebras than $\mathcal{F}(E)$. To show that $\mathcal{F}$ is final, we need only show that given any $\mathcal{L}$-algebra $\mathcal{A}$, there is a homomorphism from $\mathcal{A}(E)$ to $\mathcal{F}(E)$ for every $E \in \mathcal{L}$. This is easy since every ground Σ_E-equation that does *not* hold in $\mathcal{F}(E)$ is necessarily inconsistent with some (maximal) extension of E, and therefore could no hold in $\mathcal{A}(E)$. $\dashv$

Theorem 4 often needs to be applied in an indirect way. For instance, suppose we wish to consider the language of all finite theories over a fixed finite signature. Such a language would be a more realistic representation of evolutionary possibilities than a language which includes infinite theories. However, maximal extensions in this case are often infinite, and hence this language cannot satisfy condition (1) of Theorem 4. We thus need a final model for a *sublanguage* of a language to which Theorem 4 is applicable. In category theoretic terms, a final object is a *limit*. We therefore need a construction for categories of language models that preserves limits — the standard construction of this kind being a *right adjoint*. In general, suppose we have a language $\mathcal{L}$ which is known to have a final model, and a language $\mathcal{L}'$ for which we wish to derive a final model. Suppose CL and CL' are the (functor) categories of all models of $\mathcal{L}$ and $\mathcal{L}'$. What we need is an adjoint pair of functors $H \dashv G$, where $H : CL' \to CL$ and $G : CL \to CL'$. Given this, if $\mathcal{F}$ is the final object in CL, then $G(\mathcal{F})$ would be the fina object in CL'. Since we are working with preorder categories, the usual conditions for the adjunction between such a pair of functors reduce to a particularly simple form called a *Galois connection* [ML]. Here is a statement of the theorem adapted for our purposes:

Theorem 5 (Galois Connection Theorem) *Let P and Q be two preorders and $H : P \to Q$, $G : Q \to P$ two order preserving functions. Then H (regarded as a functor) is a left adjoint of G iff for all $p \in P$ and $q \in Q$, $H(p) \geq q$ iff $p \geq G(q)$. Such a pair H and G is called a* **Galois connection** *from P to Q.*

We now consider a practically important case in which such a Galois connection exists.

Theorem 6 *Suppose a final model $\mathcal{F}$ exists for a language $\mathcal{L}$ and $\mathcal{L}'$ is a sublanguage of $\mathcal{L}$ such that every theory in $\mathcal{L}$ is the union of (signatures and equations from) a directed set of theories in $\mathcal{L}'$. Then the reduct of $\mathcal{F}$ to $\mathcal{L}'$ is a final model for $\mathcal{L}'$.*

Proof (**sketch**) Since model functors preserve order, the set of algebras in any model of $\mathcal{L}'$ corresponding to a directed set of theories is also directed, and therefore the union of the corresponding congruences is BASE-preserving. We can now define the functor $H : CL' \to CL$. Let $\mathcal{A} \in CL'$ be any model of $\mathcal{L}'$. $H(\mathcal{A})$ is a model of $\mathcal{L}$ which extends $\mathcal{A}$ in a natural way. Thus, for each $E \in \mathcal{L}'$, $H(\mathcal{A})(E) = \mathcal{A}(E)$. If $E \notin \mathcal{L}'$, it is the union of a directed subset of $\mathcal{L}'$, and $H(\mathcal{A})(E)$ is the quotient of I_E by the BASE-preserving union of the corresponding congruences. The reverse functor $G : CL \to CL'$ simply maps a model of $\mathcal{L}$ to its reduct to $\mathcal{L}'$. Given arbitrary models $\mathcal{A} \in CL$ and $\mathcal{B} \in CL'$, it is easy to check that $H(\mathcal{B}) \leq \mathcal{A}$ iff $\mathcal{B} \leq G(\mathcal{A})$, thus establishing the Galois connection, and showing that H and G together constitute a pair of adjoints. ⊣

4 Examples

We now consider a number of languages to illustrate the application of the ideas and results of the last section. These examples show that the evolutionary space used in relativizing the (observable) semantics of individual theories can have a profound effect on the content of the semantics. The language context defines the expected range of validity of assertions about particular specifications. The larger the language the fewer the valid assertions.

Throughout the following, the extension structure in all language examples will include **all permissible** *extensions, unless otherwise mentioned.* We assume a fixed arbitrary BASE theory for all languages as well.

Example A $\mathcal{U}$ = *the language of all* finite BASE-*preserving theories* E *(with* $\Gamma \supseteq \Sigma_E$*). (Recall that* Γ *is our overall "large" singature, containing infintely many symbols of all possible arities and co-arities.)*

Example B *For each fixed signature* $\Sigma \subseteq \Gamma$, *let* $\mathcal{L}_\Sigma$ = *be the language of all* finite and infinite BASE-*preserving* Σ-*theories* E.

Example C *For each such* Σ, *let* $\mathcal{M}_\Sigma$ = *be the language of all* finite BASE-*preserving* Σ-*theories* E.

Lemma 7 *The final* $\mathcal{U}$-*algebra exists and is identical to the initial* $\mathcal{U}$-*algebra.*

Proof The proof hinges on the fact that each theory E of $\mathcal{U}$ is finite, so there are new symbols in Γ which can be used to build contexts to separate any given pair of ground terms, provided the pair is not provably equal in E. That is, let t and u be Σ_E-terms of the same sort such that $t \neq_E u$. Suppose v and w are two Σ_{BASE}-terms that correspond to different observable values. Now let g be a new symbol that is not in Σ_E and let $E' = \{g(t) = v,\ g(u) = w\}$. Clearly, the function g provides a context in the extension E' of E to separate t and u. Therefore $t \not\approx_E u$. This implies $t \approx_E u \iff t =_E u$, i.e., the only $\mathcal{U}$-algebra is the initial one. ⊣

To put this in perspective, recall that we are interested in language models as a way of making the semantics of individual theories more realistic by taking account of their possible evolution. The result above implies that if evolutionary possibilities include the addition of arbitrary new symbols, then the only results guaranteed to hold in all extensions are those which are valid in the initial algebra. Therefore, results obtained via reasoning techniques based on final algebra semantics, such as proof by consistency [KMb], are not necessarily carried over to arbitrary extensions if they introduce new symbols. This conclusion does not change even if the language is restricted to only those theories which are semantically "unambiguous"; i.e., those which have final algebras in the usual sense.

In contrast, in any language where extensions are not allowed to introduce new symbols, the final model, when it exists, is usually *different* from the initial model, and moreover, assigns the final algebra to any theory for which such an algebra exists.

We shall use the following result, which is of independent interest because it shows that the idea of taking final models of languages really is a generalization of ordinary final algebra semantics.

Theorem 8 *Let $\mathcal{L}$ be any language in which $E \leq E'$ implies that $\Sigma_E = \Sigma_{E'}$. If the final $\mathcal{L}$-algebra exists, and E is a theory in $\mathcal{L}$ such that the final interpretation of E exists, then the interpretation assigned to E in the final $\mathcal{L}$-algebra is exactly the final interpretation of E.*

Proof It is enough to show that if two ground Σ_E-terms t_1 and t_2 have the same interpretation in the final algebra F for E then they are inseparable. We show the contrapositive of this. Suppose that t_1 and t_2 are separable; i.e., there is an extension E' of E and a context $C[\cdot]$ such that $C[t_1] =_{E'} v$ and $C[t_2] =_{E'} w$, where v and w are $\Sigma_{\mathbf{BASE}}$-terms that represent distinct observable values. Since $I_{E'}$ is a model for E as well, $F(C[t_1]) = F(v)$ and $F(C[t_2]) = F(w)$ and hence $F(t_1) \neq F(t_2)$. ⊣

We now consider Examples B and C from above. Example C represents the situation where the *signature* of a specification has reached final form, but the *equational constraints* on observable behavior are still evolving. The incomplete array specification in the introduction is an example of this situation, and we shall return to this below. If one had to choose a fixed universe of evolutionary possibilities to generalize final algebra semantics, this class of languages would be a very good choice for two reasons:

1. These languages represent a sufficiently large portion of the evolutionary space for a specification to make them useful, while at the same time being small enough to retain the observable character of the resulting semantics.

2. There is a final model for each language in this class, and therefore the generalization would be universally applicable.

To show that final models exist for Example C, we first need to consider Example B. This class of languages allows *infinite* theories. Because this class has maximal theories, it is easier to work with. Then we shall deduce the existence of final models for Example C using Theorem 6.

Lemma 9 *For each $E \in \mathcal{L}_\Sigma$ there exists some (not necessarily unique) $M \in \mathcal{L}_\Sigma$ such that $M \geq E$ and M is maximal.*

Proof Define a weakly increasing sequence $\langle E_i : i \in \omega \rangle$, by recursion as follows: Let $E_0 = E$. Given E_i, let

$$E_{i+1} = \begin{cases} E_i \cup \{e_i\} & \textit{if } I_{E_i}/|e_i| \textit{ is } \textsc{base} \textit{ preserving} \\ E_i & \text{otherwise} \end{cases}$$

where $|e_i|$ is the congruence on $T_{\Sigma_{E_i}}$ generated by equation e_i.

Let $M = \lim_{i \to \infty} E_i$. So $M \in \mathcal{L}_\Sigma$ and $M \geq E_0 = E$. We claim that M is maximal. If not, let N be a theory in $\mathcal{L}_\Sigma$ such that $N > M$, and fix some ground Σ_E-equation e which holds in I_N but not I_M. Then for some i, $e = e_i$. But $E_i/|e|$ is BASE-preserving, since $I_N|\Sigma_E$ is a quotient of it. Thus, $E_{i+1} = E_i \cup |e|$. So e holds in M and this is a contradiction. ⊣

Corollary 10 *For any signature Σ, there is a final $\mathcal{L}_\Sigma$-algebra for the language $\mathcal{L}_\Sigma$ of all (finite and infinite) theories with signature Σ.*

Proof (sketch) In this language, E is maximal iff I_E has no proper quotients that preserve the BASE. Therefore I_E is the only minimal BASE-preserving Σ-algebra satisfying E. The rest follows by Theorem 4. ⊣

Corollary 11 *For any signature Σ, there is a final $\mathcal{M}_\Sigma$-algebra for the language $\mathcal{M}_\Sigma$ of all* **finite** *theories with signature Σ.*

Proof Languages with all finite and infinite theories for a fixed signature are just "completions" (in the usual sense of cpo completion) of corresponding languages with only finite theories. The corollary therefore follows from the previous corollary and Theorem 6. Note that the algebras assigned to finite theories by $\mathcal{M}_\Sigma$ are the same as those assigned by $\mathcal{L}_\Sigma$. ⊣

For convenience, we shall speak of the final model for $\mathcal{M}$ instead of the final model for $\mathcal{M}_\Sigma$ for a given signature Σ. The signature involved will be clear from the context.

Example D *The incomplete specification for arrays discussed in the introduction.* EXT *has a single equation,* $\mathsf{val}(x, \mathsf{alt}(y, z, a)) = \mathsf{if}\ \mathsf{eq}(x, y)\ \mathsf{then}\ z\ \mathsf{else}\ \mathsf{val}(x, a)$.

Here, the algebra assigned by the final model for $\mathcal{M}$ would create new values of the sort Nat for expressions of the form $\mathsf{val}(x, \mathsf{empty})$, in a sense "suspending judgement" on them, while validating the assertion that distinctions based on obsolete values at an index are immaterial, i.e.,

$$\mathsf{alt}(x, y, \mathsf{alt}(x, z, a)) = \mathsf{alt}(x, y, a) \ .$$

The (semantic) creation of new values of sort **Nat** would be considered undesirable by some authors, but it is not very damaging for reasoning about equational assertions. In any case, these values would not be considered observable in our framework.

Our last set of examples is suggested by the work of Kapur-Musser [KMa] on completable theories.

Definition Given a BASE theory and a theory E, an algebra $A \in \mathbf{Norm}(E)$ is *standard* if for every term $t \in T_{\Sigma_E}$ of observable sort, there is a term $v \in T_{\text{BASE}}$ such that $A(t) = A(v)$. In other words, A is standard if the carriers of A for observable sorts do not contain new "non-standard" (and therefore non-observable) values. The theory E is *completable* if there is a standard algebra in $\mathbf{Norm}(E)$.

The *idea* of completability is to exclude theories which do not have sufficiently complete extensions, because in some sense incomplete theories of this kind are doomed to ultimate "failure". Practically, it turns out that techniques for reasoning with ambiguous specifications using proof by consistency work out much better if all theories are assumed to be completable. The precise *definition* of completability is tricky because there are theories which have *infinite* sufficiently complete extensions but no finite ones. (see Example H below). We therefore deviate from the original definitions of [KMa] and use a semantic definition based on a notion of *standard* algebras, which we believe is more general without damaging the usefulness of the concept.

The following example shows that our method of constructing models of languages is more general than considering the standard extensions of a completable theory.

Example E *A theory E which is* not completable, *yet where the final model for $\mathcal{M}$ exists and is the intuitively correct model. The* BASE *contains two symbols* $0 :\rightarrow \mathsf{Num}$ *and* $\mathsf{succ} : \mathsf{Num} \rightarrow \mathsf{Num}$. *Thus the terms* $\mathsf{succ}^k(0)$, *for* $k \geq 0$, *are distinct observable values. E adds the symbol* $\mathsf{pred} : \mathsf{Num} \rightarrow \mathsf{Num}$ *and equations*

$$\mathsf{succ}(\mathsf{pred}(x)) = x$$
$$\mathsf{pred}(\mathsf{succ}(x)) = x.$$

Every Σ_E-term is equivalent either to 0 *or to a unique term of the form* $\mathsf{succ}^k(0)$ *or* $\mathsf{pred}^k(0)$ *for some* $k \geq 1$. *It is obvious that I_E itself is* BASE*-preserving, but no proper quotient of I_E is* BASE*-preserving. I_E is therefore the final algebra for E, and by Theorem 8, is assigned to E by $\mathcal{M}$.*

Here are some examples of languages of completable theories.

Example F $\mathcal{Z}_\Sigma$ = *the language of all (finite and infinite)* **completable** *theories with signature* Σ.

Example G $\mathcal{N}_\Sigma$ = *the language of of all* **finite completable** *theories with signature* Σ

Lemma 12 *For each* $E \in \mathcal{Z}_\Sigma$ *there exists some (not necessarily unique)* $M \in \mathcal{Z}_\Sigma$ *such that* $M \geq E$ *and* M *is maximal.*

Proof Very similar to the proof of Lemma 9, except that the condition for $E_{i+1} = E_i \cup \{e_i\}$ includes the stipulation that $E_i \cup \{e_i\}$ must be completable. ⊣

Corollary 13 *For any signature Σ, there is a final $\mathcal{N}_\Sigma$-algebra.*

Proof The proof proceeds by using Theorems 4 and 6 as before. ⊣

The next example shows the definition of completability is delicate. This example is not needed in the sequel, and in fact, it is not really an example concerning final algebras. Still, it is of interest because it shows (for the first time, as far as we know) that completability is different from finite completability.

Example H *A theory which is completable by an infinite set of equations but by no finite set.*

E has a single-sorted signature Σ_E with a constant 0 and two unary operators r and l, and the empty set of equations. Σ_E-terms can obviously be written as strings of r and l terminated by 0. If p and q are strings of r and l alone, then let $p : q$ denote their concatenation. For the visible terms, take all those of the form $v : l^n0$ where $|v| < n$ and $n \geq 0$. I.e., the tail of l's is longer than what precedes it. For example, the following are visible: $l0$, 0, $rll0$, $rllrlllll0$.

The first thing to notice is that this theory is completable. To see this, call a term **critical** if it is not visible, but its maximal subterm is visible. Now consider the set equations $\{t = 0 \mid t \text{ is critical}\}$. We claim that under this set, every term is provably equivalent to a visible term. Note that if t is equivalent to a visible term, then lt and rt are equivalent either visible or critical terms. Since 0 is visible, the claim follows by induction on the length of a term t.

Of course, it must also be checked that this infinite set of equations does not identify any visible terms. To see this, note that the set of equations with the given orientation form a canonical (confluent terminating) rewriting system in which each visible term is in normal form and each non-visible term has a unique *redex*,a critical subterm. Each non-visible term can therefore be reduced to a visible one through a unique sequence of rewriting steps, replacing its critical subterm by 0 at each step. In other words, each non-visible term has a unique visible term as its normal form. Since the system is canonical and visible terms are normal forms, this proves that no two visible terms are provably equal.

Now that we know that the system is completable via an infinite set of equations, we claim that no finite set can do this. First, we claim that there is no consistent equation $f(x) = g(x)$, where f and g are strings of l and r, if $f \neq g$. To see this, suppose that f and g are different, and let n be the larger of the lengths of f and g. It is a consequence of this equation that $f(l^{n+1}0) = g(l^{n+1}0)$. But these two are distinct visible terms. Second, for the same reason, there no consistent equations $f(x) = g(y)$ involving two different variables. Third, there are no consistent equations $f(x) = t$ involving variables on only one side because it would follow from such an equation that $f(x) = f(f(x))$, and this contradicts our point above.

Suppose that the original theory were finitely completable. Then we know that all of the equations must be equalities of ground terms $t = u$. Let all of the (finitely many) terms that appear in these equations be listed as $t_1, \ldots, t_n$. Let t^* be any non-visible term which does not have any t_i as a subterm. Such a t^* always exists by an easy combinatorial argument. Now the original finite set of equations cannot equate t^* to any visible term. This proves that the system above is not finitely completable. This completes our examination of Example H.

Our last example shows the difference between the semantics induced by the language $\mathcal{N}_\Sigma$ of all finite completable theories vs. the language $\mathcal{M}_\Sigma$ of all finite theories for a given signature. (This is the motivating example from [KMa]).

Example I *The final $\mathcal{N}_\Sigma$ model and the final $\mathcal{M}_\Sigma$-models may differ. Let* BASE *be a theory of natural numbers with only*

$$\mathsf{0} :\to \mathsf{Nat}$$
$$\mathsf{succ} : \mathsf{Nat} \to \mathsf{Nat}$$

and no equations. Now let E be a theory which introduces three new symbols

$$\mathsf{plus} : \mathsf{Nat} \times \mathsf{Nat} \to \mathsf{Nat}$$
$$\mathsf{double} : \mathsf{Nat} \to \mathsf{Nat}$$
$$\mathsf{k} :\to \mathsf{Nat}$$

subject to the following three equations:

$$\begin{aligned}
\mathsf{plus}(0, x) &= x \\
\mathsf{plus}(\mathsf{succ}(x), y) &= \mathsf{succ}(\mathsf{plus}(x, y)) \\
\mathsf{double}(0) &= 0 \\
\mathsf{double}(\mathsf{succ}(x)) &= \mathsf{succ}(\mathsf{succ}(\mathsf{double}(x)))
\end{aligned}$$

If $\mathcal{G}$ and $\mathcal{F}$ are the final models of $\mathcal{N}_{\Sigma_E}$ and $\mathcal{M}_{\Sigma_E}$ respectively, then the assertion $\mathsf{double}(x) = \mathsf{plus}(x, x)$ holds in $\mathcal{G}(E)$ but not in $\mathcal{F}(E)$. Th reason is that one may equate $\mathsf{double}(\mathsf{k}) = \mathsf{succ}(0)$ in an algebra in $\mathsf{Norm}(E)$ but such an algebra cannot be standard. No *completable* extension of E can therefore *imply* such equations which violate the required assertion.

Finality is perhaps the simplest embodiment of a kind of intensionality that contrasts with the extensionality of initial algebra semantics. It is necessary to consider different languages $\mathcal{L}$ in order to get intuitively satisfying semantics. This is the common thread in all of our examples. It seems unlikely that some fixed absolute restriction will be satisfactory in all situations, and relativization allows a kind of "semantic engineering" to adjust the class $\mathcal{Z}$ in a variety of ways. To make a related point, we showed in Example A that finality reduces to initiality if correct implementations are given sufficient liberty to enrich the signature. So if observable semantics is to offer something new, one must consider different languages $\mathcal{L}$ of reasonable models.

5 Related Work

Most previous results in the literature regarding the existence of final algebras can be understood in the framework of Section 3 using the characterization of final models based on inseparability. Wand [Wan] considers the problem in the abstract setting of algebraic theories, and his notion of extension does not even require extensions to share any sorts with the base theories. However, for most practical purposes, Wand's framework consists of a singleton language consisting of what he calls the *extension*, with his *base* theory serving as the BASE. His Λ-faithfulness condition amounts to assuming that the extension preserves the BASE of observable values, and his Λ-fullness condition is equivalent to sufficient completeness. Given the Λ-fullness condition, it is easy to show that inseparability is transitive, and hence the language has a final model. Theorem 2 then implies that the (obviously maximal) single theory in the language has a final algebra.

Kapur-Musser [KMa] consider ideas similar to those embodied in Theorem 2 in the context of inductive reasoning. Their treatment appears to be implicitly directed towards the language of all theories for a fixed signature — they do not articulate a notion of language. In later work [KMa], they have chosen to restrict their language to completable theories. The connections between their work and ours are not entirely clear to us, although final models can be derived for both their languages within our framework.

The idea of language models described here can also be applied to equational languages of "executable specifications" such as the language of regular systems [HO] and the equational subset of Miranda[4] [Tur]. The algebras in this case need to be "computable" with respect to observable values since the specifications are meant to be effective. A final model always exists under these constraints, and other interesting issues such as limiting completeness arise. [Tha] describes the relevant results.

6 Conclusions

We have generalized final algebra semantics, based on the observation that specifications are intended to support reasoning about classes of models. We introduced a notion of language to formalize ranges of applicability, showed how to lift the notions of initial and final models from theories to languages, and gave existence theorems for final models. We presented several examples to illustrate the broad applicability of observable semantics. It can be used in situations involving incomplete specifications, and it works even when there are special requirements such as completability, the possibility (or impossibility) that extensions add new operations, etc. As a result, we believe that the method of observable semantics is a meaningful basis for reasoning in realistic situations.

---

[4]Miranda is a trademark of Research Software Limited.

References

[ADJ] J. A. Goguen, J. W. Thatcher, E. G. Wagner, and J.B. Wright (ADJ). An initial algebra approach to the specification, correctness, and implementation of abstract data types. In R. T. Yeh, editor, *Current Trends in Programming Methodology IV*, Prentice-Hall, 1978.

[GH] J. V. Guttag and J. J. Horning. The algebraic specification of abstract data types. *Acta Informatica*, 10(1):27–52, 1978.

[HO] M. C. Hoffman and M. J. O'Donnell. Programming with equations. *ACM TOPLAS*, 4(1):83–112, 1982.

[Kam] S. Kamin. Final data types and their specifications. *ACM TOPLAS*, 5(1):97–121, 1983.

[KMa] D. Kapur and D. R. Musser. Inductive reasoning with incomplete specifications. In *Proceedings of the Symposium on Logic in Computer Science*, pages 367–377, 1986.

[KMb] D. Kapur and D. R. Musser. Proof by consistency. *General Electric Corporate Research and Development Report 84GEN008*, 1984.

[Mil] R. Milner. Fully abstract models of typed λ–calculi. *Theoretical Computer Science*, 4:1–22, 1977.

[ML] S. Mac Lane. *Categories for the Working Mathematician.* Springer-Verlag, 1971.

[Tha] S. R. Thatte. Full abstraction and limiting completeness in equational languages. *Theoretical Computer Science*, 1989.

[Tur] D. A. Turner. Miranda: a non-strict functional language with polymorphic types. In *LNCS 201*, Springer-Verlag, 1985.

[Wan] M. Wand. Final algebra semantics and data type extensions. *Journal of Computer and System Sciences*, 19(1):27–44, 1977.

Termination, Deadlock and Divergence*

L. Aceto
M. Hennessy
Computer Science
School of Mathematical and Physical Studies
University of Sussex
Falmer, Brighton BN1 9QH, England

Abstract

In this paper we introduce a process algebra which incorporates explicit representations of successful termination, deadlock and divergence, and analyze its semantic theory. We give both an operational and a denotational semantics for the language and show that they agree. The operational theory is based upon a suitable adaptation of the notion of bisimulation preorder. The denotational semantics for the language is given in terms of the initial continuous algebra which satisfies a set of equations E, CI_E. We show that CI_E is fully abstract with respect to our choice of behavioural preorder.

1 Introduction

In this paper we wish to develop a theory for a process algebra which incorporates some explicit representation of termination, deadlock and divergence. We develop both an operational theory based on bisimulations, [Pa81], and an equational theory similar to those for **CCS**, **ACP**, [HM85], [H88b], [BK85].

The theory of **ACP**, [BK84], [BK85], deals with deadlock explicitly by introducing into the signature of the calculus a distinguished constant symbol δ. Deadlock can also occur directly in processes. If p can only perform actions from the set H then the process $\partial_H(p)$ is considered to be the same as the deadlocked process δ. But **ACP**, at least in its original formulation, does not have an explicit representation of successful termination.

On the other hand **CCS**, [Mil80], has a single "terminated" process, *nil*, which stands for both successful termination and deadlock. This choice is justified by the

*This work has been supported by a grant from the United Kingdom Science and Engineering Research Council.

fact that in **CCS** these two kinds of termination are experimentally indistinguishable, due to the restricted form of sequential composition, *action-prefixing*, present in the calculus. As **ACP** allows sequential composition this is no longer the case. Consider, for example, the process $nil; p$. Since nil is successfully terminated $nil; p$ can perform any action which p may perform. On the other hand, it is natural to assume that the process $\delta; p$ is deadlocked and will never perform any action. Thus, in the presence of sequential composition, there is an observable difference between nil and δ.

We will express desirable properties of processes by means of equations. For example

$$\delta; x = \delta$$

represents the fact that a deadlocked process can never proceed, and

$$nil; x = x$$

the fact that nil is a properly terminated process. Many of the equations for our language are already well known either from **CCS** or **ACP**. However, the presence of the terminated process nil invalidates some of those from **ACP**. The equation

$$(x + y); z = x; z + y; z$$

is part of the theory of **ACP**, [BK85], but is not valid for our language, at least in its general form. In fact, if x is δ and y is nil then, assuming that $\delta + nil = \delta$, the left-hand side is equal to $\delta; z$, i. e. δ, whereas the righthand side is equal to $nil; z + \delta; z$, i. e. $z + \delta$. If z is a non-trivial process it is then reasonable to assume that δ and $z + \delta$ are different processes.

We will also have within our language processes which may diverge internally. We let Ω be a process which can only diverge internally. Using the usual notation for recursive terms this could also be represented by $rec\,x.\ \tau; x$, where τ is an internal unobservable move. Obviously we would expect nil and Ω to be different processes and we will also demand that δ and Ω be different. The latter requirement is less defensible, but we are motivated by the *information-theoretic* view of computation as advocated by Scott. Here the process which can only diverge, Ω, contains no information and is therefore considered less than any other process. There is some information available about the process δ, namely that it is deadlocked; so Ω and δ should be considered different. In the presence of Ω, and in particular taking Scott's approach to semantics, it is natural to express our theory in terms of *inequations*. One inequation is

$$\Omega \leq x,$$

and more generally the equations given above could be viewed as shorthand for two inequations, $t = u$ representing $t \leq u$ and $u \leq t$.

The main purpose of this paper is to show that an adequate semantic theory for a process algebra which contains divergence, termination and deadlock can be constructed using a suitable set of inequations, E. More specifically we propose as a denotational semantics the initial continuous algebra generated by E, CI_E, [GTWW77], [Gue81]. This is in contrast to [BK82] where metric spaces are used for this purpose in place of continuous partial orders. The advantage of the latter is that all of the usual operators found in process algebra may be interpreted, whereas using metric spaces we can only readily interpret operators which are contractive. For instance, unguarded recursive definitions give rise to operators which are not contractive; in addition to this drawback, silent actions and abstraction operators have never been dealt with satisfactorily in this framework. Moreover, we can apply the existing and well understood theory of algebraic cpo's, for example to show the existence of CI_E and to derive useful proof techniques such as Scott Induction, [LS87].

In order to show that CI_E is a reasonable model we develop a behavioural or observational view of processes and prove that this coincides with the interpretation given by CI_E. This is given in terms of a variation on bisimulation equivalence, [Pa81]. To take divergence into account we generalize bisimulation equivalence, $\approx$, to a preorder $\sqsubseteq$ which is often called pre-bisimulation preorder. Intuitively, $p \sqsubseteq q$ means that p and q are bisimilar except that at times p may diverge more frequently than q; in the absence of divergence $p \sqsubseteq q$ will imply $p \approx q$. This type of behavioural relation has been studied in [HP80], [Mil81], [Wal87], [Ab87a,b]. Here we modify it to take into consideration termination and deadlock and show that two processes are behaviourally related with respect to this new relation if and only if they are related in the equational model CI_E. In other words, CI_E is *fully abstract* with respect to this new behavioural preorder. There may be other fully abstract models, but CI_E is distinguished by being initial in the category of fully abstract models. In fact, it is initial in the category of models which are consistent with the behavioural preorder.

We now give a brief outline of the remainder of the paper. In §2 we define the language whose semantic properties will be investigated in the paper. The language is endowed with both an operational and a denotational semantics. The operational semantics is defined in §2.1 following standard lines by means of Plotkin's *Structural Operational Semantics* (**SOS**), [Mil80], [Pl81]. Section 2.1 also introduces several definitions and notational conventions which will be used throughout the paper. The denotational semantics for the language is given in §2.2. The definition is based on the well-known techniques of *Initial Algebra Semantics*, [GTWW77], [Gue81]; as already mentioned, we propose as a denotational model for the language the initial continuous Σ-algebra which satisfies a set of equations E, CI_E. The following sections are entirely devoted to showing that CI_E is indeed a reasonable denotational model for our language. As argued by Milner, [Mil83], operational semantics should be the touchstone for assessing mathematical models for concurrent languages. The agreement between denotational models and operational ones is called *full abstraction* in [Mil77], [Pl77], [HP79]. In this paper we follow Milner and Plotkin's paradigm and justify the choice of our denotational model by showing that CI_E is fully abstract with respect to a natural notion of an operational or behavioural preorder over our language. The behavioural preorder is introduced in §3, where several constraints which behavioural relations have to meet in order to be related to denotational ones are also discussed. In particular, it is argued that, in order to be related to $\leq_E$, a behavioural preorder should be *finitely approximable* ([H81],[Ab87b]) and *closed with respect to all contexts.*

We then state the main result of the paper, namely that our choice of behavioural preorder, $\sqsubseteq^c_\omega$, which possesses these two properties, coincides with $\leq_E$ over our language. The proof of this result, which requires considerable analysis of $\sqsubseteq^c_\omega$, will be given in the full version of the paper.

We end with a conclusion in which we discuss the results of the paper and relationships with related work.

2 The Language

Let *Act* be a countable set of atomic action symbols. It is assumed that *Act* comes equipped with a bijection $\bar{\cdot} : Act \rightarrow Act$ which is its own inverse. The set *Act* will be called the set of *observable actions* and will be ranged over by $a, b, \ldots$. Let τ and δ be two distinguished symbols not occurring in *Act*. The symbol τ will stand for an internal, unobservable action; these actions will occur when processes communicate with each other. $Act_\tau =_{def} Act \cup \{\tau\}$ will be called the set of *actions* and will be ranged over by $\mu, \gamma \ldots$. The symbol δ will stand for a *deadlocked process*, a process that cannot perform any move but is not successfully terminated. Successful termination will be denoted by the constant symbol *nil*.

The set of constant symbols in the process algebra we will consider is completed by the symbol Ω; Ω will stand for a process that can internally diverge. Alternatively, one may think of Ω as the totally undefined process, the process about which the environment has no information at all. Ω is not deadlocked and has not successfully terminated. The process combinators used to build new systems from existing ones will be the following:

- $+$ for *nondeterministic choice*,
- ; for *sequential composition*,
- $|$ for *parallel composition*,
- $\partial_H(\cdot)$ for the *encapsulation operator.*

Formally:

Definition 2.1 *For each $n \in \omega$ let Σ_n, the set of operation symbols of arity n, be defined as follows:*

- $\Sigma_0 = \{nil, \delta, \Omega\} \cup Act_\tau$
- $\Sigma_1 = \{\partial_H(\cdot) \mid H \subseteq Act \wedge H = \overline{H}\}$
- $\Sigma_2 = \{+, ;, |\}$
- $\Sigma_n = \emptyset$, *for each* $n > 2$.

The signature Σ is defined as $\Sigma = \bigcup_{n \geq 0} \Sigma_n$.

Let Var *be a countable set of variables, ranged over by $x, y \ldots$. The syntax of* recursive terms *over Σ is then defined by*

$$t ::= f(t_1, \ldots, t_k)\,(f \in \Sigma_k) \mid x \mid rec\,x.\ t.$$

We assume the usual notions of free and bound variables in terms, with $rec\,x.\ \_$ as the binding constructor. The set of recursive terms over Σ will be denoted by $REC_\Sigma(\mathtt{Var})$ and will be ranged over by $t, u, \ldots$. The set of closed recursive terms *over Σ will be denoted by REC_Σ and will be ranged over by $p, q, p' \ldots$. The set of* syntactically finite processes *(i. e. those not involving occurrences of $rec\,x.\ t$) will be denoted by $FREC_\Sigma$ and will be ranged over by $d, e, d' \ldots$.*

Notationally, all the binary operators will be used in infix form, with the assumption that ; binds stronger than |, which in turn binds stronger than +. The constructor $rec\,x.\ \_$ will have the lowest precedence among all the operators.

2.1 The operational semantics

The operational semantics for the language REC_Σ consists of three different components. The first is an interpretation of REC_Σ as a labelled transition system in Plotkin's **SOS** style, [Mil80], [Pl81]. This associates with each action symbol μ a binary infix relation. Intuitively, $p \xrightarrow{\mu} q$ means that p may perform the action μ and thereby be transformed into q. The second is a *successful termination predicate*, $\surd$, which will be written in a postfix manner. Intuitively, $p\surd$ if p has terminated successfully, which will mean, among other things, that p cannot perform any further actions. We would expect $nil\surd$ but not $\Omega\surd$, not $a;p\surd$ and not $\partial_{\{a\}}(a;p)\surd$. The final component is a *convergence predicate*, $\downarrow$. Intuitively, $p\downarrow$ means that the set of actions which p can initially perform is fully specified. It will turn out that $nil\downarrow$ but not $\Omega\downarrow$.

Definition 2.2 *Let $\surd$ be the least subset of REC_Σ which satisfies:*

i) $nil \in \surd$,

ii) $p \in \surd$ *implies* $\partial_H(p) \in \surd$,

iii) *$p \in \surd$ and $q \in \surd$ imply $p+q, p;q, p|q \in \surd$,*

iv) *$t[rec\,x.\ t/x] \in \surd$ implies $rec\,x.\ t \in \surd$.*

In what follows we will write $p\surd$ iff $p \in \surd$. Note that the process $nil + \delta$ is not considered successfully terminated. Intuitively, the process is "stagnating" on a branch of its computation and the environment has no way of discarding this branch.

Definition 2.3 *Let $\downarrow$ be the least subset of REC_Σ which satisfies*

i) *$nil \downarrow, \delta \downarrow, \mu \downarrow$*

ii) *$p \downarrow$ implies $\partial_H(p) \downarrow$*

iii) *$p \downarrow, q \downarrow$ imply $(p+q) \downarrow, (p|q) \downarrow$*

iv) *$t[rec\,x.\ t/x] \downarrow$ implies $rec\,x.\ t \downarrow$*

v) *$p\surd, q \downarrow$ imply $(p;q) \downarrow$*

vi) *$\neg(p\surd), p \downarrow$ imply $(p;q) \downarrow$.*

Intuitively, $p \downarrow$ iff p is a completely specified process, i. e. if we can expand the recursive definition of p a finite number of times to obtain at the top level all the possible moves of p. Clause vi) of the definition of the predicate $\downarrow$ deserves some comment. It expresses the intuition that, if p is not successfully terminated, $p;q$ is a completely specified process if p is; in fact, in this case the set of initial moves of $p;q$ coincides with that of p. $\uparrow$, the *divergence* predicate, will denote the complement of $\downarrow$, i. e. $p \uparrow$ iff $\neg(p \downarrow)$.

Example 2.1 *The following processes are divergent:*

- $rec\,x.\ a + x$
- $rec\,x.\ a;x + \Omega$
- $rec\,x.\ a;x + rec\,x.\ a + a|x$.

The predicate $\downarrow$ is used to detect a form of "syntactic divergence". Roughly, $p \uparrow$ if p contains unguarded recursive definitions, [Mil80], or unguarded occurrences of the divergent process Ω. One can show that $p\surd$ implies $p \downarrow$ using induction on the proof of $p\surd$. Of course the converse is not true; for instance, $a;q \downarrow$ but $a;q \notin \surd$.

Definition 2.4 *For each $\mu \in Act_\tau$, let $\xrightarrow{\mu}$ be the least binary relation on REC_Σ which satisfies the following axiom and rules:*

1. $\mu \xrightarrow{\mu} nil$

2. $p \xrightarrow{\mu} p'$ *implies* $p + q \xrightarrow{\mu} p'$, $q + p \xrightarrow{\mu} p'$

3. $p \xrightarrow{\mu} p'$ *implies* $p; q \xrightarrow{\mu} p'; q$

4. $p\surd$, $q \xrightarrow{\mu} q'$ *imply* $p; q \xrightarrow{\mu} q'$

5. $p \xrightarrow{\mu} p'$ *implies* $p|q \xrightarrow{\mu} p'|q$, $q|p \xrightarrow{\mu} q|p'$

6. $p \xrightarrow{a} p'$, $q \xrightarrow{\bar{a}} q'$ *imply* $p|q \xrightarrow{\tau} p'|q'$

7. $p \xrightarrow{\mu} p'$, $\mu \notin H$ *imply* $\partial_H(p) \xrightarrow{\mu} \partial_H(p')$

8. $t[rec\,x.\ t/x] \xrightarrow{\mu} p'$ *implies* $rec\,x.\ t \xrightarrow{\mu} p'$.

For any p, let $Sort(p) = \{\mu \in Act_\tau \mid \exists \sigma \in Act_\tau^\star, q \in REC_\Sigma : p \xrightarrow{\sigma\mu} q\}$, where, for $\sigma \in Act_\tau^\star$, $\xrightarrow{\sigma}$ is defined in the natural way. One can check that, for each p, $Sort(p)$ is finite. That is, according to the terminology of [Ab87a,b], the transition system $\langle REC_\Sigma, Act_\tau, \longrightarrow\rangle$ is *sort finite*. Some of our results will depend on this fact.

The three concepts defined above take no account of the special nature of τ. Following Milner [Mil80], τ is meant to be an internal invisible action. We now define three weaker versions of $\xrightarrow{\mu}$, $\surd$ and $\downarrow$ which use this assumption.

Let $\stackrel{\mu}{\Longrightarrow}$ denote $(\xrightarrow{\tau})^\star \circ \xrightarrow{\mu} \circ (\xrightarrow{\tau})^\star$. So $p \stackrel{\mu}{\Longrightarrow} q$ means that p may evolve to q performing the action μ and possibly silent moves. We will also use the relation $\stackrel{\varepsilon}{\Longrightarrow}$, defined as $(\xrightarrow{\tau})^\star$.

Let $Stable(p) = \{q \mid p \stackrel{\varepsilon}{\Longrightarrow} q \text{ and } q \not\xrightarrow{\tau}\}$. Then the weak counterpart to $\surd$ is defined by

$$p\surd\!\!\surd \text{ if, for each } q \in Stable(p),\ q\surd.$$

For example $nil\surd\!\!\surd$, $\tau + \delta\surd\!\!\surd$, but not $\delta\surd\!\!\surd$. This relation is characterized by:

$$p\surd\!\!\surd \Longleftrightarrow \begin{cases} i) & p \not\xrightarrow{\tau} \text{ and } p\surd, \text{ or} \\ ii) & p \xrightarrow{\tau} \text{ and, for each } q,\ p \xrightarrow{\tau} q \text{ implies } q\surd\!\!\surd. \end{cases}$$

Note that $rec\,x.\ \tau; x + a\surd\!\!\surd$ which is somewhat anomalous. However, we will only apply the "weak tick" predicate $\surd\!\!\surd$ to processes which cannot perform an infinite sequence of τ-actions. Such processes are semantically divergent, which brings us to our final weak predicate. Let $\Downarrow$ be the least predicate over REC_Σ which satisfies

$$p\downarrow \text{ and (for each } q,\ p \xrightarrow{\tau} q \text{ implies } q\Downarrow) \text{ imply } p\Downarrow.$$

Intuitively, $p\Downarrow$ means that p cannot perform τ-actions indefinetely and a syntactically divergent process cannot be reached by performing these actions. Formally, one can prove

$$p\Downarrow \Leftrightarrow p \not\xrightarrow{\tau^\omega} \text{ and } p \stackrel{\varepsilon}{\Longrightarrow} q \text{ implies } q\downarrow.$$

Note also that $p\surd$ implies $p \Downarrow$. This follows because we already know that $p\surd$ implies $p \downarrow$ and one can also show that it implies $p \xrightarrow{\mu}$ for no μ, including τ.

In the semantic preorder to be defined in §3 we will use versions of $\Downarrow$ which are parameterized by actions:

- $p \Downarrow \tau$ if $p \Downarrow$
- $p \Downarrow a$ if $p \Downarrow$ and, for each q, $p \stackrel{a}{\Longrightarrow} q$ implies $q \Downarrow$.

2.2 Denotational semantics

As pointed out in the introduction, the main purpose of this paper is to show that an adequate semantic theory for the process algebra described in the previous section can be constructed using a suitable set of inequations, E. Following [CN76], [GTWW77], [Gue81], [H88a], we propose as a denotational semantics for REC_Σ the initial continuous algebra generated by a set of equations E, CI_E. We recall that a continuous Σ-algebra is a Σ-algebra whose carrier is an algebraic cpo and whose operations are interpreted as continuous functions. Homomorphisms between continuous Σ-algebras are continuous Σ-homomorphisms and are always strict.

In order to show that CI_E is a reasonable model for REC_Σ, in subsequent sections we will develop a behavioural theory of processes and prove that this corresponds to the interpretation given by CI_E. In other words, CI_E is fully-abstract with respect to the behavioural preorder we will introduce in the next section. We assume the reader is familiar with the basic notions of continuous algebras (see, e. g. , the above quoted references); however, in what follows we give a quick overview of the way a denotational semantics can be given to $REC_\Sigma(\mathtt{Var})$ following the standard lines of algebraic semantics, [Gue81]. The interested reader is invited to consult [H88a] for an explanation of the theory.

Let Σ be the signature introduced in definition 2.1 and A be any Σ-cpo. A denotational semantics for the language $REC_\Sigma(\mathtt{Var})$ is given by the mapping

$$A[\![\cdot]\!] : REC_\Sigma(\mathtt{Var}) \rightarrow [ENV_A \rightarrow A],$$

where $ENV_A = [\mathtt{Var} \rightarrow A]$ is the set of A-environments, ranged over by the metavariables $\rho, \rho' \ldots$. As usual, $\rho[x \rightarrow a]$ will denote the environment which is defined as follows

$$\rho[x \rightarrow a](y) = \begin{cases} a & \text{if } x = y \\ \rho(y) & \text{otherwise.} \end{cases}$$

For completeness sake we define $A[\![\cdot]\!]$ by structural induction on recursive terms as follows:

i) $A[\![x]\!]\rho = \rho(x)$

ii) $A[\![f(t_1, \ldots, t_k)]\!]\rho = f_A(A[\![t_1]\!]\rho, \ldots, A[\![t_k]\!]\rho)$ $(f \in \Sigma_k)$

iii) $A[\![rec\, x.\, t]\!]\rho = Y\lambda a.\, A[\![t]\!]\rho[x \to a]$,

where Y denotes the least fixed-point operator.

Note that for each $p \in REC_\Sigma$, $A[\![p]\!]\rho$ does not depend on the environment ρ. The denotation of a closed term p will be denoted by $A[\![p]\!]$ and we write $p \leq_A q$ iff $A[\![p]\!] \leq_A A[\![q]\!]$ and $p =_A q$ iff $p \leq_A q$ and $q \leq_A p$.

As already pointed out, a natural choice of A would be the initial Σ-cpo CI_E in the class of Σ-cpo's which satisfy some set of equations, or inequations, E defined over the signature Σ. The equations that we will consider will express desirable properties of processes; many of them are already well known from **CCS** or **ACP**. Some of the equations which are part of the theory of **ACP** have had to be modified due to the fact that our calculus is richer. For example, **ACP** does not deal explicitly with successful termination and divergence. Let $\mathcal{C}(E)$ denote the category of Σ-cpo's which satisfy the equations in Figure 1 and continuous Σ-homomorphisms. The following result is then standard, [CN76], [GTWW77], [Gue81], [H88a].

Proposition 2.1 *$\mathcal{C}(E)$ has an initial object CI_E.*

3 The Behavioural Semantics

3.1 The behavioural preorder

This section is devoted to an operational preorder which will be the behavioural counterpart of the denotational relation $\leq_E$ (the ordering relation in the initial model CI_E) over REC_Σ. The existence of such a behavioural preorder, defined using a by now well-established mathematical tool, will reinforce CI_E as a reasonable model for the language REC_Σ.

The behavioural preorder will be defined using a variation of bisimulation equivalence, [Pa81], [Mil83], suitable for our language REC_Σ. Let *Rel* denote the set of binary relations over REC_Σ. We define a functional $\mathcal{F} : Rel \to Rel$ as follows:

given $\mathcal{R} \in Rel$, $p\mathcal{F}(\mathcal{R})q$ iff, for each $\mu \in Act_\tau$,

i) if $p \stackrel{\mu}{\Longrightarrow} p'$ then, for some q', $q \stackrel{\hat{\mu}}{\Longrightarrow} q'$ and $p'\mathcal{R}q'$

ii) if $p \Downarrow \mu$ then

  a) $q \Downarrow \mu$

  b) if $q \stackrel{\mu}{\Longrightarrow} q'$ then, for some p', $p \stackrel{\hat{\mu}}{\Longrightarrow} p'$ and $p'\mathcal{R}q'$

iii) if $p \Downarrow$ then $p\surd \Leftrightarrow q\surd$.

A1 $x + y = y + x$

A2 $x + (y + z) = (x + y) + z$

A3 $x + x = x$

A4 $x + nil = x$

A5 $x + \delta = x$ if $x \notin \surd$

B1 $x; nil = x = nil; x$

B2 $\delta; x = \delta$

B3 $x; (y; z) = (x; y); z$

B4 $(x + y); z = x; z + y; z$ if $x, y \notin \surd$

C1 $\delta | \delta = \delta$

C2 Let $x \equiv \sum_{i \in I} \mu_i; x_i\{+\Omega\}$,

$$\delta | x = x | \delta = \begin{cases} \delta\{+\Omega\} & \text{if } I = \emptyset \\ \sum_{i \in I} \mu_i; (\delta | x_i)\{+\Omega\} & \text{otherwise} \end{cases}$$

Exp Let $x \equiv \sum_{i \in I} \mu_i; x_i\{+\Omega\}$
and $y \equiv \sum_{j \in J} \gamma_j; y_j\{+\Omega\}$,

E1 $\partial_H(nil) = nil$

E2 $\partial_H(\delta) = \delta$

E3 $\partial_H(\mu) = \begin{cases} \delta & \text{if } \mu \in H \\ \mu & \text{otherwise} \end{cases}$

E3 $\partial_H(x; y) = \partial_H(x); \partial_H(y)$

E5 $\partial_H(x + y) = \partial_H(x) + \partial_H($

Ω1 $\Omega \leq x$

Ω2 $\tau; (x + \Omega) \leq x + \Omega$

Ω3 $\partial_H(\Omega) \leq \Omega$

Ω4 $\Omega; x \leq \Omega$

T1 $\mu; \tau = \mu$

T2 $\tau; x + x = \tau; x$

T3 $\mu; (x + \tau; y) =$
$\mu; (x + \tau; y) + \mu; y$

$$x | y = \sum_{i \in I} \mu_i; (x_i | y) + \sum_{j \in J} \gamma_j; (x | y_j) + \sum_{(i,j): \mu_i = \overline{\gamma_j}} \tau; (x_i | y_j)\{+\Omega\}$$

Note: The summation notation in axiom **Exp** is justified by axioms **A1**-**A5**. In axiom **Exp** an empty sum is understood as *nil*, $\{+\Omega\}$ indicates that Ω is an optional summand of a term and Ω is a summand of the right hand side iff it is either a summand of x or of y.

Figure 1: The set of inequations E

The notation $\hat{}$ is used to simplify the definition: $\hat{\tau}$ stands for ε and $\hat{a}$ stands for a.

The functional $\mathcal{F}$ is one of the methods for adapting the usual defining functional of bisimulation equivalence. A number of variations are discussed in [Wal87], [Ab87a,b]. There are also a number of ways of defining a behavioural preorder using $\mathcal{F}$. An established method is to take $\sqsubseteq$ to be the largest relation $\mathcal{R} \in Rel$ such that $\mathcal{R} \subseteq \mathcal{F}(\mathcal{R})$, [Mil83], [Mil88]. This relation is easily seen to be a preorder, i. e. a reflexive and transitive relation, and is, in fact, the maximum fixed-point of the equation $\mathcal{R} = \mathcal{F}(\mathcal{R})$. The preorder $\sqsubseteq$ also satisfies many of the properties that we have already discussed in the introduction. For example, for every $p \in REC_\Sigma$, $\Omega \sqsubseteq p$; also $\delta; p \simeq \delta$ and $nil; p \simeq p$, where $\simeq$ is the kernel of $\sqsubseteq$, i. e. $\simeq = \sqsubseteq \cap \sqsubseteq^{-1}$. The processes nil and δ are incomparable with respect to $\sqsubseteq$. In fact, it is easy to see that $nil \Downarrow$ and $\delta \Downarrow$, but $nil \surd\!\!\!/$ whereas $\delta \notin \surd\!\!\!/$. Note also that $\delta + \Omega \simeq \Omega$. This follows from the definition but it is also perfectly reasonable. Intuitively, we would expect that, for every action a, $\delta + \Omega \sqsubseteq \delta + a$. But $\delta + a \simeq a$, so that $\delta + \Omega$ should be less than a for each a; the only such process is Ω.

Clause *iii*) in the definition of $\mathcal{F}(\mathcal{R})$ takes care of deadlock considerations and there are a number of equivalent ways of stating it. Suppose we say that

$$p \textit{ must terminate} \text{ if } p \Downarrow \text{ and } p \stackrel{\varepsilon}{\Longrightarrow} p' \not\xrightarrow{} \text{ implies } p' \surd.$$

Then *iii*) could be replaced by:

iiia)′ p *must terminate* implies q *must terminate*
iiib)′ $p \Downarrow$, q *must terminate* imply p *must terminate*.

Alternatively, suppose we say that p is *deadlocked* if $p \Downarrow$, $p \xrightarrow{\mu}$ for no μ, but $p \notin \surd$ and p *may deadlock* if $p \stackrel{\varepsilon}{\Longrightarrow} p'$ for some p' such that p' is deadlocked. Then clause *iii*) could also be replaced by:

iiia)′ p *may deadlock* implies q *may deadlock*
iiib)′ $p \Downarrow$, q *may deadlock* imply p *may deadlock*.

However, replacing clause *iii*) with clauses such as

$$\text{if } p \Downarrow \text{ then } p\surd \text{ iff } q\surd$$

or

$$\text{if } p \Downarrow \text{ then } p \text{ is deadlocked iff } q \text{ is deadlocked}$$

would lead to a different semantic preorder. The terms $\tau; \delta$ and δ would be distinguished as would $a; \tau; \delta$ and $a; \delta$. Since $a; \tau$ and a are identified this would mean that the revised semantic preorder would not be preserved by ;.

An alternative method for using $\mathcal{F}$ to obtain a behavioural preorder is to apply it inductively as follows:

- $\sqsubset\!\!\!\sim_0 = REC_\Sigma \times REC_\Sigma$ (the top element in the lattice $(Rel, \subseteq)$)
- $\sqsubset\!\!\!\sim_{n+1} = \mathcal{F}(\sqsubset\!\!\!\sim_n)$

and finally $\sqsubset\!\!\!\sim_\omega = \bigcap_{n\geq 0} \sqsubset\!\!\!\sim_n$.

The two relations $\sqsubset\!\!\!\sim$ and $\sqsubset\!\!\!\sim_\omega$ are in general different. For example, [Ab87b], take the synchronization trees p and q defined as follows:

$$p \equiv a^\omega + \Omega,\ q \equiv \sum_{k\in\omega} a^k\delta + \Omega.$$

Then it is easy to see that $p \sqsubset\!\!\!\sim_\omega q$, but $p \not\sqsubset\!\!\!\sim q$. Two equivalent terms in our language are $rec\,x.\ a; x+\Omega$ and $\partial_{\{\alpha,\beta\}}(rec\,x.\ \alpha; a; \overline{\alpha}|(\overline{\alpha}+\overline{\beta})|\beta; x)$, respectively. All the properties of $\sqsubset\!\!\!\sim$ discussed above are also true of $\sqsubset\!\!\!\sim_\omega$. In deciding which preorder to use, we will take into account the type of semantic model we discussed in the previous section. We wish to define a behavioural preorder $\lesssim$ which satisfies

$$p \lesssim q \iff p \leq_E q, \tag{1}$$

where $p \leq_E q$ means $CI_E[\![p]\!] \leq CI_E[\![q]\!]$, for the set of inequations E in Figure 1. This requirement induces certain constraints on $\lesssim$, the most important of which is called *finite approximability*. For any binary relation $\mathcal{R}$ over REC_Σ, let $\mathcal{R}^F$ be defined by:

$$p\mathcal{R}^F q \text{ if, for every finite term } d,\ d\mathcal{R}p \text{ implies } d\mathcal{R}q.$$

We say that $\mathcal{R}$ is *finitely approximable* (*fa*) if $\mathcal{R} = \mathcal{R}^F$. This intuitively means that $\mathcal{R}$ is essentially determined by how it behaves on finite terms. By the general construction of CI_E, [H88a], it follows that $\leq_E$ is *fa* and therefore, to meet (1), we must also choose a behavioural preorder which is also *fa*. The above example shows that $\sqsubset\!\!\!\sim$ is not *fa*, as $p \sqsubset\!\!\!\sim^F q$ but $p \not\sqsubset\!\!\!\sim q$.

There is one further complication caused by requirement (1). The relation $\leq_E$ is, by definition, *closed with respect to all contexts*. To explain this we need some notation. For any binary relation $\mathcal{R}$ over REC_Σ, let $\mathcal{R}$ be extended to $REC_\Sigma(\mathtt{Var})$ by:

$$t\mathcal{R}u \text{ if, for every closed substitution } \rho,\ t\rho\mathcal{R}u\rho.$$

For any $\mathcal{R}$ over $REC_\Sigma(\mathtt{Var})$ define the new relation $\mathcal{R}^c$ by:

$$t\mathcal{R}^c u \text{ if, for every context } C[\cdot] \text{ such that } C[t] \text{ and } C[u] \text{ are closed, } C[t]\mathcal{R}C[u].$$

Then $\mathcal{R}$ is said to be closed with respect to contexts if $\mathcal{R} = \mathcal{R}^c$. By construction, it follows that $\leq_E$ is closed with respect to contexts. However, this is not true of $\sqsubset\!\!\!\sim$ or $\sqsubset\!\!\!\sim_\omega$. The usual counterexample associated with the **CCS** + operator, [Mil80], works:

$$a \sqsubset\!\!\!\sim_\omega \tau; a \text{ but } a + b \not\sqsubset\!\!\!\sim_\omega \tau; a + b.$$

We may sum up this discussion by saying that in order to reflect the semantic ordering $\leq_E$ behaviourally, it is necessary to choose a behavioural preorder which is both finitely approximable and preserved by contexts. One can show that $\sqsubseteq_\omega$ is *fa* and therefore it is appropriate to take as our behavioural preorder $\sqsubseteq_\omega^c$, its closure with respect to all contexts. The proof that $\sqsubseteq_\omega$ is *fa* depends on the fact that our operational semantics is sort finite. In a transition system which is not sort finite $\sqsubseteq_\omega$ may not be *fa*. For instance, consider the following synchronization trees from [Ab87b]:

$$\begin{aligned} p &\equiv a(\textstyle\sum_{n\in\omega} b_n nil + \Omega) + \Omega \\ q &\equiv \textstyle\sum_{n\in\omega} a(\sum_{m\in\omega-\{n\}} b_m nil + \Omega) + \Omega, \end{aligned}$$

where, for each $n \neq m$, $b_n \neq b_m$. Then $p \sqsubseteq^F q$, but $p \not\sqsubseteq_2 q$.

We may now state the main theorem of the paper.

Theorem 3.1 *For any* $p, q \in REC_\Sigma$, $CI_E[\![p]\!] \leq CI_E[\![q]\!] \Longleftrightarrow p \sqsubseteq_\omega^c q$.

The proof may be found in the full version of the paper. The central parts of the proof are to establish the following two results:

1. $d \sqsubseteq_\omega^c q$ implies $CI_E[\![d]\!] \leq CI_E[\![q]\!]$, for $d \in FREC_\Sigma$ and $q \in REC_\Sigma$, and

2. $\sqsubseteq_\omega^c$ is finitely approximable.

Both these results require a reformulation of $\sqsubseteq_\omega^c$ into a more manageable form. Then the proof of 1 follows along similar lines to the proof of corresponding results in [H88a], whereas the proof of 2 requires a characterization of $\sqsubseteq_\omega^c$ in terms of modal properties of processes, as in [HM85] and [Ab87b], which depends on the sort-finiteness of the labelled transition system semantics for the language REC_Σ.

4 Conclusion

In this paper we have developed a semantic theory for a process algebra which incorporates some explicit representation of successful termination, deadlock and divergence. The process algebra that we have considered has been endowed with both an operational and a denotational semantics and the two semantic views of processes have been shown to agree. Namely, we have shown that the denotational model that we have proposed in the paper, the initial continuous algebra which satisfies a set of equations CI_E, is fully abstract with respect to a natural operational preorder over the language. As pointed out in [H81], [H88a], our choice of a denotational semantics for the language studied in this paper gives us a complete axiomatic proof system (albeit a non recursively enumerable one) for closed terms of the language. Moreover, as our denotational model is based upon the well known theory of algebraic cpo's, rather than metric spaces as in [BK82], we may obtain effective proof systems for the language by using induction rules such as Scott Induction and Fixed-Point Induction, [LS87].

The language we have considered in this paper incorporates features from **CCS** and **ACP**. It extends **ACP** by allowing an explicit representation of successful termination and divergence; moreover, our language allows for general recursive definitions. The auxiliary operators which **ACP** uses to axiomatize $|$ (namely, $\|\!\_$, for *left-merge*, and $|_c$, for the *communication merge*) could be added to our language without affecting the results of the paper. The same can be said of Hennessy's left-merge, $\not|$, [H88b]. The language extends **CCS** as it allows general sequential composition and an explicit representation of deadlock (as opposed to successful termination). However, the signature of **CCS** contains a family of *relabelling operators* $\_[R]$, where $R : Act_\tau \longrightarrow Act_\tau$ is a function such that $R(\bar{a}) = \overline{R(a)}$ and $R(\tau) = \tau$. The introduction of such an operator in the signature of our language would cause some problems. To see that, we recall that our results about the finite approximability of the behavioural preorder $\sqsubseteq^c_\omega$ depend on the sort-finiteness of our transition system semantics for the language. However, if Act is infinite this is no longer the case. To show this, consider an enumeration $\{a_0, a_1, \ldots, a_i, \ldots\}$ of the set of observable actions Act. Using the enumeration of Act, we may define a relabelling S such that $S(a_i) = a_{i+1}$, for each $i \in \omega$. Take the process p defined as follows, [Ab87b]:

$$p \equiv rec\, x.\ a_0 + x[S].$$

Then it is easy to see that the unguarded recursive definition and the generality of S give rise to a process which is not sort-finite. In fact, $p \xrightarrow{a_i}$ for each $i \in \omega$. As a consequence of these observations, our behavioural preorder would not be finitely approximable and CI_E would not be fully abstract with respect to it.

However, it may be argued that one rarely, if ever, needs relabelling operators of such a generality. In practice, relabelling functions are usually assumed to be constant on all but finitely many actions in Act_τ. If we allow only this kind of relabellings in our language then the resulting transition system semantics will again be sort-finite, [Ab87a,b], and thus all of the results of the paper will carry through to this extended language.

We end this conclusion with a brief comparison with related work. Several term model constructions, [Mil77], for **CCS** and **SCCS**-like languages have been proposed in the literature. See for example [HP80], [H81]. In each of these papers, a denotational semantics is given to the languages considered by means of the initial continuous algebra which satisfies a set of equations E. The denotational model is then shown to be fully abstract with respect to a behavioural preorder. In [DH84] the authors show how, for the *testing equivalences* they introduce, the denotational models have a natural representation in terms of a particular class of trees, the *acceptance trees* of [H85]. In [Ab87b], the author takes a language-independent standpoint and analyzes the general relationships between strong prebisimulation, $\sqsubseteq$, over transition systems and its finitary part, $\sqsubseteq^F$. The author also shows how his general results may be used to obtain a fully abstract model with respect to (the finitary part of) strong prebisimulation over a version of **SCCS**, [Mil83], with only finite summations and relates his model to the one in [H81]. In [Wal87] a behavioural relation similar to $\sqsubseteq$ is studied and applied to **CCS**; complete axiomatizations are given for finite and regular processes. In many ways the present paper may be considered as an extension of this work, employing ideas from [HP80]. It provides the first comprehensive treatment of a weak version of prebisimulation and, in addition, it establishes a mathematical setting within which the notions of termination, divergence and deadlock may be compared and contrasted.

The dichotomy deadlock/successful termination has been dealt with in a different fashion in **CSP** [BHR84], [Hoare85] and the latest papers on **ACP** [BG87a,b]. Both these process algebras introduce an explicit constant standing for successful termination, `SKIP` in **CSP** and ε in **ACP**. These constants obey the following operational rules:

- $\mathtt{SKIP} \stackrel{\surd}{\longrightarrow} \mathtt{STOP}$, and
- $\varepsilon \stackrel{\surd}{\longrightarrow} \delta$,

where `STOP` and δ are the constants used to denote deadlock in **CSP** and **ACP**, respectively. The intuition captured by the above-given rules is that successful termination is an action in the behaviour of a process, the action processes perform when they terminate. On the other hand, a deadlocked process like `STOP` or δ is one that cannot perform any move, not even a successful termination one. This is reflected in the equational laws satisfied by, e.g., ε in **ACP**. For instance, in the equational theory of **ACP** with the empty process ε, the equation

$$\delta + \varepsilon = \varepsilon$$

replaces our $\delta + nil = \delta$. Indeed, in that theory δ always gets cancelled in a sum context, i.e. the equation

$$x + \delta = x$$

holds without any conditions on x. However, the equation

$$x + \varepsilon = x$$

no longer holds (contrary to what happens for our *nil*).

The introduction of a special termination action is somewhat artificial and complicates the operational semantics. The approach we have followed in this paper is instead based upon the intuition that both deadlocked processes and successfully terminated ones do not perform any move and that the only way of behaviourally distinguishing them is to observe their behaviour in contexts built using sequential composition. However, we can simulate the **ACP** theory of ε in our framework by introducing a special action, $\surd$, and defining ε to be $\surd;\delta$. In this way we would obtain the **ACP** laws for ε. Alternatively, we could revise the language by replacing *nil* with ε. Our results carry through to the revised language after simple modifications to the operational semantics, the set of equations E, the behavioural preorder and the modal logic. These changes are needed in order to take into account the different nature between ε and *nil*. This shows that the proof techniques employed in the paper to prove our full-abstraction result are indeed quite general and easily adapted to capture different notions of successful termination in a language.

5 References

[Ab87a] S. Abramsky, Observation Equivalence as a Testing Equivalence, TCS 53, pp. 225-241, 1987

[Ab87b] S. Abramsky, A Domain Equation for Bisimulation, Imperial College Technical Report, 1987

[BG87a] J. C. M. Baeten and R. J. van Glabbeek, Abstraction and Empty Process in Process Algebra, Report CS-R8721, CWI Amsterdam, 1987 (to appear in Fundamenta Informaticae)

[BG87b] J. C. M. Baeten and R. J. van Glabbeek, Merge and Termination in Process Algebra, Proceedings 7th Conference on Foundations of Software Technology and T.C.S. (K.V. Nori ed.), LNCS 287, pp. 153-172, Springer Verlag, 1987

[BHR84] S. D. Brookes, C. A. R. Hoare and A. W. Roscoe, A Theory of Communicating Sequential Processes, JACM 31,3, pp. 560-599, 1984

[BK82] J. A. Bergstra and J. W. Klop, Fixed Point Semantics in Process Algebra, Report IW 206/82, Centre for Mathematics and Computer Science, Amsterdam, 1982

[BK84] J. A. Bergstra and J. W. Klop, Process Algebra for Synchronous Communication, Information and Control, 60, pp. 109-137, 1984

[BK85] J. A. Bergstra and J. W. Klop, Algebra of Communicating Processes with Abstraction, TCS 37, 1, pp. 77-121, 1985

[CN76] B. Courcelle and M. Nivat, Algebraic Families of Interpretations, Proceedings 17th IEEE Symposium on Foundations of Computer Science, 1976

[DH84] R. De Nicola and M. Hennessy, Testing Equivalences for Processes, TCS 34,1, pp. 83-134, 1987

[GTWW77] J. A. Goguen, J. W. Thatcher, E. G. Wagner and J. B. Wright, Initial Algebra Semantics and Continuous Algebras, JACM 24,1, pp. 68-95, 1977

[Gue81] I. Guessarian, *Algebraic Semantics*, Lecture Notes in Computer Science vol. 99, Springer-Verlag, Berlin, 1981

[H81] M. Hennessy, A Term Model for Synchronous Processes, Information and Control 51,1, pp. 58-75, 1981

[H85] M. Hennessy, Acceptance Trees, JACM 32,4, pp. 896-928, 1985

[H88a] M. Hennessy, *Algebraic Theory of Processes*, MIT Press, 1988

[H88b] M. Hennessy, Axiomatising Finite Concurrent Processes, SIAM Journal on Computing, October 1988

[HM85] M. Hennessy and R. Milner, Algebraic Laws for Nondeterminism and Concurrency, JACM 32,1, pp. 137-161, 1985

[Hoare85] C. A. R. Hoare, *Communicating Sequential Processes*, Prentice-Hall, 1985

[HP79] M. Hennessy and G. Plotkin, Full Abstraction for a Simple Parallel Programming Language, Proc. MFCS, Lecture Notes in Computer Science vol. 74, Springer-Verlag, 1979

[HP80] M. Hennessy and G. Plotkin, A Term Model for CCS, Proceedings 9th MFCS, Lecture Notes in Computer Science vol. 88, Springer-Verlag, 1980

[LS87] J. Loeckx and K. Sieber, *The Foundations of Program Verification (2nd edition)*, Wiley-Teubner Series in Computer Science, 1987

[Mil77] R. Milner, Fully Abstract Models of Typed Lambda-Calculi, TCS 4, pp. 1-22, 1977

[Mil80] R. Milner, *A Calculus of Communicating Systems*, Lecture Notes in Computer Science Vol. 92, Springer-Verlag, 1980

[Mil81] R. Milner, A Modal Characterization of Observable Machine-Behaviour, Proc. 6th CAAP, Lecture Notes in Computer Science vol. 112, pp. 23-34, Springer-Verlag, 1981

[Mil83] R. Milner, Calculi for Synchrony and Asynchrony, TCS 25, pp. 267-310, 1983

[Mil88] R. Milner, Operational and Algebraic Semantics of Concurrent Processes, LFCS Report Series, ECS-LFCS-88-46, February 1988 (to appear as a chapter of the *Handbook of Theoretical Computer Science*)

[Pa81] D. Park, Concurrency and Automata on Infinite Sequences, Lecture Notes in Computer Science vol. 104, Springer-Verlag, 1981

[Pl77] G. Plotkin, LCF Considered as a Programming Language, TCS 5, pp. 223-255, 1977

[Pl81] G. Plotkin, A Structural Approach to Operational Semantics, Report DAIMI FN-19, Computer Science Dept. , Aarhus University, 1981

[St87] C. Stirling, Modal Logics for Communicating Systems, TCS 49, pp. 311-347, 1987

[Wal87] D. Walker, Bisimulation Equivalence and Divergence in CCS, LFCS Report Series, ECS-LFCS-87-29, June 1987 (extended abstract in Proc. LICS 1988)

A Category-theoretic Semantics for Unbounded Indeterminacy

Prakash Panangaden*
Computer Science Dept.
Upson Hall
Cornell University
Ithaca, NY 14850
prakash@cs.cornell.edu

James R. Russell†
Computer Science Dept.
Upson Hall
Cornell University
Ithaca, NY 14850
jrr@cs.cornell.edu

Abstract

In this paper we give a category-theoretic semantics for a simple imperative language featuring unbounded indeterminacy. This semantics satisfies the categorical analogues of continuity and has the meaning of while loops defined as colimits of ω-diagrams. Furthermore, it collapses via an abstraction function to a semantics that is fully abstract, and coincides with the operational semantics. The abstraction function is the only discontinuous function appearing in our semantics.

1 Introduction

In the recent book "Fairness" by Nissim Francez [Fra86] the following quote appears:

> One of the more challenging problems in modeling fairness properties is to reconcile infinite fair behaviors with approximations and limits.

The presence of fairness usually points to the presence of unbounded indeterminacy [AO83]. In a recent paper, Apt and Plotkin [AP86] showed that it is impossible to have a continuous, least fixed-point, fully abstract semantics using domain theory. This proof does not depend on the details of any particular powerdomain construction. We show that by using category theoretic methods developed by Lehmann [Leh76], one can get a "continuous" semantics in which the meanings of while loops are given by colimits of ω-diagrams. The semantics that we provide is adequate, though not fully abstract. However it collapses, via an abstraction function, to a semantics that coincides with operational equality and *is* fully abstract. Apt and Plotkin [AP86] give a semantics for this language that is fully abstract

*Supported by NSF grant CCR 8818979
†Supported in part by an NSF graduate fellowship

but it is not continuous. The failure of continuity in our case is isolated to the abstraction function.

Lehmann's category theoretic approach to powerdomains was based on the idea that morphisms in a category could convey a more precise notion of approximation than partial orders could. He developed category-theoretic generalizations of many standard domain-theoretic concepts including the Smyth powerdomain [Smy78]. He did not use these constructions for defining the semantics of any languages. Several years later, Samson Abramsky [Abr83] sketched a semantics for an applicative language with unbounded indeterminacy and investigated mathematical properties of the resulting powerdomains. The language that he studied had its operational semantics defined via a term rewriting system. Our study is in the context of an imperative language, i.e. one with an updateable store. Our work gives a more detailed study of the relationship between the operational semantics and the denotational semantics. Abramsky's study of this relationship is done by defining an appropriate operational preorder on computation sequences, along lines suggested by Boudol [Bou80], and relating the this operational preorder to the categorical approximation. Boudol's analysis of the operational semantics of an applicative language could probably be mimicked in our setting but we felt that our treatment of the operational semantics would be more perspicuous to most readers. Our work clearly owes much to Abramsky's treatment of the subject though the details are developed differently.

We view our work as the beginning of a study of category-theoretic techniques for modeling unbounded indeterminacy. Recent work has shown that there are more "vicious" forms of unbounded indeterminacy than the one described here or in Apt and Plotkin. In particular we know that various commonly considered dataflow primitives, such as fair merge, are not even monotone in a suitable sense [PS87, PS88]. We are interested in developing fixed-point theories for reasoning about such primitives as well.

The rest of this paper is organized as follows. Section 2 describes the language and its operational semantics. In section 3 we describe categorical powerdomains; this material is essentially a summary of the relevant parts of Lehmann's thesis [Leh76]. Sections 4 and 5 describe the semantics and provide some examples. In section 6 we establish the relationship between the operational semantics and the denotational semantics.

2 Operational Semantics of the Language

We describe a simple language for non-determinism based on the one presented in Apt and Plotkin [AP86]. Our language is a simplified version of that one, the state consisting of the value of a single integer variable, usually called x. It should be clear that the treatment extends readily to any 'flat' domain of states.

The Atomic Commands are the ones that change the state, i.e. set the value of x.

$$A \ ::= \ x := n \mid x := x - 1 \mid x := x + 1 \mid x :=?$$

The intended meaning of $x :=?$ is non-deterministic assignment to x of any value from the underlying domain (which in this case is the integers). The details of the Boolean expressions are not important, since they are intended to be side-effect free. We assume at least the ability to detect when x is 0.

$$B ::= x = 0 \mid x \neq 0 \mid \cdots$$

Finally, we have the following Commands:

$$S ::= A \mid \mathsf{skip} \mid S_1; S_2 \mid \mathsf{if}\ B\ \mathsf{then}\ S_1\ \mathsf{else}\ S_2\ \mathsf{fi} \mid \mathsf{while}\ B\ \mathsf{do}\ S\ \mathsf{od}$$

We give the operational semantics via a one-step transition function **Com** $\times$ **States** $\rightarrow$ **Com** $\times$ **States** $\cup$ **States**. We assume the existence of a boolean evaluation function $\mathcal{B}[\![\cdot]\!]$: **Bexp** $\rightarrow$ (**States** $\rightarrow$ $\{\mathsf{true}, \mathsf{false}\}$); we explicitly assume that the evaluation of boolean expressions terminates. In what follows we use σ to range over **States**.

$\langle x := n,\, \sigma\rangle \rightarrow n$
$\langle x := x - 1,\, \sigma\rangle \rightarrow \sigma - 1$
$\langle x := x + 1,\, \sigma\rangle \rightarrow \sigma + 1$
$\langle x :=?,\, \sigma\rangle \rightarrow n\ \forall n \in D$
$\langle \mathsf{skip},\, \sigma\rangle \rightarrow \sigma$
$\langle S_1; S_2,\, \sigma\rangle \rightarrow \langle S_1'; S_2,\, \sigma'\rangle$ if $\langle S_1,\, \sigma\rangle \rightarrow \langle S_1',\, \sigma'\rangle$
$\langle S_1; S_2,\, \sigma\rangle \rightarrow \langle S_2,\, \sigma'\rangle$ if $\langle S_1,\, \sigma\rangle \rightarrow \sigma'$
$\langle \mathsf{if}\ B\ \mathsf{then}\ S_1\ \mathsf{else}\ S_2\ \mathsf{fi},\, \sigma\rangle \rightarrow \langle S_1,\, \sigma\rangle$ if $\mathcal{B}[\![B]\!]\sigma = \mathsf{true}$
$\langle \mathsf{if}\ B\ \mathsf{then}\ S_1\ \mathsf{else}\ S_2\ \mathsf{fi},\, \sigma\rangle \rightarrow \langle S_2,\, \sigma\rangle$ if $\mathcal{B}[\![B]\!]\sigma = \mathsf{false}$
$\langle \mathsf{while}\ B\ \mathsf{do}\ S\ \mathsf{od},\, \sigma\rangle \rightarrow \langle S; \mathsf{while}\ B\ \mathsf{do}\ S\ \mathsf{od},\, \sigma\rangle$ if $\mathcal{B}[\![B]\!]\sigma = \mathsf{true}$
$\langle \mathsf{while}\ B\ \mathsf{do}\ S\ \mathsf{od},\, \sigma\rangle \rightarrow \sigma$ if $\mathcal{B}[\![B]\!]\sigma = \mathsf{false}$

We define the notation $\langle S,\, \sigma\rangle \uparrow$ to mean that there exists an infinite sequence of transitions

$$\langle S,\, \sigma\rangle = \langle S_0,\, \sigma_0\rangle \rightarrow \langle S_1,\, \sigma_1\rangle \rightarrow \langle S_2,\, \sigma_2\rangle \rightarrow \cdots,$$

and we define the operational meaning function $Op[\![]\!]$ by

$$Op[\![S]\!]\sigma \stackrel{\text{def}}{=} \{\sigma' | \langle S,\, \sigma\rangle \rightarrow^* \sigma'\} \cup \{\perp | \langle S,\, \sigma\rangle \uparrow\}.$$

3 Categorical Powerdomains

Powerdomains were originally introduced as the domain-theoretic analogues of powersets and were intended to be used for semantic treatments of indeterminacy [Plo76, Smy78]. The approach was generalized to categories by Lehmann [Leh76]. Lehmann's approach was to use categories as the semantic spaces rather than domains. His idea was that a 'more detailed' notion of approximation between elements can be expressed by using morphisms between objects in a category than by using partial orders. Dually, one may view traditional domain theory as a special case of a category theoretic approach in which the homsets are at most singletons.

Abramsky [Abr83] used Lehmann's construction to model unbounded indeterminacy. A categorical approach, but with a different construction for the powerdomain, was also used by Panangaden [Pan85] to model dataflow networks with fair merge. Recently it has been realized that one can use categories to model lambda calculi [Coq88].

Though Lehmann's original construction was for domains that were generalized to categories, in this paper we assume that the underlying domain is a partial order. As already noted, a poset can be viewed as a category in which the arrows are unique; we will often adopt this view of our domain when discussing the powerdomain construction.

Given a domain D the construction of what we will call the *categorical powerdomain* of D, $CP(D)$, is straightforward. The objects of the category are taken to be multisets - sets with repetitions - of the elements of D. Multisets are represented as tagged sets, i.e. each element of a multiset M is written d_α, where $d \in D$, and α is chosen from some (uncountable) set Γ of tags so that no two elements of M have the same tag. Arrows $G : A \longrightarrow B$ are defined such that for each $b_\beta \in B$, G uniquely associates with b_β an $a_\alpha \in A$ and an arrow $a \longrightarrow b$ of D. In our case where D is a poset, arrows are always unique and G need only associate with each $b_\beta \in B$ an $a_\alpha \in A$ with $a \sqsubseteq b$. Hence, we will use the following definition of an arrow G:

$$G \subseteq A \times B \text{ such that } \begin{array}{l} i)\ \langle a_\alpha, b_\beta \rangle \in G \Rightarrow a \sqsubseteq b\ \ \& \\ ii)\ \forall b_\beta \in B \exists! a_\alpha \in A . \langle a, b \rangle \in G. \end{array}$$

Composition of arrows and identity arrows in $CP(D)$ are straightforward.

It is important to note that two multisets with the same number of copies of the same element, but possibly different tags, are isomorphic as objects of $CP(D)$. Since objects need only be specified up to isomorphism, the actual tagging used is unimportant. Furthermore, the constructions described in the rest of this paper are sufficiently clear that explicitly stating the tagging is more cumbersome than illuminating, and hence the tagging details are largely omitted.

The categorical analog of a continuous function is a functor that preserves colimits of ω-diagrams. The definition of f a functor automatically assures the analog of monotonicity - i.e. if there is an arrow $a \xrightarrow{r} b$, then there is an arrow $f(a) \xrightarrow{f(r)} f(b)$.

We now present two theorems about $CP(D)$ originally due to Lehmann. These establish that $CP(D)$ has the properties that one associates with a complete partial order.

Theorem 1. $\{\perp\}$ is the initial object.

Proof: For all $d \in D$ we have $\perp \sqsubseteq d$, hence for any object B of $CP(D)$ there is a unique arrow $\{\perp\} \longrightarrow B$ given by $G = \{\langle \perp, b \rangle | b \in B\}$.

Theorem 2. All ω-diagrams have colimits

Proof (sketch): Consider a diagram $X_0 \xrightarrow{G_0} X_1 \xrightarrow{G_1} X_2 \xrightarrow{G_2} \cdots$.

Let S be the collection of all chains $Q = \{q_0, q_1, \ldots\}$ s.t. for all i we have $q_i \in X_i$ and $\langle q_i, q_{i+1} \rangle \in X_i$. We define the colimiting object X^* by $X^* = \mathrm{colim}(X_i) =$

$\biguplus_{Q \in S} \sqcup Q$. The colimiting arrows $X_i \xrightarrow{G_i^*} X^*$ are given by $G_i^* = \biguplus_{Q \in S}\{\langle x, \sqcup Q\rangle | x \in X_i \cap Q\}$. We omit the details of commutativity and couniversality.

Note that in the above (and in the sequel) we use the 'tagged union' symbol $\biguplus$ for unions that yield multisets. The intention is that the union guarantees that each element of the result has a unique tag, and hence multiple copies of the same element will not be identified.

4 Semantics of the Language

In this section we use the categorical powerdomain $CP(D)$ defined above to give a denotational semantics to our language. First, we need to define our domains. Our base domain D will be an unrelated set of states (which in this case is the integers ω). However, since non-termination is a possibility in the language, we must consider $CP(D_\perp)$, the categorical powerdomain of D with $\perp$ added. Our semantics will then be a function $\mathbf{Com} \to (D \to CP(D_\perp))$, i.e. the meaning of commands will be given as ω-colimit preserving functors from D to $CP(D_\perp)$.

Before we can give the semantics, there are three auxiliary functors we must define:

Singleton $\{\cdot\}$: This takes an element d of D to the object $\{d\}$ (the singleton multiset containing one copy of d) of $CP(D_\perp)$. It takes an arrow $a \longrightarrow b$ in D to the arrow $\{a\} \xrightarrow{G} \{b\}$ given by $G = \{\langle a, b\rangle\}$. It is easily seen that this is ω-colimit preserving.

Lifting $(\cdot)^\dagger$: This is a functor between the functor categories $(D \to CP(D_\perp))$ and $(CP(D_\perp) \to CP(D_\perp))$. Note that $\perp \notin D$, so a functor $f : (D \to CP(D_\perp))$ is not defined on $\perp$, while multisets $A \in CP(D_\perp)$ may contain $\perp$. For this reason it is necessary that lifting specify explicitly the action of $f^\dagger$ and $\eta^\dagger$ on $\perp$.

$$f^\dagger(A) \stackrel{\text{def}}{=} \biguplus_{a \in A} f(a) \uplus \biguplus_{\perp \in A}\{\perp\}$$

$$f^\dagger(G : A \longrightarrow B) \stackrel{\text{def}}{=} \biguplus_{\substack{\langle a, b\rangle \in G \\ a \neq \perp}} f(a \longrightarrow b) \uplus \biguplus_{\substack{\langle \perp, b\rangle \in G \\ b \neq \perp}} \biguplus_{y \in f(b)} \langle \perp, y\rangle \uplus \biguplus_{\langle \perp, \perp\rangle \in G} \{\langle \perp, \perp\rangle\}$$

The above describes what lifting does to objects (which in this case are functors). We now describe what lifting does to arrows (which are natural transformations). Recall that natural transformations are maps from objects to arrows.

Given $\eta : f \longrightarrow g$, we have $\eta^\dagger : f^\dagger \longrightarrow g^\dagger$, given by

$$\eta^\dagger(A) \stackrel{\text{def}}{=} \biguplus_{\substack{a \in A \\ a \neq \perp}} \eta(a) \uplus \biguplus_{\perp \in A} \{\langle \perp, \perp\rangle\}.$$

Two important properties of lifting are that $g^\dagger$ is ω-colimit preserving if g is, and that $(\cdot)^\dagger$ is itself an ω-colimit preserving functor.

Sequential composition $\cdot ; \cdot$: $f; g$ is shorthand for $g^\dagger \circ f$, and as such it inherits the desirable ω-colimit preservation properties from lifting.

We now describe a semantic function $\mathcal{D}[\![\cdot]\!] : \mathbf{Com} \to (D \to CP(D_\perp))$. Note that since D is completely 'flat', when it is interpreted as a category the only arrows are the identities. Thus, when we describe the functors in the semantics we will specify only their action on objects, since functors by definition preserve identity arrows. If we were to give a semantics for a non-flat domain, we would be required to explicitly specify the action of the semantic functors on the arrows of the domain, since in general this is not trivial. We use the variable d to refer to elements of D.

$\mathcal{D}[\![x := ?]\!] = \lambda d.D$
$\mathcal{D}[\![x := c]\!] = \lambda d.\{c\}$
$\mathcal{D}[\![x := x - 1]\!] = \lambda d.\{d - 1\}$
$\mathcal{D}[\![x := x + 1]\!] = \lambda d.\{d + 1\}$
$\mathcal{D}[\![\mathsf{skip}]\!] = \lambda d.\{d\}$
$\mathcal{D}[\![S_1; S_2]\!] = \mathcal{D}[\![S_1]\!]; \mathcal{D}[\![S_2]\!]$
$\mathcal{D}[\![\mathsf{if}\ B\ \mathsf{then}\ S_1\ \mathsf{else}\ S_2\ \mathsf{fi}]\!] = \lambda d.\begin{cases} \mathcal{D}[\![S_1]\!]d & \text{if } \mathcal{B}[\![B]\!]d = \mathsf{true} \\ \mathcal{D}[\![S_2]\!]d & \text{if } \mathcal{B}[\![B]\!]d = \mathsf{false} \end{cases}$

We want to define the meaning of a while loop as a colimit, and since the meaning is a functor, it must be the colimit of a diagram in the functor category $D \to CP(D_\bot)$. This is the category whose objects are functors $D \to CP(D_\bot)$ and whose arrows are natural transformations. The initial object is the functor $\lambda d.\{\bot\}$, as is easily verified. We will call this Ω. Define

$\mathcal{D}[\![\mathsf{while}\ B\ \mathsf{do}\ S\ \mathsf{od}]\!] = \mathrm{colim}(W_0 \xrightarrow{\eta_0} W_1 \xrightarrow{\eta_1} W_2 \xrightarrow{\eta_2} \cdots)$

where W_i and η_i are defined as follows:

$W_0 = \Omega = \lambda d.\{\bot\}$

For $n \geq 1$: $W_n = \lambda d.\begin{cases} (\mathcal{D}[\![S]\!]; W_{n-1})d & \text{if } \mathcal{B}[\![B]\!]d = \mathsf{true} \\ \mathcal{D}[\![\mathsf{skip}]\!]d & \text{if } \mathcal{B}[\![B]\!]d = \mathsf{false} \end{cases}$

$\eta_0 x : W_0 x \longrightarrow W_1 x = \begin{cases} \biguplus_{x' \in \mathcal{D}[\![S]\!]x}\{\langle\bot,\ \bot\rangle\} & \text{if } \mathcal{B}[\![B]\!]x = \mathsf{true} \\ \{\langle\bot,\ x\rangle\} & \text{if } \mathcal{B}[\![B]\!]x = \mathsf{false} \end{cases}$

For $n \geq 1$: $\eta_n x : W_n x \longrightarrow W_{n+1} x =$

$$\begin{cases} \biguplus_{\substack{x' \in \mathcal{D}[\![S]\!]x \\ x' \neq \bot}} \eta_{n-1} x' \uplus \biguplus_{\bot \in \mathcal{D}[\![S]\!]x} \{\langle\bot,\ \bot\rangle\} & \text{if } \mathcal{B}[\![B]\!]x = \mathsf{true} \\ \{\langle x,\ x\rangle\} & \text{if } \mathcal{B}[\![B]\!]x = \mathsf{false} \end{cases}$$

The colimit is determined pointwise; i.e. for all d, $\mathcal{D}[\![\mathsf{while}\ B\ \mathsf{do}\ S\ \mathsf{od}]\!]d = \mathrm{colim}(W_0 d \xrightarrow{\eta_0 d} W_1 d \xrightarrow{\eta_1 d} W_2 d \xrightarrow{\eta_2 d} \cdots)$.

5 Some Examples

Example 1.

$\mathcal{D}[\![\mathsf{while\ true\ do\ skip\ od}]\!]x =$

$$\begin{aligned} &\mathrm{colim}((\lambda d.\{\bot\})x \longrightarrow (\lambda d.\{d\}; \lambda d.\{\bot\})x \longrightarrow (\lambda d.\{d\}; \lambda d.\{d\}; \lambda d.\{\bot\})x \longrightarrow \cdots) \\ = \ &\mathrm{colim}(\{\bot\} \longrightarrow \{\bot\} \longrightarrow \{\bot\} \longrightarrow \cdots) \\ = \ &\{\bot\} \end{aligned}$$

Thus, $\mathcal{D}[\![\text{while true do skip od}]\!] = \lambda d.\{\bot\} = \Omega$. This is a pleasant property, since it means we can replace Ω in the definition of $\mathcal{D}[\![\text{while } B \text{ do } S \text{ od}]\!]$ by $\mathcal{D}[\![\text{loop}]\!]$, where loop $\equiv$ while true do skip od.

Example 2. $\mathcal{D}[\![x := 0]\!]d = \{0\}$

Example 3.

$$\begin{aligned}
&\mathcal{D}[\![\text{while } x > 0 \text{ do } x := x - 1 \text{ od}]\!]d = \\
&\quad \mathrm{colim}(\overbrace{\{\bot\} \longrightarrow \{\bot\} \longrightarrow \cdots \{\bot\}}^{d \text{ times}} \longrightarrow \{0\} \longrightarrow \{0\} \longrightarrow \cdots) \\
&= \{0\}
\end{aligned}$$

Thus, $\mathcal{D}[\![\text{while } x > 0 \text{ do } x := x - 1 \text{ od}]\!] = \lambda d.\{0\} = \mathcal{D}[\![x := 0]\!]$.

Example 4. $\mathcal{D}[\![x :=?; \text{while } x > 0 \text{ do } x := x - 1 \text{ od}]\!] =$

$$\begin{aligned}
\mathrm{colim}(\ &\{\bot,\ \bot,\ \bot,\ \ldots\ \} \\
&\{0,\ \bot,\ \bot,\ \ldots\ \} \\
&\{0,\ 0,\ \bot,\ \ldots\ \} \\
&\qquad \cdots) \\
= \{0,0,0,\ldots\}&
\end{aligned}$$

Of course, we should have expected this, since

$$\begin{aligned}
\mathcal{D}[\![x :=?; \text{while } x > 0 \text{ do } x := x - 1 \text{ od}]\!] &= \mathcal{D}[\![x :=?; x := 0]\!] \\
&= (\lambda d.\{0\})^{\dagger} \circ (\lambda d.D) \\
&= \biguplus_{d \in D} \{0\}.
\end{aligned}$$

This may seem unnatural, but it is a necessary effect of our approach that $\mathcal{D}[\![x :=?; \text{while } x > 0 \text{ do } x := x - 1 \text{ od}]\!]$ is different from $\mathcal{D}[\![x := 0]\!]$. The use of multisets and functors provides a more detailed description of the approximations, but at the same time can increase the cardinality of the approximate objects to account for non-deterministic behavior. The semantics of programs over the determinate portion of our language (i.e. without "$x :=?$") will always be a singleton, and will satisfy all the usual semantic relations.

The previous example also serves to point out how our semantics differs from those considered by Apt and Plotkin. In their proof they consider the same statement as above, but in a domain theoretic setting, without arrows for approximation or multiset objects, they show

$$\begin{aligned}
\mathcal{D}[\![x :=?; \text{while } x > 0 \text{ do } x := x - 1 \text{ od}]\!] &= \bigsqcup\Big\{\{0,\bot\},\{0,\bot\},\{0,\bot\},\ldots\Big\} \\
&= \{0,\bot\}.
\end{aligned}$$

This clearly does not agree with the operational behavior, therefore full abstraction must fail. In our semantics, $\bot$ is not a possibility in the limit, and hence except for the multiplicity of the result, we can achieve full abstraction. This is the subject of the next section.

6 Relationship with the Operational Semantics

In this section we begin by investigating properties of the operations semantics $Op[\![]\!]$, ultimately proving its compositionality. We then demonstrate that an abstraction of the denotational semantics $\mathcal{D}[\![]\!]$ exactly coincides with $Op[\![]\!]$. From these, the full-abstraction of the abstracted semantics follows as a corollary.

Lemma 1. For all statements S and states d,

$$Op[\![S]\!]d = \bigcup_{\substack{S', d' \text{ s.t.} \\ \langle S, d\rangle \to \langle S', d'\rangle}} Op[\![S']\!]d' \cup \{d' | \langle S, d\rangle \to d'\}.$$

Proof: Straightforward from the transition relations.

Lemma 2. For all statements S_1, S_2, and states d,

$$Op[\![S_1; S_2]\!]d = \bigcup_{\substack{d' \in Op[\![S_1]\!]d \\ d' \neq \bot}} Op[\![S_2]\!]d' \cup \{\bot | \bot \in Op[\![S_1]\!]d\}.$$

Proof: Straightforward from the transition relations.

Theorem 3. If $Op[\![S]\!] = Op[\![S']\!]$, then for all contexts $C[\cdot]$, $Op[\![C[S]]\!] = Op[\![C[S']]\!]$.

Proof: The proof is a structural induction on the context $C[\cdot]$.

Case $C \equiv \text{if } B \text{ then } [\cdot] \text{ else } T \text{ fi}$: From Lemma 1 it follows that, for all d,

$$\begin{aligned}
Op[\![\text{if } B \text{ then } S \text{ else } T \text{ fi}]\!]d &= \begin{cases} Op[\![S]\!]d & \text{if } \mathcal{B}[\![B]\!]d = \text{true} \\ Op[\![T]\!]d & \text{if } \mathcal{B}[\![B]\!]d = \text{false} \end{cases} \\
\text{(by hypothesis)} &= \begin{cases} Op[\![S']\!]d & \text{if } \mathcal{B}[\![B]\!]d = \text{true} \\ Op[\![T]\!]d & \text{if } \mathcal{B}[\![B]\!]d = \text{false} \end{cases} \\
\text{(by lemma)} &= Op[\![\text{if } B \text{ then } S' \text{ else } T \text{ fi}]\!]d
\end{aligned}$$

Case $C \equiv \text{if } B \text{ then } [\cdot] \text{ else } T \text{ fi}$: Similar to above.

Case $C \equiv [\cdot]; T$: We know from Lemma 2 that

$$\begin{aligned}
Op[\![S; T]\!]d &= \bigcup_{\substack{d' \in Op[\![S]\!]d \\ d' \neq \bot}} Op[\![T]\!]d' \cup \{\bot | \bot \in Op[\![S]\!]d\} \\
\text{(by hyp.)} &= \bigcup_{\substack{d' \in Op[\![S']\!]d \\ d' \neq \bot}} Op[\![T]\!]d' \cup \{\bot | \bot \in Op[\![S']\!]d\} \\
\text{(by lemma)} &= Op[\![S'; T]\!]d
\end{aligned}$$

Case $C \equiv T; [\cdot]$: As above, we know that

$$\begin{aligned} Op[\![T; S]\!]d &= \bigcup_{\substack{d' \in Op[\![T]\!]d \\ d' \neq \bot}} Op[\![S]\!]d' \cup \{\bot \mid \bot \in Op[\![T]\!]d\} \\ \text{(by hyp.)} &= \bigcup_{\substack{d' \in Op[\![T]\!]d \\ d' \neq \bot}} Op[\![S']\!]d' \cup \{\bot \mid \bot \in Op[\![T]\!]d\} \\ \text{(by lemma)} &= Op[\![T; S']\!]d \end{aligned}$$

Case $C \equiv \textsf{while } B \textsf{ do } [\cdot] \textsf{ od}$: We show that for all d,

$$Op[\![\textsf{while } B \textsf{ do } S \textsf{ od}]\!]d \subseteq Op[\![\textsf{while } B \textsf{ do } S' \textsf{ od}]\!]d.$$

By symmetry, the reverse must also be true, and the desired equality follows.

Case 1: If $\mathcal{B}[\![B]\!]d = \textsf{false}$, then

$$Op[\![\textsf{while } B \textsf{ do } S \textsf{ od}]\!]d = Op[\![\textsf{while } B \textsf{ do } S' \textsf{ od}]\!]d = \{d\}$$

Case 2: Suppose $\mathcal{B}[\![B]\!]d = \textsf{true}$, and let $d' \in Op[\![\textsf{while } B \textsf{ do } S \textsf{ od}]\!]d$, $d' \neq \bot$. Then $\exists n \geq 1$ and a sequence $d = d_0, d_1, \ldots, d_n = d'$ such that

$$\begin{aligned} &d_{i+1} \in Op[\![S]\!]d_i \text{ for } i = 0, \ldots, n-1, \\ &\mathcal{B}[\![B]\!]d_i = \textsf{true} \text{ for } i = 0, \ldots, n-1, \\ &\text{and } \mathcal{B}[\![B]\!]d_n = \textsf{false}. \end{aligned}$$

By the hypothesis, this means

$$\begin{aligned} &d_{i+1} \in Op[\![S']\!]d_i \text{ for } i = 0, \ldots, n-1, \\ &\mathcal{B}[\![B]\!]d_i = \textsf{true} \text{ for } i = 0, \ldots, n-1, \\ &\text{and } \mathcal{B}[\![B]\!]d_n = \textsf{false}, \end{aligned}$$

which is equivalent to

$$d' \in Op[\![\textsf{while } B \textsf{ do } S' \textsf{ od}]\!]d.$$

Case 3: Suppose $\mathcal{B}[\![B]\!]d = \textsf{true}$, and $\bot \in Op[\![\textsf{while } B \textsf{ do } S \textsf{ od}]\!]d$. Then there exists a sequence $d = d_0, d_1, \ldots$ such that

$$\begin{aligned} &d_{i+1} \in Op[\![S]\!]d_i \text{ for all } i, \\ &\text{and } \mathcal{B}[\![B]\!]d_i = \textsf{true} \text{ for all } i. \end{aligned}$$

By the hypothesis, this means

$$\begin{aligned} &d_{i+1} \in Op[\![S']\!]d_i \text{ for all } i, \\ &\text{and } \mathcal{B}[\![B]\!]d_i = \textsf{true} \text{ for all } i, \end{aligned}$$

or equivalently,

$$\bot \in Op[\![\text{while } B \text{ do } S' \text{ od}]\!]d.$$

∎

Definition Let $P_S(D_\bot)$ be the Smyth powerdomain of D. Viewed as a category, the objects of $P_S(D_\bot)$ are sets of elements of D, and there is an arrow $A \longrightarrow B$ if and only if $A \sqsubseteq_S B$. $P_S(D_\bot)$ is the result of collapsing the objects of $CP(D_\bot)$ from multisets to sets, and of collapsing parallel arrows to a single one in $P_S(D_\bot)$. We define $ab : CP(D_\bot) \rightarrow P_S(D_\bot)$ as the obvious abstraction functor, taking multiset objects of $CP(D_\bot)$ to their corresponding set objects of $P_S(D_\bot)$, and taking arrows of $CP(D_\bot)$ to the corresponding arrows of $P_S(D_\bot)$. ab is easily seen to preserve composition and identity arrows.

Note that ab is not an ω-colimit preserving functor (as can be seen by applying ab to the sets in Example 4). This is the only aspect of our semantics that is not "continuous".

Theorem 4. $ab(\mathcal{D}[\![S]\!]) = Op[\![S]\!]$ for all commands S.

Proof: The proof is a structural induction by cases.

Case $S \equiv A$: For the atomic commands A, it is trivially the case that

$$Op[\![A]\!]d = ab(\mathcal{D}[\![A]\!]d) \text{ for all } d \text{ in } D.$$

Case $S \equiv \text{skip}$: Also trivially,

$$Op[\![\text{skip}]\!]d = \{d\} = ab(\mathcal{D}[\![\text{skip}]\!]d) \text{ for all } d \text{ in } D.$$

Case $S \equiv \text{if } B \text{ then } S_1 \text{ else } S_2 \text{ fi}$: From Lemma 1 it follows that, for all d,

$$\begin{aligned}
Op[\![\text{if } B \text{ then } S_1 \text{ else } S_2 \text{ fi}]\!]d &= \begin{cases} Op[\![S_1]\!]d & \text{if } \mathcal{B}[\![B]\!]d = \text{true} \\ Op[\![S_2]\!]d & \text{if } \mathcal{B}[\![B]\!]d = \text{false} \end{cases} \\
\text{(by ind. hyp.)} &= \begin{cases} ab(\mathcal{D}[\![S_1]\!]d) & \text{if } \mathcal{B}[\![B]\!]d = \text{true} \\ ab(\mathcal{D}[\![S_2]\!]d) & \text{if } \mathcal{B}[\![B]\!]d = \text{false} \end{cases} \\
&= ab(\mathcal{D}[\![\text{if } B \text{ then } S_1 \text{ else } S_2 \text{ fi}]\!]d)
\end{aligned}$$

Case $S \equiv S_1; S_2$: From the definition of sequential composition we have

$$\begin{aligned}
\mathcal{D}[\![S_1; S_2]\!]d &= (\mathcal{D}[\![S_2]\!])^\dagger(\mathcal{D}[\![S_1]\!]d) \\
&= \biguplus_{\substack{d' \in \mathcal{D}[\![S_1]\!]d \\ d' \neq \bot}} \mathcal{D}[\![S_2]\!]d' \uplus \biguplus_{\bot \in \mathcal{D}[\![S_1]\!]d} \{\bot\}.
\end{aligned}$$

This means that, for all d,

$$\begin{aligned}
ab(\mathcal{D}[\![S_1; S_2]\!]d) &= \bigcup_{\substack{d' \in ab(\mathcal{D}[\![S_1]\!]d)\\ d' \neq \bot}} ab(\mathcal{D}[\![S_2]\!]d') \cup \{\bot \mid \bot \in ab(\mathcal{D}[\![S_1]\!]d)\}\\
\text{(by ind. hyp.)} &= \bigcup_{\substack{d' \in Op[\![S_1]\!]d\\ d' \neq \bot}} Op[\![S_2]\!]d' \cup \{\bot \mid \bot \in Op[\![S_1]\!]d\}\\
\text{(by Lemma 2)} &= Op[\![S_1; S_2]\!]d
\end{aligned}$$

<u>Case $S \equiv$ while B do S od:</u> This case proceeds via two lemmas.

Lemma 3. For all d, $Op[\![\text{while } B \text{ do } S \text{ od}]\!]d \subseteq ab(\mathcal{D}[\![\text{while } B \text{ do } S \text{ od}]\!]d)$

Proof: Let $d' \in Op[\![\text{while } B \text{ do } S \text{ od}]\!]d$.

Case 1: If $\mathcal{B}[\![B]\!]d = \text{false}$, then $d = d'$, and

$$\begin{aligned}
\mathcal{D}[\![\text{while } B \text{ do } S \text{ od}]\!] &= \operatorname{colim}(W_0 d \to W_1 d \to \ldots)\\
&= \operatorname{colim}(\{\bot\} \to \{d\} \to \ldots)\\
&= \{d\},
\end{aligned}$$

so $d' \in ab(\mathcal{D}[\![\text{while } B \text{ do } S \text{ od}]\!]d)$.

Case 2: Suppose $\mathcal{B}[\![B]\!]d = \text{true}$, and $d' \neq \bot$. Then $\exists n \geq 1$ and a sequence $d = d_0, d_1, \ldots, d_n = d'$ such that

$d_{i+1} \in Op[\![S]\!]d_i$ (equivalently $\langle S,\, d_i\rangle \to^* d_{i+1}$) for $i = 0, \ldots, n-1$,
$\mathcal{B}[\![B]\!]d_i = \text{true}$ for $i = 0, \ldots, n-1$,
and $\mathcal{B}[\![B]\!]d_n = \text{false}$.

Now note that if $\mathcal{B}[\![B]\!]d_i = \text{true}$, then for any $j \geq 1$

$$\begin{aligned}
W_j d_i &= (\mathcal{D}[\![S]\!]; W_{j-1})d_i\\
&= \biguplus_{\substack{e \in \mathcal{D}[\![S]\!]d_i\\ e \neq \bot}} W_{j-1}e \uplus \{\bot \mid \bot \in \mathcal{D}[\![S]\!]d_i\},
\end{aligned}$$

so given that $d_{i+1} \in Op[\![S]\!]d_i$ implies that $d_{i+1} \in \mathcal{D}[\![S]\!]d_i$ by the induction hypothesis, we have that

$$W_{j-1}d_{i+1} \subseteq W_j d_i, \text{ for } i < n.$$

Also, $\mathcal{B}[\![B]\!]d_n = \text{false}$ implies that

$$W_j d_n = \{d_n\} \text{ for any } j > 0.$$

This, together with the chain $d = d_0, d_1, \ldots, d_n = d'$ defined above, give us

$$\{d'\} = \{d_n\} = W_1 d_n \subseteq W_2 d_{n-1} \subseteq \cdots \subseteq W_n d_1 \subseteq W_{n+1} d_0 = W_{n+1} d.$$

Now, if we look at the definition of the natural transformation η_j, for $j > 0$, we see that

$$\begin{aligned} d' \neq \perp,\ d' \in W_{n+1} &\Rightarrow d' \in W^* d \\ &\Leftrightarrow d' \in \mathcal{D}[\![\textsf{while } B \textsf{ do } S \textsf{ od}]\!]d \\ &\Rightarrow d' \in ab(\mathcal{D}[\![\textsf{while } B \textsf{ do } S \textsf{ od}]\!]d). \end{aligned}$$

Case 3: Suppose $\mathcal{B}[\![B]\!]d = \textsf{true}$, and $d' = \perp$. Then there exists a sequence $d = d_0, d_1, \ldots$ such that

$$\begin{aligned} &d_{i+1} \in Op[\![S]\!]d_i \text{ for all } i, \\ &\text{and } \mathcal{B}[\![B]\!]d_i = \textsf{true} \text{ for all } i. \end{aligned}$$

We wish to show that $\mathcal{D}[\![\textsf{while } B \textsf{ do } S \textsf{ od}]\!]d$ contains $\perp$, or (by unfolding the definition) that the diagram

$$W_0 d_0 \overset{\eta_0 d_0}{\longrightarrow} W_1 d_0 \overset{\eta_1 d_0}{\longrightarrow} W_2 d_0 \overset{\eta_2 d_0}{\longrightarrow} \cdots$$

contains an infinite chain $\{\perp\} \longrightarrow \{\perp\} \longrightarrow \cdots$.

Noting that by the induction hypothesis, $d_{i+1} \in Op[\![S]\!]d_i$ for all i, it is a straightforward induction to show that, for any $n \geq 0$

$$\eta_n d_j : W_n d_j \longrightarrow W_{n+1} d_j \ni \langle \perp, \perp \rangle \text{ for all } j \geq 0.$$

$\mathcal{B}[\![B]\!]d_j = \textsf{true}$ for all $j \geq 0$ implies

$$\eta_0 d_j : W_0 d_j \longrightarrow W_1 d_j \ni \langle \perp, \perp \rangle \text{ for all } j \geq 0$$

from the definition of η_0. For $n > 1$, we have that, for all $j \geq 0$

$$\begin{aligned} \eta_n d_j : W_n d_j \longrightarrow W_{n+1} d_j \quad &\ni \biguplus_{x \in \mathcal{D}[\![S]\!]d_j} \eta_{n-1} x \\ (\text{since } d_{j+1} \in \mathcal{D}[\![S]\!]d_j) \quad &\ni \eta_{n-1} d_{j+1} \\ (\text{by induction}) \quad &\ni \langle \perp, \perp \rangle. \end{aligned}$$

Thus,

$$\begin{aligned} d' = \perp \ &\in \ \mathrm{colim}(W_0 d_0 \overset{\eta_0 d_0}{\longrightarrow} W_1 d_0 \overset{\eta_1 d_0}{\longrightarrow} W_2 d_0 \overset{\eta_2 d_0}{\longrightarrow} \cdots) \\ &= \ \mathcal{D}[\![\textsf{while } B \textsf{ do } S \textsf{ od}]\!]d, \end{aligned}$$

and we have $d' \in ab(\mathcal{D}[\![\textsf{while } B \textsf{ do } S \textsf{ od}]\!]d)$. ∎

Lemma 4. For all d, $ab(\mathcal{D}[\![\textsf{while } B \textsf{ do } S \textsf{ od}]\!]d) \subseteq Op[\![\textsf{while } B \textsf{ do } S \textsf{ od}]\!]d$

Proof: Let $d' \in \mathcal{D}[\![\textsf{while } B \textsf{ do } S \textsf{ od}]\!]d$. Then

$$d' \in \mathrm{colim}(W_0 d \rightarrow W_1 d \rightarrow \cdots) \Rightarrow d' = \sqcup Q,$$

where Q is a chain through the diagram, as defined in Theorem 2. From the definition of the W_i and the η_i, we know that the only distinct elements of Q are $\perp$ and d'. Thus, $d' \in Q$, which implies $d' \in W_n d$ for some finite n. Let m be the least such index. Now, assuming $d' \neq \perp$, there is a sequence $d = d_0, d_1, \ldots, d_m = d'$ such that

$$\begin{aligned}&d_{i+1} \in \mathcal{D}[\![S]\!]d_i \text{ for } i = 0, \ldots, m-1,\\ &\mathcal{B}[\![B]\!]d_i = \mathsf{true} \text{ for } i = 0, \ldots, m-1,\\ &\text{and } \mathcal{B}[\![B]\!]d_m = \mathsf{false}.\end{aligned}$$

But this means that (by induction hypothesis)

$$\begin{aligned}&d_{i+1} \in Op[\![S]\!]d_i \text{ for } i = 0, \ldots, m-1,\\ &\mathcal{B}[\![B]\!]d_i = \mathsf{true} \text{ for } i = 0, \ldots, m-1,\\ &\text{and } \mathcal{B}[\![B]\!]d_m = \mathsf{false}.\end{aligned}$$

This we know is equivalent to $d' = d_m \in Op[\![\mathsf{while}\ B\ \mathsf{do}\ S\ \mathsf{od}]\!]d$.

If $d' = \perp$, then there is an infinite sequence $d = d_0, d_1, \ldots$ such that

$$\begin{aligned}&d_{i+1} \in \mathcal{D}[\![S]\!]d_i \text{ for all } i,\\ &\text{and } \mathcal{B}[\![B]\!]d_i = \mathsf{true} \text{ for all } i.\end{aligned}$$

But this means that (by induction hypothesis)

$$\begin{aligned}&d_{i+1} \in Op[\![S]\!]d_i \text{ for all } i,\\ &\text{and } \mathcal{B}[\![B]\!]d_i = \mathsf{true} \text{ for all } i,\end{aligned}$$

which implies $\perp = d' \in Op[\![\mathsf{while}\ B\ \mathsf{do}\ S\ \mathsf{od}]\!]d$. ∎

Corollary 1 (Full-abstraction) For all statements S and S' we have

$$ab(\mathcal{D}[\![S]\!]) = ab(\mathcal{D}[\![S']\!])$$

if and only if

$$Op[\![C[S]]\!] = Op[\![C[S']]\!] \text{ for all contexts } C[\cdot].$$

Proof: $ab(\mathcal{D}[\![S]\!]) = ab(\mathcal{D}[\![S']\!])$ is equivalent to $Op[\![S]\!] = Op[\![S']\!]$ by Theorem 4. If we take $C[\cdot]$ to be the empty context, we see that $Op[\![C[S]]\!] = Op[\![C[S']]\!]$ implies that $Op[\![S]\!] = Op[\![S']\!]$; Theorem 3 gives us the reverse implication.

7 Conclusions

The result in section 6 is the main result of this paper. The main achievement is that the semantics generalizes the usual notion of continuous least fixed-point semantics. The failure of continuity, as required by Apt and Plotkin's result, is isolated to the abstraction functor.

The framework we have developed is general enough to accommodate domains that are not 'flat', and even Lehmann's categorically generalized domains. We are developing a semantics for a language with infinite output streams, together with a full-abstraction result for this setting. Additional future work includes a detailed comparison of this approach with strictly domain theoretic approaches to the same problem.

References

[Abr83] S. Abramsky. On semantic foundations for applicative multiprogramming. In J. Diaz, editor, *Proceedings of the Tenth International Conference On Automata, Languages And Programming*, pages 1–14, New York, 1983. Springer-Verlag.

[AO83] K. R. Apt and E.-R. Olderog. Proof rules and transformations dealing with fairness. *Sci. Comput. Prog.*, 3:65–100, 1983.

[AP86] K. R. Apt and G. D. Plotkin. Countable nondeterminism and random assignment. *Journal Of The ACM*, 33(4):724–767, 1986.

[Bou80] G. Boudol. *Semantique Operationalle et Algebrique Des Programmes Recursifs Non-Deterministes.* PhD thesis, University de Paris VII, 1980. These d'Etat.

[Coq88] T. Coquand. Categories of embeddings. In *Proceedings of the Third IEEE Symposium on Logic In Computer Science*, 1988.

[Fra86] N. Francez. *Fairness.* Springer-Verlag, 1986.

[Leh76] D. Lehmann. *Categories for Fixed-point Semantics.* PhD thesis, University of Warwick, 1976.

[Pan85] P. Panangaden. Abstract interpretation and indeterminacy. In *Proceedings of the 1984 CMU Seminar on Concurrency*, pages 497–511, 1985. LNCS 197.

[Plo76] G. D. Plotkin. A powerdomain construction. *SIAM Journal of Computing*, 5(3):452–487, 1976.

[PS87] P. Panangaden and V. Shanbhogue. On the expressive power of indeterminate primitives. Technical Report 87-891, Cornell University, Computer Science Department, November 1987.

[PS88] P. Panangaden and E. W. Stark. Computations, residuals and the power of indeterminacy. In Timo Lepisto and Arto Salomaa, editors, *Proceedings of the Fifteenth ICALP*, pages 439–454. Springer-Verlag, 1988. Lecture Notes in Computer Science 317.

[Smy78] M. B. Smyth. Powerdomains. *Journal of Computer and System Sciences*, 16:23–36, 1978.

Algebraic Types in PER Models

by J.M.E. Hyland[1],

Department of Pure Mathematics and Mathematical Statistics,
16 Mill Lane, Cambridge CB2 1SB, England

E.P. Robinson[2],

Department of Computing and Information Science,
Queen's University, Kingston, Ontario, K7L 3N6, Canada

and G. Rosolini

Dipartimento di Matematica,
Università degli Studi, 43100 Parma, Italy

ABSTRACT

Huet has conjectured that the interpretations of a class of types (the "algebraic types") in the PER model on the natural numbers for the second-order lambda calculus are in a certain sense the initial algebras. In this paper we examine several different PER models, and show that Huet's conjecture holds in each.

Introduction

If you are given a model of the polymorphic lambda calculus (or of anything else for that matter), the first question you are likely to ask is "how good is it?" For many programming languages this might translate straight into a technical question about whether the model is fully abstract with respect to some operational semantics. For the strongly normalizing second-order lambda calculus, this particular question degenerates, and becomes simply the problem of characterizing the equational theory of the model. Seen from a slightly different point of view, it is also a question about

[1]The authors would like to acknowledge the support of the SERC during the preparation of this paper. The second author would also like to acknowledge the support of the Queen's University ARC and the Natural Sciences and Engineering Research Council (Canada), and the third that of the M.P.I.

[2]Currently at the School of Cognitive and Computing Sciences, University of Sussex, Brighton, BN1 9QH, England

how close the model you are given is to the term model for the theory. Specifically, how close it is to the term model for the bare theory, with no additional types, and no extra equations between terms. In this case there is, however, another important question that we can ask (even if we cannot yet precisely formulate it): "Are all the polymorphic values parametric?" (*cf.* [Rey83, Fre89b]).

One rather crude way of measuring this is by examining the interpretations of the polymorphic natural numbers

$$\Pi X.\,[X \to X] \to [X \to X]$$

and the polymorphic booleans

$$\Pi X.\,[X \to [X \to X]].$$

In a model close to the term model or in a parametric model, one might expect that these interpretations would contain, in some suitable sense, only the closed terms of given type in the calculus. (Indeed Freyd, following Reynolds, has proposed this as part of a series of tests for the inherent parametricity of a model [Fre89b]). It is a straightforward consequence of normalization (*cf.* [Gir72, Gir71]), that the only closed terms of these types are the polymorphic Church numerals

$$\Lambda X.\ \lambda fx.\ f^n(x)$$

in the first case, and the two elements

$$\Lambda X.\ \lambda xy.\ x, \qquad \Lambda X.\ \lambda xy.\ y$$

in the second. In the first case the result goes back to [Gir72], but the second seems to be folklore, and the earliest explicit reference to it that we can find is in [BB85]! A few remarks seem to be in order:

1) This little syntactic result is irrelevant to the major proof-theoretic concerns of [Gir72]. Even the results on *representability* of functions (which have been reworked recently by [Sta81], [SFO83], and [Lei83]) do not require one to show explicitly that there are no "non-standard" terms of type $\Pi X.\,[X \to X] \to [X \to X]$ (in much the same way that results on the representability of numeric functions in the untyped lambda calculus are not invalidated by the existence of lambda terms other than numerals).

2) In so far as these results are relevant to the implementation of programming languages (*cf. e.g.* [Fai86]), it is the syntactic result which matters and not the semantic ones we present here. However, as we will explain in section 2, we can give a semantic proof of the syntactic result, by considering the special case of the PER-model generated by an open term algebra.

3) We do not in this paper explicitly consider the other obvious way of comparing a model with the term model; we do not consider the theory generated by the model. It is however worth noting that the results in this area are somewhat ambivalent.

Results such as those presented in this paper show that certain types contain only elements corresponding to closed terms, and that no closed terms of these types are unnecessarily identified. However, the types involved are all of rather a low level, and at higher levels certain non-trivial equations hold in all PER models. An example is given at the end of the paper.

The syntactic result of [BB85] is considerably more general than the two results mentioned above. If we rewrite the polymorphic natural numbers and booleans in the extended calculus with product, then the connection between the two types becomes clearer. The natural numbers become $\Pi X.\, [X \times [X \to X] \to X]$, and the booleans $\Pi X.\, [X \times X \to X]$. We can see that they both fall into the same pattern: given certain operations concerning the parametric type X we have to produce a value of type X. In the first case the data we have are a value of type X and a unary function $X \to X$, and in the second two values of type X.

Böhm and Berarducci extend the syntactic characterization above to types derived from a general many-sorted algebraic signature (and, though this will not concern us, slightly beyond). Let Σ be a signature in the sense of many-sorted algebra. Thus Σ is given by a collection of basic sorts $A_1, \ldots, A_n$, and a collection of basic operations $f_1, \ldots, f_m$, corresponding to functions

$$f_i : A_{i1} \times \ldots \times A_{iN(i)} \to A_{r(i)}.$$

We assume that constants are given by nullary operations. Closed terms of each type are built up inductively in the usual way: if $\sigma_1, \ldots, \sigma_{N(i)}$ are closed terms of types $A_1, \ldots, A_{N(i)}$ respectively, then $f_i(\sigma_1, \ldots, \sigma_{N(i)})$ is a closed term of type $A_{r(i)}$. We recall that the closed terms form the initial algebra for the signature.

Corresponding to each basic sort A_i we define types $\mathcal{A}_i^{\times}$ and $\mathcal{A}_i$ in the extended and the pure second-order lambda calculus respectively:

$$\mathcal{A}_i^{\times} = \Pi A_1 \ldots A_n.\ (\, \Sigma^{\times}(A_1, \ldots, A_n) \to A_i\,),$$

(we have overloaded the sort symbols, using them also as type variables).
Here $\Sigma^{\times}(A_1, \ldots, A_n)$ is a type which encodes the signature Σ:

$$\begin{aligned}
&(A_{11} \times \ldots \times A_{1N(1)} \to A_{r(1)}) \\
&\times (A_{21} \times \ldots \times A_{2N(2)} \to A_{r(2)}) \\
&\times \ldots \\
&\times (A_{m1} \times \ldots \times A_{mN(m)} \to A_{r(m)})
\end{aligned}$$

$\mathcal{A}_i$ is the curried version of $\mathcal{A}_i^{\times}$

$$\begin{aligned}
\Pi A_1 \ldots A_n.\ \ & [\, (A_{11} \to \ldots \to A_{1N(1)} \to A_{r(1)}) \\
& \to\ (A_{21} \to \ldots \to A_{2N(2)} \to A_{r(2)}) \\
& \to\ \ldots \\
& \to\ (A_{m1} \to \ldots \to A_{mN(m)} \to A_{r(m)}) \\
& \to\ \ A_i\,].
\end{aligned}$$

Note that there are terms in the extended calculus which define isomorphisms between the types $\mathcal{A}_i$ and the types $\mathcal{A}_i^\times$, and hence that $\mathcal{A}_i$ is isomorphic to $\mathcal{A}_i^\times$ in any model. Note also that a value of type $\mathcal{A}_i^\times$ is a function which takes a Σ-algebra $B = (B_1, \ldots, B_n, e_1, \ldots, e_m)$ as parameter, and returns a value of type B_i. Thus we shall refer to these types as *algebraic types*. In this paper, we use the types $\mathcal{A}_i$ of the pure calculus rather than the types $\mathcal{A}_i^\times$. The only reason for taking this option is that the use of this representation rather than the other seems to make our calculations slightly simpler.

Böhm-Berarducci show that any closed term of type $\mathcal{A}_i$ is reducible to one of the form $\Lambda A_1, \ldots, A_n\ \lambda f_1, \ldots, f_m.\ \sigma$, where σ is a closed Σ-term of type A_i (this time overloading function symbols as value variables).

In particular we obtain the results above about $\Pi X. [X \to X] \to [X \to X]$ and $\Pi X. [X \to [X \to X]]$ as special cases. Both correspond to one-sorted theories. The polymorphic natural numbers arise from the theory with one constant 0, and one unary operation s, and the polymorphic booleans arise from a signature with no operations but with two constants, 0 and 1. We can also see that there are no closed terms of type $\Pi X. X$, and that the only closed term of type $\Pi X. X \to X$ is the polymorphic identity.

In this paper we shall look at two different "PER" models for the second-order calculus, and show that for any signature Σ the algebraic types $\mathcal{A}_i$ are interpreted by the carriers for the initial Σ-algebra in PER. In particular, the elements of the algebraic types correspond to the closed terms of the free calculus. Our major result is for the standard PER model on the natural numbers, thus proving a conjecture of Huet, as well as showing that the PER model satisfies at least this portion of Freyd's criteria for inherent parametricity.

As is by now well-known, a PER model can be viewed as a small complete category of sets inside a realizability topos (*cf.* [Hyl87], [HRR89]). From this point of view, the polymorphic types are interpreted as a product in the topos. In the case of the algebraic types this product is quite simply the product of all Σ-algebras in the category of types.

1 A recap on PER models

Let A be a partial combinatory algebra (also called a partial Schönfinkel algebra). Thus A is given by a set A, together with a partial binary operation $\cdot$, representing functional application, together with two elements S and K. (As usual, we shall adopt the convention that application associates to the left, and drop the use of the symbol $\cdot$. Thus instead of $((\mathsf{S} \cdot X) \cdot Y) \cdot Z$, we shall simply write $\mathsf{S}XYZ$.) In order to have a partial combinatory algebra, we require also that for all X, Y in A, both $\mathsf{K}XY$ and $\mathsf{S}XY$ are always defined, and that for all X, Y, Z

$$\begin{aligned} \mathsf{K}XY &= X \\ \text{and} \quad \mathsf{S}XYZ &= (XZ)(YZ) \end{aligned}$$

where the equality means that if one side is defined, then so is the other, and they are equal.

The instances with which we shall be particularly concerned are the partial combinatory algebra N of integers with Kleene application, and the total combinatory algebra Λ of β-equivalence classes of open untyped lambda terms (the "free" lambda algebra on countably many generators). However, there are many other interesting Schönfinkel algebras around. These include algebras of closed lambda terms, and algebras of functions recursive with respect to some oracle, as well as the algebras arising from domain-theoretic models of the untyped lambda calculus, such as D_∞ or $\mathsf{P}\omega$.

Since A contains combinators S and K it enjoys a form of combinatory completeness, and in particular has a notion of pairing. Specifically, there is an element Pair of A such that $\mathsf{Pair}XY$, for which we shall write $\langle X, Y\rangle$, is always defined. Moreover there are elements P_0 and P_1 such that $\mathsf{P}_0\langle X, Y\rangle = X$ and $\mathsf{P}_1\langle X, Y\rangle = Y$.

Given such an A we can construct a model PER(A) of the second-order lambda calculus. The types of the model are the partial equivalence relations (or *per's*) on A. That is to say they are symmetric and transitive, but not necessarily reflexive, relations. Given a per R, we shall refer to the set of x such that for some y, xRy as the *domain* of R.

In the course of the paper we shall make considerable use of per's P with the property that xPy iff x is in the domain of P and $x = y$. We shall call such per's *canonically projective* (*cf.* [RR88]). In particular if $U \subseteq \mathsf{A}$, then we shall refer to *the canonically projective per on domain* U.

Given two per's R and S, a map from R to S is given by an element ϕ of A, such that

1. ϕx is defined for all x in the domain of R,
2. for all x, y in A, if xRy then $\phi x\, S\, \phi y$.

The intuition is that ϕ induces a map between the quotients of A by R and S. Accordingly, we specify that two elements ϕ and ψ induce the same map from R to S if (they both induce maps from R to S and) for all x, y such that xRy, $\phi x\, S\, \psi y$.

Given types (per's) R and S we can now interpret the product type $R \times S$ as the per whose domain is the set of pairs $\langle x, y\rangle$ where x is in the domain of R and y is in the domain of S, and where

$$\langle x, x'\rangle\ R \times S\ \langle y, y'\rangle \text{ iff } xRy \text{ and } xSy'$$

The function space type $[R \to S]$ is interpreted as the per whose domain is the set of ϕ inducing a function from R to S, and where

$$\phi[R \to S]\psi \text{ iff } \phi \text{ and } \psi \text{ induce the same map from } R \text{ to } S.$$

These definitions give the cartesian closed category structure on PER(A), and hence we have an interpretation of the first-order typed lambda calculus. Given a type expression $F[X]$, we interpret it as the family of types $F[R]$ as R varies through PER(A), and similarly for type expressions with more than one free variable. Finally, the polymorphic types are interpreted by intersection:

$$x\ \{\Pi X.\, F[X]\}\ y \text{ iff for all } R \text{ in PER(A)},\ x\ F[R]\ y.$$

There is more than one way to handle the presentation of the interpretation of terms in this model, and we leave it to the reader to pick his or her favourite approach.

It is by now well-known that a PER model constructed in this way can also be viewed as a small complete category of sets inside the realizability topos $\mathcal{A}$ generated from A (*cf.* [Hyl87], [HRR89]). From this point of view, the polymorphic types are interpreted by a product in the topos. The results we shall present below can also be read in this topos-theoretic setting. We shall show that for the toposes $\mathcal{L}$ (constructed from Λ) and $\mathcal{E}\!f\!f$, algebraic types form the carriers of the appropriate initial algebra in the topos.

2 PER(Λ)

Let Λ be the combinatory algebra of open untyped lambda terms (with respect to $\beta\eta$-equivalence). Let us suppose that the algebraic type $\mathcal{A}_i$ is interpreted in PER(Λ) by the per $[\![\mathcal{A}_i]\!]$. The characterization of these types is due to Eugenio Moggi.

2.1 PROPOSITION (Moggi, *cf.* [BC88]) $[\![\mathcal{A}_i]\!]$ is the canonically projective per (*cf.* section 1) whose domain is

$$\{\lambda f_1 \ldots f_m.\, \sigma \mid \sigma \text{ is a closed } \Sigma\text{-term of type } A_i\ \}.$$

Proof. Given τ in the domain of $[\![\mathcal{A}_i]\!]$, choose distinct variables $x_1 \ldots x_m$ not free in τ, and for $i = 1 \ldots n$ define the per X_i to be the canonically projective per on

$$\{\sigma[\vec{x}/\vec{f}] \mid \sigma \text{ is a closed } \Sigma\text{-term of type } A_i\ \}.$$

These types carry a Σ-algebra structure in which, when we follow through the implications of currying all our functions, the interpretation of f_i is induced by x_i.

If we specialize τ at $X_1, \ldots, X_n$, and then apply the resulting function to the variables $x_1, \ldots, x_m$, we must get some element $\sigma[\vec{x}/\vec{f}]$ of X_i. Thus

$$\tau x_1 \ldots x_m = \sigma[\vec{x}/\vec{f}]$$

and hence

$$\tau = \lambda f_1 \ldots f_m.\, \sigma.$$

In particular τ is closed.

We have thus established that the per $[\![\mathcal{A}_i]\!]$ is on the correct domain. It remains to show that it is canonically projective. Suppose $\tau\ [\![\mathcal{A}_i]\!]\ \tau'$, then

$$\begin{aligned} \tau x_1 \ldots x_m &= \sigma[\vec{x}/\vec{f}] \\ \tau' x_1 \ldots x_m &= \sigma'[\vec{x}/\vec{f}]. \end{aligned}$$

Since X_i is canonically separated, $\sigma = \sigma'$, and thus by the η rule $\tau = \tau'$. □

2.2 COROLLARY The algebra $[\![\mathcal{A}(\Sigma)]\!]$ is the initial Σ-algebra in the realizability topos associated to Λ, and hence is also the initial Σ-algebra in PER(Λ).

The proof of this is left to the reader. Those not interested in topos theory may wish to note that the second statement admits of a simple direct proof.

If, instead of PER(Λ), we take the per model on the algebra Λ^β of β-equivalence classes of open untyped terms, then we can still obtain a result analogous to proposition 2.1.

2.3 PROPOSITION If the algebraic type $\mathcal{A}_i$ is interpreted in PER(Λ^β) by the per $[\![\mathcal{A}_i]\!]_\beta$, then $[\![\mathcal{A}_i]\!]_\beta$ is isomorphic to the canonically projective per on

$$\{\lambda f_1 \ldots f_m.\ \sigma \mid \sigma \text{ is a closed } \Sigma\text{-term of type } A_i\ \}.$$

Proof. The proof is much as before, except for a slight difficulty introduced by the fact that we can no longer use

$$\tau x_1 \ldots x_m = \sigma[\vec{x}/\vec{f}]$$

to conclude that

$$\tau = \lambda f_1 \ldots f_m.\ \sigma.$$

Note, however, that if $\tau x_1 \ldots x_m = \tau' x_1 \ldots x_m$, then $\tau[\![\mathcal{A}_i]\!]_\beta \tau'$, and hence that

$$\lambda x.\ \lambda f_1 \ldots f_m.\ x f_1 \ldots f_m$$

induces an isomorphism from $[\![\mathcal{A}_i]\!]_\beta$ to the per required. □

To conclude the section we give an alternative proof of the result of Böhm-Berarducci.

2.4 PROPOSITION (Böhm-Berarducci) Any closed term of type $\mathcal{A}_i$ is $\beta\eta$-equivalent to one of the form

$$\Lambda X_1 \ldots X_n.\ \lambda f_1 \ldots f_m.\ \sigma$$

where σ is a closed Σ-term of type A_i.

Proof. Let T be a closed term of type $\mathcal{A}_i$, and τ the lambda term obtained from it by erasing the type information. Consider the interpretation of T in PER(Λ) at the algebra on types $X_1 \ldots X_n$ defined as above. This is $\tau x_1 \ldots x_m$. Hence $\tau x_1 \ldots x_m$ reduces to $\sigma[\vec{x}/\vec{f}]$. But any reduction of $\tau x_1 \ldots x_m$ lifts to a reduction of $T[X_1 \ldots X_n] x_1 \ldots x_m$ (where the X_i are type variables). Thus we conclude that

$$T[X_1 \ldots X_n]\, x_1 \ldots x_m \quad \text{reduces to} \quad \sigma[\vec{x}/\vec{f}]$$

and hence that

$$T = \Lambda X_1 \ldots X_n.\ \lambda f_1 \ldots f_m.\ \sigma$$

as required. □

It is interesting to compare the proof given by Böhm and Berarducci with this one. Böhm and Berarducci use the strong normalization of polymorphic lambda terms, and then a simple argument as to the structure of a normal form of the required type. Strong normalization is itself proved by a kind of realizability argument (essentially realizability using the combinatory algebra of strongly normalizable untyped terms), as was pointed out by Tait ([Tai75]). Here, we use a less sophisticated realizability to show normalizability, and to characterize normal forms, for a very restricted class of types.

3 A simple case

We now try to prove results analogous to those in the previous section for PER models over more general partial combinatory algebras. Such an algebra does not necessarily contain anything that we can use as a variable, and so we have to attempt a different line of proof.

We shall begin by considering the simplest non-trivial case—the polymorphic booleans, $\Pi X. [X \to [X \to X]]$. We recall that this corresponds to a one-sorted signature with no basic operations but two constants 0 and 1.

Let θ be an element of our algebra A contained in the interpretation of the type $\mathcal{A}_i$, and let us suppose that we are given a per B, together with two elements b_0 and b_1. We want to examine the interpretation of $\theta[B]b_0b_1$. The first stage of the proof is to realize that this has to be either b_0 or b_1. Indeed, since the interpretation of $\theta[B]b_0b_1$ is given by the value of $\theta b_0 b_1$, we can look at the per B' whose only elements are b_0 and b_1, and in which elements are related iff they are equal. We now use the fact that $\theta b_0 b_1 = \theta[B']b_0b_1$ must be an element of B'. Hence it is either b_0 or else b_1.

We shall see later that this stage of the proof generalizes nicely to arbitrary algebraic types. The next stage is however more problematic. We have to show that θ is uniformly defined. If $\theta[F]f_0f_1$ is f_0, for one (suitable) per F and elements f_0 and f_1, then $\theta[B]b_0b_1$ is b_0 for all per's B and elements b_0 and b_1.

To do this, we first fix our reference per F to have only the elements 0 and 1, $0 \neq 1$, and in which 0 is not related to 1. We shall suppose, with no real loss of generality, that $\theta[F]01 = 0$. Now let's turn our attention back to B, b_0 and b_1. In order to show that θ is uniform we have to show that $\theta b_0 b_1 = b_0$. The easy case is when the sets $\{b_0, b_1\}$ and $\{0, 1\}$ are disjoint.

In this case, we glue the per's B' and F together, to get a per B'' in which 0 is related to b_0, and 1 is related b_1. Now, as far as B'' is concerned, 0 is the same element as b_0, and 1 is the same element as b_1. It follows that $\theta b_0 b_1 = \theta[B'']b_0 b_1$ is the same element of B'' as $\theta 01 = \theta[B'']01 = 0$. So $\theta b_0 b_1 = b_0$, as required.

Now, if $\{b_0, b_1\}$ and $\{0, 1\}$ are not disjoint, we simply pick $b'_0 \neq b'_1$ disjoint from both, and apply the argument above twice.

This shows that in any PER model the interpretation of the polymorphic booleans contains only the two polymorphic projections. (The result of course generalizes to all algebraic signatures which contain only constants).

To recap, the first stage of the proof was to cut down a large per (or in general, algebra) B to a small algebra B' which still contained all the elements we were interested in (in general this is the subalgebra of reachable elements). This stage generalizes, as we shall see in the next section. The second stage was to show that as well as behaving as expected on each individual algebra, our function behaved uniformly on all algebras. We managed this via a gluing construction involving disjoint per's. This construction does not generalize well. There seem to be two separate problems. One is getting the algebra operations on the glued per's, and the other is finding a disjoint algebra to glue with in the first place.

4 Some general remarks

This section contains an account of that fragment of the theory for arbitrary algebraic types that holds in general.

As before, let Σ be a signature, and $(\mathcal{A}_i)_{i\in\{1\ldots n\}}$ its associated family of algebraic types. Note that if $\mathcal{M}$ is any model of the second-order lambda calculus whatever, then the interpretations $[\![\mathcal{A}_i]\!]$ in $\mathcal{M}$ of the types $\mathcal{A}_i$ form in a canonical way the carriers for a Σ-algebra structure. In this structure the operation ϕ_i is given by the interpretation in $\mathcal{M}$ of the term

$$\begin{aligned}\lambda\, x_1 : \mathcal{A}_{i1} \;\ldots\; x_{N(i)} : \mathcal{A}_{iN(i)}.\;\; &\Lambda\, B_1 \ldots B_n.\; y_1 \ldots y_m. \\ &y_i \;\; (x_1[B_1 \ldots B_n]\, y_1 \ldots y_m) \ldots (x_{N(i)}[B_1 \ldots B_n]\, y_1 \ldots y_m).\end{aligned}$$

Let us call this algebra $\mathcal{A}(\Sigma)$. Now, if B is any other Σ-algebra in $\mathcal{M}$ (with carriers $B_1, \ldots, B_n$ and operations $\psi_1, \ldots, \psi_m$ say), then there is a canonical Σ-algebra homomorphism from $\mathcal{A}(\Sigma)$ to B. In general, this is given by the functions $\theta_i : [\![\mathcal{A}_i]\!] \to B_i$, where θ_i is the interpretation in $\mathcal{M}$ of

$$\lambda a : \mathcal{A}_i.\; a[B_1 \ldots B_n]\, \psi_1 \ldots \psi_m.$$

In other words $\theta_i(a)$ is the interpretation of a at the Σ-algebra B. In the case of a PER model, where we know that $\mathcal{A}(\Sigma)$ is the product of all Σ-algebras, this homomorphism is just the projection at B.

If $\mathcal{M}$ should happen to contain an initial Σ-algebra ($\mathcal{F}$ say), then we have also a map $\mathcal{F} \to \mathcal{A}(\Sigma)$, exhibiting $\mathcal{F}$ as a retract of $\mathcal{A}(\Sigma)$. We want to show that this retraction is an isomorphism. For this we use the following lemma.

4.1 LEMMA Suppose $\mathcal{M} = \mathrm{PER}(\mathsf{A})$ is an arbitrary PER model for the second-order lambda calculus, and that Σ is a signature in the sense of many-sorted logic. Then

(i) $\mathcal{M}$ contains an initial Σ-algebra $\mathcal{F}$. (Moreover, $\mathcal{F}$ is also the initial Σ-algebra in the associated realizability topos.)

(ii) $\mathcal{A}(\Sigma)$ is canonically isomorphic to $\mathcal{F}$ iff for any $a \in \mathcal{A}_i$ there is a Σ-term σ of type A_i such that

$$a = \Lambda A_1 \ldots A_n.\ \lambda\phi_1 \ldots \phi_m.\ \sigma.$$

Proof. To prove the first assertion, take some suitable Gödel encoding of the syntax of Σ into the Church numerals of the partial combinatory algebra, and take the carriers of $\mathcal{F}$ to be (the canonically projective per's whose underlying sets are) the images of the closed terms of the appropriate sort under this encoding. The operations of the algebra are now given by the functions which construct compound terms out of their components, and are therefore recursive. Given any other Σ-algebra B in the topos, there is a homomorphism from $\mathcal{F}$ to B, given by decoding the elements of $\mathcal{F}$ as closed Σ-terms, and then giving the interpretation of the term in B. This operation is recursive in codes for the operations on B, and hence is an internal homomorphism. It is the unique homomorphism from $\mathcal{F}$ to B since the carriers of $\mathcal{F}$ are canonically separated, and their underlying sets give the free algebra in *Sets*.

To prove the second assertion, note that if a is any element of $\mathcal{A}_i$, then if there is a closed Σ-term σ such that $a = \Lambda A_1 \ldots A_n.\ \lambda\phi_1 \ldots \phi_m.\ \sigma$, then since the intepretation of $a[\mathcal{F}]$ is essentially σ, such a σ is necessarily unique. We thus have $\mathcal{F}$ as a retract of $\mathcal{A}_i$, where the inclusion sends σ to $\Lambda A_1 \ldots A_n.\ \lambda\phi_1 \ldots \phi_m.\ \sigma$, and the retraction sends a to $a[\mathcal{F}]$. The second assertion of the lemma now says that this retraction is an isomorphism if and only if it is surjective. □

To sum up, if $\mathcal{M}$ is a PER model, PER(A), then we have the following situation (treating the model as an internal category of sets in the realizability topos):

1. The product of all Σ-algebras in PER(A), is itself a Σ-algebra in PER(A), and is given by taking the canonical Σ-algebra structure on the interpretations of the algebraic types associated to Σ.

2. PER(A) also contains an initial Σ-algebra.

3. There is thus a canonical map from the initial algebra to the product algebra.
4. There is also a canonical map from the product algebra to the initial algebra (the projection).

We want to show that these two maps are inverse isomorphisms.

We recall that in any model $\mathcal{M}$, a value of type $[\![\mathcal{A}_i]\!]$ can be regarded as a function taking a Σ-algebra B as parameter, and producing a value of type B_i. An important feature of PER models is that the value produced does not depend on the carriers $B_1 \ldots B_n$, but only on the names of the operations of B. More formally, we have

4.2 Lemma Suppose B and C are Σ-algebras in a PER model $\mathcal{M} = \mathrm{PER}(\mathsf{A})$, where $B = (B_1, \ldots, B_n, e_1, \ldots, e_m)$ and $C = (C_1, \ldots, C_n, e_1, \ldots, e_m)$. Then if $a \in [\![\mathcal{A}_i]\!]$,

$$\begin{aligned} a[B_1 \ldots B_n]\, e_1 \ldots e_m &= a[C_1 \ldots C_n]\, e_1 \ldots e_m \\ &= a e_1, \ldots, e_m. \end{aligned}$$

Recall that we want to show that a is $\Lambda A_1 \ldots A_n.\ \lambda\langle\phi_1, \ldots, \phi_m\rangle.\ \sigma$, where σ is a closed Σ-term of type A_i. The lemma above allows us to restrict the class of Σ-algebras at which we have to look.

Suppose $\mathcal{M} = \mathrm{PER}(\mathsf{A})$ is the PER model on a partial combinatory algebra A. Given elements $e_1, \ldots, e_m$ of A which induce the operations on some Σ-algebra $B = (b_1, \ldots, b_n, e_1, \ldots, e_m)$ in $\mathcal{M}$, we define the Σ-algebra $\mathcal{O}(e_1, \ldots, e_m)$ (the *orbit* of $e_1 \ldots e_m$), with carriers $\mathcal{O}_i(e_1, \ldots, e_m)$, where $\mathcal{O}_i$ is the subset of the carrier of B_i consisting of the interpretations of closed Σ-terms, and in which the operations are given by $e_1, \ldots, e_m$. If $e_1 \ldots e_m$ do not induce the operations on any Σ-algebra, then $\mathcal{O}(e_1, \ldots, e_m)$ is undefined.

It is easy to see that the definition of $\mathcal{O}(e_1, \ldots, e_m)$ does not depend on the algebra B; $\mathcal{O}_i$ consists of the elements $(\lambda\phi_1 \ldots \phi_m.\ \sigma)\, e_1 \ldots e_m$ such that σ is a Σ-term of type A_i. Moreover $\mathcal{O}$ is defined iff all such elements are defined. It is also clear that $\mathcal{O}$ is in some sense a minimal Σ-algebra. This intuition can be made precise using the notion of *inclusion* ([CFS87]). We recall that an inclusion is a map of per's induced by the identity function on A. The category of Σ-algebras in $\mathrm{PER}(\mathsf{A})$ with homomorphisms given by inclusions is a poset, and the minimal elements are the algebras $\mathcal{O}(e_1, \ldots, e_m)$.

4.3 Lemma If $B = (B_1, \ldots, B_n, e_1, \ldots, e_m)$ is a Σ-algebra in $\mathrm{PER}(\mathsf{A})$, and $a \in [\![\mathcal{A}_i]\!]$, then

$$a[B_1 \ldots B_n]\, e_1 \ldots e_m = a[\mathcal{O}_1 \ldots \mathcal{O}_n]\, e_1 \ldots e_m,$$

and in particular $a[B_1 \ldots B_n]\, e_1 \ldots e_m \in \mathcal{O}_i$.

As a direct consequence we have:

4.4 COROLLARY Given an element a of $[\![\mathcal{A}_i]\!]$, then for any Σ-algebra B, where $B = (B_1, \ldots, B_n, e_1, \ldots, e_m)$ there is a closed Σ-term σ of type A_i (possibly depending on $e_1, \ldots, e_m$, but certainly independent of $B_1, \ldots, B_n$), such that

$$a[B_1 \ldots B_n]\, e_1 \ldots e_m = (\lambda\phi_1 \ldots \phi_m.\, \sigma)\, e_1 \ldots e_m.$$

This corollary is the analogue of showing that if θ is a polymorphic boolean, then $\theta[B]b_0b_1$ has to be either b_0 or b_1. It is the local version of the result we really want. So, it now remains to show that the σ whose existence is guaranteed by this corollary can be chosen independently of $e_1 \ldots e_m$. Note that in order to do this we need only look at the Σ-algebras $\mathcal{O}(x_1, \ldots, x_m)$. The crucial technical result we shall need is, however best expressed for more general algebras. We first make a definition.

Definition Suppose A is a partial combinatory algebra, and U and V are subsets of A, then we say that U and V are *recursively disjoint* if there is some element ϕ of A such that for all $u \in U$, $\phi u = \ulcorner 0 \urcorner$, and for all $v \in V$, $\phi v = \ulcorner 1 \urcorner$.

4.5 LEMMA Suppose that $\theta : B \to C$ is a homomorphism of Σ-algebras, where $B = (B_1, \ldots, B_n, e_1, \ldots, e_m)$, and that $C = (C_1, \ldots, C_n, d_1, \ldots, d_m)$, and for each i the domains of the per's B_i and C_i are recursively disjoint. Then for any $a \in [\![\mathcal{A}_i]\!]$

$$\{a[C_1 \ldots C_n]\, d_1 \ldots d_m\} \quad C_i \quad \{\theta_i\, (a[B_1 \ldots B_n]\, e_1 \ldots e_m)\},$$

i.e. a evaluated at C is the image under θ of a evaluated at B.

Proof. We define for each i, a per $B_i \otimes C_i$, whose domain is $B_i \cup C_i$, and where $n\ B_i \otimes C_i\ m$ iff either
(*i*) $n\ B_i\ m$ or else $n\ C_i\ m$
or (*ii*) $(\theta_i n)\ C_i\ m$ or else $n\ C_i\ (\theta_i m)$.
(We are gluing B_i to C_i along θ_i.) Since B_i and C_i are recursively disjoint for each i, the types $B_i \otimes C_i$ carry a natural Σ-algebra structure. We can define this in two different ways, either by means of functions $e'_1 \ldots e'_m$, which restrict to $e_1 \ldots e_m$ on B, or else via $d'_1 \ldots d'_m$, which restrict to $d_1 \ldots d_m$ on C. We cannot do both simultaneously, since any constants must come either from B or from C.

Since $e'_1 \ldots e'_m$ and $d'_1 \ldots d'_m$ both define the same algebra structure on $B \otimes C$ we have

$$a[B_1 \otimes C_1 \ldots B_n \otimes C_n]\, e'_1 \ldots e'_m =_{B_i \otimes C_i} a[B_1 \otimes C_1 \ldots B_n \otimes C_n]\, d'_1 \ldots d'_m.$$

Since e'_i restricts to e_i on B, we have

$$a[B_1 \ldots B_n]\, e_1 \ldots e_m \quad =_{B_i} \quad a[B_1 \ldots B_n]\, e'_1 \ldots e'_m,$$

and similarly

$$a[C_1 \ldots C_n]\, d_1 \ldots d_m \quad =_{C_i} \quad a[C_1 \ldots C_n]\, d'_1 \ldots d'_m.$$

Now apply lemma 4.2, and we obtain

$$a[B_1 \dots B_n]\, e_1 \dots e_m \quad =_{B_i \otimes C_i} \quad a[C_1 \dots C_n]\, d_1 \dots d_m,$$

from which the result follows. □

This lemma is only useful if we can find an algebra disjoint from the one which we wish to study. However, for the case A = N it is easy to find algebras which fill up the whole space available. Consider the polymorphic integers, corresponding to a one-sorted algebra with one constant and a single unary function. If for our constant we take 0, and for our function the successor function, then we obtain an algebra which takes up the whole of $\mathbb{N}$. To get round this we use a continuity argument which will allow us to compress our algebra so that it takes up only a finite amount of space, thus allowing us easily to find algebras disjoint from it. Unfortunately our argument does not work for general combinatory algebras, and so fails for general PER models.

5 PER(N)

We can express the continuity argument we need in very simple, though slightly abstract terms. For readers unhappy with this, a direct proof of the central proposition is given in an appendix. Again, let Σ be any many-sorted algebraic signature, and A any partial combinatory algebra. Then the Σ-algebras in PER(A) form a metric space (Σ-*Alg*) in which

$$d(B, C) \leq 1/n$$

if and only if the interpretation of σ in B is the same as the interpretation of σ in C for all Σ-terms σ of size not greater than n. This is true also in the internal topos-theoretic sense, with a slight delicacy—the metric function

$$d : \Sigma\text{-}Alg \times \Sigma\text{-}Alg \to \mathrm{Q}$$

is not represented internally. However, the predicate

$$d(B, C) \leq q$$

on Σ-*Alg* $\times$ Σ-*Alg* $\times$ Q is, and this will suffice for our purposes.

It is easy to see that Σ-*Alg* is complete with respect to this "metric", and that it is the completion of its subspace of finite Σ-algebras. Indeed, suppose that $B = (B_1, \dots, B_n, b_1, \dots, b_m)$ is an arbitrary Σ-algebra, then for each i such that the orbit $\mathcal{O}_i(b_1, \dots, b_m)$ is non-empty we can pick a "canonical" element c_i of $\mathcal{O}_i(b_1, \dots, b_m)$. Now, given an integer l, we can find elements b_{li} of A, such that

$$b_{li}\, z_1 \dots z_{N(i)} \quad = \quad \begin{cases} b_i\, z_1 \dots z_{N(i)} & \text{if each } z_i \text{ is } [\![\sigma]\!]_b (\text{where } [\![\sigma]\!]_b = (\lambda f_1 \dots f_m.\ \sigma)\ b_1 \dots b_m), \text{ for some closed term } \sigma \text{ of size less than } l \\ c_i & \text{otherwise} \end{cases}$$

Moreover, we can find the b_{li} recursively in l. Note that if for some i the orbit $\mathcal{O}_i(b_1, \ldots, b_m)$ is empty, then no closed term of any type can depend on a term of type A_i, and hence that empty orbits cause no problems. Clearly, for each l, the orbits $\mathcal{O}_i(b_{l1}, \ldots, b_{lm})$ are defined, and are finite subsets of the $\mathcal{O}_i(b_1, \ldots, b_m)$. Furthermore, the Σ-algebras $\mathcal{O}(b_{l1}, \ldots, b_{lm})$ form a Cauchy sequence (even an internal Cauchy sequence) whose limit is B.

In the case $\mathsf{A} = \mathsf{N}$, this means that Σ-*Alg* is a complete separable metric space.

We can also represent the collection of closed terms of sort A_i internally in the topos, via Gödel enumeration. Abusing notation, let A_i be the canonically projective per whose domain is the set of codes for closed Σ-terms of type A_i. We regard A_i as a discrete metric space. Whatever partial combinatory algebra A we use, it is always separable, and has a linear order (for example that inherited from the order on $\mathbb{N}$).

Now let a be a value of type $[\![\mathcal{A}_i]\!]$. We can use a to define a map

$$\phi_a : \Sigma\text{-}Alg \to A_i.$$

$\phi_a(B)$ is the least σ in A_i such that a evaluated on B gives the intepretation of σ (*cf.* corollary 4.4). For a suitable choice of the linear order on A_i, the function ϕ_a is represented internally in the topos.

In the case of the effective topos ($\mathsf{A} = \mathsf{N}$), we can use a well-known theorem due to Čeitin and independently Moschovakis ([Če62], *cf.* [Bee85]) on the continuity of effective operations in the presence of Markov's principle. Expressed in somewhat more topos-theoretic terms than usual, this states

5.1 THEOREM If in the effective topos $\mathcal{E}\!f\!f$, f is a function from a complete separable metric space X to a separable metric space Y, then f is pointwise continuous, *i.e.* given a fixed $x \in X$, the formula

$$\forall \epsilon > 0.\ \exists \delta > 0.\ d_X(x, x') < \delta \ \rightarrow\ d_Y(f(x), f(x')) < \epsilon$$

is satisfied.

The continuity result we need is an immediate corollary of this theorem:

5.2 PROPOSITION Suppose a is in the domain of $[\![\mathcal{A}_i]\!]$, and $\mathcal{O}(x_1, \ldots, x_m)$ is defined. Then there is a number h, such that for all $l \geq h$ we have $a\, x_{l1} \ldots x_{lm} = a\, x_1 \ldots x_m$.

(Here, we are using juxtaposition to denote Kleene application, and associating to the left as usual; thus $ax_1 \ldots x_m$ means $\{\ldots\{\{a\}\, x_1\} \ldots\}\, x_m$.)

In the appendix we shall give a proof of this proposition which does not depend on Čeitin's theorem. Unfortunately, both this more concrete proof, and the proof of Čeitin's theorem itself, seem to depend on properties of the partial combinatory algebra N which are not completely general.

We now conclude our proof that the algebraic types are interpreted as initial Σ-algebras in the case $\mathsf{A} = \mathsf{N}$.

We first note that $\mathbb{N}$ is recursively isomorphic to $\{n \in \mathbb{N} \mid n = k \bmod m\}$ for any m, and k, and hence that we can find a copy of the initial Σ-algebra disjoint from any finite set.

Let $\mathcal{F} = (F_1, \ldots, F_n, y_1, \ldots, y_m)$ be our designated initial algebra, and suppose that

$$a[F_1 \ldots F_n]\, y_1 \ldots y_m = [\![\sigma]\!]_y.$$

Then, given a Σ-algebra $B = (B_1, \ldots, B_n, b_1, \ldots, b_m)$ (without loss of generality we can suppose that $B = \mathcal{O}(b_1, \ldots, b_m)$), we can find an h such that

$$a[F_1 \ldots F_n]\, y_{h1} \ldots y_{hm} = a[F_1 \ldots F_n]\, y_1 \ldots y_m$$

and also

$$a[B_1 \ldots B_n]\, b_{h1} \ldots b_{hm} = a[B_1 \ldots B_n]\, b_1 \ldots b_m.$$

Now pick a copy $\mathcal{F}'$ of the initial algebra avoiding both the finite sets $\mathcal{O}(y_{h1}, \ldots, y_{hm})$ and $\mathcal{O}(b_{h1}, \ldots, b_{hm})$, and apply lemma 4.5 twice to show that

$$a[B_1 \ldots B_n]\, b_1 \ldots b_m = [\![\sigma]\!]_b.$$

Thus we have shown:

5.3 THEOREM When $\mathsf{A} = \mathsf{N}$, the algebra $\mathcal{A}(\Sigma)$ is the initial Σ-algebra in $\mathcal{E}\!f\!f$, and hence $\mathcal{A}(\Sigma)$ is also the initial Σ-algebra in PER(N).

6 Conclusion

We have achieved our purpose in proving the theorem above. However, some concluding remarks appear to be in order. The proof we have given seems to use rather a lot of machinery, for a fairly small result. In particular, it is disturbing that it will only work for a relatively small class of Schönfinkel algebras. It is possible, however, that these results do inevitably depend on the algebra chosen. The small amount of experimental evidence available would tend to support this view. Apart from the proof for the algebra of open lambda terms, due to Moggi and outlined above, we have a proof due to Peter Freyd of the result for the polymorphic natural numbers. This proof avoids the explicit use of a general continuity principle by using an elegant combinatorial trick. However, it also works only for the PER model on the natural numbers. More recently, Freyd has also been able to show that the interpretation of the types $\Pi X.[[A \to X] \to X]$ for an arbitrary per A, is isomorphic to A. The proof he has given here also works only for PER models based on a relatively restricted class of algebras (again including the Kleene algebra on the natural numbers) [Fre89a].

Finally, we should return to the relationship between PER models and the free term model. We stated above that any PER model satisfies equations which are not provable in the calculus, and which therefore do not hold in the free term model. We can now give an example. If we let P be the type $\Pi X.X \to X$, $Q = P \to P$ be the type of endomorphisms of P, and S be the "double negation" $\Pi X.[[Q \to X] \to X]$ of Q, then it is well-known that there is a closed term of type S which is not formally evaluation at any closed term of type Q (the term $\Lambda X.\ \lambda f : Q \to X.\ f(\lambda a : P.\ (a[X \to P](\lambda x : X.\ \Lambda Y.\ \lambda y : Y.\ y)(f(\lambda p : P.p))))$ will do). It follows that in any extensional model, either we violate parametricity, in that S is not isomorphic to Q, or there is a non-standard element of Q, or else some equation holds in the model which does not hold in the term model. In the particular case of PER models, we know that P contains only the polymorphic identity. It follows that Q and S are also singletons, and that the term we have given is equated in the model to evaluation at the identity map on P.

Appendix

In this appendix we give a direct proof of the crucial continuity result proposition 5.2. We have chosen to use a proof style modelled on Gandy's proof of the Kreisel-Lacombe-Shoenfield theorem. This gives a very slick proof in which we use the second recursion theorem to pull assorted useful integers out of a definition whose construction is about as obvious as a magician's hat. There is always an alternative to using the recursion theorem in this way. Most often, it is a longer and equally unintuitive combinatorial proof. Readers are referred to the proof of Čeitin's theorem in [Bee85] for an example.

The structure of this proof is such that it is more convenient to use the interpretations of the type $\mathcal{A}_i^\times$ of the extended calculus, rather than the types $\mathcal{A}_i$. Recall that if $B = (B_1, \ldots, B_n, b_1, \ldots, b_m)$ is a Σ-algebra, then a value a of type $[\![\mathcal{A}_i^\times]\!]$ interpreted at $B_1, \ldots, B_n$ takes the tuple $\langle b_1, \ldots, b_m\rangle$ as parameter, and produces a value of type B_i as result. (We contrast $a\,[B_1 \ldots B_n]\,\langle b_1, \ldots, b_m\rangle$ with $a[B_1 \ldots B_n]\,b_1 \ldots b_m$). As before, we can define a series of finite approximants $\mathcal{O}^\times(b_{l1}, \ldots, b_{lm})$ to B, and our proposition becomes:

5.2 **Proposition** Suppose a is in the domain of $[\![\mathcal{A}_i^\times]\!]$, and $\mathcal{O}^\times(x_1, \ldots, x_m)$ is defined. Then there is a number h, such that for all $l \geq h$ we have

$$a\,\langle x_{l1}, \ldots, x_{lm}\rangle = a\,\langle x_1, \ldots, x_m\rangle.$$

Proof. Let $\Phi(g, y, y', m)$ be the formula

$$\exists w, w' \leq m[T(g, y, w)\ \wedge\ T(g, y', w')\ \wedge\ Uw = Uw']$$

where T is Kleene's T-predicate, and U its output function. Use the recursion theorem to find an integer e_j such that

$$\{e_j\}\langle z_1,\ldots,z_{N(j)}\rangle = \begin{cases} \{x_j\}\langle z_1,\ldots,z_{N(j)}\rangle & \text{if some } z_i = [\![\sigma]\!]_x \text{ for some } \sigma \text{ of size } r\text{, where } \Phi(a,\langle x_1,\ldots,x_m\rangle,\langle e_1,\ldots,e_m\rangle,r) \\ \{x_k\}\langle z_1,\ldots,z_{N(j)}\rangle & \text{if for some } r\text{, some } z_i \text{ is of the form } [\![\sigma]\!]_x\text{, where } \sigma \text{ is a term of size } r\text{, but where any such } r \text{ satisfies } \Phi(a,\langle x_1,\ldots,x_m\rangle,\langle e_1,\ldots,e_m\rangle,r) \text{ and where } k \text{ is the least integer such that } \Phi(a,\langle x_1,\ldots,x_m\rangle,\langle e_1,\ldots,e_m\rangle,k) \text{ and } \{a\}\langle x_1,\ldots,x_m\rangle \neq \{a\}\langle x_{k1},\ldots,x_{km}\rangle \end{cases}$$

First note that there must be an r such that $\Phi(a,\langle x_1,\ldots,x_m\rangle,\langle e_1,\ldots,e_m\rangle,r)$. If we suppose that there is not, then $\{e_i\} = \{x_i\}$ on $\mathcal{O}$. This however implies that $\{a\}\langle x_1,\ldots,x_m\rangle = \{a\}\langle e_1,\ldots,e_m\rangle$, using the functionality of a on the Σ-algebra $\mathcal{O}$. Thus there is an r such that $\Phi(a,\langle x_1,\ldots,x_m\rangle,\langle e_1,\ldots,e_m\rangle,r)$, a contradiction. We can now let h be the least integer such that $\Phi(a,\langle x_1,\ldots,x_m\rangle,\langle e_1,\ldots,e_m\rangle,h)$.

To conclude the proof, let n be the least integer greater than or equal to h such that $\{a\}\langle x_{n1},\ldots,x_{nm}\rangle \neq \{a\}\langle x_1,\ldots,x_m\rangle$, if such exists. For such an n we have that for each i, $\{x_{ni}\} = \{e_i\}$ on $\mathcal{O}(x_{n1},\ldots,x_{nm})$. This implies that $\{a\}\langle x_{n1},\ldots,x_{nm}\rangle = \{a\}\langle e_1,\ldots,e_m\rangle$, again using the functionality of a. But since we know that the formula $\Phi(a,\langle x_1,\ldots,x_m\rangle,\langle e_1,\ldots,e_m\rangle,h)$ actually holds, we also have $\{a\}\langle e_1,\ldots,e_m\rangle = \{a\}\langle x_1,\ldots,x_m\rangle$, which thus leads to a further contradiction. We conclude that for all $n \geq h$, we have $\{a\}\langle x_{n1},\ldots,x_{nm}\rangle = \{a\}\langle x_1,\ldots,x_m\rangle$, as required. □

References

[BB85] C. Böhm and A. Berarducci. Automatic synthesis of typed λ-programs on term algebras. *Theor. Comp. Sci.*, 39:135–154, 1985.

[BC88] V. Breazu-Tannen and T. Coquand. Extensional models for polymorphism. *Theor. Comp. Sci.*, 59:85–114, 1988.

[Bee85] M.J. Beeson. *Foundations of Constructive Mathematics.* Springer-Verlag, Berlin, 1985.

[CFS87] A. Carboni, P.J. Freyd, and A. Scedrov. A categorical approach to realizability and polymorphic types. In *Proceedings of the 3rd A.C.M. Work-*

shop on Mathematical Foundations of Programming Language Semantics, Springer-Verlag, Berlin, 1987.

[Če62] G.S. Čeitin. Algorithmic operators in constructive metric spaces. *Trudy Mat. Inst. Steklov*, 67:295–361, 1962. English translation in A.M.S. Translations (2) **64**, 1-80.

[Fai86] J. Fairbairn. A new type-checker for a functional programming language. *Science of Computer Programming*, 6:273–290, 1986.

[Fre89a] P.J. Freyd. personal communication, 1989.

[Fre89b] P.J. Freyd. Structural polymorphism. distributed notes, 1989.

[Gir71] J-Y. Girard. Une extension de l'interprétation de gödel. In J.E. Fenstad, editor, *Proc. of the Second Scandinavian Logic Symposium*, pages 63–92, North-Holland Publishing Company, Amsterdam, 1971.

[Gir72] J-Y. Girard. *Interprétation fonctionelle et élimination des coupures de l'arithmétique de l'ordre supérieur.* PhD thesis, Paris, 1972.

[HRR89] J.M.E. Hyland, E.P. Robinson, and G. Rosolini. The discrete objects in the effective topos. *Proceedings of the London Mathematical Society*, 1989. to appear.

[Hyl87] J.M.E. Hyland. A small complete category. In *Proc. of the Conference on Church's Thesis: Fifty Years Later*, 1987.

[Lei83] D. Leivant. Reasoning about functional programs and complexity classes associated with type disciplines. In *24th Annual IEEE Symposium on the Foundations of Computer Science, Tucson, Arizona*, pages 460–496, 1983.

[Rey83] J.C. Reynolds. Types, abstraction, and parametric polymorphism. In R.E.A. Mason, editor, *Information Processing '83*, pages 513–523, North Holland, 1983.

[RR88] E.P. Robinson and G. Rosolini. *Colimit completions and the effective topos.* Technical Report 34, Università degli Studi, Parma, 1988.

[SFO83] D. Leivant S. Fortune and M.J. O'Donnell. The expressiveness of simple and second-order type structures. *J. A.C.M.*, 30:151–185, 1983.

[Sta81] R. Statman. Number theoretic functions computable by polymorphic programs. In *Proc. of the 22nd IEEE symposium on the Foundations of Computer Science*, 1981.

[Tai75] W.W. Tait. A realizability interpretation of the theory of species. In R. Parikh, editor, *Proceedings of the Boston Logic Colloquium*, pages 240–251, Springer-Verlag, Berlin, 1975.

Pseudo-Retract Functors for Local Lattices and Bifinte L-Domains

Elsa L. Gunter
Department of Computer and Information Science,
University of Pennsylvania,
Philadelphia, PA 19104-6389

Abstract

Recently, a new category of domains used for the mathematical foundations of denotational semantics, that of L-domains, has been under study. In this paper we consider a related category of posets, that of *local lattices*. First, a completion operator taking posets to local lattices is developed, and then this operator is extended to a functor from posets with embedding-projection pairs to local lattices with embedding-projection pairs. The result of applying this functor to a local lattice yields a local lattice isomorphic to the first; this functor is a pseudo-retract.

Using the functor into local lattices, a continuous pseudo-retraction functor from ω-bifinite posets to ω-bifinite L-domains can be constructed. Such a functor takes a universal domain for the ω-bifinite posets to a universal domain for the ω-bifinite L-domains. Moreover, the existence of such a functor implies that, from the existence of a saturated universal domain for the ω-algebraic bifinites, we can conclude the existence of a saturated universal domain for the ω-bifinite L-domains.

1 Introduction

In the search for structures for use as mathematical foundations of the denotational semantics of programming languages, attention has for several years been focused on the category of bounded-complete (ω-algebraic) directed-complete posets. In particular, the requirement was made that every bounded subset of the poset should have a least upper bound. If one thinks of the points in the poset as being partial information, and of one point as bieng greater than another if it contains more information, then we are insisting that every demonstrably consistent set of information should have a join in the whole poset of information.

However, recently a new class of posets has been studied as an alternative, namely the (ω-algebraic) L-domains. An L-domain is a directed-complete poset (a cpo) with the property that the set of elements below each element forms a complete lattice. So how do L-domains differ from bounded-complete cpo's? In an L-domain, bounded sets need not have joins in the poset as a whole. We only require that such a bounded set have a unique minimal upper bound below any given upper bound. Thinking in terms of information, we require a demonstrably consistent collection of information have a join relative to a given witness of this consistency, relative to each particular way we have more complete information. The idea of L-domains was introduced by Achim Jung in his investigations into extensions of Smyth's Theorem [Jun88b, Jun88a]. They were independently discovered by Thierry Coquand as a special instance of his categories of embeddings [Coq88].

For the first part of this paper, we will focus our attention on a class of posets more general than the L-domains. The class we shall focus attention on is that of *local lattices*. A local lattice is a poset with the requirement that each principal ideal (the set of points below a given one) is a complete lattice. From the point of view of computation this is an unusual class to be looking at because we are not making the requirement that our posets be directed-complete. However, in the end we will restrict to finite posets (and then go to the ω-bifinites), and finite local lattices are the same thing as finite L-domains.

2 From Posets to Local Lattices

A classic example of a completion operator on posets is given by the following:

Definition 2.1 Let P be a poset. Given $X \subseteq P$, let

$$\underline{X} = \{y \mid y \leq x \text{ for all } x \in X\} = \{y \mid y \leq X\},$$

i.e., the set of all lower bounds for X in P. Let

$$M(P) = \{\underline{X} \mid X \subseteq P\}$$

ordered by subset inclusion (*i.e.* $\underline{X} \leq \underline{Y}$ iff $\underline{X} \subseteq \underline{Y}$). Let $j : P \to M(P)$ by $j(x) = \underline{\{x\}} = \downarrow x$ the principal ideal generated by x. ∎

This construction is known as the MacNeille completion. Some properties which this construction has are:

1. $M(P)$ is a complete lattice.

2. $M(P) \cong P$ iff P is a complete lattice.

3. j is an injective function which preserves and reflects order, and which preserves all existing meets and joins.

(See Theorem III.3.11 in [Joh82], for example.) In this section, we shall develop a similar construction for local lattices.

Definition 2.2 A poset L is called an *local lattice* if for every $x \in L$ the principal ideal generated by x (given by $\downarrow x = \underline{x}$) forms a complete lattice under the restricted ordering. ∎

Thus we have that an L-domain is a cpo which is also a local lattice. A fairly typical example of a local lattice is the following:

Example 2.3

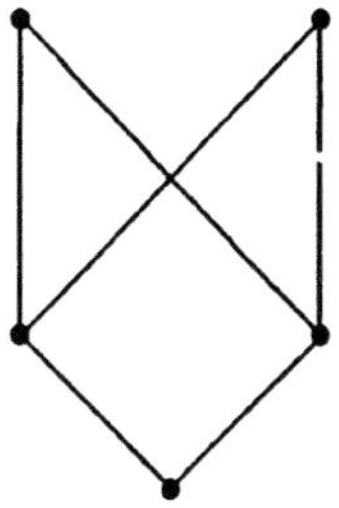

Notice that, while it is true that any bounded subset of a local lattice must have a meet, as the previous example shows, it need not be the case that a bounded subset have a join (*i.e.* a local lattice need not be bounded complete). However, what will be the case is that below any upper bound for a given bounded subset there will exist a unique minimal upper bound. That is, while we do not have absolute joins for bounded subsets, we do have "joins" for bounded subsets relative to given upper bounds.

Definition 2.4 Let P be a poset, X a bounded subset of P, and u an upper bound for X. Then v is a *join for X relative to u* if v is the join of X in the poset $\downarrow u$ with the restricted order relation. ∎

Note that an element v is the join of a bounded set X iff v is the join of X relative to every upper bound for X.

The following gives us a useful way of thinking about local lattices.

Lemma 2.5 *For any poset P the following are equivalent:*

1. *Every non-empty bounded subset has a meet.*

2. *For each bounded subset and each upper bound of it, there exists a join for the bounded subset relative to the upper bound.*

3. *P is a local lattice.* ∎

A complete lattice is a poset in which every subset has a meet. One way of thinking about the MacNeille completion is that it parsimoniously adds meets for subsets of the original poset. (For any subset $X \subseteq P$, the new set $\underline{X}$ will be the meet of $j^*(X) = \{j(x) \mid x \in X\}$ in $M(P)$.) Given the above lemma, a first attempt at turning a poset into a local lattice might be to add, in a similar fashion, meets for bounded subsets. This construction does not quite work. One of the difficulties is that, in adding meets for all bounded subsets of the original poset, you may create new bounded subsets without meets. While some sense can be made out of iterating the process (by injecting each partial result into the MacNeille completion of the original poset), there are other difficulties which arise. (Ultimately, we are aiming for a functor from posets to local lattices and this process just does not mesh well with functions.) A more fruitful approach lies in attempting to add relative joins for bounded subsets in a reasonably parsimonious fashion, and this is the approach we shall pursue.

But how frugal is frugal? If we only include one minimal upper bound for each bounded subset of our original poset, then we will turn it into a bounded-complete poset, and not just a local lattice. On the other hand, if we try to add one minimal upper bound for each existing upper bound of a given bounded subset, we will obviously just get a big mess. The solution is to amalgamate all the upper bounds of a given bounded subset which must end up having the same minimal upper bound in the local lattice we are trying to build, and then for each such amalgamation add one relative join.

Definition 2.6 Let P be a poset. Given a bounded subset X of P, define

$$\overline{X} = \{y \mid y \geq x \text{ for all } x \in X\} = \{y \mid y \geq X\}.$$

That is, $\overline{X}$ is the set of upper bounds for X in P, or the *ceiling* of X in P. For each bounded subset $X \subseteq P$ define an equivalence relation $\sim$ on $\overline{X}$ by the following:

1. If a and b are elements of $\overline{X}$ and $\{a, b\}$ is bounded in P, then $a \sim b$.

2. If $a \sim b$ and $b \sim c$, then $a \sim c$.

That is, we take the transitive closure of the relation of consistency. We say that a and b are *linked in* $\overline{X}$ if $a \sim b$. Let us write $[a]_X$ for the equivalence class of a in $\overline{X}$ under this equivalence relation. ∎

If we view a poset as a graph where there is an edge between two points precisely when one is greater than the other, then $[a]_X$ is just the connected component of $\overline{X}$ which contains a. Two elements a and b of $\overline{X}$ are linked iff there exists a chain $a_0, \ldots, a_n$ in $\overline{X}$ such that $a = a_0$, $b = a_n$, and for all i, $0 \leq i < n$ either $a_i \leq a_{i+1}$ or $a_i \geq a_{i+1}$. As a consequence, the notion of being linked is preserved by monotone functions.

Lemma 2.7 *Let P and Q be posets, and suppose $f : P \to Q$ is a monotonic function. If X is a bounded subset of P and u and v are two upper bounds which are linked in $\overline{X}$, then $f(u)$ and $f(v)$ are linked in $\overline{Y}$ for all subsets Y of Q such that $f^*(\overline{X}) \subseteq \overline{Y}$ (where $f^*(\overline{X}) = \{f(w) \mid w \in \overline{X}\}$).* ∎

As the next lemma shows us, these equivalence classes do unite upper bounds which must have the same relative join.

Lemma 2.8 *Let X be a bounded subset of a local lattice L, and suppose y and z are two upper bounds of X which are linked in $\overline{X}$. Then a join for X relative to y is a join for X relative to z.* ▮

Using equivalence classes of minimal upper bounds, we are now in a position to define our construction.

Definition 2.9 For any poset P, define

$$L(P) = \{A \subseteq P \mid A = [a]_X,\ X \subseteq P \text{ bounded},\ a \in \overline{X}\}.$$

Define a partial ordering $\sqsubseteq$ on $L(P)$ by superset containment. That is, say that $A \sqsubseteq B$ if $A \supseteq B$. ▮

The ordering on $L(P)$ is really just the same as the ordering on the upper or Smyth power domain. Note that every element of $L(P)$ is an upwards-closed subset of P. With this ordering on $L(P)$ we are in a position to see that our operator L does as it is supposed to do and turns a poset into a local lattice.

Theorem 2.10 *For any poset P, the poset $L(P)$ is a local lattice.* ▮

If $L(P)$ is to be thought of as the completion of P with respect to being a local lattice, we need to be able to view P as living in $L(P)$. That is, we need an injective function $\iota_P : P \to L(P)$ such that $\iota_P(p) \leq \iota_P(q)$ iff $p \leq q$.

Definition 2.11 For any poset P define $\iota_P : P \to L(P)$ by $\iota_P(p) = [p]_{\{p\}} = \uparrow p$, the principal filter generated by p. ▮

Theorem 2.12 *The function $\iota_P : P \to L(P)$ is an injective function which preserves and reflects order, (that is, $\iota_P(p) \leq \iota_P(q) \Leftrightarrow p \leq q$), and preserves all existing meets and joins.* ▮

If $L(P)$ is to be thought of as a completion of P with respect to being a local lattice, in addition to having ι_P, we need to know that if P is already a local lattice, then L doesn't change it (up to isomorphism).

Theorem 2.13 *For any poset P we have $P \cong L(P)$ iff P is a local lattice.* ▮

It might be nice if we had that L were a minimal completion operator. That is, we might like to have that if P is a poset, M is local lattice, and $j : P \to M$ is an injection preserving and reflecting order and preserving all existing meets and joins, then there exists a monotonic injection $\eta : L(P) \to M$ such that $j = \eta \circ \iota_P$. (Actually, we would want that η should be an embedding (see the next section)). Unfortunately, this is too much to hope for. However, we do get the following factorization result.

Proposition 2.14 *Given a poset P, a local lattice M, and a monotonic function $f : P \to M$, there exists a monotonic function $\eta : L(P) \to M$ such that $f = \eta \circ \iota_P$. Moreover, if $\nu : L(P) \to M$ is another monotonic function such that $f = \nu \circ \iota_P$, then $\eta(A) \leq \nu(A)$ for all $A \in L(P)$.* ▮

3 Dealing with Functions

So far we have built an operator L which takes a poset and turns it into a local lattice, and if the poset is already a local lattice, it leaves it alone (up to isomorphism). Moreover, we have a way of viewing P as living in $L(P)$ via our monotone injective function ι_P. But what we are after is more, namely that L be (extended to) a functor from an appropriate category of posets to a correspondingly appropriate category of local lattices. For both categories the morphisms we will be interested in are embedding-projection pairs.

Definition 3.1 Given posets P and Q, a monotone function $\epsilon : P \to Q$ is an *embedding* if there exists a monotone function $\rho : Q \to P$ such that $\rho \circ \epsilon = \mathrm{id}_P$ and $\epsilon \circ \rho(q) \leq q$ for all $q \in Q$. The function ρ is called a *projection* from Q to P. ∎

It should be noted that an embedding is an injection and a projection is a surjection. It is a fairly straight-forward and well-known fact (see Proposition 0.3.2 in [GHK*80], for example) that an embedding uniquely determines its corresponding projection, and likewise a projection uniquely determines its corresponding embedding. Therefore, instead of referring to the category of posets with embedding-projection pairs, we could equally well refer to the category of posets with embeddings or to the category of posets with projections.

In order to extend L to a functor from the category of posets with embedding-projection pairs ($\mathbf{PO}^{ep}$) to the category of local lattices with embedding-projection pairs ($\mathbf{LL}^{ep}$), we first need to extend L to act on embedding-projection pairs.

For sake of convenience, if $f : X \to Y$, let $f^* : 2^X \to 2^Y$ denote the function $f^*(U) = \{f(u) \mid u \in U\}$.

Definition 3.2 Let P and Q be posets, and suppose that $(\epsilon : P \to Q, \rho : Q \to P)$ is an embedding-projection pair. Define $(L(\epsilon) : L(P) \to L(Q), L(\rho) : L(Q) \to L(P))$ by

$$L(\epsilon)(A) = [\epsilon(a)]_{\epsilon^*(\underline{A})} \qquad \text{some } a \in A$$

and

$$L(\rho)(B) = [\rho(b)]_{\rho^*(\underline{B})} \qquad \text{some } b \in B.$$ ∎

That $L(\epsilon)$ and $L(\rho)$ are well-defined functions, *i.e.*, that their definition is independent of the choices of a and b, is given to us by Lemma 2.7. The next thing we need to know is that if (ϵ, ρ) is an embedding-projection pair, then so is $(L(\epsilon), L(\rho))$.

Lemma 3.3 *If P and Q are posets and $(\epsilon : P \to Q, \rho : Q \to P)$ is an embedding-projection pair, then $(L(\epsilon), L(\rho))$ is also an embedding-projection pair.* ∎

Just showing that L takes embedding-projection pairs to embedding-projection pairs is not all that we must do. It is important to verify that L also preserves the identity morphisms and that it preserves composition.

Theorem 3.4 *Let* $\mathbf{PO}^{ep}$ *be the category of posets and embeddings-projection pairs (where the arrow points in the direction of the embedding). Let* $\mathbf{LL}^{ep}$ *be the category of local lattices with embeddings-projection pairs. Then* $L : \mathbf{PO}^{ep} \to \mathbf{LL}^{ep}$ *is a functor.* ∎

We have more from L, namely that it commutes with the action of the ι_P's.

Lemma 3.5 *For every* P *and* $Q \in \mathbf{PO}^{ep}$, *and* $(\epsilon, \rho) \in \mathrm{Hom}_{\mathbf{PO}^{ep}}(P, Q)$ *we have*

$$L(\epsilon) \circ \iota_P = \iota_Q \circ \epsilon \qquad \textit{and} \qquad \iota_P \circ L(\rho) = \rho \circ \iota_Q.$$

That is, the following diagrams commute:

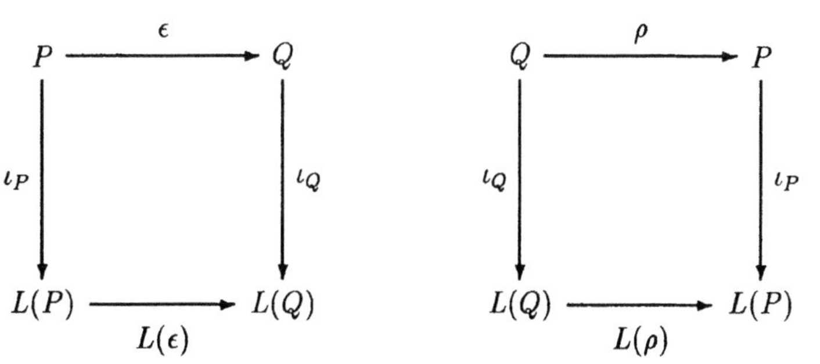

∎

Definition 3.6 Let $\mathbf{C}$ be a category and let $\mathbf{B}$ be a full subcategory. A functor $F : \mathbf{C} \to \mathbf{B}$ is a *pseudo-retraction* if there exists a natural isomorphism from the identity functor on $\mathbf{B}$ to F restricted to $\mathbf{B}$, *i.e.* if there exists a family of isomorphisms $i_B : B \cong F(B)$, one for each object in $\mathbf{B}$, such that for all arrows $f : B_1 \to B_2$ in $\mathbf{B}$, we have $F(f) \circ i_{B_1} = i_{B_2} \circ f$. ∎

Corollary 3.7 *The functor* $L : \mathbf{PO}^{ep} \to \mathbf{LL}^{ep}$ *is a pseudo-retraction.* ∎

4 Restricting to the Bifinites

So far we have been looking at the categories of posets and of local lattices. However, for a class of posets to be used as a mathematical foundation for the denotational semantics of programming languages (in order to sensibly model computation), one usually restricts to the class of directed-complete posets (cpo's). Unfortunately, the functor L which we have built does not take cpo's to cpo's; it does not in general preserve directed-completeness. However, it does preserve finiteness.

Recall that an *ω-bifinite domain* is a directed-complete poset which is an ω-bilimit (directed colimit) of an ω-chain in the category $\mathbf{DCPO}_\perp^{ep}$ of directed-complete posets with continuous embedding-projection pairs of finite posets with least element. (An ω-chain in a category $\mathbf{C}$ is a functor $F : \omega \to \mathbf{C}$ from the ordinal ω viewed as a category.) In [GJ88] it is observed that the subcategories of ω-bifinite domains ($\omega\mathbf{B}^{ep}$) and ω-bifinite L-domains ($\omega\mathbf{BL}^{ep}$) are closed under the formation of bilimits of ω-chains in the category of directed complete posets with continuous embedding-projection pairs. If we restrict our functor L to the finite posets, we may use it to construct a functor $\overline{L}$ from the ω-bifinite posets to the ω-bifinite L-domains which has much the same behavior as L.

Although we only intend to use the next in this paper in the the context of categories of posets with embedding-projection pairs, it makes sense in a much broader setting, and we shall state it in that broader setting.

Definition 4.1 Let $\mathbf{C}$ be a category with colimits of ω-chains and let $\mathbf{B}$ be a non-empty full subcategory of $\mathbf{C}$. We shall say that $\mathbf{B}$ is an *ω-closed* subcategory of $\mathbf{C}$ if $\mathbf{B}$ is closed under the formation of colimits of ω-chains and if every object A of $\mathbf{C}$ is an object of $\mathbf{B}$ whenever there exists an object B of $\mathbf{B}$ and an arrow $f : A \to B$ in $\mathbf{C}$. ▮

The following lemma is certainly a previously-known fact. However, the author could find no direct reference to it.

Lemma 4.2 *Let $\mathbf{B}$ be an ω-closed subcategory of $\omega\mathbf{B}^{ep}$, and let $\mathbf{A}$ be the intersection of $\mathbf{B}$ with the finite posets (i.e. the full subcategory of $\mathbf{B}$ whose objects are exactly the finite posets of $\mathbf{B}$). Let $I : \mathbf{A} \to \mathbf{B}$ be the inclusion functor. Suppose that F is a functor from $\mathbf{A}$ to any ω-complete category $\mathbf{C}$. Then there exists a functor $\widehat{F} : \mathbf{B} \to \mathbf{C}$ which preserves ω-colimits and whose restriction to $\mathbf{A}$, i.e., $\widehat{F} \circ I$, is naturally isomorphic to F. Moreover, any other functor $G : \mathbf{B} \to \mathbf{C}$ which preserves ω-colimits and whose restriction to $\mathbf{A}$ is naturally isomorphic to F is naturally isomorphic to $\widehat{F}$.* ▮

Corollary 4.3 *Let $\mathbf{B}$ be an ω-closed subcategory of $\omega\mathbf{B}^{ep}$, and let $\mathbf{A}$ be the intersection of $\mathbf{B}$ with the finite posets, $\mathbf{PO}^{ep}_{<\omega}$. Let $I : \mathbf{A} \to \mathbf{B}$ be the inclusion functor of $\mathbf{A}$ into $\mathbf{B}$, and let $J : \mathbf{PO}^{ep}_{<\omega} \to \omega\mathbf{B}^{ep}$ be the inclusion functor of finite posets with embedding-projection pairs into the bifinites. Let $G : \mathbf{PO}^{ep}_{<\omega} \to \mathbf{A}$ be a pseudo-retraction functor. Then there exists a pseudo-retraction $\overline{G} : \omega\mathbf{B}^{ep} \to \mathbf{B}$ which preserves ω-bilimits and whose restriction to the finite posets, $\overline{G} \circ J$, is naturally isomorphic to $I \circ G$. Pictorially, we have that the following diagram commutes up to natural isomorphism.*

$$\begin{array}{ccc} \mathbf{PO}^{ep}_{<\omega} & \xrightarrow{\quad G \quad} & \mathbf{A} \\ {\scriptstyle J}\big\downarrow & & \big\downarrow{\scriptstyle I} \\ \omega\mathbf{B}^{ep} & \xrightarrow[\quad \overline{G} \quad]{} & \mathbf{B} \end{array}$$

▮

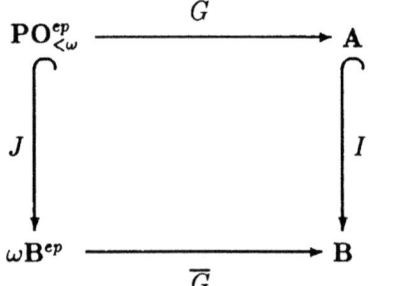

In [GJ88], it is also shown that the category of ω-bifinite L-domains has a universal domain.

Definition 4.4 Let **C** be a category. An object $\mathcal{U}$ is *universal* in **C** if for every object A of **C**, there is a (not necessarily unique) arrow $f : A \to \mathcal{U}$. ∎

Notice that it is immediate from the definitions, that if $\mathcal{U}$ is universal in a category **C** and $F : \mathbf{C} \to \mathbf{B}$ is a pseudo-retraction, then $F(\mathcal{U})$ is universal in **B**. That a universal domain $\mathcal{U}$ exists for the ω-bifinites was shown by Gunter in [Gun87]. Therefore, we have

Corollary 4.5 *Let $\mathcal{U}$ be a universal domain for the ω-bifinites, and let $\overline{L}$ be a continuous extension of the functor L restricted to the finite posets. Then $\overline{L}(\mathcal{U})$ is a universal domain in the category of ω-bifinite L-domains.* ∎

5 Functors and Universal Domains

As we have already seen, pseudo-retraction functors provide us with an easy means to conclude that a subcategory of ω-bifinites has a universal domain. In [GJ88] the notion of a specific kind of universal domain, that of a *fully-saturated* universal domain, was introduced. As was shown in that paper, fully-saturated universal domains have the advantage that, when they exist, they are unique up to isomorphism. In the remainder of this paper, we shall show that pseudo-retraction functors also provide us with a means to conclude that a subcategory of ω-bifinites has a fully-saturated universal domain.

We begin with some preliminary definitions and facts.

Recall that an object B is said to be finite in a category **C** provided that for every functor $F : \mathbf{I} \to \mathbf{C}$ with colimit (k_I, K), if there exists a arrow $f : B \to K$, then there exists an object I of **I** and an arrow $h : B \to F(I)$ such that $f = k_I \circ h$. This is the categorical equivalent of an element being finite (or compact) in a poset. In the various categories in which we are interested, the finite objects are just those of finite cardinality.

Definition 5.1 (Gunter and Jung) An arrow $f : A \to B$ in a category **C** is an *increment* if for every pair of arrows h, g in **C** with $f = h \circ g$ either h or g is an isomorphism. A category **C** is *incremental* if

1. **C** has an initial object,

2. **C** has colimits of ω-chains,

3. every object A of **C** is a colimit of an ω-chain (A_i, a_{ji}) where A_0 is initial, each A_i is finite (in the category **C**), and each arrow $a_{i,i+1} : A_i \to A_{i+1}$ is an increment. ∎

Lemma 5.2 (Gunter and Jung) *Let* $\mathbf{C}$ *be an incremental category and let* $\mathbf{B}$ *be an* ω*-closed full subcategory of* $\mathbf{C}$. *Then* $\mathbf{B}$ *is an incremental category.* ▮

In particular, we have that $\omega\mathbf{BL}^{ep}$ is incremental, since $\omega\mathbf{B}^{ep}$ is.

Definition 5.3 (Gunter and Jung) Let $\mathbf{C}$ be an incremental category and let A be an object in $\mathbf{C}$. An object A^+ together with an arrow $s : A \to A^+$ is a *saturation* of A in $\mathbf{C}$ if, for every increment $f : B \to B'$ and arrow $g : B \to A$, there exists an arrow $h : B' \to A^+$ such that $h \circ f = s \circ g$. That is, there exists an h which makes the following diagram commute.

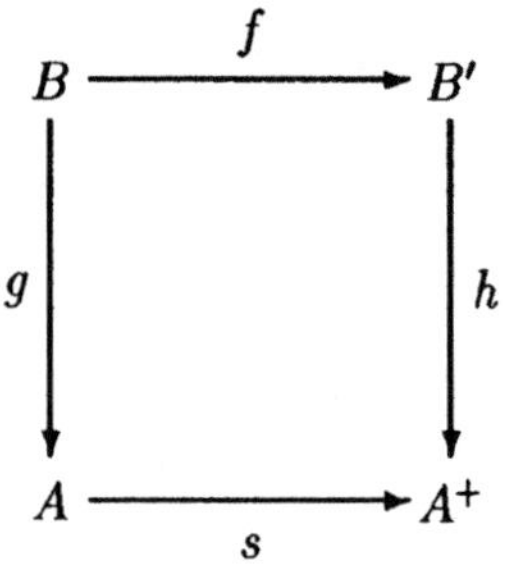

The category $\mathbf{C}$ is said to have *finite saturations* if for every finite object A in $\mathbf{C}$ there exists a saturation $s : A \to A^+$ such that A^+ is finite. ▮

The following fact about saturations follows immediately from the definition.

Lemma 5.4 *Let* A, A^+ *be objects in an incremental category* $\mathbf{C}$, *and let* $f : A \to A^+$ *be a saturation of* A. *Then, for every object* C *in* $\mathbf{C}$, *if there exists an arrow* $g : A^+ \to D$, *then* $g \circ f : A \to D$ *is another saturation of* A *and if there exists an arrow* $h : C \to A$, *then* $f \circ h : C \to A^+$ *is a saturation of* C. ▮

In general, the image of a finite saturation under a functor will not be a finite saturation. Even if the functor is a pseudo-retraction, the image of a finite saturation will not generally be a finite saturation. However, as the next lemma shows, a pseudo-retraction will carry some finite saturations to finite saturations.

Lemma 5.5 *Let* $\mathbf{C}$ *be an incremental category and let* $\mathbf{B}$ *be an* ω*-closed full subcategory of* $\mathbf{C}$. *Let* $F : \mathbf{C} \to \mathbf{B}$ *be a pseudo-retraction of* $\mathbf{C}$ *to* $\mathbf{B}$. *If* A *is an object in* $\mathbf{B}$ *and* $s : A \to A^+$ *is a saturation of* A *in* $\mathbf{C}$, *then* $F(s) : F(A) \to F(A^+)$ *is a saturation of* $F(A)$ *in* $\mathbf{B}$. ▮

Definition 5.6 (Gunter and Jung) Let $\mathbf{B}$ be an ω-closed full sub-category of $\omega\mathbf{B}^{ep}$. An object U of $\mathbf{B}$ is *fully saturated* in $\mathbf{B}$ if for every pair of objects M, N and arrows $f : M \to U$ and $g : M \to N$ in $\mathbf{B}$ there exists an arrow $h : N \to U$ such that $f = h \circ g$. Thus, we have the following commutative diagram:

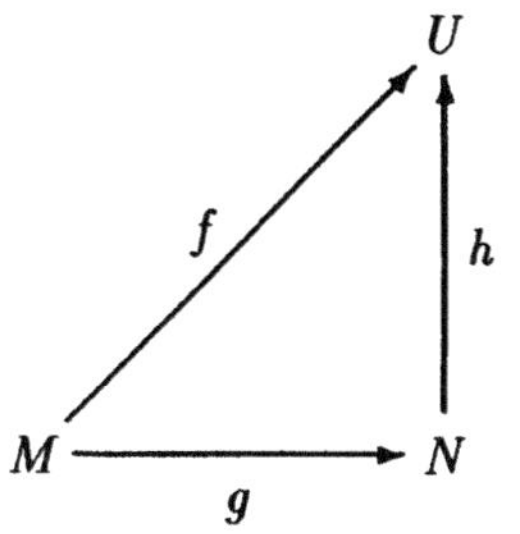

∎

From the definition we can see that a fully saturated object is a universal domain. More then that, it was shown in [GJ88] that any fully saturated object is unique up to isomorphism. Also in that paper a means for constructing a fully saturated object was given.

Theorem 5.7 (Gunter and Jung) *Suppose that* $\mathbf{B}$ *is a closed full sub-category of* $\omega\mathbf{B}^{ep}$*. Let* (S_i, s_{ij}) *be an* ω*-chain in* $\mathbf{B}$ *where* S_0 *is initial, each* S_i *is finite, and each* $s_{i+1,i} : S_i \rightarrow S_{i+1}$ *is a saturation in* $\mathbf{B}$*, and let* $\mathcal{U}$ *be the bilimit of this* ω*-chain. Then* $\mathcal{U}$ *is fully saturated in* $\mathbf{B}$*.* ∎

Now, a pseudo-retraction functor does not preserve arbitrary finite saturations , so just taking the image of an ω-chain of finite saturations and then taking its bilimit need not get us a fully saturated domain. (In fact, in the case of $\overline{L}$ it will not.) However, a pseudo-retraction does give us a way of constructing an ω-chain of finite saturations out of the image of an ω-chain of finite saturations.

Lemma 5.8 *Let* (S_i, s_{ji}) *be an* ω*-chain in* $\omega\mathbf{B}^{ep}$ *where* S_0 *is initial, each* S_i *is finite, and each* $s_{i,i+1} : S_i \rightarrow S_{i+1}$ *is a saturation. Let* $\mathbf{B}$ *be an* ω*-closed full subcategory of* $\omega\mathbf{B}^{ep}$*, and suppose there exists a pseudo-retraction* $F : \omega\mathbf{B}^{ep} \rightarrow \mathbf{B}$ *which takes finite objects of* $\omega\mathbf{B}^{ep}$ *to finite objects of* $\mathbf{B}$*. Then there exists a subchain* (S_{n_i}, s_{n_j,n_i}) *(with* $n_i > n_j$ *for all* $i > j$*) such that* $F(S_{n_0})$ *is initial in* $\mathbf{B}$*, and a sequence of embeddings* $e_i : F(S_{n_i}) \rightarrow S_{n_{i+1}}$ *such that* $e_i \circ F(s_{n_{i-1},n_i}) = s_{n_i,n_{i+1}} \circ e_{i-1}$ *and such that* $F(e_i) : F^2(S_{n_i}) \rightarrow F(S_{n_{i+1}})$ *is a saturation in* $\mathbf{B}$*. Pictorially, we have*

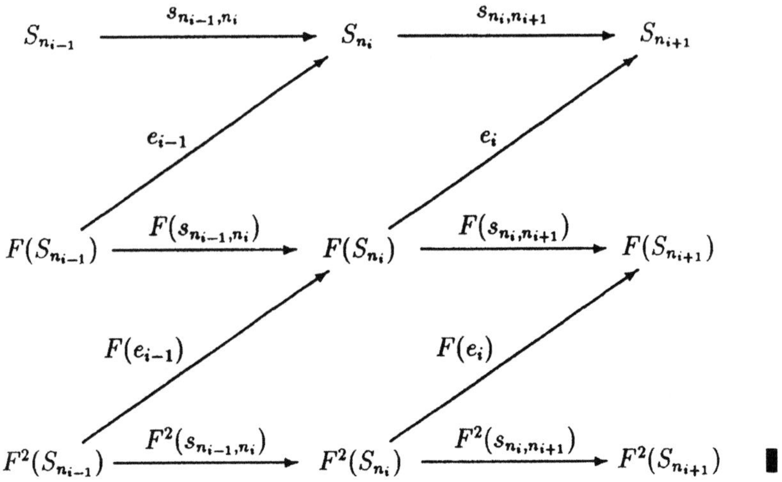

Theorem 5.9 *Let* $\mathbf{B}$ *be an* ω*-closed full subcategory of* $\omega\mathbf{B}^{ep}$*, and suppose there exists a pseudo-retraction* $F : \omega\mathbf{B}^{ep} \to \mathbf{B}$ *which takes finite objects of* $\omega\mathbf{B}^{ep}$ *to finite objects of* $\mathbf{B}$*. Then there exists a fully saturated object in* $\mathbf{B}$*.* ▮

Corollary 5.10 *There exists a fully saturated object for the category of* ω*-bifinite L-domains.* ▮

References

[Coq88] Thierry Coquand. Categories of embeddings. In Y. Gurevich, editor, *Logic in Computer Science*, pages 256–263, IEEE Computer Society, July 1988.

[GHK*80] G. Gierz, K. H. Hofmann, K. Keimel, J. D. Lawson, M. Mislove, and D. S. Scott. *A Compendium of Continuous Lattices.* Springer-Verlag, 1980.

[GJ88] Carl A. Gunter and Achim Jung. Coherence and consistency in domains (extended outline). In Y. Gurevich, editor, *Logic in Computer Science*, pages 309–319, IEEE Computer Society, July 1988.

[Gun87] Carl A. Gunter. Universal profinite domains. *Information and Computation*, 72:1–30, 1987.

[Joh82] Peter T. Johnstone. *Stone Spaces.* Cambrigde University Press, 1982.

[Jun88a] Achim Jung. *Cartesian Closed Categories of Algebraic CPO's.* Technical Report 1110, Technische Hochschule Darmstadt, Fachbereich Mathematik, 1988.

[Jun88b] Achim Jung. *Cartesian Closed Categories of Domains.* PhD thesis, Darmstadt Technische Hochschule, 1988.

L-domains and Lossless Powerdomains

Radhakrishnan Jagadeesan *
Computer Science Department,
Cornell University

Abstract

The category of L-domains was discovered by A. Jung while solving the problem of finding maximal cartesian closed categories of algebraic CPO's and continuous functions. In this note we analyse properties of the lossless powerdomain construction, that is closed on the algebraic L-domains. The powerdomain is shown to be isomorphic to a collection of subsets of the domain on which the construction was done. The proof motivates a certain finiteness condition on the inconsistency relations of elements. It is shown that all algebraic CPO's D whose basis $B(D)$ has property M satisfy the condition. In particular, the coherent L- domains satisfy the condition.

1 Introduction

Recent work by A. Jung and C. Gunter shows that the L-domains discovered independently by A. Jung[6] and T. Coquand[1] form an interesting cartesian closed category. P. Buneman proposed the lossless powerdomain construction that is closed on L-domains. Buneman's construction was based on intuitions from databases. We were studying the construction as a possible candidate for modelling oracleisable indeterminacy. The present paper investigates some purely mathematical questions that arose from the study.

A representation theorem for the lossless powerdomain is proved for the lossless powerdomain in Section 3. The proof suggests some additional natural conditions on L-domains. The proof uses the notion of finite separability, the ability to separate elements of the sets that constitute the powerdomain by disjoint basic Scott open sets (this is made precise in Section 3). This suggests that a natural condition to consider is that the inconsistency relation have finite witnesses. Section 4 discusses this notion of finite inconsistency. Discussions with C. Gunter and E. Gunter revealed the relationship of this property to the coherent L-domains[4].

*Research supported in part by NSF grant CCR 8818979

2 Preliminaries

This section outlines the basic definitions and facts that are used in the note.

A subset X of a partially ordered set is directed iff it is non-empty and every pair of elements in X has an upper bound in X. A partial order D is said to be (directed) complete if every directed subset X of D has a least upper bound in D. We shall only consider CPO's with a least element, which will be denoted $\perp$. An element d of D is said to be compact if for every directed set D such that $d \sqsubseteq \bigsqcup D$, there is an element $x \in D$ such that $d \sqsubseteq x$. The set of all elements of D greater than a compact element d is denoted by $d \uparrow$. The set of all elements of D less than an element x is denoted by $x \downarrow$. A CPO D is said to be algebraic if every element is the lub of a directed set of compact elements. The set of compact elements of an algebraic CPO D is denoted by $B(D)$. Step functions of the form $d \searrow e$ where d and e are compact elements of D and E respectively are compact elements of the function space $D \rightarrow E$. The set of minimal upper bounds of a subset A of D that are below elements of a subset B of D is denoted by $mub_A\ (B)$.

An algebraic CPO D is said to be an algebraic L-domain when any of the following equivalent conditions hold [6].

1. For each $x \in D$, the set $x \downarrow \bigcap B(D)$ is a $\vee$-semilattice with smallest element.

2. For each upper bound x of a finite subset A of $B(D)$, there is a unique minimal upper bound of A below x.

3. For any finite subset A of $B(D)$, $mub_A\ (D) = mub_{(mub_A\ (D))}\ (D)$

3 The Lossless Powerdomain

Peter Buneman discovered the lossless powerdomain construction [3] that is closed on L-domains.

Definition 1 *Let D be an L-domain. Define the preorder $P_L(D)$as follows:*

$$|P_L(D)| = \{(e_1 \ldots e_n) \mid (\forall 1 \le i \le n)\ [e_i \in B(D)] \ \wedge\ (\forall 1 \le i,j \le n)\ [i \ne j \Longrightarrow mub_{(e_i,\, e_j)}\ (D) = \phi]\}$$

and the elements of $|P_L(D)|$ are ordered by the Egli-Milner ordering, $\sqsubseteq_{EM}$.

Actually $P_L(D)$ is a partial order. The lossless powerdomain $\overline{P_L(D)}$ is constructed by ideal-completion of $P_L(D)$.

Lemma 1 *$\overline{P_L(D)}$is an L-domain if D is.*

Proof: It suffices to prove that $(\forall c \in P_L(D))\ [c \downarrow \bigcap P_L(D)]$ is an $\vee$- semilattice under the partial order $\sqsubseteq_{EM}$. Let

$$a = (e_1 \ldots e_n),\ b = (d_1 \ldots d_m),\ c = (f_1 \ldots f_p),\ a, b \sqsubseteq_{EM} c$$

Then, $\bigsqcup\{a,b\}$ under c is given by,

$$\bigsqcup\{a,b\} = \bigcup\{mub_{(e_i,d_j)}(c) \mid 1 \le i \le n, 1 \le j \le m,\ e_i \uparrow \bigcap d_j \uparrow \bigcap c \ne \phi\}$$

The verification that the above definition is correct is quite easy. ∎

It is shown that the ideals in the lossless powerdomain are representatives of their fringe sets. Fringe sets are the sets generated by following the partial order arrows among the elements of the ideal. The following definition captures the idea of "following arrows".

Definition 2 *A GENERATOR d over a ideal $I \in \overline{P_L(D)}$ is a function $d:\ I \to D$ such that :*

- $i \in I \Longrightarrow d\,(i) \in i$
- $(\,i \in I \wedge\ j \in I\ \wedge i \sqsubseteq_{EM} j) \Longrightarrow d\,(i) \sqsubseteq d\,(j)$

The following lemma is an easy consequence of the definition.

Lemma 2 *If d is a GENERATOR over I, $d\,(I)$ is a directed set in D.*

Definition 3 *Let $I \in \overline{P_L(D)}$. The set generated by I is*

$$S_I = \{\ \bigsqcup d\,(I) \mid\ d \textit{ is a Generator on } I\}$$

Since the lossless powerdomain embodies finite branching only, one expects the sets generated to be Scott-compact. The proof requires the following lemma, that is a consequence of Rudin's lemma [2].

Lemma 3 *Let $I \in \overline{P_L(D)}$. Let $\tilde{I}$ be cofinal in I. Let P be a predicate defined on D such that*

1. $(\forall i \in\ \tilde{I})\ (\exists e \in\ i)\ [P(e)]$
2. $(d \in i \in \tilde{I}\ \wedge e \in j \in \tilde{I}\ \wedge i \sqsubseteq_{EM} j\ \wedge d \sqsubseteq e\ \wedge P(e)) \Longrightarrow P(d)$

Then there is a generator d over I such that $(\forall i \in \tilde{I})\ [P\ (d\ (i))]$.

Lemma 4 *If $I \in \overline{P_L(D)}$, S_I is non-empty and Scott-compact.*

Proof:

1. Define a predicate P on D by

$$P(e) \Longleftrightarrow\ (\exists i \in I)\ [e \in i]$$

¿From lemma 3, we get a generator on D. Hence S_I is non-empty.

2. Let $\{x_\alpha\}$ be a net in S_I. Define a predicate P on D as follows:

$$P(e) \Longleftrightarrow (\exists i \in I)\ [e \in i\ \wedge\ e \uparrow \bigcap \{x_\alpha\} \text{ is cofinal}\ ]$$

From lemma 3 , we get a generator d. Consider $\bigsqcup d\ (I)$. Any neighbourhood $e\uparrow$ of $\bigsqcup d\ (I)$ has non-empty intersection with $d\ (I)$, and consequently is cofinal in $\{x_\alpha\}$. Hence $\bigsqcup d\ (I)$ is an accumulation point of $\{x_\alpha\}$ ∎

However, not all non-empty Scott-compact sets are generated by some ideal in the lossless powerdomain. Consider the L-domain in Fig 1. The set $\{x, y\}$ cannot be generated by any ideal in $\overline{P_L(D)}$even though the set $\{x, y\}$ is finite and hence compact in the Scott Topology. This observation motivates the following definition.

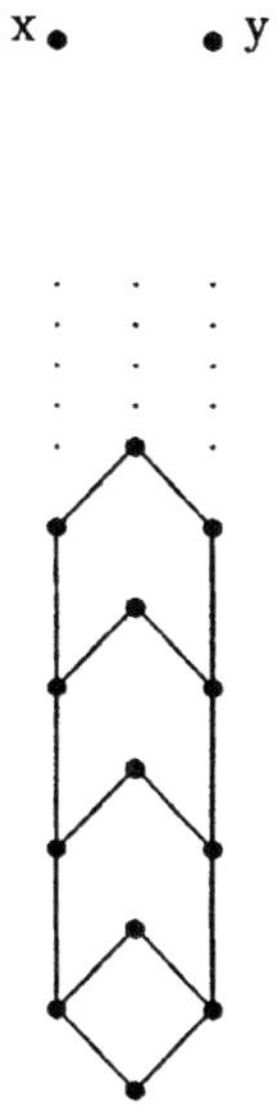

Fig. 1

Definition 4 $S \subseteq D$ *is* FINITELY SEPARABLE *if*

$$(\forall S_{fin} = \{x_1 \ldots x_n\},\ S_{fin} \subseteq S)\ (\exists\,(e_1 \ldots e_m) \in P_L(D))\ [n \leq m\ \wedge\ (\forall 1 \leq i \leq n)(e_i \sqsubseteq x_i)\ \wedge\ (e_1 \ldots e_m) \sqsubseteq_{EM} S]$$

The following lemma is an easy consequence of the above definition.

Lemma 5 *If* $I \in \overline{P_L(D)}$, S_I *is* FINITELY SEPARABLE.

It turns out that one can generate all *FINITELY SEPARABLE*, Scott-compact sets.

Definition 5 *Let D be an L-domain. Then Ψ_D is the set of all Scott-compact finitely-separable subsets of D ordered by $\sqsubseteq_{EM}$*

Note that Ψ_D is a partial order. Now, we define maps between $\overline{P_L(D)}$and Ψ_D in a natural manner, and show that the definitions do indeed constitute an order isomorphism between the partial orders, giving us the required representation theorem.

Lemma 6 *Define* $\Phi : \overline{P_L(D)} \rightarrow \Psi_D$ *by*

$$\Phi\,(I) = S_I$$

Then, Φ *is monotone.*

Proof: Let $I_1 \in \overline{P_L(D)}, I_2 \in \overline{P_L(D)}, I_1 \subseteq I_2$. Let $\Phi\,(I_1) = S_{I_1}, \Phi\,(I_2) = S_{I_2}$.

- Let $x \in S_{I_1}$
  $\Longrightarrow (\exists d)\,[\bigsqcup d\,(I_1) = x]$, where d is generator on I_1. Define

  $$\tilde{I_2} = \{i |\ i \in I_2\ \wedge (\exists j \in I_1)\,[j \sqsubseteq_{EM} i]$$

  Note that $\tilde{I_2}$ is cofinal in I_2. Define predicate P by

  $$P(e) \Longleftrightarrow (\exists i,\ j)\,[i \in \tilde{I_2}\ \wedge\ e \in i\ \wedge\ j \in I_1\ \wedge\ d\,(j) \sqsubseteq e\ \wedge \tilde{e} \in i\ \wedge i \in \tilde{I_2}]$$

  The generator $\tilde{d}$ on I_2 given by lemma 4 satisfies $x \sqsubseteq \bigsqcup \tilde{d}$
  $\Longrightarrow (\forall x \in S_{I_1})\,(\exists y \in S_{I_2})\,[x \sqsubseteq y]$

- Let $x \in S_{I_2}$
  $\Longrightarrow (\exists \tilde{d})\,[\bigsqcup \tilde{d}\,(I_2) = x]$, where $\tilde{d}$ is a generator on I_2. Restriction of $\tilde{d}$ to I_1 gives a generator d on I_2 such that $d\,(I_1) \subseteq \tilde{d}\,(I_2)$. Hence, we have $\bigsqcup d\,(I_1) \sqsubseteq \bigsqcup \tilde{d}\,(I_2)$. Hence, we have

  $$(\forall y \in S_{I_2})\,(\exists x \in S_{I_1})\,[x \sqsubseteq y] \quad \blacksquare$$

Lemma 7 *Let* $S \in \Psi_D$, $(e_1 \ldots e_n) \sqsubseteq_{EM} S$, $(d_1 \ldots d_m) \sqsubseteq_{EM} S$. *Then*

$$(\forall 1 \le i \le n,\ 1 \le j \le m)\,[mub_{(e_i,\, d_j)}\,(S) \textit{ is finite }]$$

Proof: Let $1 \le i \le n,\ 1 \le j \le m$. Consider the basic open cover of S consisting of

- $\tilde{e} \uparrow,\ \tilde{e} \in mub_{(e_i,\, d_j)}\,(S)$
- $d_{\tilde{j}} \uparrow,\ \tilde{j} \ne j,\ 1 \le \tilde{j} \le m$
- $e_{\tilde{i}} \uparrow,\ \tilde{i} \ne i,\ 1 \le \tilde{i} \le n$

This open cover has a finite subcover. Since all members of $(e_1, \ldots e_n)$ and $(d_1, \ldots d_m)$ are pairwise inconsistent, we deduce that $mub_{(e_i,\, d_j)}\,(S)$ is finite. $\blacksquare$

Lemma 8 *Define* $\tilde{\Phi} : \Psi_D \rightarrow \overline{P_L(D)}$ *by*

$$\tilde{\Phi}\,(S) \;= \{\,(e_1 \ldots e_m)|\,(e_1 \ldots e_m) \sqsubseteq_{EM} S \;\wedge (e_1 \ldots e_m) \in P_L(D)\}$$

Then, $\tilde{\Phi}$ is monotone.

Proof:

- We have to first show that $\tilde{\Phi}$ is well defined. Let $S \in \Psi_D$. Then we have
  - $\{\bot\} \in \tilde{\Phi}\,(S)$. So, $\tilde{\Phi}\,(S)$ is non-empty.
  - $(e_1 \ldots e_n) \sqsubseteq_{EM} S,\ (d_1 \ldots d_m) \sqsubseteq_{EM} S \Longrightarrow$ $\bigcup\{\{mub_{(e_i,\ d_j)}\,(S)\}\ |1 \leq i \leq n,\ 1 \leq j \leq m,\ e_i \uparrow \bigcap d_j \uparrow \bigcap S \neq \phi\} \sqsubseteq_{EM} S$. Hence, $\tilde{\Phi}\,(S)$ is directed.
  - $\tilde{\Phi}\,(S)$ is obviously downward closed.

  Hence, $\tilde{\Phi}\,(S)$ is an element of $\overline{P_L(D)}$.
- $S_1 \sqsubseteq_{EM} S_2 \Longrightarrow \tilde{\Phi}\,(S_1) \subseteq \tilde{\Phi}\,(S_2)$. Hence, $\tilde{\Phi}$ is monotone. ∎

Lemma 9 $\tilde{\Phi} \circ \Phi \;= Id$

Proof: Note that it suffices to prove that

$$(\forall i \in P_L(D))\ [i \in I \Longleftrightarrow i \sqsubseteq_{EM} S_I]$$

The 'if' part is obvious. For the reverse direction, consider

$(e_1 \ldots e_m) \in P_L(D)\ \wedge\ (e_1 \ldots e_m) \sqsubseteq_{EM} S_I$
$\Longrightarrow (\exists x_1 \ldots x_m \in S_I)\ (\forall 1 \leq k \leq m)\ [e_k \sqsubseteq x_k]$

Let $d_1 \ldots d_m$ be the generators on I corresponding to $x_1 \ldots x_m$. Since $d_k\ (I)$ is a directed set for all $1 \leq k \leq m$, we have

$$(\forall 1 \leq k \leq m)\ (\exists \tilde{e_k}, i_k)\ [\tilde{e_k} \in i_k \in I\ \wedge d_k\ (i_k) = \tilde{e_k}\ \wedge e_k \sqsubseteq \tilde{e_k}\ ]$$

Since I is directed, $(\exists i \in I)\ [i_1 \ldots i_m \sqsubseteq_{EM} i]$. Note that $(\forall 1 \leq i \leq m)\ (\exists \tilde{e_i} \in i)\ [e_i \sqsubseteq \tilde{e_i}]$. Define

$$\tilde{I} = \{j|\ j \in I \wedge i \sqsubseteq_{EM} j\}$$

Note that $\tilde{I}$ is cofinal in I.

- We have

$$(\forall j \in \tilde{I})\ (\forall 1 \leq k \leq m)\ (\exists \tilde{e_k} \in j)\ [e_k \sqsubseteq \tilde{e_k}]$$

- We need to prove that $(\exists j \in \tilde{I})\ (\forall \tilde{e} \in j)\ (\exists 1 \leq k \leq m)\ [e_k \sqsubseteq \tilde{e}]$ Suppose not. Define a predicate P by:

$$P(e) \Longleftrightarrow (\exists j \in \tilde{I})\ [e \in j\ \wedge e_1 \not\sqsubseteq e \ldots e_m \not\sqsubseteq e]$$

P satisfies the conditions of lemma 3. The generator yielded by lemma 3 gives an element in S_I that is not greater than any of $e_1 \ldots e_m$. This is a contradiction since $(e_1 \ldots e_m) \sqsubseteq_{EM} S_I$

Hence, we have an element of I above $(e_1 \ldots e_m)$. The result follows by downward closure of I. ∎

Lemma 10 $\Phi \circ \tilde{\Phi} = Id$

Proof: Let $I = \tilde{\Phi}\,(S)$, where $S \in \Psi_D$.

1. Let d be a generator over I. We shall prove by contradiction that $(\exists x \in S)\ [\bigsqcup d\,(I) \sqsubseteq x]$. Suppose not. We have the cover

   $$\{e \uparrow \mid (\exists i \in I)\ [e \in i\ \wedge\ e \neq d\,(i)]\}$$

   Since S is Scott-compact, there is a finite sub-cover

   $$\{e_k \uparrow \mid (\exists i_k \in I)\ [e_k \in i_k\ \wedge\ e \neq d\,(i_k)\ \wedge\ 1 \leq k \leq n]\}$$

   Since I is directed,

   $(\exists i \in I)\ [i_1 \ldots i_k \sqsubseteq_{EM} i]$
   $\Longrightarrow \{e \uparrow \mid e \in (i - \{d\,(i)\})\}$ is a cover of S

   This is a contradiction, since $i \sqsubseteq_{EM} S$ means that there is an element in S greater than $d\,(i)$

2. Let $x \in S$. From definition of $\tilde{\phi}\,(S)$,

   $$(\forall i \in \tilde{\phi}\,(S))\ (\exists e_i \in i)\ [e_i \sqsubseteq x]$$

   Also, the e_i's are unique and form a directed set. Hence, a generator d_x can be defined in the obvious way such that $\bigsqcup d_x\,(I) \sqsubseteq x$.

3. Let $x, y \in S\ \wedge x \neq y$. Then from finite separability of S $d_x \neq d_y$, where d_x, d_y are the generators on I defined as above.

4. Now we shall show that $\bigsqcup d_x\,(I) = x$. Let $e \sqsubseteq x$. Consider the cover

   $$\{b \uparrow \mid b \in i \in I,\ b \neq d_x\,(i)\}\ \cup \{e \uparrow\}$$

   Note that the above is a cover because of finite separability. Since S is Scott-compact, we have a finite sub-cover

   $$\{b_k \uparrow \mid b_k \in i_k \in I,\ b_k \neq d_x\,(i_k),\ 1 \leq k \leq n\}\ \cup \{e \uparrow\}$$

Since I is directed, $(\exists i \in I)\ [i_1 \ldots i_n \sqsubseteq_{EM} i]$. Hence, we note that

$$\{b \uparrow \mid b \in i,\ b \neq d_x\ (i)\} \bigcup \{e \uparrow\}$$

is a cover. Hence, we have

$$(\forall y \in S)\ [d\ (i) \sqsubseteq y \Longrightarrow e \sqsubseteq y]$$

Hence, we deduce that we have the open cover consisting of

- $\tilde{e}$, where $\tilde{e} \in i - \{d\ (i)\}$
- $\tilde{m}$, where $\tilde{m} \in mub_{(d\ (i),\, e)}\ (S)$

The above open cover has a finite sub-cover. Hence we deduce that $mub_{(d\ (i),\, e)}\ (S)$ is finite. Hence $\tilde{i}$ defined as

$$\tilde{i} = mub_{(d\ (i),\, e)}\ (S) \bigcup (i - \{d\ (i)\}$$

satisifies $\tilde{i} \sqsubseteq_{EM} S\ \wedge \tilde{i} \in P_L(D)$. Hence, $e \sqsubseteq \bigsqcup d_x\ (I)$

The above shows that the generators on I generate precisely the elements of S. Hence,

$$\Phi \circ \tilde{\Phi} = Id. \quad \blacksquare$$

Theorem 1 $\overline{P_L(D)}$*is isomorphic to* Ψ_D

4 Finitely Detectable Inconsistency

Consider the elements $x,\ y$ in Figure 1. Every pair of finite elements $e_x,\ e_y$ below $x,\ y$ respectively have upper-bounds. However $x,\ y$ do not have an upper bound. One might demand that the inconsistency relation have finite witnesses to make it continuous. The above discussion motivates the following definition.

Definition 6 *An algebraic CPO D is said to have property FI (for finitely detectable inconsistency) if*
$(\forall x,\ y \in D)\ [mub_{(x,\, y)}\ (D) = \phi \Longrightarrow (\exists e_x,\ e_y\ compact)$
$[e_x \sqsubseteq x\ \wedge e_y \sqsubseteq y\ \wedge\ mub_{(e_x,\, e_y)}\ (D) = \phi;]$

It is easy to check that all Scott domains have the above property. C. Gunter observed that all coherent L-domains [4] have property FI. The following lemma due to A. Jung [5] enables us to prove a stronger result.

Lemma 11 *An algebraic CPO D is Lawson-compact if and only if B (D) has property M.*

Lemma 12 *Let D be an algebraic CPO such that every finite subset of $B(D)$ has a complete finite set of minimal upper bounds. (i.e $B(D)$ has Property M). Then, D has property FI.*

Proof: Let x, $y \in D$. Let $\{d_j | \ j \in I_x\}$ and $\{e_i | \ i \in I_y\}$ be the compact elements approximating x, y respectively, where I_x and I_y are index sets. Furthermore, let us assume that

$$(\forall d_j, e_i)\ [j \in I_x \ \wedge \ i \in I_y \Longrightarrow mub_{(d_j, e_i)}(D) \neq \phi]$$

Let $C = \{\{e_i \uparrow\} | \ i \in I_y \} \ \cup \ \{\{d_j \uparrow\} | \ j \in I_x\}$. Then C is a collection of closed sets in the Lawson topology on D satisfying the finite intersection condition. Result follows from the compactness of the Lawson Topology on D. ∎

In particular all SFP objects D have property FI.

Acknowledgements

I would like to thank Prof. Panangaden for invaluable guidance. Indeed, most of the ideas in the representation theorem arose during discussions with him. The original conjecture that all coherent L-domains satsify property FI is due to Prof. Gunter. The section on finitely detectable inconsistency is based on discussions with Prof. Gunter and Prof. Elsa Gunter during their visit to Cornell. I would like to thank Prof. A. Jung for pointing out an error in the original version of this note and for motivating the general topological structure of the proof of Lemma 12.

References

[1] T. Coquand. Categories of embeddings. *Logic In Computer Science*, 1988.

[2] G. Gierz, J.D. Lawson, and A. Stralka. Quasicontinuous posets. *Houston Journal of Mathematics 9*, 1983.

[3] C. A. Gunter. Private Communication, 1988.

[4] C. A. Gunter and A. Jung. Coherence and consistency in domains. In *Logic in Computer Science*, 1988.

[5] A. Jung. Private Communication, 1988.

[6] A. Jung. Cartesian closed categories of algebraic cpo's. Technical report, Technische Hochshule Darmstadt, Fachbereich Mathematik, 1988.

Does "N+1 Times" Prove More Programs Correct Than "N Times"?

Ana Pasztor *
Florida International University
School of Computer Science, University Park
Miami, FL 33199

Abstract

Slightly modifying Burstall-Manna-Waldinger's Intermittent Assertions program verification method (also called SOMETIME method and denoted here by *Bur*), Sain[21] obtained the so called SOME OTHER TIME method and showed that it proves strictly more programs correct than the SOMETIME method and, on the other hand, proves less than or equally many programs correct as the method obtained from *Bur* by imposing the axioms of successor on time.

The present paper answers a question raised by R. H. Thomason in 1984 by proving that between the SOME OTHER TIME method and the SOMETIME method with the successor axioms there is an **infinite** hierarchy of program verification methods called N TIMES systems, ($N \geq 1$).

1 Sometime Logics of Programs. The Problem

The framework of our discussion will be first order temporal logic. Syntactically, the formulas of first order temporal logic differ from the classical first order formulas in that they contain two types of variables, the so called gobal, or timeless, and the so called local, or time-dependent variables. They also contain so called modalities, like, in our case, *Fst* ("first"), *Nxt* ("next"), and *Smt* ("Sometime"). For example, the sentence "the temperature changes" is expressed in temporal logic as the formula $\exists x(x = \ tempr \wedge Smt(x \neq tempr))$, or, the sentence "the temperature is increasing" is expressed as the formula $\exists x(x = tempr \wedge Nxt(x < tempr))$, x being a timeless, and *tempr* a time-dependent variable.

*Research supported by NSF Grant CCR-8807155

Time-dependent or local variables (like *tempr*) behave in formulas as *constants*, yet they change in time. At each time point however, they are *constant.*

First we define the language L_d^B of the so called SOMETIME logic of type d. Throughout the rest of the paper d will denote an arbitrary but fixed signature (i.e. similarity type). F_d denotes the set of all first order formulas of type d (without modalities!). The set F_d^B of all formulas of L_d^B consists of all usual first order formulas of type d of F_d (whose variables are the global or timeless variables), expanded with new individual constant symbols y_i $(i \in \omega)$. In addition, if, say, ϕ and ψ are in F_d^B, then $Fst\phi$, $Nxt\phi$, $Smt\phi$, $\neg\phi$, $\phi \wedge \psi$ and $\exists x\phi$ (x being a global variable) are also in F_d^B.

The *models* of the SOMETIME logic of type d are the usual Kripke models satisfying the modal induction scheme $Ind(B)$ defined as follows:

$$Ind(B) \equiv \left\{[Fst\varphi \wedge Alw(\varphi \rightarrow Nxt\varphi)] \rightarrow Alw\varphi : \varphi \in F_d^B\right\}.$$

We will view here the Kripke models $\mathcal{M}$ for F_d^B as 2-sorted models and write them in the following form: $\mathcal{M} = \langle \underset{\sim}{T}, \underset{\sim}{D}, y_i\rangle_{i\in\omega}$, where $\underset{\sim}{T} = \langle T, 0, suc\rangle$ is the time structure with T the set of time instances of $\mathcal{M}$, $0 \in T$ and $suc : T \rightarrow T$; $\underset{\sim}{D}$ is a model of type d whose underlying set is D, and $y_i : T \rightarrow D$ for every $i \in \omega$. (Sometimes we will make no difference in notation between a model or a structure $\underset{\sim}{D}$ and its underlying set, denoting both by D.) At a time instance $t \in T$, our "world" or "state" is the classical model $\underset{\sim}{A}_t = \langle \underset{\sim}{D}, y_i(t)\rangle_{i\in\omega}$, where $y_i(t) \in D$ is the interpretation of the constant symbol y_i of F_d^B. So, $\underset{\sim}{A}_{t_1}$ and $\underset{\sim}{A}_{t_2}$ can differ only in the interpretations of the constant symbols y_i. (The standard notation for our Kripke model $\mathcal{M}$ would be $\langle T, E_0, E_1, E_2; A_t\rangle_{t\in T}$, where E_i $(i < 3)$ are the accessibility relations corresponding to our modalities.)

We define the validity relation $\models$ in the usual way through the forcing relation $\Vdash$. Let $\mathcal{M}$ be a Kripke model like defined above. If $\varphi \in F_d^B$ contains no modalities then $t \Vdash \varphi$, if $\underset{\sim}{A}_t \models \varphi$; $t \Vdash Fst\varphi$ if $0 \Vdash \varphi$; $t \Vdash Nxt\varphi$ if $suc(t) \Vdash \varphi$ and $t \Vdash Smt\varphi$ if there is a $t' \in T$ such that $t' \Vdash \varphi$ (note that t and t' are unrelated). Then $\mathcal{M} \models \varphi$ if for all $t \in T$ $t \Vdash \varphi$.

This completes the definition of the language L_d^B of the temporal SOMETIME logic of type d. Let us recall that there is a decidable inference system $\overset{dB}{\vdash}$ for the language L_d^B, such that for every $Th \cup \{\phi\} \subseteq F_d^B$, $Th \models \phi$ iff $Th \overset{dB}{\vdash} \phi$ (see e.g. Prior[16], but also Clifford[5] and Segerberg[22]).

Next we introduce our programming language and various temporal logics of *programs* to prove programs correct.

P_d denotes the set of all deterministic goto programs (i.e. flow charts) of type d using only variables x_i $(i \in \omega)$. If $p \in P_d$, φ, $\psi \in F_d$ (i.e. φ and ψ are *classical* first order formulas of type d—they contain no modalities!), then $\varphi \to \Box(p, \psi)$ denotes the *partial correctness assertion* (*pca* for short) concerning the program p with respect to input condition φ and output condition ψ. Informally it states that whenever the input stisfies φ and the program p is in a halting state, the output satisfies ψ. $\Box(p, \psi)$ abbreviates $true \to \Box(p, \psi)$. Since every *pca* $\varphi \to \Box(p, \psi)$ can be written in the form $\Box(p', \psi')$, we will henceforth only consider *pca*'s of the form $\Box(p, \psi)$. For a detailed definition of P_d the reader is referred to e.g. [3].

Next we recall from Manna-Pnueli[9] the axiom system $Ax(p)$ associated to a given $p \in P_d$. Let $\bar{x} = \langle x_0, \ldots, x_c \rangle$ contain all the variables occurring in p (including the control variable to range over command labels, such that an evaluation of $\bar{x}$ in model $\underset{\sim}{D}$ provides a *complete* description of some state of p). $Ax(p)$ states that $\bar{y} = \langle y_0, \ldots, y_c \rangle$ codes an execution sequence of p. In other words, if $\mathcal{M} = \langle \underset{\sim}{T}, \underset{\sim}{D}, y_i \rangle_{i \in \omega}$ and $\bar{y}(t) = \langle y_0(t), \ldots, y_c(t) \rangle$ for every $t \in T$, then $Ax(p)$ states that $\langle \bar{y}(t) : t \in T \rangle$ is an execution sequence (of states) of p in the *usual* sense, i.e. $t_2 = suc(t_1)$ implies that $\bar{y}(t_1)$ and $\bar{y}(t_2)$ are in the *state transition relationship* $Sttr$ associated to p in $\underset{\sim}{D}$. Formally this is formulated as $\wedge_{i \le c}(y_i(t_1)\ Sttr\ y_i(t_2))$ or $\bar{y}(t_1)\ Sttr\ \bar{y}(t_2)$, which is a classical first order formula of type d. For example, if l_m is the label of the assignment statement $x_i := \tau$ (labels are constant terms of type d), then $t \Vdash (\bar{x}\ Sttr\ \bar{y})$ iff $t \Vdash (y_c = l_{m+1} \wedge y_i = \tau \wedge \wedge_{j<c, j \ne i}(y_j = x_j))$. Now

$$Ax(p) \equiv \forall \bar{x}[\wedge_{m \le N}[(y_c = \ell_m \wedge \bar{x} = \bar{y}) \longrightarrow Nxt(\bar{x}\ Sttr\ \bar{y})] \wedge Fst\ y_c = \ell_0 \wedge \wedge_{m<n \le N} \ell_m \ne \ell_n],$$

where N is the number of commands in p and ℓ_m $(m \le N)$ are the labels in p. $Ax(p)$ is formulated in the language L_d^B of the SOMETIME Logic of type d.

This brings us to Burstall-Manna-Waldinger's Intermittent Assertions Method or SOMETIME method, which we denote by *Bur* (and whose first description appeared e.g. in Burstall[4] and independently in Andréka et al.[2]). This method provides the following inference system $\overset{Bur}{\vdash}$: Let $Th \subseteq F_d$ and let $\Box(p,\psi)$ be a *pca*. Then $Th \overset{Bur}{\vdash} \Box(p,\psi)$, iff

$$Th \cup Ax(p) \overset{dB}{\vdash} (y_c = \ell_{HALT} \rightarrow \psi(\bar{y}))$$

where c is the index of the control variable x_c, ℓ_{HALT} is the label of the HALT command of p (and is a constant term of type d) and so $y_c = \ell_{HALT}$ means that the program p is in a halting state.

Note that $Alw\varphi$ abbreviates $\neg Smt\neg\varphi$.

Sain[18] and [20] proved that $Bur \underset{\Box}{\equiv} Ind$, i.e. from the point of view of proving pca's *Bur* is complete with respect to those Kripke models whose *time structure* satisfies the full induction scheme *Ind*.

Burstall-Manna-Waldinger's Intermittent Assertions Method is a so called endogenous program verification method (see e.g. Pnueli[15] pg. 56). Endogenous program verification methods develop a logic $L(p)$ "around" the program p and prove the property in question within the logic $L(p)$. Other endogenous methods are Pnueli's Temporal Logics of Programs (c.f. Pnueli[15]), in particular his SOMETIME IN THE FUTURE method (also referred to as NEXTTIME method—c.f. Manna-Pnueli[9]), which we denote by *Pnu*.

The method *Pnu* is not based on the language L_d^B of the SOMETIME logic, but on an extension of it containing the new modality *Sfu* ("Sometimes in the future"). This language we denote by L_d^P. The syntax (i.e. the set F_d^P of formulas) of L_d^P is now defined according to the same principles as that of L_d^B. But for the models $\mathcal{M} = \langle \underset{\sim}{T}, \underset{\sim}{D}, y_i \rangle_{i\in\omega}$ they must satisfy the axioms $Afu\varphi \rightarrow AfuAfu\varphi$, $Afu\varphi \rightarrow Nxt\varphi$ and $Fst(Afu\varphi \rightarrow Alw\varphi)$, as well as the modal induction scheme $Ind(P)$ defined below:

$$Ind(P) \equiv \left\{[\varphi \wedge Afu(\varphi \rightarrow Nxt\varphi)] \rightarrow Afu\varphi : \varphi \in F_d^P\right\}.$$

As for the forcing and validity relations, we have to add the interpretation of the *Sfu* modality. Namely, $t \Vdash Sfu\phi$ iff there is a time point $t_1 \geq t$, such that $t_1 \Vdash \phi$.

Let us recall that there is a decidable inference system $\overset{dP}{\vdash}$ for the language L_d^P, such that for every $Th \cup \{\phi\} \subseteq F_d^P$, $Th \models \phi$ iff $Th \overset{dP}{\vdash} \phi$ (see e.g. Manna-Pnueli[9] with the addition of the axioms $SmtAlw\varphi \rightarrow Alw\varphi$ and $Alw\varphi \rightarrow Fst\varphi$, but also see Prior[16], Clifford[5] and Segerberg[22]).

The method *Pnu* provides the following inference system $\overset{Pnu}{\vdash}$ for proving correctness of programs: Let $Th \subseteq F_d$ and let $\Box(p,\psi)$ be a *pca*. Then $Th \overset{Pnu}{\vdash} \Box(p,\psi)$, iff

$$Th \cup Ax(p) \overset{dP}{\vdash} (y_c = \ell_{HALT} \rightarrow \psi(\bar{y})).$$

One of the most interesting results proved in e.g. Sain[18] within the framework of Nonstandard (or Absolute) Logics of Programs (NLP for short— for a survey see Pasztor[12]), is formally stated as $(\star) Bur \underset{\square}{<} Pnu$ and says that from the point of view of proving programs partially correct, Pnueli's SOMETIME IN THE FUTURE method is strictly more powerful than Burstall-Manna-Waldinger's Intermittent Assertions Method (c.f. Sain[19]).

If we have two program verification systems or methods, say A and A', then $A \underset{\square}{\leq} A'$ means that every *pca* provable by A is also provable by A'. $A \underset{\square}{<} A'$ means $A \underset{\square}{\leq} A'$ together with the existence of a *pca* provable by A' but not provable by A'. ($A \underset{\square}{\equiv} A'$ denotes $A \underset{\square}{\leq} A'$ and $A' \underset{\square}{\leq} A$.)

Németi[10] and Sain[18] proved that if they change the language L_d^B of the SOMETIME logic by imposing on the time structure $\underset{\sim}{T}$ of the models the successor axioms Ts, they obtain a new method $Bur + Ts$ (in place of Bur), which still has less program verifying power than Pnu, i.e. $Bur + Ts \underset{\square}{<} Pnu$). Németi[10] and Andréka[1] formulated the following problem:

$$Bur \underset{\square}{<} Bur + Ts?$$

In other words, when reasoning about programs, is it useful to impose the successor axioms on the time scale and thus change the semantics, or are these axioms perhaps too trivial to help in proving more programs correct? Corollary 2 of Sain[21] and BIV.1. Kovetkezmeny of Sain[20] answer the question by proving $Bur \underset{\square}{<} Bur + Ts$. The proof of this result is given by exchanging in the semantics of the language L_d^B of the SOMETIME logic the interpretation of *Smt* as follows: $t \Vdash Smt\phi$ if there is $t_1 \neq t$ such that $t_1 \Vdash \phi$. The resulting method (replacing Bur) is called SOME OTHER TIME method and is denoted by Bur^+, and the following holds:

$$(\star\star) \quad Bur \underset{\square}{<} Bur^+ \underset{\square}{\leq} Bur + Ts.$$

This result gives rise immediately to the following *new* question:

$$Bur^+ \underset{\square}{<} Ind \cup Ts?$$

i.e. is $Ind \cup Ts$ strictly more powerful for proving partial correctness assertions than Bur^{+}?

R.H. Thomason remarked during a presentation of I. Sain of her result ($\star\star$) in one of D. Scott's weekly seminars at Carnegie-Mellon University, that the *proof* of ($\star\star$) really hinges on the fact that she was able to express in Bur^{+} the new modality Twc (read TWICE, where $t \Vdash Twc\ \phi$ if there are time points t_1 and t_2 such that $t_1 \neq t_2$ and $t_1,\ t_2 \Vdash \phi$), together with the axioms

$$Twc\varphi \leftrightarrow Twc\, Nxt\varphi \vee (Fst\varphi \wedge Smt\, Nxt\varphi)$$

governing the behavior of the successor function, and that in reality she was proving $Bur \underset{\square}{<} Twc$, i.e. that the TWICE method is strictly more powerful than the SOMETIME method in proving partial correctness assertions.

Thomason also raised the question whether the THREE TIMES (denoted $3*$) modality together with axioms stating an even "better" behavior of the successor function (i.e. more like that of the *usual* successor function) on time would further increase the reasoning power. He went even further and asked the very general question (which finally brings us to the main results of the *present* paper) formulated below. For every $n \geq 1$ we define the $n*$ system ("N TIME system") to be much like the SOMETIME system augmented with a new modality $n*$ and the following two axiom schemes governing the behavior of the successor function:

$$\begin{aligned} n * Nxt\,\varphi &\rightarrow n * \varphi \\ n * \varphi &\rightarrow [n * Nxt\,\varphi \vee (Fst\,\varphi \wedge Smt\, Nxt^{n-1}\,\varphi)]. \end{aligned}$$

(Note that with increasing n, the successor function will behave more and more like the *usual* successor function on time.) Is it then true that

$$n * \underset{\square}{<} (n+1) * ?$$

In other words, do the $n*$ systems provide an *infinite hierarchy* of program verification methods whose powers lie strictly between Bur and $Bur + Ts$?

It is in the present paper that this question is answered *for the first time.* The answer is YES, the $n*$ systems *do* provide such an infinite hierarchy.

Before concluding this section let us remark that all the above mentioned results belong to the field of *the comparative study of program verification methods* and, as so many other interesting results, are due to the use of Nonstandard (or Absolute) Logics of Programs. Contrary to a general belief, NLP has a great variety of logics, some making explicit use of time, but others being based on weak second order logics, relational run semantics of programs, etc. (see e.g. Andréka- Németi-Sain[3], Hájek[6], Makowsky- Sain[7] and [8], or Pasztor[11] to mention only a few).

2 N TIMES LOGICS. The Main Result.

The modalities we are going to work with are *Fst*, *Nxt*, *Smt* and $n*$ ($n \geq 1$).

The language of our N TIMES Logic of type d is defined as $L_d^{n*} = \langle F_d^{n*}, M_d^{n*}, \models \rangle$. F_d^{n*} consists of all formulas of the language L_d^B of the SOMETIME logic, expanded by the new modality $n*$ (i.e. if $\varphi \in F_d^{n*}$, then $n * \varphi \in F_d^{n*}$, too). M_d^{n*} is the class of all models of L_d^{n*} and consists of all Kripke models $\mathcal{M} = \langle \underset{\sim}{T}, \underset{\sim}{D}, y_i \rangle_{i \in \omega}$ as defined for the language L_d^B which satisfy the two axiom schemes $Ax_1(n*)$ and $Ax_2(n*)$, as well as the induction scheme $Ind(n*)$ defined below:

$Ax_1(n*) \equiv \{n * Nxt\varphi \rightarrow n * \varphi \,:\, \varphi \in F_d^{n*}\}$

$Ax_2(n*) \equiv \{n * \varphi \rightarrow [n * Nxt\varphi \vee (Fst\varphi \wedge Smt\, Nxt^{n-1}\varphi)] \,:\, \varphi \in F_d^{n*}\}$

$Ind(n*) \equiv \{[Fst\varphi \wedge Alw(\varphi \rightarrow Nxt\varphi)] \rightarrow Alw\varphi : \varphi \in F_d^{n*}\}$.

Let $Ax(n*) = Ax_1(n*) \cup Ax_2(n*)$.

The validity relation $\models$ is the same as for L_d^B, expanded by the following definition: $t \Vdash n*\varphi$ if there are n different time points $t_0, t_1, \ldots, t_{n-1} \in T$ such that $t_i \Vdash \varphi$ ($i < n$).

Theorem 1: There is a decidable inference system $\overset{n*td}{\vdash}$ for the language L_d^{n*} such that for every $Th \subseteq F_d^{n*}$ and $\varphi \in F_d^{n*}$ we have $Th \models \varphi$ iff $Th \overset{n*td}{\vdash} \varphi$.

Proof: The idea of the proof is to reduce (or translate) the language L_d^{n*} to the complete classical 3- sorted language $L_{td} = \langle F_{td}, M_{td}, \models \rangle$ defined in e.g. Andréka-Németi-Sain[2] (but also see Andréka[1], Németi[10], Pasztor[12] or e.g. Sain[17]) by a total computable function $\Theta : F_d^{n*} \rightarrow F_{td}$ such that for every $\varphi \in F_d^{n*}$ and every $\mathcal{M} \in M_d^{n*}$, $\mathcal{M} \models \varphi$ iff $\mathcal{M} \models \Theta(\varphi)$.

□ Theorem 1.

Remark: There *is* also a *Hilbert style* inference system for L_d^{n*}, but it is the topic of a follow up paper to define it and prove the Main Result of the present paper on pure syntactical level.

Let $p \in P_d$, $\psi \in F_d$, $Th \subseteq F_d$ (without modalities!). We say that *the pca* $\Box(p,\psi)$ *is provable from* Th *by the* $n*$ *system* $\overset{n*}{\vdash}$, in symbols $Th \overset{n*}{\vdash} \Box(p,\psi)$, if

$$Th \cup Ax(p) \overset{n*td}{\vdash} (y_c = \ell_{HALT} \rightarrow \psi(\bar{y}))$$

where c is the index of the control variable x_c, ℓ_{HALT} is the label of the HALT command of p (and is a constant term of type d) and so $y_c = \ell_{HALT}$ means that the program p is in a halting state.

Completeness Lemma: Let $p \in P_d$, $Th \subseteq F_d$ and $\psi \in F_d$. Then for any $n \geq 1$, $Th \overset{n*}{\vdash} \Box(p,\psi)$ iff for every Kripke model $\mathcal{M}$ of type d, $\mathcal{M} \models Th \cup Ax(n*) \cup Ax(p) \cup Ind(n*)$ implies

$$\mathcal{M} \models (y_c = \ell_{HALT} \rightarrow \psi(\bar{y})).$$

Proof: see Pasztor[13].

The Main Result of the present paper is formulated in

Theorem 2: For every $n \geq 1$

$$\overset{n*}{\vdash} \underset{\Box}{<} \overset{(n+1)*}{\vdash}.$$

3 The Proof of Theorem 2

Let d be the signature consisting of the constant symbols $\mathbf{0}$, Q, q, r, f and h, the unary function symbol g, the unary relation symbols P and S, and the binary relation symbol R.

Let p be the program of type d of Figure 1. For all natural numbers $k \geq 0$ and $n \geq 1$, we define D_k^n to be the model of type d of Figure 2 (top part).

We will prove the following, from which Theorem 2 follows:

Theorem 3: For every $n \geq 1$,

$$Th(D_0^n) \overset{(n+1)*}{\vdash} \Box(p, false), \quad \text{but} \quad Th(D_0^n) \overset{n*}{\nvdash} \Box(p, false).$$

(For every model D, $Th(D)$ denotes its first order theory).

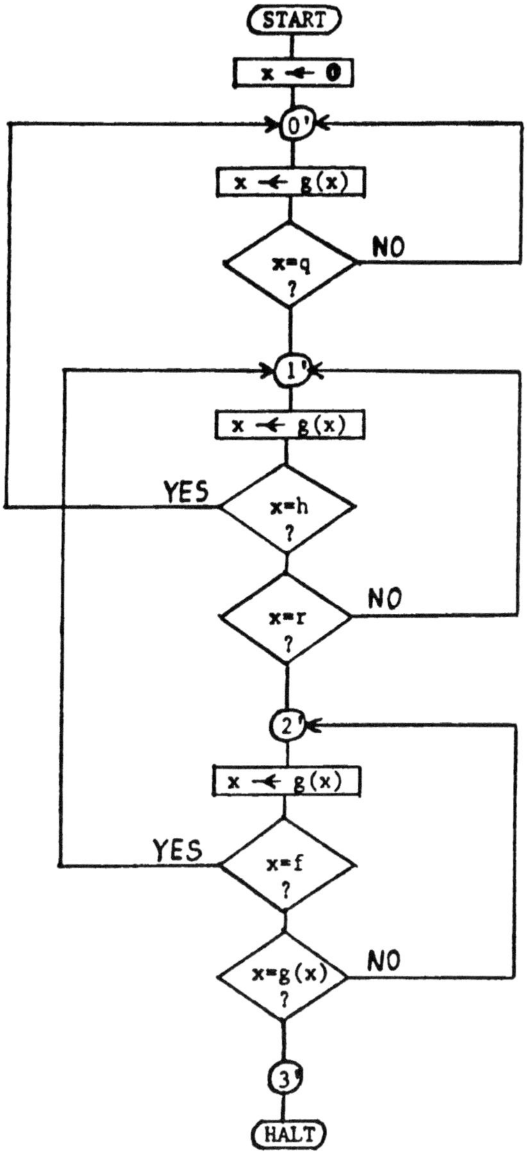

Figure 1: The program p of type d. The labels $0'$, $1'$, $2'$ and $3'$ denote constant terms of type d.

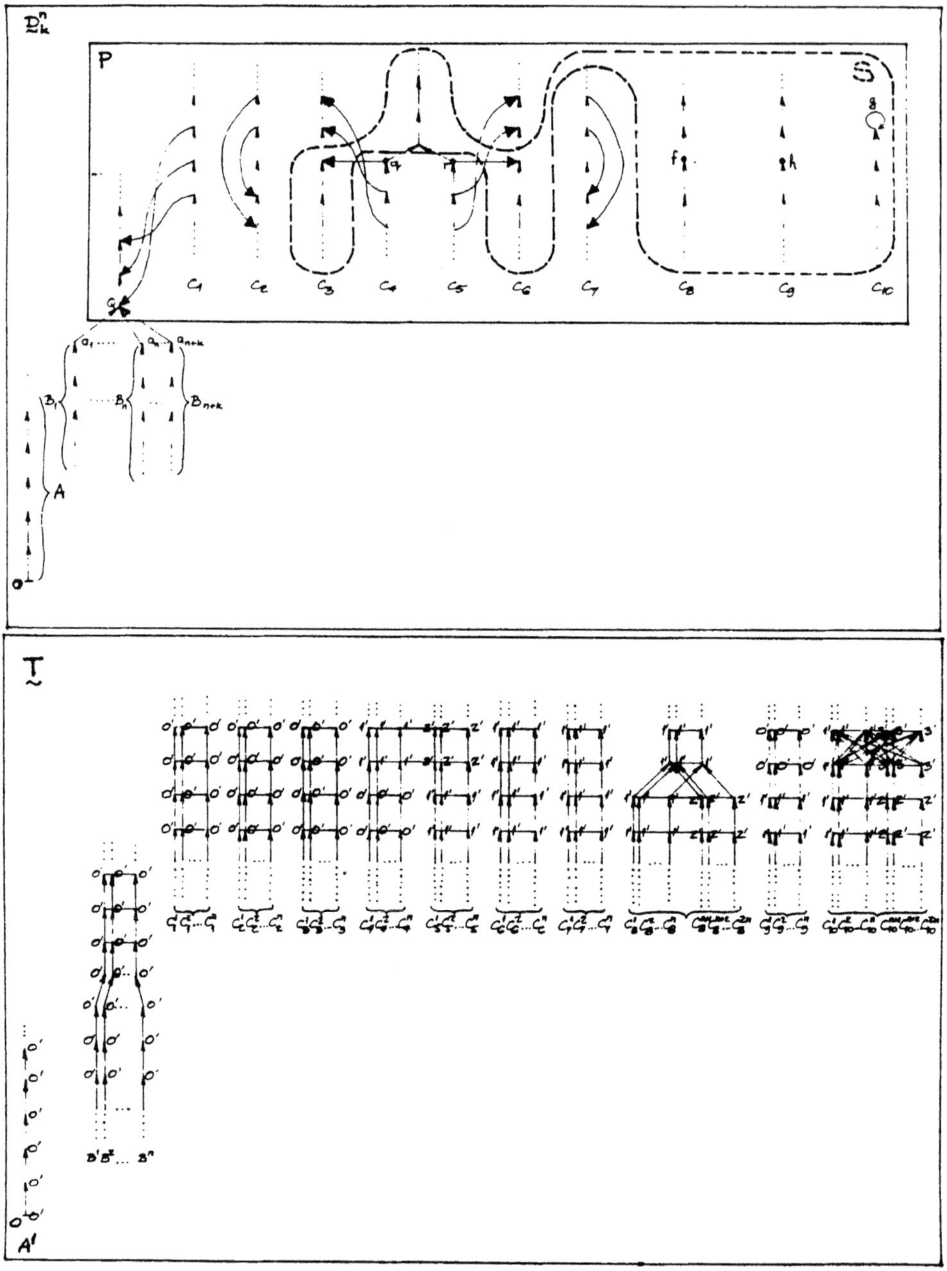

Figure 2: In T the labels on the time points define $y_1(=y_c)$, while y_0 assigns A^1 to A, B^i to $\hat{B}_i = B_i \cup \{Q, g(Q), g(g(Q)), \ldots\}$ $(1 \leq i \leq n)$ and C_i^j to C_i $(1 \leq i \leq 10, 1 \leq j \leq 2n)$ in the obvious way (following the run of the program p). The straight arrows in T define the successor function. The horizontal line between two time points t_1 and t_2 is to indicate that $y_0(t_1) = y_0(t_2)$.

Proof: (I) We will first prove that for every $n \geq 1$, $Th(D_0^n) \overset{(n+1)*}{\vdash} \Box(p, false)$, which is a direct consequence of

Proposition 1: Let $n \geq 1$ and $k \geq 0$. For every Kripke model $\mathcal{M}$ of type d, if

$$\mathcal{M} \models Th(D_k^n) \cup Ind((n+1)*) \cup Ax(p) \cup Ax((n+1)*)$$

and if $|\{i : \mathcal{M} \models Smt\, y_0 = a_i\}| \leq n$, then

$$\mathcal{M} \models \neg Smt\, y_1 = 3'.$$

Remark: Intuitively $|\{i : \mathcal{M} \models Smt\, y_0 = a_i\}| \leq n$ means that "p will run through at most n B_i's in $\mathcal{M}$", and $\mathcal{M} \models \neg Smt\, y_1 = 3'$ that "p will never reach HALT in $\mathcal{M}$", i.e. $\mathcal{M} \models \Box(p, false)$.

Proof of Proposition 1: We will do induction on n.

(IH) Induction hypothesis: Suppose that for every $m < n$ and every $k \geq 0$, if $\mathcal{M}$ is a Kripke model of type d and

$$\mathcal{M} \models Th(D_k^m) \cup Ind((m+1)*) \cup Ax(p) \cup Ax((m+1)*)$$

and $|\{i : \mathcal{M} \models Smt\, y_0 = a_i\}| \leq m$, then

$$\mathcal{M} \models \neg Smt\, y_1 = 3'.$$

Now assume $\mathcal{M} \models Th(D_k^n) \cup Ind((n+1)*) \cup Ax(p) \cup Ax((n+1)*)$ and $|\{i : \mathcal{M} \models Smt\, y_0 = a_i\}| \leq n$, but $\mathcal{M} \models Smt\, y_1 = 3'$. We will derive a contradiction. But first we have to state a series of lemmas.

Lemma 1: $A \models Smt\, y_0 = x$, i.e. for every $n \in \omega$

$$\mathcal{M} \models Smt\, y_0 = g^n 0.$$

Lemma 2: $A \models \neg 2 * y_0 = x$.

Remark 1: As a consequence of Lemma 1 and 2, for every $\varphi \in F_d^{n*}$ we can define a formula $\varphi' \in F_d^{(n+1)*}$ such that $\mathcal{M} \models (\varphi \longleftrightarrow \varphi')$ in the following way:

1. If φ contains no $n*$ modality, then $\varphi' = \varphi$.
2. Suppose we have already defined φ'. Then $(n*\varphi)' = (n+1)*\varphi' \vee [\bigvee_{k\leq n}(n+1)*(y_0 = g^k\mathbf{0} \vee \varphi')]$.

We will not distinguish in notation between φ and φ'.

Lemma 3: $\mathcal{M} \models Ax(n*) \cup Ind(n*)$.

Lemma 4: For all $i \leq n+k$, $B_i \models \neg 2 * y_0 = x$.

Lemma 5: $|\{i : \mathcal{M} \models Smt\, y_0 = a_i\}| = n$.

As a direct consequence of Lemma 4 and 5, we can now state

Lemma 6: $\mathcal{M} \models n * y_0 = Q \wedge \neg(n+1) * y_0 = Q$.

Lemma 7: $\mathcal{M} \models Alw\forall x((y_0 = x \wedge P(x)) \rightarrow n * y_0 = x)$.

Lemma 8: $\mathcal{M} \models Alw\forall x((y_0 = x \wedge S(x) \rightarrow (n+1) * y_0 = x)$

Lemma 9: $(C_4 \cap \overline{S^{\mathcal{M}}}) \cup (C_5 \cap \overline{S^{\mathcal{M}}}) \models \neg(n+1)*y_0 = x$ (where $\overline{S^{\mathcal{M}}}$ is the complement of $S^{\mathcal{M}}$).

Finally, using the previous results, we can prove

Lemma 10: $\mathcal{M} \models Alw\forall x_1((y_0 = x_1 \wedge \neg S(x_1) \wedge P(x_1) \wedge \neg(n+1) * y_0 = x_1) \rightarrow \neg\exists x_2[R(x_1, x_2) \wedge (Smt\, y_0 = x_2 \rightarrow (n+1) * y_0 = x_2)])$.

proof: Let φ denote the formula $\forall x_1((y_0 = x_1 \wedge \neg S(x_1) \wedge P(x_1) \wedge \neg(n+1) * y_0 = x_1) \longrightarrow \Psi)$, where Ψ denotes $\neg\exists x_2[R(x_1, x_2) \wedge (Smt\, y_0 = x_2 \rightarrow (n+1)*y_0 = x_2)]$. We prove $\mathcal{M} \models Alw\varphi$ by using $\mathcal{M} \models Ind((n+1)*)$. Obviously $\mathcal{M} \models Fst\varphi$, since $\mathcal{M} \models Fst\, y_0 = \mathbf{0}$ and $\mathbf{0}^{\mathcal{M}} \notin P^{\mathcal{M}}$. Suppose $t \Vdash \varphi$ for some time point $t \in T$. We want to prove $suc(t) \Vdash \varphi$. For that suppose $suc(t) \Vdash (y_0 = x_1 \wedge \neg S(x_1) \wedge P(x_1) \wedge \neg(n+1)y_0 = x_1)[a]$ for some $a \in D_k^n$. Let $t \Vdash (y_0 = x)[b]$. Then $g^{\mathcal{M}}(b) = a$ and we have the cases listed below.

case 1) $b \in \bar{S}^{\mathcal{M}} \cap P^{\mathcal{M}}$ and $t \Vdash \neg(n+1)*(y_0 = x)[b]$. Then, since $t \Vdash \varphi$, we have $t \Vdash \Psi[b]$. Therefore $b \notin C_4 \cup C_5$, since by Lemma 8 and $\mathcal{M} \models Ax(n+1)*)$, $C_3 \cup C_6 \models (n+1)*y_0 = x$, and so if $b \in C_4 \cup C_5$ were true, $t \Vdash \neg\Psi[b]$ would hold by $\mathcal{M} \models Th(D_k^n)$.

case 1.1) If $b \in C_0$, then obviously $suc(t) \Vdash \Psi[a]$, so $suc(t) \Vdash \varphi$.

case 1.2) If $b \in C_1$, then $suc(t) \Vdash \Psi[a]$, since $C_0 \models \neg(n+1) * y_0 = x$. (Remember that by Lemma 7, $\mathcal{M} \models \neg(n+1) * y_0 = Q$, so by $\mathcal{M} \models Ax((n+1)*) \cup Ind((n+1)*) \cup Th(D^n_k)$, $C_0 \vdash \neg(n+1) * y_0 = x$.) This proves $suc(t) \Vdash \varphi$.

case 1.3) If $b \in C_2$ (then $a \in C_2$ and) then $suc(t) \Vdash \Psi[a]$, for if $suc(t) \Vdash \neg\Psi[a]$, i.e. $suc(t) \Vdash [R(x_1, x_2) \wedge (Smt\, y_0 = x_2 \rightarrow (n+1) * y_0 = x_2)][a, a']$ for some $a' \in C_2$, then by $\mathcal{M} \models Ax(p) \cup Ind((n+1)*) \cup Th(D^n_k) \cup \{Fst\, y_1 = 0', Smt\, y_1 = 3'\}$ we can prove $\mathcal{M} \models Smt(y_0 = x)[a']$, which implies $\mathcal{M} \models (n+1) * (y_0 = x)[a']$, but then, by $\mathcal{M} \models Ax((n+1)*)$, $\mathcal{M} \models (n+1) * (y_0 = x)[a]$, contradicting $suc(t) \Vdash \neg(n+1) * (y_0 = x)[a]$.

case 1.4) If $b \in C_7$, then $suc(t) \Vdash \Psi[a]$ for the same reasons as in case 1.3.

case 1.5) As we showed above $C_3 \cup C_6 \models (n+1) * y_0 = x$, therefore $a \notin C_3 \cup C_6$.

case 2) $b \notin \bar{S}^{\mathcal{M}} \cap P^{\mathcal{M}}$ or $\mathcal{M} \models (n+1) * (y_0 = x)[b]$.

case 2.1) If $\mathcal{M} \models (n+1)(y_0 = x)[b]$, then $\mathcal{M} \models (n+1) * Nxt(y_0 = x)[a]$ (by $\mathcal{M} \models Th(D^n_k) \cup Ax(p)$) and then, by $\mathcal{M} \models Ax((n+1)*)$, $\mathcal{M} \models (n+1) * (y_0 = x)[a]$ contradicting our assumptions.

case 2.2) If $b \in S^{\mathcal{M}}$, then, since $a \notin C_3 \cup C_6$ (see case 1.5 above), $a \in S^{\mathcal{M}}$, too, contradicting our assumptions.

case 2.3) If $b \notin P^{\mathcal{M}}$, then $a = Q^{\mathcal{M}}$, so $suc(t) \Vdash \Psi[a]$ by $\mathcal{M} \models Th(D^n_k)$. This proves $suc(t) \Vdash \varphi$.

□ Lemma 10

But now notice that using Lemma 8 and 9, $Ax((n+1)*)$ and the fact that $\mathcal{M} \models Smt\, y_0 = q$ (by $\mathcal{M} \models Ax(p) \cup Ind((n+1)*) \cup \{Smt\, y_1 = 3' \wedge Fst\, y_1 = 0'\}$), we can see that for any time point t for which $t \Vdash y_0 = q$, $t \not\Vdash \forall x_1((y_0 = x_1 \wedge \neg S(x_1) \wedge P(x_1) \wedge \neg(n+1) * y_0 = x_1) \rightarrow \neg\exists x_2[R(x_1, x_2) \wedge (Smt\, y_0 = x_2 \rightarrow (n+1) * y_0 = x_2)])$ since $\mathcal{M} \models (y_0 = x_1 \wedge \neg S(x_1) \wedge P(x_1) \wedge \neg(n+1) * y_0 = x_1 \wedge \exists x_2[R(x_1, x_2) \wedge (Smt(y_0 = x_2) \rightarrow (n+1) * y_0 = x_2)])[q]$. This contradicts Lemma 10.

Therefore we conclude that our assumption $\mathcal{M} \models Smt\, y_1 = 3'$ was false.

□ Proposition 1

If we choose $k = 0$ in Proposition 1, then we can certainly imply (using the Completeness Lemma) $Th(D_0^n) \overset{(n+1)*}{\vdash} \Box(p, false)$. This concludes the proof of part **(I)**.

(II) We will now prove that for every $n \geq 1$

$$Th(D_0^n) \overset{n*}{\not\vdash} \Box(p, false).$$

Let therefore $\mathcal{M}$ be the Kripke model of Figure 2 with $k = 0$. One can easily see that $\mathcal{M} \models Th(D_0^n)$ and that $\langle y_0, y_1 \rangle$ is a terminating run of p in $\mathcal{M}$, which implies $\mathcal{M} \models Ax(p)$ and $\mathcal{M} \models \mathit{Smt}\, y_1 = 3'$. It is left to prove $\mathcal{M} \models Ax(n*) \cup Ind(n*)$.

Lemma 11: $\mathcal{M} \models Ax_1(n*)$.

proof: We have to prove that for every $\varphi \in F_d^{n*}$, $\mathcal{M} \models n * Nxt\varphi \to n * \varphi$.

For this, let $\varphi = \varphi(x_1, \ldots, x_m) \in F_d^{n*}$ and assume $\mathcal{M} \models (n * Nxt\varphi(x_1, \ldots, x_m))[a_1, \ldots, a_m]$ for some $a_1, \ldots, a_m \in D$. Then there are n different time points $t_1, \ldots, t_n$ in T, such that $suc(t_1), \ldots, suc(t_n) \Vdash \varphi[a_1, \ldots, a_m]$. The definition of T is such that $suc(t_1), \ldots, suc(t_n)$ are also all different, except when for some i, j and l in $\{1, \ldots, n\}$ $t_i \in C_8^l$, $t_j \in C_8^{n+l}$, $suc(t_i) = suc(t_j)$ and $y_0(suc(t_i)) = y_0(suc(t_j)) = f^{\mathcal{M}}$. Suppose that $(t_{i_1}, t_{j_1}), \ldots, (t_{i_k}, t_{j_k})$ are all such pairs with $t_{i_p} \in C_8^{l_p}$ and $t_{j_p} \in C_8^{n+l_p}$ for some $l_p \in \{1, \ldots, n\}$ such that $suc(t_{i_p}) = suc(t_{j_p})$ and $y_0(suc(t_{i_p})) = y_0(suc(t_{j_p})) = f^{\mathcal{M}}$ ($p \in \{1, \ldots, k\}$, t_{i_p}, $t_{j_p} \in \{t_1, \ldots, t_n\}$). For each $t_{j_p} \in C_8^{n+l_p}$ ($p \in \{1, \ldots, k\}$), we find $l'_p \in \{1, \ldots, n\}$, such that for no i $t_i \in C_8^{l'_p}$ or $t_i \in C_8^{n+l'_p}$ ($i \in \{1, \ldots, n\}$), and such that if $p \neq p'$ (p, $p' \in \{1, \ldots, k\}$), then $l'_p \neq l'_{p'}$, and then replace $t_{j_p} \in C_8^{n+l_p}$ by the time point $t^{j_p} \in C_8^{n+l'_p}$ for which $y_0(suc(t^{j_p})) = f^{\mathcal{M}}$. Notice that for each $p \in \{1, \ldots, k\}$, $suc(t_{i_p}) \neq suc(t^{j_p})$, moreover, $suc(t^{j_p}) \Vdash \varphi[a_1, \ldots, a_m]$, since there is an obvious automorphism of $\mathcal{M}$ interchanging $C_8^{n+l_p}$ with $C_8^{n+l'_p}$ and leaving everything else fixed.

This way we have arrived at n different time points $t'_1, \ldots, t'_n$, such that $suc(t'_1), \ldots, suc(t'_n)$ are also all different and $suc(t'_i) \Vdash \varphi[a_1, \ldots, a_m]$ for all $i \in \{1, \ldots, m\}$.

This proves $\mathcal{M} \models (n * \varphi)[a_1, \ldots, a_m]$.

$\Box$ Lemma 11

Lemma 12: $\mathcal{M} \models Ax_2(n*)$.

proof: We have to prove that for every $\varphi \in F_d^{n*}$, $\mathcal{M} \models (n * \varphi \rightarrow n * Nxt\varphi \vee [Fst\varphi \wedge SmtNxt^{n-1}\varphi])$. For this suppose that $\mathcal{M} \models (n * \varphi)[a_1, \ldots, a_m]$ for some $\varphi = \varphi(x_1, \ldots, x_m) \in F_d^{n*}$ and $a_1, \ldots, a_m \in D$. Then there are n different time points $t_1, \ldots, t_n$, such that $t_i \Vdash \varphi[a_1, \ldots, a_m]$ $(i \in \{1, \ldots, n\})$. Since *suc* is a function on T and every time point except 0 has a "predecessor", either the n time points $t_1, \ldots, t_n$ have n different "predecessors" and then $\mathcal{M} \models (n * Nxt\varphi)[a_1, \ldots, a_m]$, or one of the time points is 0 and any of the others has infinitely many "predecessors", in which case $\mathcal{M} \models (Fst\varphi \wedge SmtNxt^{n-1}\varphi)[a_1, \ldots, a_m]$.

□ Lemma 12

Lemma 13: $\mathcal{M} \models Ind(n*)$.

proof: The proof makes use of the ingenious ultrapower technique developed in Andreka-Nemeti-Sain [3].

Suppose that for some $\varphi \in F_d^{n*}$, $\mathcal{M} \models \neg([Fst\varphi \wedge Alw(\varphi \rightarrow Nxt\varphi)] \rightarrow Alw\varphi)$, i.e. $\mathcal{M} \models Fst\varphi \wedge Alw(\varphi \rightarrow Nxt\varphi)$, but $\mathcal{M} \models \neg Alw\varphi$. This means that there is a time point t, for which $t \not\Vdash \varphi$. Let us fix this time point t. By systematically using the construction of T , we prove that such a time point t can not exist in T. Let U be a nonprincipal ultrafilter on ω and let $\mathcal{M}^+$ denote ${}^\omega\mathcal{M}/U$, i.e. the ωth ultrapower of $\mathcal{M}$ modulo U. We define an automorphism α on $\mathcal{M}^+$ such that for some time point t^+ of $\mathcal{M}^+$ corresponding to the time point t, $t^+ \Vdash \varphi$, while $\alpha(t^+) \not\Vdash \varphi$. This contradicts the fact that α is an automorphism. So t can not exist.

□ Lemma 13

This concludes the proof of part **(II)** then. □ Theorem 3

Obviously, by the definition of the relation $\underset{\Box}{\leq}$ between verification system, Theorem 3 immediately implies our Main Result: Theorem 2.

4 Conclusion and Open Problem

In the previous section we have proved that for every $n \geq 1$, $\overset{n*}{\vdash} \underset{\Box}{\leq} \overset{(n+1)*}{\vdash}$. In fact, as mentioned in the Introduction, for $n = 1$ Sain[21] has already proved this result.

It is also clear that for every $n \geq 1$, $\overset{n*}{\vdash} \underset{\Box}{\leq} Ind \cup Ts$, since in the presence of the successor axioms on time, we can certainly express the N TIMES modality together with $Ax(n*)$ (recall that *Ind* denotes full induction on the time stucture). So what we have proved is this:

$$Bur \underset{\Box}{\equiv} \overset{1*}{\vdash} \underset{\Box}{\lneq} \overset{2*}{\vdash} \underset{\Box}{\lneq} \overset{3*}{\vdash} \underset{\Box}{\lneq} \cdots \underset{\Box}{\lneq} \overset{n*}{\vdash} \underset{\Box}{\lneq} \overset{(n+1)*}{\vdash} \underset{\Box}{\lneq} \cdots \underset{\Box}{\leq} Ind \cup Ts.$$

Very naturally then the following *new* question arises: If we define a new system $\cup n*$ with all $n*$ modalities present and postulating all $Ax(n*)$ schemes, is it true that $\cup n * \underset{\Box}{\equiv} Ind \cup Ts$?

Acknowledgements

I would like to thank Istvan Németi for the helpful and encourageing discussions on the topic of the present paper. I would also like to thank Ildiko Sain for helping me understand her paper[21] and her dissertation.

References

[1] H. Andréka, Sharpening the characterization of the power of Floyd's method, in: *Logic of Programs and their Applications*, ed.: A. Salwicki (Proc. Conf. Poznan 1980), Lecture Notes in Computer Science 148, Springer-Verlag 2983, pp. 1-26.

[2] H. Andréka, K. Balogh, K. Lábadi, I. Németi, P. Tóth, Plan for improving a working program verifier program, Software Dept. of Computing Center of the Heavy Industries, Preprint 1974, Budapest (in Hungarian).

[3] H. Andréka, I. Németi and I. Sain, A complete logic for reasoning about programs via nonstandard model theory, Parts I-II, *Theoretical Computer Science* 17 (1982), pp. 193-212 and pp. 259-278.

[4] R.M. Burstall, Program proving as hand simulation with a little induction, *IFIP Congress, Stockholm, August 3- 10, 1974.*

[5] J. Clifford,Tense logic and the logic of change, *Logique et Analyse* 34 (1966), pp 219-230.

[6] P. Hájek, Some conservativeness results for nonstandard dynamic logic, in: *Algebra, Combinatronics and Logic in Computer Science*, eds.: J. Demetrovics, G. Katona, A. Salomaa (Proc. conf. Győr Hungary 1983) Colloq. Math. Soc. J. Bolyai Vol. 42, North-Holland, 1986, pp. 443-449.

[7] J.A. Makowsky and I. Sain, On the equivalence of weak second order and nonstandard time semantics for various program verification systems, *Proceedings of the first annual IEEE Symposium on Logic in Computer Science*, Cambridge, MA, June 1986, pp. 293-300.

[8] J.A. Makowsky and I. Sain, Weak second order characterizations of various program verification systems, Technical Report #457, Technion–Israel Institute of Technology, Comp. Sci. Dept., June 1987. Submitted to Theoretical Computer Science.

[9] Z. Manna and A. Pnueli, The modal logic of programs, in: *International Colloquium on Automata, Languages and Programming '79*, Graz, Lecture Notes in Computer Science 71, Springer-Verlag, 1979, pp. 385-409.

[10] I. Németi, Nonstandard dynamic logic, in: *Logics of Programs*, ed.: D. Kozen, (Proc. Conf. New York 1981) Lecture Notes in Computer Science 131, Springer-Verlag, 1982, pp. 311-348.

[11] A. Pasztor, Recursive programs and denotational semantics in absolute logics of programs. Technical Report of Florida International University, School of Comp. Sci., #FIU- SCS-87-1, to appear in *Theoretical Computer Science.*

[12] A. Pasztor, Nonstandard Algorithmic and Dynamic Logic, in: *J. Symbolic Computation* 2 (1986), pp. 59-81.

[13] A. Pasztor, An Infinite Hierarchy of Program Verification Methods, Proceedings of Workshop on Many–Sorted Logic and its Applications in Computer Science, Sept., 12–14, 1988, Leeds, England, Academic Press (ed. J.V. Tucker), to appear.

[14] A. Pnueli, The temporal semantics of concurrent programs, *Theoretical Computer Science* 13 (1981) pp. 45-60.

[15] A. Pnueli, The temporal logic of programs, *18th Annual Symposium on Foundations of Computer Science*, 1977, pp. 46-57. Revised version: Preprint of the Weizman Institute of Science, May 1981.

[16] A.N. Prior, Past, present and future, Oxford University Press, 1967.

[17] I. Sain, Structured nonstandard dynamic logic, *Zeitschrift für Math. Logic und Grundlagen der Math.* Heft 3, 1984, pp. 481-497.

[18] I. Sain, Relative program verifying powers of the various temporal logics, *Information and Control*, to appear. An extended abstract of this is [19].

[19] I. Sain, The reasoning powers of Burstall's (modal logic) and Pnueli's (temporal logic) program verification methods, in: *Logics of Programs*, ed.: R. Parikh (Proc. Conf. Brooklyn USA 1985) Lecture Notes in Computer Science 193, Springer-Verlag, pp. 302-319.

[20] I. Sain, Dynamic logic with nonstandard model theory, Dissertation, Hungarian Academy of Sciences, Budapest, 1986 (in Hungarian).

[21] I. Sain, Is "SOME OTHER TIME" sometimes better than "SOMETIME" in proving partial correctness of programs?, to appear in a special vol. of *Studia Logica* on nonstandard methods edited by M.M. Richter and M.E. Szabo.

[22] K. Segerberg, On the logic of tomorrow, *Theoria* 33 (1967), pp 45-52.

An Algebraic Formulation For Data Refinement

A. J. POWER

Department of Mathematics and Statistics
Case Western Reserve University
Cleveland, Ohio 44106
U.S.A.

Abstract. We analyze recent work on an algebraic formulation for data refinement. Hoare's principal mathematical constructions are reviewed. Then, they are mildly reformulated and unified in terms of two principal category theoretic notions: those of an enriched category and monad, also known as a triple. The requisite definitions and theory are given, together with several examples to illustrate precisely how Hoare's work, including his main theorem may be seen in this light.

1. INTRODUCTION

In his paper *Data Refinement in a categorical setting* [6], Hoare uses some familiar notions of category theory in order to give an abstract formulation for a series of results about data refinement. He gives a succession of separate statements with separate proofs; he realizes that there is a common thread to many of his arguments; and he explicitly asks "Why do the algebraic proofs work out so easily?" Here in, we give a partial answer to that question by formulating Hoare's constructions in a unified manner, in terms of a known and developed abstract theory, the theory of monads.

Hoare's paper, as I understand it, contains two fundamental ideas. First, it gives a precise, mathematical formulation of the term *data refinement* to encapsulate the notion that it is a formal method for the stepwise development of large programs and systems. Second, it investigates the question: for which generators is data refinement a valid method for program development? In Section 2 herein, we review Hoare's principal mathematical construction in detail.

Next, we attempt to re-express Hoare's constructions in terms of a known theory. Two principal category theoretic constructions seem to be ideally suited. First is the notion of *enriched category*. We develop the basic theory in Section 3; the standard text for this theory is Kelly [7]. The second construction is that of a *monad*, also known as a *triple*. We develop its basic theory in Section 4. There are several texts on monad theory: one such text is Barr and Wells [1]; but the fundamental result of Section 4 lies in Kelly and Power [8].

Sections 3 and 4, as explained therein, do encapsulate most of Hoare's basic constructions. However, as they stand, they do not address the notion of *upward simulation*, which is central to this paper. However, using somewhat more sophisticated category theoretic machinery, we can do that too. That forms the content of Section 5.

Finally, in Section 6, we can explain how Hoare's main theorem can be seen in a more general category theoretic light.

1. HOARE'S CONSTRUCTIONS

Given a strong typed programming language, i.e., one in which every term has a unique type, we may form a category L:

the objects of L are data types

arrows are programs

composition is the familiar sequential composition of programming, denoted by semicolon.

We may introduce a partial order on $\leq$ on Arr L, the set of arrows of L, such that

if $p \leq q$, then dom $p =$ dom q and cod $p =$ cod q

$\leq$ is preserved by composition, i.e. if $p \leq q$, then $p;r \leq q;r$ and $r;p \leq r;q$ whenever the composites are defined.

The statement $p \leq q$ is to mean p *is an improvement upon* q. For instance, it could mean that whenever q terminates, p must also terminate and give the same result as q or a better one. Another way to interpret $p \leq q$ is that every result that p can give is the same as, or better than, some result that q can give (but q can give a wider range of different results). So, p is more predictable, more controllable, and more deterministic than q.

An *interpretation* of the language is a functor $G: L \to M$ preserving $\leq$, where M is some *mathematical category.*

Given functors $F, G: L \to M$, an *upward simulation* from F to G consists of the assignment, to each object A of L, of a map $\alpha A: FA \to GA$, such that for any map $p: A \to B$ in L,

$$\begin{array}{ccc} FA & \xrightarrow{\alpha A} & GA \\ {\scriptstyle Fp}\downarrow & \leq & \downarrow{\scriptstyle Gp} \\ FB & \xrightarrow[\alpha B]{} & GB \end{array},$$

i.e. $Fp; \alpha B \leq \alpha A; Gp$.

Observe that every natural transformation is an upward simulation. An upward simulation represents the idea that F is more concrete that G. The collection $\{\alpha A: FA \to GA\}$ is a collection of abstraction functions, which establish the correctness of the more concrete interpretation F. The aim is to prove the above inequality holds merely for the primitive operations invoked by the program, then to deduce it for any program written using those primitives.

Hoare also introduces *downward simulations* and *total simulations*, both of which are variants of the above.

The notion of *generator* is first defined, then modified, in the course of Hoare's paper. Initially, it is defined as a partial function from Arr L to Arr L, and later it is generalised. However, in fact, the term is used only for partial functions with a very limited collection of possible domains. So, here, I will give a formulation that seems to include all of Hoare's examples, but which allows easier reformulation in abstract terms: a *generator* is a (total) function "g": Mod$(X, L) \to$ Arr L, where Mod(X, L) is the set of $\leq$-preserving functors from X to L, and X can be any of

1 the unit category
2 the discrete category on two objects
2 the arrow category, i.e.,
the category with two objects and two non-identity arrows pictured as

$$\downarrow \leq \downarrow$$

or similar constructions, which we shall describe precisely in Section 4.

A generator is subject to axioms of the form

$$\phi(p) = \psi(p) \quad \text{or} \quad \phi(p) \leq \psi(p) \ .$$

where $\phi(p)$ and $\psi(p)$ are arrows in L generated by applications of "g" and composition, taking domains and codomains, etcetera: again, Section 4 will make this precise.

An explicit example of a generator is "abort":$\mathrm{Mod}(1, L) \to \mathrm{Arr}\ L$, with axioms such that

1) cod("abort") $= 1_{\mathrm{Mod}(1,L)} =$ dom("abort"), where $1_{\mathrm{Mod}(1,L)}$ is the identity function on the set $\mathrm{Mod}(1, L)$,
2) given $p \in \mathrm{Arr}\ L$,

$$p; \text{"abort"}(\mathrm{cod}\ p) \leq \text{"abort"}(\mathrm{dom}\ p); p \ .$$

Clearly, "abort" is to give the "abort" operator on a data type, and the axioms yield that non-termination after performing inputs and outputs of p is no worse than immediate non-termination, e.g. CHAOS in CSP.

Other examples are discriminated unions, or binary coproducts, or variant records in PASCAL; binary products, or smash products, or record declarations in PASCAL, where projections are called field names in PASCAL and cdr in LISP; higher order function spaces; and functors, natural transformations, bifunctors, contravariant functors, and zero-morphisms.

A generator "g" is to have a "mathematical meaning", which is to be a function $g\colon \mathrm{Mod}(X, M) \to \mathrm{Arr}\ M$ satisfying the axioms. Hoare studies those interpretations G such that

$$G(\text{"}g\text{"}p) = g(Gp) \ ,$$

where $p \in \mathrm{Mod}(X, L)$, and similarly for upward simulations.

One fundamental idea is that $\mathrm{Arr}\ L$ has a subset $\mathrm{Arr}\ L_0$ containing just the primitive data types and the built-in operations: one hopes to give interpretations and upward simulations only on L_0, demand that interpretations and upward simulations respect the generators as above, and then conclude that the interpretations and upward simulations extend consistently to L.

3. ENRICHED CATEGORIES

First, we seek to describe Hoare's categories with $\leq$, and $\leq$-preserving functors, as an instance of a general well known theory.

DEFINITION 3.1. *Let $\mathcal{V}$ be a cartesian closed category. A $\mathcal{V}$-**category** $\mathcal{A}$ consists of*

- *a set* ob $\mathcal{A}$ *of objects*
- *for* $A, B \in \text{ob } \mathcal{A}$, *an object* $\mathcal{A}(A, B)$ *of* $\mathcal{V}$
- *for* $A, B, C \in \text{ob } \mathcal{A}$, *arrows*

$$M : \mathcal{A}(A, B) \times \mathcal{A}(B, C) \to \mathcal{A}(A, C) \ ,$$

called composition, and

$$j_A : 1 \to \mathcal{A}(A, A) \ ,$$

called an identity, such that

$$(1 \times M); M = (M \times 1); M : \mathcal{A}(A, B) \times \mathcal{A}(B, C) \times \mathcal{A}(C, D) \to \mathcal{A}(A, D),$$

i.e. the arrows in $\mathcal{V}$ *from* $\mathcal{A}(A, B) \times \mathcal{A}(B, C) \times \mathcal{A}(C, D)$ *to* $\mathcal{A}(A, D)$ *given by* $(1 \times M); M$ *and* $(M \times 1); M$ *are equal; and similarly,*

$$(j_A \times 1); M = 1 = (1 \times j_B); M : \mathcal{A}(A, B) \to \mathcal{A}(A, B).$$

This definition appears in Kelly [7], where it is given in a more general setting: the most natural formulation for enriched category theory allows $\mathcal{V}$ to be a symmetric monoidal closed category, of which cartesian closed categories are an important, special case.

DEFINITION 3.2. *A $\mathcal{V}$-**functor** S from $\mathcal{A}$ to $\mathcal{B}$ consists of*

- *a function* ob $S : \text{ob } \mathcal{A} \to \text{ob } \mathcal{B}$
- *for each* $A, B \in \text{ob } \mathcal{A}$, *an arrow* $S_{AB} : \mathcal{A}(A, B) \to \mathcal{B}(SA, SB)$ *in* $\mathcal{V}$,

such that

$$(S \times S); M = M; S : \mathcal{A}(A, B) \times \mathcal{A}(B, C) \to \mathcal{A}(A, C)$$

and

$$j_A; S = j_{SA} : 1 \to \mathcal{B}(SA, SA).$$

With the evident composition, we have a category $\mathcal{V}$–*Cat*.

An example of the power of this analysis is given by the following theorem, which is a fundamental result of the theory, appearing in Kelly [7]:

THEOREM 3.3. *If $\mathcal{V}$ is a complete and cocomplete cartesian closed category, then so is $\mathcal{V}$–Cat.*

EXAMPLE 3.4. Let $\mathcal{V}$ = *Posets*, the cartesian closed category of partially ordered sets and order-preserving functions. Then, any of Hoare's categories with $\leq$ is precisely a $\mathcal{V}$-category, and an interpretation is precisely a $\mathcal{V}$-functor: Hoare's conditions restrict $\leq$ to a partial order on each homset and make composition a map of partial orders; everything else is automatic. Moreover, since $\mathcal{V}$–*Cat* is cartesian closed, we can discuss the *Posets*–category $[L \Rightarrow M]$. The arrows of this *Posets*–category are precisely natural transformations between interpretations. This observation is crucial to Section 6.

More sophistication is required to handle upward simulations. There is another "monoidal biclosed" structure on *Posets–Cat* in addition to the cartesian closed structure, and the "right internal hom" for that yields precisely the upward simulations. This appears in Section 5 (cf. Gray [5]).

4. MONADS

Section 3 formulated Hoare's underlying structure as an instance of a more general, already developed theory. Now, we seek to formulate his construction along the same lines. To do this, we invoke the theory of monads.

DEFINITION 4.1. *Given a category C, a* **monad** *on C is a triple $T = (T, \eta, \mu)$, consisting of*

- *a functor $T : C \to C$*
- *a natural transformation $\eta : 1 \to T$, and*
- *a natural transformation $\mu : T^2 \to T$,*

such that $\mu T; \mu = T\mu; \mu : T^3 \to T$ and $\eta T; \mu = 1 = T\eta; \mu : T \to T$.

We henceforth make the following blanket assumption:

ASSUMPTION 4.2. *C is locally finitely presentable (which we may abbreviate as l.f.p.), and T is finitary.*

These are size conditions, which keep our constructions in a precise sense "essentially algebraic". Definitions appear in several works such as [4] and [8]. It is not necessary to understand these conditions in order to understand what follows.

DEFINITION 4.3. *Given a monad T on C, a T–***algebra** *is a pair $(A, a : TA \to A)$ in C, such that $\eta A; a = 1 : A \to A$ and $\mu A; a = Ta; a : T^2 A \to A$.*

*A T–***algebra map** *from (A, a) to (B, b) is an arrow $f : A \to B$ in C such that*

$$\begin{array}{ccc} TA & \xrightarrow{Tf} & TB \\ {\scriptstyle a}\downarrow & & \downarrow{\scriptstyle b} \\ A & \xrightarrow[f]{} & B \end{array}$$

commutes. With the evident composition, we have a category T–Alg.

An indication of the power of this notation can be seen in the following theorem, which appears in Barr and Wells' [1]:

THEOREM 4.4.

(1) *The underlying functor $U : T\text{-}Alg \to C$ has a left adjoint, i.e. free algebras exist.*
(2) *T–Alg is complete and cocomplete.*
(3) *Given monads T and S and a functor $W : T\text{–}Alg \to S\text{–}Alg$ such that*

$$W; U_S = U_T : T\text{-}Alg \to C \ ,$$

W has a left adjoint, i.e. T-algebras free with respect to S– Alg exist.

We now need a theorem to explain how Hoare's constructions are related to monads. That theorem exists and appears in [8] in a general setting. Many examples of its application, with varying degrees of detail, appear in [3].

NOTATION 4.5. *Let $|C_{\text{f.p.}}|$ denote the set of finitely presented objects of C. (A definition of finitely-presented appears in [4] and [8]. In the case of interest to us, where $C = Posets\text{-}Cat$, $|C_{\text{f.p.}}|$ contains all finite objects of C plus those objects given by taking closure under coequalizers. For our purposes, it is sufficient to know that it includes all finite objects.)*

DEFINITION 4.6. *A* **theory** (S,E) **of operations and equations on** C *is given as follows:*

$S : |C_{\text{f.p.}}| \to C$ *is a function. S is often 0 on all but a small, finite number of values. $S(X)$ is called the* **object of basic operations of arity** X.

An $\boldsymbol{S}$**–algebra** *consists of an object L of C together with a function $(\ )_L : C(X,L) \to C(S(X),L)$ for each X, where $C(X,L)$ denotes the set of arrows from X to L in C.*

The **object of derived operations of arity** X, *denoted by $F(S)(X)$, is defined by*

$$F(S)(X) = \operatorname*{colim}_{n \to \infty} S_n(X),$$

where

$S_0(X) = X$,

$S_{n+1}(X) = X = \sum_Y C(Y, S + n(X)) \cdot S(Y)$, *and*

$C(Y, S_n(X)) \cdot S(Y)$ *denotes the coproduct in C of $C(Y, S + n(X))$ copies of $S(Y)$.*

Every S–algebra may be extended inductively to an $F(S)$-algebra

$$(\ )_L : C(X,L) \to C(F(S)(X),L)$$

in a canonical way: given $\phi \in C(X,L)$, define $(\phi_L)_{n+1} : S_{n+1}(X) \to L$ inductively by defining

- $(\phi_L)_{n+1}|_X = \phi$ *on the X-component of $S_{n+1}(X)$,*
- $(\phi_L)_{n+1}|_{S(Y)} = (\psi; (\phi_L)_n)_L$ *on the ψth copy of $S(Y)$ in $C(Y, S_n(X)) \cdot S(Y)$, given $\psi : Y \to S_n(X)$.*

The **equations** E *of (S,E) assigns to each arity X, a subobject $E(X)$ of $F(S)(X) \times F(S)(X)$.*

An $(\boldsymbol{S},\boldsymbol{E})$**–algebra** *is an S-algebra $(L, (\ )_L)$ such that the two composites*

$$C(X,L) \to C(F(S)(X),L) \rightrightarrows C(E(X),L)$$

given by the two projections of $E(X)$ onto $F(S)(X)$, are equal. (In the language of generalised elements, it is equivalent to say that for each $(\tau, \tau') \in E(X)$, and each $\phi \in C(X,L)$, we have $\phi_L(\tau) = \phi_L(\tau')$.)

An **(S,E)–algebra map** *is an arrow $F : L \to M$ in C such that for each $X \in |C_{\text{f.p.}}|$ and each $\phi \in C(X,L)$, we have $\phi_L; F = (\phi; F)_M$.*

With the evident composition, we have a category $(S,E)\text{–}Alg$.

THEOREM 4.7. *Given an l.f.p. category C, a category is of the form T–Alg for a finitary monad T on C if and only if it is, up to equivalence, of the form (S,E)–Alg for some theory (S,E) of operations and equations on C.*

The above theorem is proved in the enriched setting in [8]; the above is obtained by putting $\mathcal{V} = Sets$ in the main theorem of [8].

EXAMPLE 4.8. Let $C = Posets$–Cat. Then, all of Hoare's constructions exhibit L and M as T-algebras and the interpretations of interest as T-algebra maps. For instance, pursuing the "abort" example of Section 2, we want

$$\text{"abort"} : C(1,L) \to C(2,L)$$

and $d : C(2,L) \to C(Z,L)$, where

$$Z = \quad \begin{array}{ccc} & \xrightarrow{\alpha_1} & \\ \alpha_3\big\downarrow & \leq & \big\downarrow\alpha_4 \\ & \xrightarrow[\alpha_2]{} & \end{array}$$

with equations to make

$$\text{cod("abort")} = 1_L = \text{dom("abort")}$$

and, for each $\phi \in C(2,L)$,

$$d(\phi) = \quad \begin{array}{ccc} & \xrightarrow{\phi_1} & \\ \phi_3\big\downarrow & \leq & \big\downarrow\phi_4 \\ & \xrightarrow[\phi_2]{} & \end{array}$$

equations to make $\phi_1 = \text{"abort"}(\text{dom}\ \phi)$, $\phi_2 = \text{"abort"}(\text{cod}\ \phi)$, and $\phi_3 = \phi_4 = \phi$.

At this point, with a little experience, it becomes clear that the above can be formalised. As usual with formal descriptions, the details are sometimes complex. See [3] for a list of examples. In this case, we put

$$S(X) = \begin{cases} 2 & \text{if } X = 1 \\ Z & \text{if } X = 2 \\ 0 & \text{otherwise} . \end{cases}$$

Then

$$\begin{aligned} S_1(X) &= X + \sum_Y C(Y,X) \cdot S(Y) \\ &= X + C(1,X) \cdot 2 + C(2,X) \cdot Z . \end{aligned}$$

Given $\phi : X \to L$, $(\phi_L)_1 : S_1(X) \to L$ is

- ϕ on the first component,
- "abort"$(\phi(x))$ on the xth copy of $\mathit{2}$, where x is an object of X,
- $d(\phi(p))$ on the pth copy of Z, where p is an arrow in X.

So, in the case that $X = 1$, we have $S_1(1) = 1{+}\mathit{2}{+}Z$, ϕ may be regarded as an object A of L, and $(\phi_L)_1$ consists of A, the arrow "abort"(A) and a diagram of shape Z in L. So, we define $E(1) \to S_1(1) \times S_1(1)$ to be the map $1 + 1 \to S_1(1) \times S_1(1)$ with first component $(1, \mathrm{dom}(\mathit{2})) : 1 \to S_1(1)$ and a second component $(1, \mathrm{cod}(\mathit{2})) : 1 \to S_1(1) \times S_1(1)$.

We also have $S_1(\mathit{2}) = \mathit{2} + C(1,\mathit{2}) \cdot \mathit{2} + C(\mathit{2},\mathit{2}) \cdot Z$. We may write $S_1(\mathit{2}) = \mathit{2} + \mathit{2}_{\mathrm{dom}} + \mathit{2}_{\mathrm{cod}} + Z_{\mathrm{id}} + Z_{\mathrm{dom}} + Z_{\mathrm{cod}}$, where Z_{dom} is the copy of Z corresponding to the arrow dom : $1 \to \mathit{2}$ in C, and similarly for Z_{cod}. Given $\phi : \mathit{2} \to L$, ϕ may be regarded as an arrow p in L. Then, $(\phi_L)_2$ consists of p, the arrows "abort"(dom p) and "abort"(cod p), the square $d(p)$, and two other diagrams of shape Z in L. So, we define $E(\mathit{2}) \to S_1(\mathit{2}) \times S_1(\mathit{2})$ to be the map $\mathit{2}{+}\mathit{2}{+}\mathit{2}{+}\mathit{2} \to S_1(\mathit{2}) \times S_1(\mathit{2})$ with first component given by (x,y): $\mathit{2} \to \mathit{2}_{\mathrm{dom}} \times Z_{\mathrm{id}}$, where $x = \mathrm{id}_{\mathit{2}}$ and y is the inclusion onto α_1; second component given by the identity from $\mathit{2}$ to $\mathit{2}_{\mathrm{cod}}$ and the inclusion onto α_2; third component given by the identity from $\mathit{2}$ to $\mathit{2}$ and the inclusion onto α_4.

For all other X, $E(X) = 0$.

Hoare's central example of a generator is that which equips a category L with an endofunctor H. In order to formalize that, we want

$\mathrm{ob}H : C(1, L) \to C(1, L)$

$\mathrm{arr}H : C(\mathit{2}, L) \to C(\mathit{2}, L)$, and

$\mathrm{comp}H : C(\mathit{3}, L) \to C(\mathit{3}, L)$,

where $\mathit{3}$ is the category given by three objects 0, 1, and 2, with one map from m to n whenever $m \leq n$. We denote such a map by mn.

We must add equations to ensure that H preserves domains and codomains, that H preserves identities, and so that comp$H$ tells us precisely that H preserves composition. So, we put

$$S(X) = \begin{cases} 1 & \text{if } X = 1 \\ \mathit{2} & \text{if } X = \mathit{2} \\ \mathit{3} & \text{if } X = \mathit{3} \\ 0 & \text{otherwise}. \end{cases}$$

Then, $S_1(X) = X + \sum_Y C(Y, X) \cdot S(Y) = X + C(1, X) \cdot 1 + C(\mathit{2}, X) \cdot \mathit{2} + C(\mathit{3}, X) \cdot \mathit{3}$.

Given $\phi : X \to L$, $(\phi_L)_1 : S_1(X) \to L$ is

- ϕ on the first component,
- ob$H(\phi(x))$ on the xth copy of 1, where x is an object of X,
- arr$H(\phi(p))$ on the pth copy of $\mathit{2}$, where p is an arrow in X,
- a commutative triangle comp$H(\phi(f), \phi(g))$ on the (f,g)th copy of $\mathit{3}$, where (f,g) is a pair of composable arrows in X.

So, in the case $X = \mathit{2}$, we have

$$S_1(\mathit{2}) = \mathit{2} + 1_{\text{dom}} + 1_{\text{cod}} + \mathit{2}_{\text{id}} + \mathit{2}_{\text{dom}} + \mathit{2}_{\text{cod}} + C(\mathit{3}, \mathit{2}) \cdot \mathit{3},$$

using the same conventions as the previous example. Given $\phi : \mathit{2} \to L$, ϕ may be regarded as an arrow p in L. Then, $(\phi_L)_2$ consists of p, ob$H$(dom p), ob$H$(cod p), arr$H(p)$, and some other diagrams. So, we define $E(\mathit{2}) \to S_1(\mathit{2}) \times S_1(\mathit{2})$ to be the map $1 + S_1(\mathit{2}) \times S_1(\mathit{2})$ with first component given by $(x, y) \colon 1 \to 1_{\text{dom}} \times \mathit{2}_{\text{id}}$ where x is the identity and y choses the domain; second component given by the identity into 1_{cod} and choosing the codomain of $\mathit{2}_{\text{id}}$. Thus we have forced H to preserve domains and codomains.

For $X = 1$, we have

$$S_1(1) = 1 + 1 + \mathit{2} + \mathit{3},$$

ϕ may be regarded as an object A of L, and $(\phi_L)_1$ consists of A, ob$H(A)$, arr$H(1_A)$, and one other diagram. So, we define $E(1) \to S_1(1) \times S_1(1)$ to be the map $\mathit{2} \to S_1(1) \times S_1(1)$ whose first component choses the second copy of 1 in the above list and whose second component is the identity into $\mathit{2}$. Thus we have forced H to preserve identities.

Finally, for $X = \mathit{3}$, we have

$$S_1(\mathit{3}) = \mathit{3} + C(1, \mathit{3}) \cdot \mathit{3} + \sum_{m \leq n} \mathit{2}_{mn} + \mathit{3}_{id} + \text{ several more copies of } \mathit{3},$$

where $\mathit{2}_{mn}$ is the copy of $\mathit{2}$ corresponding to the map $mn : \mathit{2} \to \mathit{3}$ selecting the mn map in $\mathit{3}$. Given $\phi : \mathit{3} \to L$, ϕ may be regarded as a composable pair (f, g) in L. Then $(\phi_L)_{\mathit{3}}$ consists of (f, g), arr$H(f)$, arr$H(g)$, arr$H(f; g)$, a commutative triangle comp$H(f, g)$ and several other diagrams in L. We define $E(\mathit{3}) \to S_1(\mathit{3}) \times S_1(\mathit{3})$ to be the map $\mathit{2} + \mathit{2} + \mathit{2} \to S_1(\mathit{3}) \times S_1(\mathit{3})$ with first component given by $(x, y) : \mathit{2} \to \mathit{2}_{01} \times \mathit{3}_{\text{id}}$ where x is the identity and y is the inclusion into the 01 map of $\mathit{3}_{\text{id}}$; second component given by the identity map into $\mathit{2}_{12}$ and the inclusion into the 12 arrow of $\mathit{3}_{\text{id}}$. Thus we have forced the comutative triangle comp$H(f, g)$ to be given by arr$H(f)$, arr$H(g)$, and arr$H(f; g)$. This means that comp$H$ is completely determined by arr$H$, and the fact that it is commutative forces H to preserve composition.

For all other X, $E(X) = 0$.

Hoare's other main example of a generator equips a category with two endofunctors and a natural transformation between them. To do this, one gives two copies of the above operations together with equations for each copy, plus data and axioms for a natural transformation: so one adds

$$\alpha : C(1, L) \to C(\mathit{2}, L),$$

and

$$\alpha_{\text{arr}} : C(\mathit{2}, L) \to C(\text{"commuting square"}, L)$$

together with axioms to make the commuting square consist of the data appropriate to the definition of natural transformation, and to make α_A have the appropriate domain and codomain for each object A of L. To do this just requires minor modifications to the above.

Dubuc and Kelly [3] and Kelly and Power [8] contain many more examples, in varying degrees of detail.

5. ENRICHMENTS

Until now, we have only considered *Posets–Cat* as an ordinary category, whose objects are *Posets*–categories and whose arrows are *Posets*–functors. However, Hoare also considers upward simulations, etcetera. This leads us to the possible enrichments of *Posets–Cat*, which in turn affect the notion of "operations and equations" on *Posets–Cat*.

There are several possible enrichments of *Posets–Cat*. Enrichment over *Cat* replaces the homset of functors from $\mathcal{A}$ to $\mathcal{B}$ by the homcategory of functors from $\mathcal{A}$ to $\mathcal{B}$ together with natural transformations between them. Enrichment over a category commonly denoted by *Gpd* replaces the homset by a different homcategory, that of functors from $\mathcal{A}$ to $\mathcal{B}$ together with natural isomorphisms between them, i.e. those natural transformations that are invertible. These cases are both within the scope of Kelly and Power [8] so they both already have an existing theory, which will be discussed in more detail in Section 6.

Alas, although it is possible to enrich *Posets–Cat* so that we replace the homset by yet another homcategory, that of functors from $\mathcal{A}$ to $\mathcal{B}$ together with upward simulations between them, this enrichment is not an instance of the work in [8]. The difficulty is that the associated tensor product is not symmetric. We can still discuss monads on *Posets–Cat* and their enriched categories of algebras, and we can still discuss presentations by operations and equations; but the general theory linking the two notions simply has not been done yet to the best of my knowledge. Partly, there have not been pressing examples; and partly, it seems to add even more difficulty to an already quite difficult job. Such an analysis may well be possible in the future, as we come to a deeper understanding of the work in [8].

We have one further remark on enrichment. Hoare has restricted his attention to *Posets*–categories and *Posets*–functors. In similar mathematical situations, this has often proved not to be the definitive formulation. In such situations, the theory must encompass a treatment of non-identity isomorphisms in order to encapsulate the examples adequately. For instance, one is generally interested in any functor that takes finite products to finite products, rather than only those that take assigned finite products to assigned finite products, as there may be several possible choices of assignment. If Hoare's work extends naturally to structures rather than properties, which is certainly what his paper suggests, the definitive treatment may well be at the level of "2–categories". Substantial work has been done at this level, much of it specifically focusing on the issue of non-identity isomorphisms; a foundational paper on this work is Blackwell, Kelly and Power [2].

6. ENRICHMENT OVER *Cat* AND *Gpd*

The theory of monads on *Posets–Cat* enriches over *Cat* in a straight-forward manner, as described in [8]. The notions relevant to "operations and equations" do likewise. The only changes are that an S*-algebra* is now a functor from the category $C(X,L)$ to the category $C(S(X),L)$, and (S,E)*-algebras* are modified accordingly; an (S,E)*-algebra map* now amounts to an arrow $F : L \to M$ in C such that not only does $\phi_L; F = (\phi; F)_M$ for each $\phi \in C(X,L)$, but also, for any arrow $\sigma : \phi \to \psi$ in $C(X,L)$, we have $\sigma_L; F = (\sigma; F)_M$, where the semicolon is now extended to mean a horizontal composite of a natural transformation with a functor. An (S,E)*-algebra 2-cell* or an (S,E)*-algebra natural transformation* is a natural transformation $\gamma : F \to G$, such that for each $\phi \in C(X,L)$, we have $\phi_L; \gamma = (\phi; \gamma)_M$. Thus, we have (S,E)-*Alg* as a *Cat*-category, or in the usual abbreviated notation, as a 2-category, and we have a *Cat*-enriched version of Theorem 4.7.

Most of Hoare's examples do enrich over *Cat* in an evident manner by simply replacing each function of the form $d : C(X,L) \to C(S(X),L)$ by an evident functor for which the function is the map on objects; and by deleting some redundant data. Alas, "abort" is an exception because the functoriality of "abort" : $C(1,L) \to C(2,L)$ gives an equality where one wants an inequality. The other exception of note is where Hoare introduces a contravariant functor on L or higher-order function spaces. It is no coincidence that, in these cases, upward simulation is not valid: if upward simulation were valid, then it is easy to show that natural transformations should be valid; but that in turn says precisely that the monad enriches over *Cat*; and it has been proved in [2] that the monad for function spaces has no *Cat*-enrichment. So, the above analysis proves that upward simulation cannot be valid in some cases where Hoare shows it is invalid.

However, all of Hoare's examples do enrich over *Gpd*, i.e. it is possible to lift each function $d : C(X,L) \to C(S(X),L)$ to a functor on the natural isomorphisms of $C(X,L)$ and to do likewise with the equations. This is essentially obvious, but even more, there is a theorem to assert that it is always possible. This supports Hoare's contention that total simulation is valid for all of his constructions: if total simulation were invalid, then replacing all inequalities by the special case of equality, it would suggest that natural isomorphism was probably also invalid, but that would contradict the fact that his constructions all enrich over *Gpd*.

Finally, we may observe that if we restrict attention to natural transformations and natural isomorphisms rather than upward simulations, etcetera, all of Hoare's constructions thus restricted, are valid where he claims so, for the following general reason, if only we assume that L_0 is a *Posets*-category (so composition is defined) and L is free on L_0, as he seems to imply.

PROPOSITION 6.1. *Let (S,E) be a Cat-enriched theory of operations and equations on the 2-category Posets-Cat. Let L be the free (S,E)-algebra on L_0; let $F,G : L \to M$ be (S,E)-algebra maps; and let $\gamma : F_0 \to G_0$ be a natural transformation, where F_0 and G_0 are the restrictions of F and G to L_0. Then, γ extends uniquely toan (S,E)-algebra natural transformation from L to M.*

PROOF: This holds simply because the left adjoint to the forgetful functor from (S,E)-*Alg* to *Posets*-*Cat* is *Cat*-enriched.

In all of his examples, Hoare gives an explicit construction of the extension of γ

to L, and his conclusion follows directly from the conclusion of the proposition. A similar proposition is true for Gpd–enrichment.

BIBLIOGRAPHY

1. M. Barr and C. Wells, *Toposes, triples, and theories, Grundlehren der Math. Wissenschaften*, Springer Verlag.
2. R. Blackwell, G. M. Kelly and A. J. Power, *Two-dimensional monad theory*, J. Pure Appl Algebra, to appear.
3. E. J. Dubuc and G. M. Kelly, *A Presentation of Topoi as Algebraic Relative to Categories or Graphs*, J. Algebra **81** (1983), 420–433.
4. P. Gabriel and F. Ulmer, *Lokal prasentierbare Kategorien*, Lecture Notes in Math. **221** (1971), Springer Verlag.
5. J. W. Gray, *Formal category theory: Adjointness for 2–Categories*, Lecture Notes in Math. **391** (1974), Springer Verlag.
6. C. A. R. Hoare, *Data refinement in a categorical setting,* draft, 1987.
7. G. M. Kelly, *Basic Concepts of Enriched Category Theory*, London Math. Soc. Lecture Notes Series **64** (1982).
8. G. M. Kelly and A. J. Power, *Algebraic structure in the enriched context,* draft, 1988.

Categorical Semantics for Programming Languages

He Jifeng and C.A.R. Hoare
Programming Research Group
Oxford University Computing Laboratory

1 Introduction

The cartesian closed monoid [6] provides an algebraic semantics for an untyped lambda calculus; this language includes a surjective pairing (*cons*), partial functions, non-strict parameter passing and recursion. The cartesian closed category [7] (with natural number objects) provides an algebraic semantics for the typed lambda calculus, which includes cartesian products, discriminated unions, total functions and iteration. Topos theory [7] extends the range of definable functions and types.

This paper applies categorical method to the definition of a range of conventional programming languages, which include conditionals, non-determinism, non-termination, strictness, recursion, higher order procedures and even communications. A good example is Dijkstra's language with predicate transformer semantics [1]. To achieve this greater generality, we use a simple kind of 2-category theoty, i.e., a category whose homsets are preordered and whose composition is monotonic.

Section 2 reviews the relevant categorical concepts. We need a preorder with bottom element ($\perp$), the *glb* (greatest least bound, $\sqcap$), and limits of ascending chains. The intended meaning of $q \sqsubseteq p$ in this order is that program p is as good as q or better, for any purpose and in any context of use. A simple 2-category has such a preorder defined on each of its homsets. The definition of a natural transformation can be weakened to that of an up-simulation, or its dual, a down-simulation (called 2-natural transformation in [2]). In order to give an algebraic semantics to a wider variety of programming languages, including strict, lazy, and non-deterministic languages, the definition of an adjunction will be weakened, so that its unit and counit are simulations. In this way, simple algebraic distinction are made between familiar functional models and more conventional languages. For example the command **abort** of Dijkstra's language is the bottom element of the preorder (the worst of all programs). This is also a *zero morphism* in Dijksstra's language, i.e., a natural transformation from the identity functor to itself. In a language like **CSP** (with communication) [4], it is an up-simulation; and in **Miranda** (with partial functions and non-strict parameter) [9], it is a down-simulation. The definition of a product can also be weakened to accommodate non-determinism.

Section 3 applies the theoretical results of section 2 to a selection of constructors in a range of familiar languages. We show that they have familiar categorical interpretations. The method exploreed in this section will be extended in a later paper [5] to provide a uniform algebraic semantics to a wider variety of programming languages, including strict, lazy and non-deterministic languages.

2 Categorical Preliminaries

This section recalls the preliminary concepts in 2-category theory [2]. We assume that familiarity with most of the basic definition of category theory [7]

A 2-category C is a category with two different compositions ; (*horizontal composition*) and ; (*vertical composition*) such that

- each of its homsets C(b, c) is a category with the vertical composition.
- identities for the horizontal composition are also idntities for the vertical composition.

- these compositions together satisfy the interchange law

$$(p ; q) ; (r ; s) \; = \; (p ; r) ; (q ; s)$$

A simple example of 2-category is a category C with a preorder $\sqsubseteq$ on hom sets and with respect to which the categorical composition ; is monotonic

$$p \sqsubseteq q \Rightarrow (p; r) \sqsubseteq (q; r) \wedge (s; p) \sqsubseteq (s; q)$$

C is called a *order enriched category* if the preorder $\sqsubseteq$ defined on homset has a bottom element $\perp$, the *glb* ($\sqcap$), and limits of ascending chains ($\sqcup_n p_n$)

A total function F from a 2-category L to a 2-category M is said to be a *2-functor* if it preserves horizontal and vertical identities and distribute through both the compositions

$$\begin{aligned} F(p ; q) &= Fp ; Fq \\ F(p ; q) &= Fp ; Fq \end{aligned}$$

The 2-functors of order enriched categories are those functors which are monotonic

$$p \sqsubseteq q \; \Rightarrow \; Fp \sqsubseteq Fq$$

Let F and G be 2-functors from an order enriched category L to an order enriched category M. Then a transformation t from G to F is said to be a *Cat-natural transformation* (abbreviated $t : F \dot{\rightarrow} G$) if for all p in L

$$Fp ; t_{\overrightarrow{p}} \; = \; t_{\overleftarrow{p}} \; ; \; Gp$$

where $\overleftarrow{p}$ and $\overrightarrow{p}$ stand for the source and target of the morphism p respectively.

Using the preorder on homsets we can construct a weaker version of a natural transformation called a *simulation*, by combining the left and right hand sides of the above defining equation with an inequality rather than equality. t is said to be a *down-simulation* (abbreviated $t : F \sqsubseteq G$) if

$$Fp; t_{\vec{p}} \sqsubseteq t_{\overleftarrow{p}} ; Gp$$

In [2] t is called a *two-natural transformation.*

Dually t is said to be a *up-simulation* (abbreviated $t : F \sqsupseteq G$) if

$$Fp; t_{\vec{p}} \sqsupseteq t_{\overleftarrow{p}} ; Gp$$

It is obvious that a Cat-natural transformation is both an up-simulation and a down-simulation.

The standard definition of an adjunction in terms of natural transformations can be generalised in a similar way. This definition says that an adjunction between categories L and M is a quadruple (F, G, ϵ, η) where $F : L \rightarrow M$ and $G : M \rightarrow L$ are functors, and and satisfy the following conditions

- $\epsilon : FG \dot{\rightarrow} I_L$.
- $\eta : I_M \dot{\rightarrow} GF$.
- $F\eta ; \epsilon F = I_F$.
- $\eta G ; G\epsilon = I_G$.

where I_L and I_M are identity functors on L and F respectively.

We define the weaker notion of a *quasi-adjunction* between two categories L and M to be a quadruple (F, G, ϵ, η) satisfying

- $\epsilon : FG \sqsupseteq I_L$.
- $\eta : I_M \sqsupseteq GF$.
- $F\eta ; \epsilon F = I_F$.

- $\eta G\,;\, G\epsilon \;=\; I_G$.

Following this way we can define *quasi-coproduct* and *quasi-product*, and the notion of *junction* can be generalised to so-called *quasi-junction.*

The uniqueness theorem for adjunctions has the following analogue for quasi-adjunction [5] where a *quasi-isomorphism* is an isomorphism defined in terms of equivalence induced by the preorder $\sqsubseteq$ rather than equality.

Theorem
Let $(F, G_1, \epsilon_1, \eta_1)$ and $(F, G_2, \epsilon_2, \eta_2)$ both be quasi-adjunctions. If they are adjunctions between wide subcategories L' and M' of L and M, then functors G_1 and G_2 are quasi-isomorphic.

3 Language Constructors

In this section we show that many constructors in a range of programming languages have familiar categorical interpretations. For simplicity, we will suppress mention of types (objects) whenever possible.

3.1 Composition

In all programming languages of interest, there exists a composition operator (denoted here as $p; q$, elsewhere $p * q$). Execution of such a composite program usually (but not always) involves execution of both of its components. In a procedural programming language like **PASCAL**, we interpret this notation as *sequential execution*: q does not start until p has successfully terminated. In a functional language it denotes functional composition. This operator is associative

$$(p\,;\,q)\,;\,r \;=\; p\,;\,(q\,;\,s)$$

It has both a left and a right unit. In Dijkstra's language [1], the unit is the command skip

$$\textbf{skip}\,;p \;=\; p;\textbf{skip} \;=\; p$$

In a typed language, the composition of programs is undefined when the type of the result of the first component differs from that expected by the second component. This can be treated in category theory by associating source and target types with each program. $(p\,;\,q)$ is then well-defined iff $\overrightarrow{p} = \overleftarrow{q}$.

In the rest of this section we assume without explicit mention that all type constraints have been observed.

A *zero* of composition (if it exists) is denoted by $\emptyset$. It is the program that fails to terminate. The defining property of the zero program is

$$p\,;\emptyset \;=\; \emptyset \;=\; \emptyset\,;\,q$$

In words, a program which starts by failing to terminate is indistinguishable from one which ends by failing to terminate.

In Dijkstra's language, this role of zero program is played by the program **abort** which is the *bottom element* in its homset. It may fail to terminate; or being non-deterministic it may do even worse: it may terminate with the wrong result, or even the right one (sometimes, just to mislead you). To specify the execution of q after termination of **abort** cannot redeem the situation, because **abort** cannot be relied on to terminate. To specify execution of p before abortion is equally ineffective, because the non-termination will make any result of executing p inaccessible and unusable. In other words, composition in Dijkstra's language is *strict* in the sense that it gives bottom if either of its arguments is bottom. The above defining equation states that zero program is a natural transformation.

A language like **CSP** [4] contains commands for input and output, which have results observable before the program terminates (or fails to do so). Consequently, the aborting command **chaos** does not satisfy the above equation. However it has the weaker property that non-termination after performing the inputs and outputs of p cannot be worse than immediate non-termination. So for **CSP**, the defining property of the aborting command must replaced by

$$\emptyset ; q \; = \; \emptyset \sqsubseteq p ; \emptyset$$

which states that $\emptyset$ is an up-simulation.

In a lazy functional programming language like **Miranda** [9], the call of a function will not evaluate an argument unless the value of the argument is actually needed during execution of the body of the function. As a result, it may terminate even when applied to a non-terminating argument. However, the wholly undefined function always fails. On the principle the failure is worse than any kind of success, the property of zero program has to be replaced by

$$p ; \emptyset \; = \; \emptyset \sqsubseteq \emptyset ; q$$

i.e., zero programs become a down-simulation in this case.

We use $p \sqcap q$ to denote the best common approximation in the $\sqsubseteq$ ordering of both p and q, if it exists. It can be defined by the single law

$$r \sqsubseteq (p \sqcap q) \;\; \textit{iff} \;\; r \sqsubseteq p \; \textit{and} \; r \sqsubseteq q$$

We are going to explore the way in which composition interacts with the $\sqcap$ operator. From the defining property of $\sqcap$ and the monotonicity of composition we can derive the following weak distributive law

$$r ; (p \sqcap q) ; s \;\sqsubseteq\; (r ; p ; s) \sqcap (r ; q ; s)$$

In Dijkstra's language (and other truly non-deterministic language like **CSP**), $\sqcap$ denotes non-determinism; and the law can be strengthened to an equation

$$r ; (p \sqcap q) ; s \;=\; (r ; p ; s) \sqcap (r ; q ; s)$$

This law states that it makes no difference whether the selection between p and q is made before execution of the first operand of a composition (e.g., at compiler time), or whether it is made (at run time) after execution of the first operand. In other words, the $\sqcap$ is a junction.

However, in a functional or deterministic language it is better to postpone the application of $\sqcap$ as long as possible, because it somehow worsens its argument. The above strengthening is not valid, instead we have

$$\begin{array}{rcl} (p \sqcap q)\,;\, s & \sqsubseteq & (p\,;\, s) \sqcap (q\,;\, s) \\ r\,;\,(p \sqcap q) & = & (r\,;\, p) \sqcap (r\,;\, q) \end{array}$$

In this case, the $\sqcap$ operator is a quasi-junction.

3.2 Disjoint Union

The coproduct (disjoint union) constructor will be denoted by an infix $+$. $b + c$ is the discriminated union type, which appears, for example, in **PASCAL** as a variant record. $(p + q)$ is a case discrimination. When applied to a value of type $(\overleftarrow{p} + \overleftarrow{q})$ it first tests which variant it comes from. If it is the first variant, then p is applied, obtaining a result of type $\overrightarrow{p}$, which is then injected into the first variant of $(\overrightarrow{p} + \overrightarrow{q})$. The treatment of the second case is similar. Thus

$$\begin{array}{rcl} \overrightarrow{(p + q)} & = & \overrightarrow{p} + \overrightarrow{q} \\ \overleftarrow{(p + q)} & = & \overleftarrow{p} + \overleftarrow{q} \end{array}$$

Furthermore, it is easy to see that the above description of the case discrimination satisfies the other defining property of a bifunctor

$$(p + q)\,;(r + s) \;=\; (p\,;r) + (q\,;s)$$

The discriminated union provides a convenient method of modelling th familiar *conditional construction* of a programming language. For exam ple, the test "*even*", which tests whether a number is odd or even, can b regarded as a function from the natural number **N** to the disjoint unio N + N. When applied to an even number , say $2n$, its result $(0, 2n)$ is th same number tagged as in the first alternative of the discriminated unior whereas an odd number $2n + 1$ is mapped into $(1, 2n+1)$, the same numbe tagged as in the second alternative. To halve a number if it is even, or ad one if it is odd, can be achieved by the composition

$$even\,;(halve\ +\ add)$$

But it still remains to map the result of this conditional from the discrim inated union N + N back to the single natural number type N. For thi we need for each type b, a *merge* operator symbolised by ∇b, which maps disjoint union $(b + b)$ onto the type b, simply by forgetting the tag whic determines from which of the two (identical) types its argument has origi nated. Thus to achieve the effect

$$\textbf{if}\, even(x)\, \textbf{then}\, x := x/2\, \textbf{else}\, x := x + 1\, \textbf{fi}$$

the conditional described above should be completed as follows

$$even\,;(halve\ +\ add)\,;\,\nabla\mathbf{N}$$

If p maps b to c, p may be applied after the merging operation ∇b, or it ma be applied to both alternatives before the merging operator ∇c; the fina result of each of these applications will be the same. Thus merging operato satisfies

$$(p + p)\,;\,\nabla\,\overrightarrow{p} \;=\; \nabla\,\overleftarrow{p}\,;p$$

The above algebraic law states that ∇ is a natural transformation betwee the identity functor and the functor that maps p to $(p + p)$.

In a programming language, there are two extreme conditions for each pair of types b and c, $true_{b,c} : b \rightarrow (b+c)$ and $false_{b,c} : c \rightarrow (b+c)$:

- $true_{b,c}$ which tags its argument as the first alternative of type $b+c$.
- $false_{b,c}$ which tags its argument as the second alternatives of type $b+c$

These are called insertion functions. Thus if $(p + q)$ is executed after $true_{\overleftarrow{p},\overleftarrow{q}}$, the first alternative p is invariably selected; so the effect is the same as if p had been applied beforehand

$$true_{\overleftarrow{p},\overleftarrow{q}}\,;\,(p+q) \;=\; p\,;\,true_{\overrightarrow{p},\overrightarrow{q}}$$

Similarly

$$false_{\overleftarrow{p},\overleftarrow{q}}\,;\,(p+q) \;=\; q\,;\,false_{\overrightarrow{p},\overrightarrow{q}}$$

Thus both condition *true* and condition *false* are natural transformations. Furthermore they satisfy

$$\begin{aligned} true_{b,b}\,;\,\nabla b &= id_b \\ falae_{b,b}\,;\,\nabla b &= id_b \\ (true_{b,c} + false_{b,c})\,;\,\nabla(b+c) &= id_{b+c} \end{aligned}$$

where id_b stands for the identity function on the type b.

However, in a non-strict programming language the discriminated union of types b and c is not simply the disjoint sum of b and c as described before, but is defined by

$$b + c \stackrel{\text{def}}{=} \{\perp\} \cup \{(x, 0) \mid x \in b\} \cup \{(y, 1) \mid y \in c\}$$

where a new element $\perp$, represents the bottom element of the union type. The program $p + q$ will map $(x, 0)$ where $x \in \overleftarrow{p}$ to $(px, 0)$, and $(y, 1)$ where

$y \in \overleftarrow{q}$ to $(qy, 1)$, and the bottom element $\perp$ to $\perp$. The merging operator ∇ will be defined by

$$\begin{aligned} \nabla b : (b + b) &\rightarrow b \\ (x, 0) &\mapsto x \\ (y, 1) &\mapsto y \\ \perp &\mapsto \perp \end{aligned}$$

In this language, $+$ is a *quasi-coproduct* [5], in a sense defined up to equivalence by the laws previously given for coproduct except that the merging operator is a downward simulation, and governed by

$$(p + p) ; \nabla \overrightarrow{p} \sqsubseteq \nabla \overleftarrow{p} ; p$$

This is because the program $(p + p) ; \nabla \overrightarrow{p}$ will map $\perp$ to $\perp$, but $\nabla \overleftarrow{p} ; p$ will not so when the program is non-strict.

3.3 Product

A similar treatment can be given to the product bifunctor $p \times q$, where programs p and q are assumed to be run in parallel without interference. The associated natural transformations are the *projections* $\pi_{b,c} : (b \times c) \rightarrow b$ and $\mu_{b,c} ; (b \times c) \rightarrow c$, and the *duplicating* operator $\Delta b : b \rightarrow (b \times b)$, which maps x of type b to the pair (x, x). In a category of total functions, they satisfy

$$\begin{aligned} (p \times q) ; \pi_{\overrightarrow{p}, \overrightarrow{q}} &= \pi_{\overleftarrow{p}, \overleftarrow{q}} ; p \\ (p \times q) ; \mu_{\overrightarrow{p}, \overrightarrow{q}} &= \mu_{\overleftarrow{p}, \overleftarrow{q}} ; q \\ \Delta \overleftarrow{p} ; (p \times p) &= p ; \Delta \overrightarrow{p} \end{aligned}$$

Let p and q be programs with $\overleftarrow{p} = \overleftarrow{q}$, we define their *product* $< p, q >$ to be a program which makes a second copy of the current argument, and execute p on one of the two copies and q on the other one, and delivers the two results as a pair. In a functional programming language with lists as a data structure, this can be defined:

$$< p, q > \stackrel{\text{def}}{=} \lambda x. cons(px, qx)$$

In a categorical setting it can be formulated by

$$< p, q > \stackrel{\text{def}}{=} \Delta \overleftarrow{p} ; (p \times q)$$

From the defining properties of Δ and bifunctor $\times$ it follows that

$$< p; q; r, \ p; s; t > \ = \ p ; < q, s > ; (r \times t)$$

This states that the product function is actually a left junction from the duplicating functor, that maps p to a pair (p, p), to the bifunctor $\times$.

But in many language the above equations do not hold. Suppose that the calculation on q fails to terminate. Then the execution of $(p \times q) ; \pi_{\overrightarrow{p}, \overrightarrow{q}}$ in a strict language like **LISP** will also fail to terminate. The program $\pi_{\overleftarrow{p}, \overleftarrow{q}} ; p$ does not involve an operation on the discarded alternative q, and will therefore terminate in cases $(p \times q) ; \pi_{\overrightarrow{p}, \overrightarrow{q}}$ will not. This can be expressed mathmatically by inequations stating that the projections π and μ are downward simulations from the product bifunctor to the bifunctor that selects one of its operands,

$$(p \times q) ; \pi_{\overrightarrow{p}, \overrightarrow{q}} \sqsubseteq \pi_{\overleftarrow{p}, \overleftarrow{q}} ; p$$
$$(p \times q) ; \mu_{\overrightarrow{p}, \overrightarrow{q}} \sqsubseteq \mu_{\overleftarrow{p}, \overleftarrow{q}} ; q$$

The strong equations, of course, remain true for a lazy functional language, in which no result is computed until it is known to be needed.

In a programming language which permits non-determinism, the duplicating operator does not satisfy the equation $\Delta\,\overleftarrow{p}\,;(p \times p) = p;\Delta\,\overrightarrow{p}$. If p is non-deterministic, the two occurrences of p on the left hand side may produce different results, even when starting with the same value. However, equal results on the left hand side are still possible (by chance, say). So the left hand side can only be inferior in the sense that it is more non-deterministic. The right hand side is still a valid optimisation, as expressed by the upward simulation property [3]

$$\Delta\,\overleftarrow{p}\,;(p \times p) \sqsubseteq p;\Delta\,\overrightarrow{p}$$

Consequently one has

$$\begin{aligned} < p;q,\, p;r > &\sqsubseteq p;< q, r > \\ < q;r,\, s;t > &= < q, s >;(r \times t) \end{aligned}$$

3.4 Higher Order Functions

As useful example of a bifunctor of mixed variance is the exponental bifunctor, denoted by $\Rightarrow$. $(b \Rightarrow c)$ is a function space of functions from b to c. $(p \Rightarrow q)$ is an operation which when applied to a function f delivers the function $(p;f;q)$ as result. So the type consistency equires that f must be in $(\overrightarrow{p} \Rightarrow \overleftarrow{q})$ and the result will be in $(\overleftarrow{p} \Rightarrow \overrightarrow{q})$. So

$$(p \Rightarrow q) : (\overrightarrow{p} \Rightarrow \overleftarrow{q}) \longrightarrow (\overleftarrow{p} \Rightarrow \overrightarrow{q})$$

Furthermore $(p \Rightarrow q);(r \Rightarrow s)$ applied to f is

$$r;(p;f;q);s = (r;p);f;(q;s)$$

which is the same as $(r;p) \Rightarrow (q;s)$ applied to f. So we deduce

$$(p \Rightarrow q)\,;(r \Rightarrow s) = (r\,;p) \Rightarrow (q\,;s)$$

In summary, the bifunctor $\Rightarrow$ is contravariant in its first operand, covariant in its second.

Consider a function $f : (b \times c) \rightarrow a$, which takes a pair of arguments. The curried version of f is the same as f, except that it takes its arguments one at time. Thus $(curry\, f) : b \rightarrow (c \Rightarrow a)$ is a function which expects an argument x of type b, and delivers as result another function from c to a. When this latter function is applied to an argument y in b, it delivers the same result as f does when applied to the pair (x, y). More simply, in symbols

$$((curry\, f)\, x)\, y) = f(x, y)$$

In category theory use of variable is forbidden; furthermore, the operator needs to be subscripted by the types of its operands and is characterized by the following laws

$$curry_{b,c,d}(f) : b \rightarrow (c \Rightarrow a) \text{ for } f : (b \times c) \rightarrow a$$
$$curry_{\overleftarrow{p},\overleftarrow{q},\overrightarrow{r}}((p \times q); f; r) = p\,; curry_{\overrightarrow{p},\overrightarrow{q},\overleftarrow{r}}(f)\,; (q \Rightarrow r)$$

This states that *curry* is a contravariant junction from the covariant bifunctor $\times$ to the mix-variant bifunctor $\Rightarrow$.

The currying operator has an inverse called *uncurrying.* Its defining properties are

$$uncurry_{b,c,d}(f) : (b \times c) \rightarrow a \text{ for } f : b \rightarrow (c \Rightarrow a)$$
$$uncurry_{\overleftarrow{p},\overleftarrow{q},\overrightarrow{r}}(p; f; (q \Rightarrow r)) = (p \times q)\,; curry_{\overrightarrow{p},\overrightarrow{q},\overleftarrow{r}}(f)\,; r$$

3.5 Recursive Programs

Let Ψ be a continuous constructor satisfying for any program p

$$\Psi(p) : \overleftarrow{p} \to \overrightarrow{p}$$

The recursive program $\mu x_{b,c}.\ \Psi(x_{b,c})$ is defined in e.g., [8] as the least uppe bound of the ascending chain

$$\emptyset_{b,c} \sqsubseteq \Psi(\emptyset_{b,c}) \sqsubseteq \Psi^2(\emptyset_{b,c}) \sqsubseteq ....$$

where $\emptyset_{b,c}$ denotes the worst program with the source type b and the targe type c.

From the property of the least upper bound operator $\sqcup_n$ we can deriv for any ascending chain $\{p_n\}$

$$\sqcup_n (p_n \,;\, q) \sqsubseteq \sqcup_n(p_n) ; q$$

This law states that the least upper bound operator is a quasi-junction.

In Dijkstra's language the loop program **do** $b \to p$ **od** is defined as th least fixed point of the recursive equation

$$x = \textbf{if } b \textbf{ then } p ; x \textbf{ else skip fi}$$

Acknowledgement

Thank the anonymous referees for their help comments on earlier version this paper. Also to the Admiral B.R. Inman Centennial Chair in Computin Theory at the University of Texas at Austin for support during the studi which led to this paper. The research was also supported in part by tl Science and Engineering Research Council of Great Britain.

References

[1] E.W. Dijkstra, *A Discipline of Programming.* Prentice-Hall, Englewood Cliffs. NJ, (1976).

[2] J.W. Gray, *Formal Category Theory: Adjointness for 2-categories.* LNM 391, Springer-Verlag, (1974).

[3] M. Hennessy, *The semantics of call-by-value and call-by-name in a non-deterministic environment.* SIAM J. Comp. (1980), 67–85.

[4] C.A.R. Hoare, *Communicating Sequential Processes.* Prentice-Hall, (1985).

[5] C.A.R. Hoare and He Jifeng, *Two-categorical Semantics for Programming Languages.* in preparation.

[6] J. Lambek and P.J. Scott, *Introduction to higher order categorical logic* Cambridge University Press, (1985).

[7] Sanders Mac Lane, *Categories for the working mathematicians.* Springer-Verlag, New York Inc. (1971).

[8] D.S. Scott, *The lattice of flow diagrams.* Symposium on Semantics of Algorithmic Languages, LNM 118, E. Engeler (ed.), (1971) 311–366.

[9] D.A. Turner, *Miranda, a non-strict functional language with polymorphic types.* LNCS 201, Springer-Verlag, (1985) 1–16.

Initial Algebra Semantics for Lambda Calculi

by John W. Gray[1]
University of Illinois at Urbana-Champaign

Abstract. Ordinary and polymorphic typed lambda calculi are constructed as initial algebras for suitable endofunctors. The semantics is realized as the unique morphism from the initial algebra to an appropriate semantic algebra. In the case of the polymorphic lambda calculus, this semantic algebra is constructed from the category **bPER**$^{\bullet}$ of pointed partial equivalence relations.

0. Introduction

In [4], the structure of the syntax and semantics of various kinds of lambda-calculi is outlined. It is shown there how the syntax of a lambda calculus can be presented as an initial algebra for an appropriate endofunctor (except in the polymorphic case where the family of terms is only a subalgebra of such an initial algebra.) However, the construction of the semantics from this observation is not pursued there. What would be desirable would be for the semantics to be given as the unique morphism from such an initial algebra to some other semantic algebra constructed from mathematical data without reference to the language itself. The reason that this is desirable is that denotational semantics means that the meaning of an expression is determined by the meaning of its parts. From this one concludes two things:

i) Meanings must be determined recursively. To determine the meaning of an expression built from two other expressions, one must first determine their meanings. These are in turn determined by the meanings of their parts. In order that this not lead to an infinite regress, expressions must ultimately be built from certain basic parts whose meanings are given explicitely.

ii) The way in which the meaning of an expression is determined by the meaning of its parts must be specified in advance in some explicit form independent of the notion of meaning. For instance, the meaning of "x + y" is given by the meanings of x and y, which might be numbers and the meaning of "+" which is a certain function "plus" of two variables defined for numbers. The meaning of "x + y" is then the value of the function "plus" for the arguments given by the meanings of x and y.

---

[1]This work was partially supported by the National Science Foundation.

The points in i) and ii) describe why semantics should be given by an initial morphism from the initial algebra to some other semantic algebra. In order for i) to make sense, expressions must be built up recursively from basic expressions; i.e., expressions form the initial algebra (or term algebra) for some rules of construction. The values of meanings must lie in some domain where these rules have some concrete interpretation; i.e., in some semantic algebra for the same rules that determine the syntax of expressions.

As will be seen, this procedure works very well for ordinary typed lambda-calculi, but it breaks down in the polymorphic case for two reasons:

i) the semantics of the polymorphic lambda calculus demands outsized products.

ii) the syntax of the polymorphic lambda calculus is given by a context dependent grammar.

Nevertheless, we carry out the development as far as possible for an arbitrary category **C** to bring out its essential simplicity. The difficulties are met by chosing **C** to be the category **pPER$^\bullet$** of pointed partial equivalence relations on the natural numbers where we constructi an associated context free syntax and by giving a correct semantics in the associated category **ω-pSET$^\bullet$**.

The description of semantics given here attempts to make explicit what is tacitly understood in other descriptions. It must be that all of these observations are well-known (possibly in other forms) but it is difficult to dig them out of the literature. In particular, Lambek and Scott [9] provides a semantics only for terms with exactly one free variable. (Of course this variable can be of product type, but that still leaves many terms uninterpreted.) Reynolds [15], [16], [17] has been a constant guide in the development presented here. A chief difference between the work here and that of Reynolds is that variables here come equipped with types, so there are no type assignments in our description. A similar study has been carried out by Oles in his thesis [12], [13]. In Oles's work, which also uses type assignments, type coercions are a central concept and introduce considerable complexity into the description. However, this paper is the first place that pointed sets, or pointed pers have been explicitly introduced.

There are three general considerations that should be mentioned:

i) A lambda-calculus consists of entities of two kinds - types and terms. Each of these is given separately as an initial algebra with the endofunctor for terms constructed from the initial algebra for types. Note that we are concerned only with *free* typed lambda-calculi here. Lambek and Scott [8] also considers the non-free case.

ii) The semantic algebras will be constructed by means of categorical data from large categories so the carriers of these algebras will in general not be small sets. They will instead be large sets (proper classes), but this causes no difficulties since the description of initiality is completely first order. The endofunctors for these algebras will be described as though they were endofunctors on the category of (small) sets, but since they are all polynomial functors (in the sense of Manes and Arbib [11]), there is no difficulty in giving them proper classes as arguments.

iii) An algebra for an endofunctor H is an object X together with a morphism from H(X) to X. (For details, see 1.2 below.) There are no equations or relations to be verified, so this situation is intrinsically simpler than the Lambek and Scott treatment where it has to be verified that a cartesian closed category has actually been constructed from a lambda calculus. In effect, the only thing that has to be verified in constructing an algebra is that one has actually given a morphism with the required domain and codomain. Since our endofunctors are on the category of sets, this means that one just has to describe a *function* with the required domain and codomain.

1. SYNTAX OF LAMBDA-CALCULI

1.1 Definition. A *typed λ-calculus* L consists of a set T of types (called a *type system* here) and a set Term_τ of terms of each type $\tau \in T$.

1.2 Types

We treat two kinds of type systems: ordinary type systems and polymorphic type systems. Both will be constructed as initial algebras for suitable endofunctors on the category **SET** of sets and functions. Recall that if H : **SET** → **SET** is an endofunctor, then an H - algebra is a set X together with a structure function h : H(X) → X. X is called the *underlying set* of the H - algebra (X, h). If (X, h) and (X', h') are H - algebras, then a *homomorphism* between them is a function f : X → X' such that h' • H(f) = f • h: i.e., such that the adjoining diagram commutes. If I is the initial H-algebra (i.e., there is a unique algebra homomorphism from it to any other H - algebra) then the structure function i : H(I) → I is an isomorphism by Lambek [8]. I is constructed as the colimit of the chain $H^k(\emptyset)$ providing H is sequentially cocontinuous. If (X, h) is any H-algebra, then the unique H-algebra homomorphism from (I, i) to (X, h) is the unique function

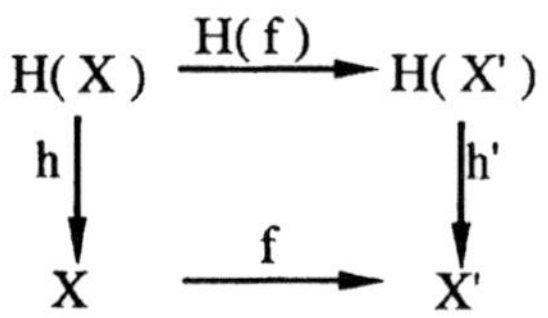

$$u : \mathrm{colim}_k H^k(\emptyset) \to X$$

whose components are given recursively by

$$u_0 : \emptyset \to X, \;\; u_{k+1} = h \bullet H(u_k) : H^{k+1}(\emptyset) \to X.$$

The kinds of endofunctors that occur here are the polynomial endofunctors that arise from algebraic signatures. (Cf. Ehrig and Mahr [2]) Consider a single sorted algebraic signature S consisting of operations $a : X^i \to X$ for $a \in A_i$, $0 \le i \le n$. Such an S determines an endofunctor H_S : **SET** → **SET** by setting

$$H_S(X) = \sum_{i=0}^{n} (A_i \times X^i)$$

where Σ means the coproduct (or disjoint union) of the indicated sets. A function h : $H_S(X) \to X$ is determined by n + 1 - cases, or components, $h_i : A_i \times X^i \to X$, and each h_i can be viewed as an A_i - indexed family of functions $h_{i,a} : X^i \to X$, for $a \in A_i$, so an H_S - algebra is exactly the same thing as an algebra (or model) for the signature S. The initial H_S - algebra is the term algebra for S. Note that $H_S(X)$ gives one step in the construction of the free (term) algebra generated by the set X, so $H_S^k(\emptyset)$ is k steps in the construction of the free algebra on no generators and $\mathrm{colim}_k H^k(\emptyset)$ is the entire free algebra on no generators.

1.2.1 **Ordinary type systems.** Let B be a set of basic types and let $H_{Type} : \mathbf{SET} \rightarrow \mathbf{SET}$ be the functor given by

$$H_{Type}(X) = B + (X \times X)$$
$$H_{Type}(f) = id_B + (f \times f).$$

Here "+" denotes the coproduct (disjoint union, or sum) of sets and × denotes the cartesian product. (As a signature, Type consists of a B - indexed family of constants and one binary operation.) An element $(x, y) \in (X \times X)$ is denoted by $(x \rightarrow y)$. Such types are called *function types*. We define the set of types T (determined by B) to be the underlying set of the initial H_{Type}-algebra. T consists of all expressions generated freely from the elements of B by the requirement that if σ and τ are in T then so is $[\sigma \rightarrow \tau]$. (See Manes and Arbib [11].) The canonical representation of this initial algebra is as binary trees whose leaves are labeled by elements of B and whose nodes are labeled by "→".

1.2.2 **Polymorphic type systems.** A *polymorphic type system* is a type system PT in which the set of basic types has the form $B = B_1 + Tv$. Tv is called the *set of type variables*. Note that B_1 may be empty. There is one additional rule: if σ is a type and t is a type variable, then $\forall t.\sigma$ is a type. In this case, PT is the initial H_{pType}-algebra for the endofunctor

$$H_{pType}(X) = B_1 + Tv + (X \times X) + (Tv \times X)$$

(As a signature, pType includes a Tv - indexed family of unary operations.) The canonical representation of this initial algebra is as trees of arity at most two whose leaves are labeled by elements of $B_1 + Tv$, whose binary nodes are labeled by "→", and whose unary nodes are labeled by $\forall t$ where t is a type variable.

There is a "free-type-variables" function fv defined on PT, constructed as follows: The set, $P_f(Tv)$, of finite subsets of Tv carries an H_{pType}-algebra structure $p : H_{pType}(P_f(Tv)) \rightarrow P_f(Tv)$ whose components are the functions

$$p_{B_1} : B_1 \rightarrow P_f(Tv) \qquad \text{where} \qquad p_{B_1}(b) = \emptyset$$
$$p_{Tv} : Tv \rightarrow P_f(Tv) \qquad \text{where} \qquad p_{Tv}(t) = \{t\}$$
$$p_{\rightarrow} : P_f(Tv) \times P_f(Tv) \rightarrow P_f(Tv) \qquad \text{where} \qquad p_{\rightarrow}(V, W) = V \cup W$$
$$p_{\forall} : Tv \times P_f(Tv) \rightarrow P_f(Tv) \qquad \text{where} \qquad p_{\forall}(t, V) = V - \{t\}$$

(Cf. Oles [12].) The unique homomorphism from the initial H_{pType}-algebra PT to $P_f(Tv)$ is denoted by $fv : PT \rightarrow P_f(Tv)$. It satisfies the equations:

$$fv(b) = \emptyset, \ \ fv(t) = \{t\},$$
$$fv(\sigma \rightarrow \tau) = fv(\sigma) \cup fv(\tau),$$
$$fv(\forall t.\sigma) = fv(\sigma) - \{t\}.$$

1.3 Terms

Corresponding to ordinary types and polymorphic types, there are two kinds of terms, ordinary terms for ordinary type systems and parametric terms for polymorphic type systems.

1.3.1 **Atoms.** Let T be a type system. For each type $\tau \in T$ there is an infinite set, Var_τ, of ordinary term variables of type τ, and a set $Const_\tau$ of constants of type τ. The set $A_\tau = Var_\tau + Const_\tau$ is called the set of *atoms* of type τ.

1.3.2 **Ordinary terms.** The set $Term_\tau$ of terms of type τ for an ordinary typed lambda calculus is constructed recursively as follows: Write $f : \tau$ or, equivalently, $f \in Term_\tau$, for "f is a term of type τ" . Then terms are specified by the following clauses.

i) If $a \in A_\tau$ then $a : \tau$.

ii) If $f : [\sigma \to \tau]$ and $g : \sigma$, then $(fg) : \tau$.

iii) If $g : \tau$ and $x \in V_\sigma$, then $(\lambda x : \sigma.\ g) : [\sigma \to \tau]$.

In terms of algebras for an endofunctor, let $\mathbf{SET}^T$ denote the product of T copies of the category of sets or, equivalently, the category of **SET**-valued functors on the discrete category determined by T. We represent an object of $\mathbf{SET}^T$ as a T-indexed family of sets, $\{X_\tau \mid \tau \in T\}$ so

$$H_{Term}(\{X_\tau \mid \tau \in T\}) = \{H_{Term}(\{X_\tau \mid \tau \in T\})_\sigma \mid \sigma \in T\}.$$

Let $H_{Term} : \mathbf{SET}^T \to \mathbf{SET}^T$ be the endofunctor whose τ'th component is given by cases as follows:

Case $\beta \in B$:

$$H_{Term}(\{X_\tau \mid \tau \in T\})_\beta = A_\beta + \Sigma_{\sigma \in T}(X_{[\sigma \to \beta]} \times X_\sigma)$$

(I.e., the only terms of type $b \in B$ are constants or variables of type b, or function applications whose codomain is of type b. These are the terms in i) and ii) above.)

Case $[\mu \to \eta]$:

$$H_{Term}(\{X_\tau \mid \tau \in T\})_{[\mu \to \eta]} = A_{[\mu \to \eta]} + \Sigma_{\sigma \in T}(X_{[\sigma \to [\mu \to \eta]]} \times X_\sigma) + (Var_\mu \times X_\eta)$$

(I.e., for a function type $[\mu \to \eta]$, there are also terms given by a variable of type μ and a term of type η. This gives the terms in iii) above.)

An H_{Term} - algebra consists of a T-indexed family of functions $h_\sigma : H_{Term}(\{X_\tau \mid \tau \in T\})_\sigma \to X_\sigma$ for all $\sigma \in T$. Finally, $\{Term_\tau \mid \tau \in T\}$ is the underlying T-indexed family of sets of the initial H_{Term}-algebra. Elements of $Term_\tau$ can be described as trees whose precise description depends on τ, and is omitted here. Note that we could work in the category $(\mathbf{SET}\downarrow T)$ of objects over T instead of $\mathbf{SET}^T$ where a more intrinsic notation could be developed. However, this probably would not clarify anything.

1.3.3 **Parametric terms.** The set $pTerm_\tau$ of terms of type τ for a polymorphic typed lambda calculus consists of all terms for the corresponding ordinary typed lambda calculus together with two more kinds of terms. The terms in $pTerm_\tau$ are called *parametric* terms since they may contain a type parameter:

iv) If $t \in Tv$ and $g : \sigma$ has the property that for all $x : \tau \in FV(g)$, $t \notin fv(\tau)$ then $\lambda t.g : \forall t.\sigma$.

v) If $f : \forall t . \sigma$ and $\tau \in T$, then $f[\tau] : [\tau/t]\sigma$

Here FV(g) means the set of ordinary free variables of g and $fv(\tau)$ means the set of free type variables of τ as defined in 1.2.2. $[\tau/t]\sigma$ means substitution of τ for t in σ as usual. Clause iv) is the non context-free clause which requires us to change from sets to pointed sets in section 4.

The endofunctor $H_{pTerm} : \mathbf{SET}^{PT} \to \mathbf{SET}^{PT}$ is defined by cases as follows:

Case $\beta \in B$:

$$H_{pTerm}(\{X_\tau \mid \tau \in T\})_\beta = A_\beta + \Sigma_{\sigma \in T}(X_{[\sigma \to \beta]} \times X_\sigma) + \Sigma_{[\tau'/t]\sigma = \beta}(X_{\forall t.\sigma})$$

Case $[\mu \to \eta]$:

$$H_{pTerm}(\{X_\tau \mid \tau \in T\})_{[\mu \to \eta]} = A_{[\mu \to \eta]} + \Sigma_{\sigma \in T}(X_{[\sigma \to [\mu \to \eta]]} \times X_\sigma) + (Var_\mu \times X_\eta) + \Sigma_{[\tau'/t]\sigma = [\mu \to \eta]}(X_{\forall t.\sigma})$$

Case $\tau = \forall t.\mu$:

$$H_{pTerm}(\{X_\tau \mid \tau \in T\})_{\forall t.\mu} = A_{\forall t.\mu} + \Sigma_{\sigma \in T}(X_{[\sigma \to \forall t.\mu]} \times X_\sigma) + \Sigma_{[\tau'/t']\sigma = \forall t.\mu}(X_{\forall t'.\sigma}) + X_\mu$$

This last case introduces inadmissible terms because of the restriction in iv), so the collection of polymorphic terms, $\{pTerm_\tau \mid \tau \in PT\}$, is only a subfamily of the underlying family of the initial H_{pTerm} - algebra. It is characterized by the property that if $\tau = \forall t.\sigma$, then for all terms of type τ of the form $\lambda t.g$, t is not a free type variable in the type of any free ordinary variable of g. (See Section 4 below.)

2. SEMANTICS OF LAMBDA-CALCULI AS INITIAL MORPHISMS

2.1 Semantics of ordinary types in a cartesian closed category

Let **C** be a cartesian closed category and let obj(**C**) be the class of objects of **C**. As in 1.2.1, let $H_{Type}(X) = B + (X \times X)$.

2.1.1 Theorem. Obj(**C**) underlies an H_{Type}- algebra.

Proof. We define an H_{Type}- algebra structure on the class obj(**C**) as follows:

$$h : B + (obj(\mathbf{C}) \times obj(\mathbf{C})) \to obj(\mathbf{C})$$

is given by two components since it is a function from a coproduct of two classes.

i) $h_B : B \to obj(\mathbf{C})$ is some user defined function.

ii) $h_\to : obj(\mathbf{C}) \times obj(\mathbf{C}) \to obj(\mathbf{C})$ is given by $h_\to(C, C') = [C \to C']$. Here $[C \to C']$ denotes the function space construction in **C**.

Then $[[\ -\ ]] : T \to obj(C)$ is the unique H_{Type}- algebra homomorphism from T to obj(C). I.e.,

$$[[\beta]] = h_B(\beta),$$
$$[[\sigma \to \tau]] = h_{\to}([[\sigma]], [[\tau]]) = [\ [[\sigma]] \to [[\tau]]\ ].$$

Note that one is free to define h_B, and $h_{\to}$ in any way at all as far as the existence of an interpretation function for types is concerned. Thus, T can be "interpreted" in any set or class equipped with a B-indexed family of elements and a single binary operation. However, such a random choice would not agree with the intended intuitive meaning of the types $[\sigma \to \tau]$. Also, it might not be extendable to an interpretation of terms. However, h_B is completely arbitrary, and the notation should reflect that the interpretation of types is a function of the chosen interpretation of basis types. For simplicity, this is not done here.

2.2 Semantics of ordinary terms in a cartesian closed category

We will use the morphisms of **C** to interpret the terms. Let $mor_C(\mathbf{C})$ denote the subclass of morphisms of **C** whose codomain is the object C in **C** and let $dom_C : mor_C(\mathbf{C}) \to obj(\mathbf{C})$ denote the domain function in **C**; i.e., if $(f : C' \to C) \in mor_C(\mathbf{C})$, then $dom_C(f) = C'$. The set of all ordinary variables of all types is denoted by $Var = \sum_{\tau \in T} Var_\tau$ and $P_f(Var)$ denotes the set of finite subsets of Var. A typical element of $P_f(Var)$ is a set of the form $\{x_i : \tau_i \mid i \in I\}$ where $\tau_i \in T$ and I is a finite set of indices. There is a function prod : $P_f(Var) \to ob(\mathbf{C})$ given by

$$prod(\{x_i : \tau_i \mid i \in I\}) = \prod_{i \in I} [[\tau_i]].$$

For each object C in **C**, consider the class M_C given by the pullback diagram:

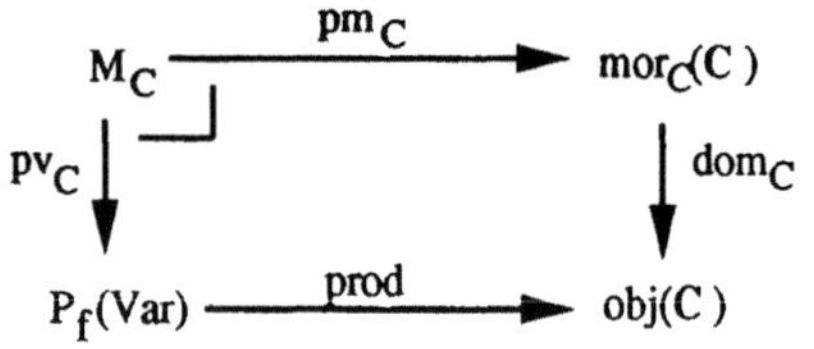

Thus, elements of M_C are pairs

$$\underline{m} = (\{x_i : \tau_i \mid i \in I\}, m : \prod_{i \in I} [[\tau_i]] \to C)$$

since for elements in the pullback, $prod(\{x_i : \tau_i \mid i \in I\}) = \prod_{i \in I} [[\tau_i]] = dom_C(m)$. Here $pm_C(\underline{m}) = m$ and $pv_C(\underline{m}) = \{x_i : \tau_i \mid i \in I\}$. Recall that functions $f : X \to M_C$ are determined by pairs of functions $f_1 : X \to P_f(Var)$ and $f_2 : X \to mor_C(\mathbf{C})$ with the property $prod \bullet f_1 = dom_C \bullet f_2$. Such an f is denoted by $<f_1, f_2>$. It is the unique function such that

$$pv_C \bullet <f_1, f_2> = f_1 \text{ and } pm_C \bullet <f_1, f_2> = f_2,$$

and is given on elements by the formula: $f(x) = (f_1(x), f_2(x))$.

The intention is to interpret $Term_\tau$ in $M_{[[\tau]]}$ via a family of functions $F_\tau : Term_\tau \to M_{[[\tau]]}$, where $[[\tau]]$ is the interpretation of the type τ constructed above in 2.1.1,. As will be seen in the proof of the following theorem, it is not possible to give a direct interpretation of $Term_\tau$ in $mor_{[[\tau]]}(C)$ since the main step of the proof, part iii), can only be carried out if one knows what the free variables giving rise to the domain of $m \in mor_{[[\tau]]}(C)$ are.

2.2.1 Theorem. The family of classes $\{M_{[[\tau]]} \mid \tau \in T\}$ underlies an H_{Term} - algebra.

Proof. For each $\tau \in T$ a function $k_\tau : H_{Term}(\{M_{[[\sigma]]} \mid \sigma \in T\})_\tau \to M_{[[\tau]]}$ must be constructed. Depending on τ, this will have two or three components, but the first two components are always the same.

i) $(k_\tau)_A : A_\tau \to M_{[[\tau]]}$ itself consists of two components since $A_\tau = Const_\tau + Var_\tau$.

a) $(k_\tau)_{Const} : Const_\tau \to M_{[[\tau]]}$ is given by

$$(k_\tau)_{Const}(c_\tau) = (\emptyset, chose_\tau(c_\tau) : 1_C \to [[\tau]]) \text{ for any } c_\tau \in Const_\tau.$$

Here, $chose_\tau : Const_\tau \to mor_{[[\tau]]}(C)$ is a user defined function, subject to the restriction that for each $c_\tau \in Const_\tau$, $chose_\tau(c_\tau) : 1_C \to [[\tau]]$. In terms of pure functions, $(k_\tau)_{Const} = \langle '\emptyset \bullet !, chose_\tau \rangle$ where, for any set X, $! : X \to 1$ is the unique function from X to the one element set 1 and $'\emptyset : 1 \to P_f(Var)$ is the "name of the empty set" of variables; i.e., the function whose only value is $\emptyset \in P_f(Var)$, where $\emptyset$ denotes the empty set. Thus, $'\emptyset \bullet ! : Const_\tau \to P_f(Var)$ is the function assigning the empty set of variables to every constant. Note that $\prod_{i \in \emptyset} [[\rho_i]] = 1_C$, where 1_C is the terminal object of **C**.

b) $(k_\tau)_{Var} : Var_\tau \to M_{[[\tau]]}$ is given by

$$(k_\tau)_{Var}(x : \tau) = (\{x : \tau\}, id_{[[\tau]]} : [[\tau]] \to [[\tau]]) \text{ for any } x : \tau \in Var_\tau.$$

In pure functional terms, $(k_\tau)_{Var} = \langle \{ - \} \bullet in_\tau, 'id_{[[\tau]]} \bullet ! \rangle$ where $in_\tau : Var_\tau \to Var$ is the τ'th coproduct inclusion, $\{ - \} : Var \to P_f(Var)$ is the singleton function taking each variable to the set consisting just of that variable, ! is as in part a), and $'id_{[[\tau]]} : 1 \to mor_{[[\tau]]}(C)$ is the name of the identity morphism of $[[\tau]]$.

ii) The second component of k_τ is a function from $\sum_{\sigma \in T} (M_{[[\sigma \to \tau]]} \times M_{[[\sigma]]})$ to $M_{[[\tau]]}$, so it has a component for each $\sigma \in T$ denoted by

$$(k_\tau)_{\sigma\to} : M_{[[\sigma \to \tau]]} \times M_{[[\sigma]]} \to M_{[[\tau]]}.$$

Let $\underline{m} = (\{x_i : \tau_i \mid i \in I\}, m : \prod_{i \in I} [[\tau_i]] \to [[\sigma \to \tau]]) \in M_{[[\sigma \to \tau]]}$

$\underline{n} = (\{x_j : \tau_j \mid j \in J\}, n : \prod_{j \in J} [[\tau_j]] \to [[\sigma]]) \in M_{[[\sigma]]}$.

Then

$$(k_\tau)_{\sigma\to} (\underline{m}, \underline{n}) = (\{x_i : \tau_i \mid i \in I\} \cup \{x_j : \tau_j \mid j \in J\},$$
$$app_{[[\sigma]],[[\tau]]} \bullet (m \times n) \bullet \langle pr_I, pr_J \rangle)$$

In more diagrammatic terms, $(k_\tau)_{\sigma\to} (\underline{m}, \underline{n})$ is determined by its two projections:

$$pv_{[[\tau]]} ((k_\tau)_{\sigma\to} (\underline{m}, \underline{n})) \in P_f(Var)$$

$$pm_{[[\tau]]} ((k_\tau)_{\sigma\to} (\underline{m}, \underline{n})) \in mor_{[[\tau]]}(C).$$

The first is given by

$$pv_{[[\tau]]} ((k_\tau)_{\sigma\to} (\underline{m}, \underline{n})) = \{x_i : \tau_i \mid i \in I\} \cup \{x_j : \tau_j \mid j \in J\}$$

and the second is the function making the diagram

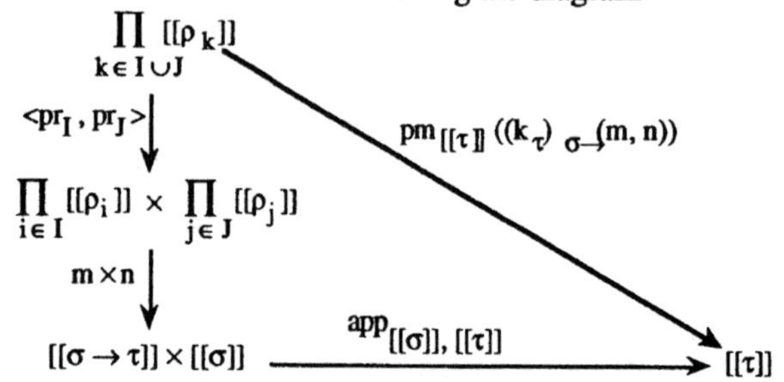

commute; i.e., it is the composed function

$$pm_{[[\tau]]}((k_\tau)_{\sigma\to}(\underline{m}, \underline{n})) = app_{[[\sigma]], [[\tau]]} \bullet (m \times n) \bullet <pr_I, pr_J>$$

where app is the "apply" operation from the cartesian closed structure of **C**. A pure functional description of $(h_\tau)_{\sigma\to}$ is very complicated to write out.

iii) Case $\tau = [\mu \to \sigma]$. Then k_τ has a third component which is a function from $Var_\mu \times M_{[[\sigma]]}$ to $M_{[[\tau]]}$ denoted by $(k_\tau)_\lambda$. Let $x : \mu \in Var_\mu$ and let

$$\underline{m} = (\{x_i : \tau_i \mid i \in I\}, m : \Pi_{i \in I} [[\tau_i]] \to [[\sigma]]) \in M_{[[\sigma]]}.$$

Then,

$$(k_\tau)_\lambda (x : \mu, \underline{m}) = (\{x_i : \tau_i \mid i \in I\} - \{x : \mu\}, \text{"curry(m)"}).$$

To explain this, as before the two projections of $(k_\tau)_\lambda$ have to be described separately. Firstly,

$$pv_{[[\tau]]}((k_\tau)_\lambda (x : \mu, \underline{m})) = \{x_i : \tau_i \mid i \in I\} - \{x : \mu\},$$

where "-" means delete $x : \mu$ from $\{x_i : \tau_i \mid i \in I\}$ if it is a member of that set; otherwise do nothing. Secondly, the term "curry(m)", i.e., $pm_{[[\tau]]}((k_\tau)_\lambda (x : \mu, \underline{m}))$ is defined as follows:

a) If $x : \mu = x_{i_0} : \tau_{i_0}$ for some $i_0 \in I$, then rewrite $\Pi_{i \in I} [[\tau_i]]$ as

$$(\Pi_{I \ni i \neq i_0} [[\tau_i]]) \times [[\tau_{i_0}]]$$

and consider

$$m \bullet a : (\Pi_{I \ni i \neq i_0} [[\tau_i]]) \times [[\tau_{i_0}]] \to [[\sigma]]$$

where a is an appropriate isomorphism given by associativity and commutativity of products. Then

$$pm_{[[\tau]]}((k_\tau)_\lambda (x : \mu, \underline{m})) = \text{curry}(m \bullet a) : \Pi_{I \ni i \neq i_0} [[\tau_i]] \to [\, [[\tau_{i_0}]] \to [[\sigma]]\,].$$

Since $[\, [[\tau_{i_0}]] \to [[\sigma]]\,] = [[\tau_{i_0} \to \sigma]] = [[\mu \to \sigma]]$, this has the correct codomain. Note that $\Pi_{I \ni i \neq i_0} [[\tau_i]] = \text{prod}(\{x_i : \tau_i \mid i \in I\} - \{x : \mu\})$ in this case.

b) If $x : \mu \neq x_i : \tau_i$ for all $i \in I$. Rewrite $\Pi_{i \in I} [[\tau_i]]$ as $\Pi_{i \in I} [[\tau_i]] \times 1$ and consider $m \bullet p : \Pi_{i \in I} [[\tau_i]] \times 1 \to [[\sigma]]$ where p is the projection from $\Pi_{i \in I} [[\tau_i]] \times 1$ to $\Pi_{i \in I} [[\tau_i]]$. Then curry $(m \bullet p) : \Pi_{i \in I} [[\tau_i]] \to [\, 1 \to [[\sigma]]\,]$. But

$$[! \to [[\sigma]]\,] : [\, 1 \to [[\sigma]]\,] \to [\, [[\mu]] \to [[\sigma]]\,] = [[\mu \to \sigma]]$$

where $! : [[\mu]] \to 1$ is the unique such morphism. Hence we can define

$$pm_{[[\tau]]}((k_\tau)_\lambda (x : \mu, \underline{m})) = [[! \to \sigma]] \bullet curry(m \bullet p) : \Pi_{i \in I} [[\tau_i]] \to [[\mu \to \sigma]].$$

Note that again $\Pi_{i \in I} [[\tau_i]] = prod(\{x_i : \tau_i \mid i \in I\} - \{x : \mu\})$.

(This is the step where one has to know what the free variables of m are. That is why we have to work with the pullbacks $M_{[[\tau]]}$ rather that directly with the classes $mor_C(C)$.)

This defines $k_\tau : H_{Term}(\{M_{[[\tau]]} \mid \tau \in T\}) \to M_{[[\tau]]}$ in all cases, so $\{M_{[[\tau]]} \mid \tau \in T\}$ carries an H_{Term} - algebra structure. Hence, there is a unique H_{Term} - algebra homomorphism

$$F : \{Term_\tau \mid \tau \in T\} \to \{M_{[[\tau]]} \mid \tau \in T\}$$

with components $F_\tau : Term_\tau \to M_{[[\tau]]}$ for each $\tau \in T$.

2.2.2 The free-variable functions. Let $Fv_\tau = pv_{[[\tau]]} \bullet F_\tau : Term_\tau \to P_f(Var)$ be the first projection of F_τ. Fv is the "free variable" function that assigns to each term its set of free variables. By construction it satisfies the recursive equations:

Case $x : \tau \in Var_\tau$;	$Fv_\tau(x : \tau) = \{x : \tau\}$.
Case $c \in Const_\tau$;	$Fv_\tau(c) = \emptyset$.
Case $f \in Term_{\sigma \to \tau}$ and $g \in Term_\sigma$;	$Fv_\tau((fg)) = Fv_{\sigma \to \tau}(f) \cup Fv_\sigma(g)$.
Case $g \in Term_\sigma$;	$Fv_{\mu \to \sigma}(\lambda x : \mu . g) = Fv_\sigma(g) - \{x : \mu\}$.

2.2.3 The semantics functions. Let $[[-]]_\tau = pm_{[[\tau]]} \bullet F_\tau : Term_\tau \to mor_{[[\tau]]}(C)$ be the second projection of F_τ. [[-]] is the *semantics* (or *interpretation*) function for terms (using the same notation as for types). By construction, if $f \in Term_\tau$, then

$$[[f]]_\tau : (\Pi_{x : \sigma \in Fv_\tau(f)} [[\sigma]]) \to [[\tau]].$$

Also by construction, it satisfies the recursive equations:

Case $x : \tau \in Var_\tau$; $\quad [[x : \tau]]_\tau = id_{[[\tau]]} : [[\tau]] \to [[\tau]]$.

Case $c \in Const_\tau$; $\quad [[c]]_\tau = chose_\tau(c_\tau) : 1_C \to [[\tau]]$.

Case $f \in Term_{\sigma \to \tau}$ $\quad [[fg]]_\tau = app_{[[\sigma]], [[\tau]]} \bullet ([[f]]_{\sigma \to \tau} \times [[g]]_\sigma) \bullet <pr_{Fv(f)}, pr_{Fv(g)}>$

and $g \in Term_\sigma$; $\quad : (\Pi_{x : \mu \in Fv(f) \cup Fv(g)} [[\mu]]) \to [[\tau]]$

Case $g \in Term_\eta$; $\quad [[\lambda x : \mu . g]]_{\mu \to \eta} =$

if $x : \mu \in Fv(g)$

then $curry([[g]]_\eta \bullet a) : \Pi_{x : \sigma \in Fv(f) - \{x : \mu\}} [[\sigma]]) \to [[\mu \to \eta]]$

else $[! \to [[\eta]]\,] \bullet curry([[g]]_\eta \bullet p) : \Pi_{x : \sigma \in Fv(f)} [[\sigma]]) \to [[\mu \to \eta]]$

Note: id, chose, app, !, a, and p are described in the proof of 2.2.1.

2.2.4 Substitution. Let f and g be terms in a typed λ-calculus and let x be a variable. If $f{:}\tau$ and $x{:}\tau$ have the same type τ, then *substitution* of $f{:}\tau$ for $x{:}\tau$ in g, denoted by $[f{:}\tau/x{:}\tau]g$, is defined. $[f{:}\tau/x{:}t]g$ has the same type as g, and $FV([f{:}\tau/x{:}\tau]g) = (FV(g) - \{x{:}\tau\}) \cup FV(f{:}\tau)$. It is given recursively by the following rules:

a) $[f{:}\tau/x{:}\tau]x = f{:}\tau$.
b) $[f{:}\tau/x{:}\tau]a = a$ for all atoms $a \neq x$.
c) $[f{:}\tau/x{:}\tau]\,(g_1\, g_2) = (([f{:}\tau/x{:}\tau]\, g_1)([f{:}\tau/x{:}\tau]\, g_2))$
d) $[f{:}\tau/x{:}\tau]\,(\lambda x{:}\tau.h) = (\lambda x{:}\tau.h)$ (i.e., don't substitute for bound variables.)
e) $[f{:}\tau/x{:}\tau]\,((\lambda y{:}\tau'.g)$

$= (\lambda y{:}\tau'.[f{:}t/x{:}\tau]\, g)$ if $y{:}\tau' \neq x{:}\tau$ and $(y{:}\tau' \notin FV(f)$ or $x{:}\tau \notin FV(g))$

$= (\lambda z{:}\tau'\,.\,[f{:}\tau/x{:}\tau]\,[z{:}\tau'/y{:}\tau']g)$ if $y{:}\tau' \neq x{:}\tau$ and $y{:}\tau' \in FV(f)$ and $x{:}\tau \in FV(g)$

where z is the first variable in some given ordering of V_τ that is not free in f or g.

For fixed $f{:}\tau$ and $x{:}\tau$, $[f{:}\tau/x{:}\tau]\colon Term_\sigma \to Term_\sigma$ for each σ. In terms of H_{Term}-algebras, substitution is not a homomorphism of the initial algebra, since an initial algebra does not admit any non-identity endomorphisms. However, it is a homomorphism from the initial algebra to another H_{Term}-algebra with the same carrier as the initial algebra. Let $Sub_\sigma = Term_\sigma$ for all $\sigma \in T$, and let $t_\sigma : H_{Term}(\{Term_\tau \mid \tau \in T\}) \to Term_\sigma$ be the structure isomorphisms for the initial algebra. Define $s(f{:}\tau, x{:}\tau)_\sigma = t_\sigma : H_{Term}(\{Sub_\mu \mid \mu \in T\})_\sigma \to Sub_\sigma$ for all $\sigma \neq \tau$ and

$$s(f{:}\tau, x{:}\tau)_\tau(g{:}\tau) = \text{if } g{:}t\rho = x{:}\tau \text{ then } f{:}\tau \text{ else } g{:}\tau.$$

This defines an H_{Term}-algebra $\{(Sub_\sigma, s(f{:}\tau, x{:}\tau)_\sigma) \mid \sigma \in T\}$ and

$$[f{:}\tau/x{:}\tau]\colon \{Term_\sigma \mid \sigma \in T\} \to \{(Sub_\sigma, s(f{:}\tau, x{:}\tau)_\sigma) \mid \sigma \in T\}$$

is the unique homomorphism from the initial algebra to this algebra. (Cf. Oles [11].)

3. Functor Category Models of Polymorphism

As in Section 2, let **C** be a cartesian closed category. If **X** is another category (in our application it will be a product of copies of |**C**|), then the functor category [**X** → **C**] need not exist as an ordinary (locally small) category since the collection of natural transformations between two functors can be a proper class. Nevertheless, one can use the definition of categories that just talks about objects and morphisms and doesn't try to collect together morphisms into hom sets, in which case there is such a functor category. It has cartesian products given by $(F \times G)(X) = F(X) \times G(X)$ so product types $\sigma \times \tau$ would have an interpretation, but there need not be function space objects. There are two ways to try to use the function space construction in **C** to interpret function space types $[\sigma \to \tau]$. One way is to assume that **X** is a discrete category (no non-identity morphisms) and then to define $[F \to G](X) = [F(X) \to G(X)]$, this being a perfectly good functor if there are no non-identity morphisms in **X**. In this case, [**X** → **C**] is cartesian closed with respect to this structure, so the ordinary typed λ-calculus can be interpreted here as usual. The other way is to replace **X** by $\mathbf{X}^{op} \times \mathbf{X}$ and change the definition of $[F \to G]$. Here we consider only the discrete case.

Let Tv be the set of type variables for a polymorphic λ-calculus and let |C| be the underlying discrete category of C. (Just forget all morphisms except identity morphisms.) $|C|^{Tv}$ can be thought of either as the Tv-fold product of |C| with itself or as the (discrete) category of functors from the discrete category with objects Tv to the discrete category |C|. In the first formulation, there are projection functors $pr_t : |C|^{Tv} \to \mathbf{C}$ for each $t \in Tv$. In either form, objects of $|C|^{Tv}$ are functions $\rho : Tv \to obj(C)$, with $pr_t(\rho) = \rho(t)$, and there are no non-identity morphisms. One can think of $|C|^{Tv}$ as type *environments* with values in |C|. We frequently use the notation $\mathbf{Env} = [Tv \to |C|\,] = |C|^{Tv}$. Note that ρ is the standard symbol for an environment. When necessary, we write $\mathbf{Env}_C$ to keep track of the category C. In what follows, we take the **X** of the preceeding paragraph to be **Env**.

3.1. Types = discrete functors. Recall from 1.2.2 that PT is the initial algebra for the endofunctor

$$H_{pType}(X) = B_1 + Tv + (X \times X) + (Tv \times X).$$

Consider the class $obj[\mathbf{Env} \to \mathbf{C}] = obj[|C|^{Tv} \to \mathbf{C}]$ of functors from **Env** to **C**. Such functors are the same as functions from **Env** to |C|; i.e., from environments to values in C, since **Env** has no non-identity morphisms. The following theorem talks about "suitable categories **C**" since this is the place where we run into the first problem mentioned in the introduction; namely,

i) the semantics of the polymorphic lambda calculus demands outsized products.

In the proof, it is pointed out that **C** = **PER** is a suitable cartesian closed category.

3.1.1 Theorem. $Obj[\mathbf{Env} \to \mathbf{C}]$ underlies an H_{pType} - algebra for suitable categories **C**.

Proof. A function of the form

$$h : H_{pType}(obj[\mathbf{Env} \to \mathbf{C}]) \to obj[\mathbf{Env} \to \mathbf{C}]$$

is a map from a coproduct of four terms to $obj[\mathbf{Env} \to \mathbf{C}]$, so it is given by four components, the first three of which are just the standard components for the semantics of types of the ordinary λ-calculus in a cartesian closed category, except for special conditions on the values of atoms.

a) $h_{B_1} : B_1 \to obj[\mathbf{Env} \to \mathbf{C}]$ is some user interpretation of the constant types as functors factoring through the category **1**.

b) $h_{Tv} : Tv \to obj[\mathbf{Env} \to \mathbf{C}]$ is given by $h_{Tv}(t) = pr_t : |C|^{Tv} \to \mathbf{C}$.

c) $h_{\to} : obj[\mathbf{Env} \to \mathbf{C}] \times obj[\mathbf{Env} \to \mathbf{C}] \to obj[\mathbf{Env} \to \mathbf{C}]$ is given by

$h_{\to}(F, G) = [F \to G]$, where $[F \to G](X) = [F(X) \to G(X)]$.

d) $h_{\forall} : Tv \times obj[\mathbf{Env} \to \mathbf{C}] \to obj[\mathbf{Env} \to \mathbf{C}]$ is given by

$h_{\forall}(t, F) = \lambda\rho.\ \Pi_{C \in |C|} F(\rho[C / t])$.

Here, $[C / t] : |C|^{Tv} \to |C|^{Tv}$ is the "update" function such that if $\rho : Tv \to |C|$, then

$\rho[C / t](t') = \text{if } t' = t \text{ then } C \text{ else } \rho(t')$.

This large product over all of the objects of **C** does not exist except for very special choices of the category **C** such as **C** = **PER**. (See Bruce and Longo [1], Hyland [5], Longo and Moggi [10], Gray [4].)

Assuming that $h_\forall$ can be defined, then [[-]] : PT → [Tv, **C**] is the unique H_{pType} - algebra homomorphism from PT to this algebra on [Tv, **C**]. Thus,

$$[[t]] = pr_t\,,$$
$$[[\sigma \to \tau]] = [\,[[\sigma]] \to [[\tau]]\,],$$
$$[[\forall t.\sigma]] = \lambda\rho.\ \Pi_{C\in|C|}[[\sigma]](\rho[C/t]).$$

3.2 Terms = natural transformations. Recall from 1.3.3 that the collection of terms for the polymorphic λ-calculus is a subset of the initial algebra for the endofunctor H_{pTerm} : $\mathbf{SET}^{PT} \to \mathbf{SET}^{PT}$. Consider the class nat[Tv, **C**] = mor(**Env** → **C**) of natural transformations between functors from **Env** to **C**. (Such natural transformations are the same as functions from the class of objects of **Env** to the class of morphisms of **C**.) As in 2.2, for any functor F : **Env** → **C**, let mor_F([**Env** → **C**]) be the class of natural transformations with codomain F and form the pullbacks M_F as in 2.2. We write n : F ⇒ G for a natural transformation from F to G.

The following result is called a "Pseudo Theorem" because this is the point where we run into the second difficulty mentioned in the introduction; namely,

ii) the syntax of the polymorphic lambda calculus is given by a context dependent grammar.

In distinction to 3.1.1, there is no suitable choice for **C**.

3.2.1 Pseudo Theorem. The family of classes $\{M_{[[\tau]]} \mid \tau \in PT\}$ underlies an H_{pTerm} - algebra for "suitable categories **C**".

Proof. For each $\sigma \in T$, a function $k_\sigma : H_{pTerm}(\{M_{[[\tau]]} \mid \tau \in PT\})_\sigma \to M_{[[\sigma]]}$ must be constructed. The first three cases are exactly the same as 2.2.1, noting that application and curry are component-wise. However, in the present case there is an extra summand whenever $\tau = [\tau'/t]\sigma$. Then k_τ has an extra component which is a function from $M_{[[\forall t.\sigma]]}$ to $M_{[[\tau]]}$ denoted by $(k_\tau)_{[\tau'/t]\sigma}$. To construct it, let

$$\underline{m} = (\{x_i : \tau_i \mid i \in I\}, m : \Pi_{i\in I}[[\tau_i]] \Rightarrow [[\forall t.\sigma]]\,) \in M_{[[\forall t.\sigma]]}).$$

Since $[[\forall t.\sigma]](\rho) = \Pi_{C\in|C|}[[\sigma]](\rho[C/t])$, the projection morphism

$$pr_{[[\tau']](\rho)} : \Pi_{C\in|C|}[[\sigma]](\rho[C/t]) \to [[\sigma]](\rho[\,[[\tau']](\rho)/t])$$

is available and $[[\sigma]]((\rho[\,[[\tau']](\rho)/t]) = [[\,[\tau'/t]\sigma\,]](\rho)$ by the substitution theorem, so we define

$$(k_\tau)_{[\tau'/t]\sigma}(\underline{m}\,) = (\{x_i : \tau_i \mid i \in I\}, pr_{[[\tau']](\rho)} \circ m : \Pi_{i\in I}[[\tau_i]] \Rightarrow [[\,[\tau'/t]\sigma\,]]\,).$$

To prepare the way for the extra case in 1.3.3, consider three constructions:

a) Call a functor F : **Env** → **C** *invariant* under [C / t] if F ∘ [C / t] = F where [C / t] : **Env** → **Env** is the endofunctor defined in 3.1.

b) If $t \in Tv$, let $pr_{t^\wedge} : |C|^{Tv} \to |C|^{Tv - \{t\}}$ be the projection morphism complimentary to pr_t : **Env** $\to$ **C**, and let obj[**Env** $\to$ **C**]$_t$ be the subclass of functors which factor through $pr_{t^\wedge}$.

c) Let $Var_t = \{x : \tau \mid t \notin fv(\tau)\}$ and let $P_f(Var_t)$ be the subset of $P_f(Var)$ consisting of finite subsets of Var_t. As in 2.2, for each object C in an arbitrary category **C**, consider the class ${}_tM_C$ given by the pullback diagram:

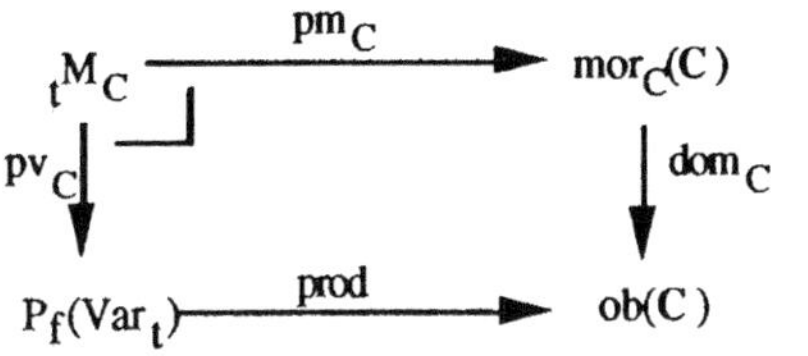

Then ${}_tM_C$ can be described as the subclass of M_C consisting of all pairs

$$\underline{m} = (\{x_i : \tau_i \mid i \in I\}, m : \textstyle\prod_{i \in I} [[\tau_i]] \to C)$$

such that for all $i \in I$, $t \notin fv(\tau_i)$. We are of course interested in the case where **C** is replaced by [**Env** $\to$ **C**].

3.2.2 Lemma. F factors through $pr_{t^\wedge}$ if and only if F : **Env** $\to$ **C** is *invariant* under $[C / t]$ for all $C \in$ **C**.

Proof. $pr_{t^\wedge}$ is clearly invariant under $[C / t]$ so any functor that factors through $pr_{t^\wedge}$ is also invariant under $[C / t]$. Conversely, choose $C_0 \in |C|$ and define $[C_0 / t]' : |C|^{Tv - \{t\}} \to |C|^{Tv}$ by the formula:

$$[C_0 / t]' (\rho'(t')) = \text{if } t' = t \text{ then } C_0 \text{ else } \rho'(t').$$

Clearly $pr_{t^\wedge} \circ [C_0 / t]' = id$ and $[C_0 / t]' \circ pr_{t^\wedge} = [C_0 / t]$. If F is invariant under $[C / t]$ for all $C \in$ **C** then it is invariant under $[C_0 / t]$ so $F = F \circ [C_0 / t] = F \circ [C_0 / t]' \circ pr_{t^\wedge} = F' \circ pr_{t^\wedge}$ where $F' = F \circ [C_0 / t]'$, so F factors through $pr_{t^\wedge}$.

3.2.3 Proposition. $[[\tau]]$ factors through $pr_{t^\wedge}$ if and only if $t \notin fv(\tau)$.

Proof. By structural induction.

3.2.4 Corollary. Let $\underline{m} = (\{x_i : \tau_i \mid i \in I\}, m : \prod_{i \in I} [[\tau_i]] \Rightarrow [[\sigma]]) \in M_{[[\sigma]]}$. Then $\underline{m} \in {}_tM_{[[\sigma]]}$ iff and only if $\prod_{i \in I} [[\tau_i]]$ is invariant under $[C / t]$ for all $C \in$ **C**

Proof. We have the following chain of equivalences:

$\underline{m} \in {}_tM_{[[\sigma]]}$
iff for all $i \in I$, $t \notin fv(\tau_i)$
iff for all $i \in I$, $[[\tau_i]]$ factors through $pr_{t^\wedge}$
iff for all $i \in I$, $[[\tau_i]]$ is invariant under $[C / t]$ for all $C \in$ **C**
iff $\prod_{i \in I} [[\tau_i]]$ is invariant under $[C / t]$ for all $C \in$ **C**

"Proof" of 3.2.1 concluded. We are now ready to define k_τ in the last case:

d) Case $\tau = \forall t.\sigma$. Then k_τ has a fourth component which is a function from $M_{[[\sigma]]}$ to $M_{[[\tau]]}$ denoted by $(k_\tau)_{\forall t.\sigma}$. Let

$$\underline{m} = (\{x_i : \tau_i \mid i \in I\}, m : \Pi_{i \in I}[[\tau_i]] \Rightarrow [[\sigma]]) \in {}_tM_{[[\sigma]]}.$$

Then $\Pi_{i \in I}[[\tau_i]](\rho[C/t]) = \Pi_{i \in I}[[\tau_i]](\rho)$ for all $C \in \mathbf{C}$ and for all $\rho \in \mathbf{Env}$.

Now $m : \Pi_{i \in I}[[\tau_i]] \Rightarrow [[\sigma]]$ is a natural transformation between functors from **Env** to $|\mathbf{C}|$ so it has components $m_\rho : \Pi_{i \in I}[[\tau_i]](\rho) \rightarrow [[\sigma]](\rho)$ for each $\rho \in \mathbf{Env}$.

In particular, for each $C \in \mathbf{C}$ it has a component

$$m_{\rho[C/t]} : \Pi_{i \in I}[[\tau_i]](\rho[C/t]) \rightarrow [[\sigma]](\rho[C/t]).$$

Now recall that

$$[[\forall t.\sigma]](\rho) = \Pi_{C \in |C|}[[\sigma]](\rho[C/t])$$

If $\underline{m} \in {}_tM_{[[\sigma]]}$, then, the domain of each $m_{\rho[C/t]}$ equals $\Pi_{i \in I}[[\tau_i]](\rho)$, so these morphisms determine an induced morphism into the product defining $[[\forall t.\sigma]]$,

$$\langle m_{\rho[C/t]} \mid C \in \mathbf{C}\rangle : \Pi_{i \in I}[[\tau_i]](\rho) \rightarrow \Pi_{C \in |C|}[[\sigma]](\rho[C/t]).$$

We define

$$(k_\tau)_{\forall t.\sigma}(\underline{m}) = (\{x_i : \tau_i \mid i \in I\}, \lambda\rho.\langle m_{\rho[C/t]} \mid C \in \mathbf{C}\rangle : \Pi_{i \in I}[[\tau_i]] \Rightarrow [[\forall t.\sigma]])$$

if $\underline{m} \in {}_tM_{[[\sigma]]}$, and $\perp$ otherwise.

Here "$\perp$" means either

a) $(k_\tau)_{\forall t.\sigma}(m)$ is a partial morphism (if that makes sense) whose domain of definition includes natural transformations whose domain in turn is a functor in $\mathrm{ob}[|\mathbf{C}|^{Tv} \rightarrow \mathbf{C}]_t$, or

b) $\mathbf{C}$ is a pointed category, e.g., a category of domains, and $\perp$ is the distinguished point in $M_{[[\tau]]}$. This, however, requires many changes; in particular, we have to give up cartesian closedness. See the discussion in Section 4 below.

The reason this is only a pseudo theorem is because in the present context there is no way to define this algebraic structure since we have neither partial morphisms nor distinguished base points. It is necessary to move to the more complex environment described in Section 4.

3.3 The free-variable functions

When k is defined, then the free variable function in 2.2.2 is clearly extended by the above definitions to

Case $g \in \mathrm{pTerm}_\sigma$	$\mathrm{Fv}(\lambda t.g) = \mathrm{Fv}(g)$
Case $f \in \mathrm{pTerms}_{\forall t.\sigma}$	$\mathrm{Fv}(f[\tau']) = \mathrm{Fv}(f)$

3.4 The semantics functions

When k is defined, then the semantics functions in 2.2.3 are clearly extended by the above definitions to

Case $g \in pTerm_\sigma$ $\quad$ $pr_C \circ [[\lambda t . f]]_\rho = [[f]]_{[C/t]\rho}$
Case $f \in pTerms_{\forall t.\sigma}$ $\quad$ $[[f[\tau]]]_\rho = pr_{[[\tau]](\rho)} \circ [[f]]_\rho$.

4. Algebras in Pointed Sets

We review here the well-known relations (from algebraic topology 25 years ago) between the category of sets and the category of pointed sets. We then show how this gives a proper semantics for terms in the polymorphic lambda calculus.

4.1 SET and bSET•

4.1.1 Pointed sets. A *pointed set* is a pair $(X, \bot_X)$ consisting of a set X and a chosen element $\bot_X \in X$, called the *base point* or *bottom* of X. There are two kinds of functions between pointed sets that it is reasonable to consider.

a) An *arbitrary* function $f : (X, \bot_X) \to (Y, \bot_Y)$ is just a function f from X to Y.

b) A *base-point preserving* function $f : (X, \bot_X) \to (Y, \bot_Y)$ is a function f from X to Y such that $f(\bot_X) = \bot_Y$.

We denote the category of pointed sets and arbitrary functions by **bSET**, the initial **b** referring to the base point. The (non-full) subcategory of pointed sets and base-point preserving functions is denoted by **bSET•**, the superscript • indicating that the functions preserve the base point.

4.1.2 Some associated functors. A *forgetful* functor is a functor that forgets part of the structure of an object. An obvious example is the underlying set functor $U^\bullet$: **bSET•** → SET given by $U^\bullet(X, \bot_X) = X$ and $U^\bullet(f) = f$. There is a left adjoint functor to $U^\bullet$ going the other way, called the *lifting* functor, $(\_)_\bot$: **SET** → **bSET•** given by the formulas:

$$(X)_\bot = (X \cup \{\bot_X\}, \bot_X)$$

where $\bot_X$ is a new point not in X; e.g., $\bot_X = \{X\}$. If $f : X \to Y$ is a function from X to Y, then $(f)_\bot : (X)_\bot \to (Y)_\bot$ is given by the formula $(f)_\bot(z) =$ *if* $(z \in X)$ *then* $f(z)$ *else* $\bot_Y$.

4.1.3 Properties of bSET•. The category **SET** is complete, cocomplete and cartesian closed. **bSET•** is complete and cocomplete, but not cartesian closed. Instead, it has another closed structure given by the *smash* product $(X, \bot_X) \# (Y, \bot_Y) = ((X \times Y)/R, [\bot_X, \bot_Y])$ where R is the smallest equivalence relation such that $(x, \bot_Y)\ R\ (\bot_X, \bot_Y)$ and $(\bot_X, y)\ R\ (\bot_X, \bot_Y)$ for all $x \in X$ and $y \in Y$. Here $[\bot_X, \bot_Y]$ is the equivalence class of $(\bot_X, \bot_Y)$ modulo R. Equivalently,

$$(X, \bot_X) \# (Y, \bot_Y) \approx ((X - \{\bot_X\}) \times (Y - \{\bot_Y\}) \cup \{\bot_{X\#Y}\}, \bot_{X\#Y})$$

where $\bot_{X\#Y}$ is a new point not in X or Y. The first representation is more useful in our context. (Cf., Gunter and Scott, [4].) The smash product is isomorphic to the ordinary product of the non-bottom points of X and Y together with a single new base point. Note that as with the cartesian product, there are (strict) projection functions from a smash product to its factors. (This is evident from the second representation.) The corresponding internal function space object for X and Y in $\mathbf{bSET}^\bullet$ is given by

$$([X \circ\!\!\rightarrow Y], \perp_{X \to Y}) = (\mathbf{bSET}^\bullet((X, \perp_X), (Y, \perp_Y)), \perp_{X \to Y}),$$

where $\perp_{X \to Y} : X \to Y$ is the constant function equal to $\perp_Y$ and $\mathbf{bSET}^\bullet((X, \perp_X), (Y, \perp_Y))$ is the set of base-point preserving functions from $(X, \perp_X)$ to $(Y, \perp_Y)$. These constructions satisfy:

$$[Z \circ\!\!\rightarrow [X \circ\!\!\rightarrow Y]] \approx [Z \# X \circ\!\!\rightarrow Y].$$

where we have omitted the base-points for readability. Since Z # - has a right adjoint, it preserves colimits and coproducts. In particular, there is the distributive law $X \# (Y + Y') \approx X \# Y + X \# Y'$ where "+" is the coaleased sum (= coproduct in $\mathbf{bSET}^\bullet$) of X and Y. Furthermore, # is directed cocontinuous; i.e., if M and M' are functors from a small directed category **D** to $\mathbf{bSET}^\bullet$, then colim (M # M') ≈ colim M # colim M'. It follows that "polynomial" endofunctors built from # and + on $\mathbf{bSET}^\bullet$ are ω - cocontinuous and hence have initial algebras.

Since the lifting functor $(\_)_\perp$ has a right adjoint, it preserves colimits and coproducts. In particular, for sets X and Y, $(X + Y)_\perp = (X)_\perp + (Y)_\perp$ where "+" on the right is the coaleased sum (= coproduct in $\mathbf{bSET}^\bullet$) of $(X)_\perp$ and $(Y)_\perp$. A direct calculation shows that $(\_)_\perp$ takes products to smash products; i.e., $(X \times Y)_\perp \approx X_\perp \# Y_\perp$.

4.2 Initial algebras in $\mathbf{bSET}^\bullet$. Call the ordinary polynomial endofunctors on **SET**, $(\times, +)$ - endofunctors, and the analogous endofunctors on $\mathbf{bSET}^\bullet$ built from # (instead of $\times$) and +, (#, +) - endofunctors. Let $F : \mathbf{SET} \to \mathbf{SET}$ be an $(\times, +)$ - endofunctor and let $F_\perp : \mathbf{bSET}^\bullet \to \mathbf{bSET}^\bullet$ denote the corresponding (#, +) - endofunctor built from F by lifting all constants in F and replacing all occurrences of $\times$ by #.

4.2.1 Proposition. If I is the initial F - algebra in **SET**, then $I_\perp$ is the initial $F_\perp$ - algebra in $\mathbf{bSET}^\bullet$.

Proof. It is evident that $F(X)_\perp \approx F_\perp(X_\perp)$, so $F(\emptyset)_\perp = F_\perp(\perp)$, and by induction that for all k, $F^k(\emptyset)_\perp \approx F_\perp{}^k(\perp)$. But then $I_\perp = (\mathrm{colim}\, F^k(\emptyset))_\perp \approx \mathrm{colim}(F^k(\emptyset)_\perp) \approx \mathrm{colim}\,(F_\perp{}^k(\perp))$, and this last term is the initial $F_\perp$ - algebra.

4.3 Syntax in $\mathbf{bPER}^\bullet$

The simplest way to treat the definition of $(k_\tau)'\forall_{t.\sigma}(m)$ in 3.2.1 and the context dependency of the definition of the collection of polymorphic terms is to work in the category $\mathbf{bSET}^\bullet$. Actually, we have to work in $\mathbf{bPER}^\bullet$ where $\mathbf{bPER}^\bullet$ is constructed from **PER** in exactly the same way that $\mathbf{bSET}^\bullet$ is constructed from **SET**, and has all the corresponding properties. Here **PER** is the category of partial equivalence relations. (Cf., [1], [4], [6], [7], [10].) A number of minor modifications are required. First we have to deal with the endofunctors and the initial algebras defining the syntax.

4.3.1 Types. The endofunctor $(H_{pType})_\perp : \mathbf{bPER}^\bullet \to \mathbf{bPER}^\bullet$ for types is given by

$$(H_{pType})_\perp(X) = (B_1)_\perp + Tv_\perp + (X \# X) + (Tv_\perp \# X).$$

where we again have omitted the base-point from the notation for the object X in $\mathbf{bPER}^\bullet$. We assume here that B_1 and Tv are objects in **PER**. $PT_\perp$ is the initial algebra in $\mathbf{bPER}^\bullet$ for this endofunctor. Its elements can be described as trees as in 1.2.2 together with a single element $\perp_{PT}$. Alternatively, if any leaf of a tree is labeled by $\perp_{B_1}$ or $\perp_{Tv}$ then that tree is equal to the element $\perp_{PT}$ (because of our representation of the smash product.)

Another way to describe this situation is to consider what it would mean to give a BNF grammar for types in $\mathbf{bSET}^\bullet$ or $\mathbf{bPER}^\bullet$. This would look like

$$Tp ::= B_1 \mid Tv \mid Tp \to Tp \mid \forall t.Tp \mid \perp,$$

except that $\perp$ cannot be substituted for Tp in either of the type constructors. Equivalently, the grammar is not given by an algebraic signature but by a specification containing the equations:

$$\perp \to Tp \;=\; Tp \to \perp \;=\; \forall t.\perp \;=\; \perp$$

4.3.2 Terms. The endofunctor $(H_{pTerm})_\perp : (\mathbf{bPER}^\bullet)^{PT_\perp} \to (\mathbf{bPER}^\bullet)^{PT_\perp}$ is defined in the same way as in 1.3.2 and 1.3.3 with certain changes as follows: First define

$$(H'_{pTerm})_\perp(\{X_\tau \mid \tau \in PT_\perp\})_\tau = (A_\tau)_\perp + \Sigma_{\sigma \in T}\,(X_{[\sigma \to \tau]} \# X_\sigma) + \Sigma_{[\tau'/t]\sigma = \tau}\,(X_{\forall t.\sigma})$$

again omitting base-points from the notation for objects X_τ in $\mathbf{bPER}^\bullet$. Then one has the same cases as before plus an extra case for $\perp_{PT}$.

Case $b \in B_1 + Tv$: $\quad (H_{pTerm})_\perp(\{X_\tau \mid \tau \in PT_\perp\})_b = (H'_{pTerm})_\perp(\{X_\tau \mid \tau \in PT_\perp\})_b$

Case $\tau = [\mu \to \sigma]$: $\quad (H_{pTerm})_\perp(\{X_\tau \mid \tau \in PT_\perp\})_{[\mu \to \sigma]} =$

$$(H'_{pTerm})_\perp(\{X_\tau \mid \tau \in PT_\perp\})_{[\mu \to \sigma]} + ((Var_\mu)_\perp \# X_\sigma)$$

Case $\tau = \forall t.\sigma$: $\quad (H_{pTerm})_\perp(\{X_\tau \mid \tau \in PT_\perp\})_{\forall t.\sigma} =$

$$(H'_{pTerm})_\perp(\{X_\tau \mid \tau \in PT_\perp\})_{\forall t.\sigma} + X_\sigma$$

Case $\tau = \perp_{PT}$: $\quad (H_{pTerm})_\perp(\{X_\tau \mid \tau \in PT_\perp\})_{\perp_{PT}} = \{\perp\}$.

Let $\{(pTerm_\tau)_\perp \mid \tau \in PT_\perp\}$ be the *initial* $(H_{pTerm})_\perp$ - algebra in $(\mathbf{bPER}^\bullet)^{PT_\perp}$ (not the subalgebra described in 1.3.3).

A description in terms of grammars here requires a separate grammar, Tr_τ, for each type τ.

Case $b \in B_1 + Tv$:

$$Tr_b ::= A_b \mid (Tr_{[v \to b]}\;\; Tr_v) \mid Tr_{\forall t.v}[\tau] \mid \perp_\tau.$$

Case $\tau = [\mu \to \sigma]$:

$$Tr_{[\mu \to \sigma]} ::= A_{[\mu \to \sigma]} \mid (Tr_{[v \to [\mu \to \sigma]]}\;\; Tr_v) \mid Tr_{\forall t.v}[\tau] \mid \lambda Var_\mu.\, T_\sigma \mid \perp_{[\mu \to \sigma]}.$$

Case $\tau = \forall t.\sigma$:

$$Tr_{\forall t.\sigma} ::= A_{\forall t.\sigma} \mid (Tr_{[v \to \forall t.\sigma]}\;\; Tr_v) \mid Tr_{\forall s.\forall t.v}[\tau] \mid \lambda t.\, Tr_\sigma \mid \perp_{\forall t.\sigma}.$$

Case $\tau = \perp_{PT}$:

$$Tr_{\perp_{PT}} ::= \perp_{\perp_{PT}}.$$

where $[\rho/t]v = \tau$ in the third clauses, and there are the same requirements on the $\bot_\tau$'s as in 4.3.1

4.4 Semantics of types in bPER•

As in 3.1, types are given by discrete functors. As before, let **Env** = [Tv → |**bPER**•|], where this is now understood as objects in **bPER**• indexed by (i.e., over) Tv. Similarly, [**Env** → **bPER**•] is understood as objects in ω-**bSET**• over the ω-**bSET**• corresponding to **Env**, whose fibres are pers. Details of the internal structure required here will be published elsewhere. For our purposes here, ω-**bSET**• has the same relation to **bPER**• as classes have to sets. (Cf. [1], [4], [10].) The base point is the constant functor $K\{\bot\}$ whose value is $\{\bot\}$. As with our extension of endofunctors on sets to classes, $(H_{pType})_\bot$ extends to an endofunctor on ω-**bSET**• given by the same expression.

4.4.1 Theorem. obj[**Env** → **bPER**•] underlies an $(H_{pType})_\bot$ - algebra in ω-**bSET**•.
Proof. Consider the functions

$$h_{B_1} : B_1 \to \text{obj}[\mathbf{Env} \to \mathbf{bPER}^\bullet]$$
$$h_{Tv} : Tv \to \text{obj}[\mathbf{Env} \to \mathbf{bPER}^\bullet]$$
$$h_{\to} : \text{obj}[\mathbf{Env} \to \mathbf{bPER}^\bullet] \times \text{obj}[\mathbf{Env} \to \mathbf{bPER}^\bullet] \to \text{obj}[\mathbf{Env} \to \mathbf{bPER}^\bullet]$$
$$h_\forall : Tv \times \text{obj}[\mathbf{Env} \to \mathbf{bPER}^\bullet] \to \text{obj}[\mathbf{Env} \to \mathbf{bPER}^\bullet]$$

constructed in 3.1.1, except that $h_\to(F, G) = [F \circ\!\!\to G]$, where $[F \circ\!\!\to G](X) = [F(X) \circ\!\!\to G(X)]$, and $[X \circ\!\!\to Y]$ is redefined up to isomorphism so that $[X \circ\!\!\to \{\bot\}] = [\{\bot\} \circ\!\!\to Y] = \{\bot\}$. The only point that needs some extra clarification is in the definition of $h_\forall$. In 3.3.1, this is given by the formula $h_\forall(t, F) = \lambda\rho.\ \Pi_{C\in|C|}F(\rho[C / t])$. Here **C** is replaced by **bPER**• so the ingredients of this formula are given by $\rho \in$ **Env** and F : **Env** → **bPER**•. Since the internal product of a family of PER's indexed by **PER** is given by their intersection, we arrive at the formula:

$$h_\forall(t, F) = \lambda\rho.\ \bigcap_{C\in|\mathbf{bPER}^\bullet|}F(\rho[C / t]).$$

These formulae extend to strict functions

$$(h_{B_1})_\bot : (B_1)_\bot \to \text{obj}[\mathbf{Env} \to \mathbf{bPER}^\bullet]$$
$$(h_{Tv})_\bot : Tv_\bot \to \text{obj}[\mathbf{Env} \to \mathbf{bPER}^\bullet]$$
$$(h_\to)_\bot : \text{obj}[\mathbf{Env} \to \mathbf{bPER}^\bullet] \# \text{obj}[\mathbf{Env} \to \mathbf{bPER}^\bullet] \to \text{obj}[\mathbf{Env} \to \mathbf{bPER}^\bullet]$$
$$(h_\forall)_\bot : Tv_\bot \# \text{obj}[\mathbf{Env} \to \mathbf{bPER}^\bullet] \to \text{obj}[\mathbf{Env} \to \mathbf{bPER}^\bullet]$$

and hence determine a function

$$h : (H_{pType})_\bot(\text{obj}[\mathbf{Env} \to \mathbf{bPER}^\bullet]) \to \text{obj}[\mathbf{Env} \to \mathbf{bPER}^\bullet].$$

Then $[[\ -\ ]] : PT_\bot \to$ obj[**Env** → **bPER**•] is the unique $(H_{pType})_\bot$ - algebra homomorphism from $PT_\bot$ to this algebra on obj[**Env** → **bPER**•]. Thus,

$$[[t]] = pr_t\ ,$$

$$[[\sigma \to \tau]] = [\, [[\sigma]] \circ\!\!\to [[\tau]] \,],$$
$$[[\forall t.\sigma]](\rho) = \lambda\rho. \bigcap_{C \in |\mathbf{bPER}^\bullet|} F(\rho[C / t]),$$
$$[[\bot_{PT}]] = K\{\bot\}.$$

4.5 Semantics of terms in bPER$^\bullet$

As in 3.2, for any functor $F : |\mathbf{bPER}^\bullet|^{Tv} \to \mathbf{bPER}^\bullet$, let

$$\mathrm{nat}[Tv, \mathbf{bPER}^\bullet]_F = \mathrm{mor}_F([|\mathbf{bPER}^\bullet|^{Tv} \to \mathbf{bPER}^\bullet])$$

be the ω- set of (internal) natural transformations with codomain F. The base point here is the unique natural transformation $!_F : K\{\bot\} \Rightarrow F$. The construction of the pullbacks M_F in 2.2 has to be modified. It is given by the pullback diagram in ω-$\mathbf{bSET}^\bullet$

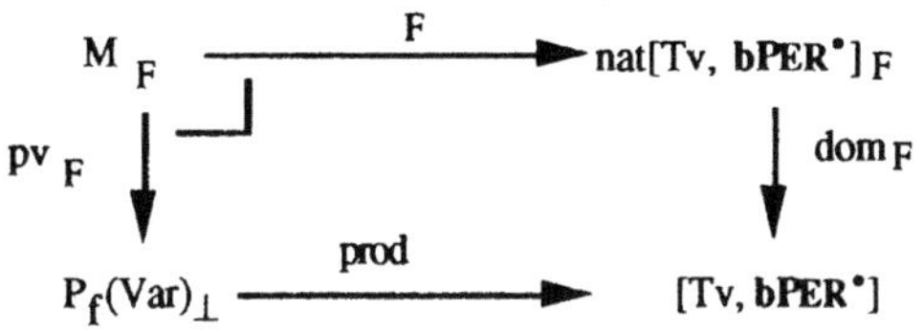

where dom_F is strict, and prod is extended to be a strict function. Since this is a pullback diagram, the description of the elements of M_F is the same as before except for the added bottom element in M_F given by the pair $(\bot, !_F)$. The sub ω- set $({}_tM_C)_\bot$ of $(M_C)_\bot$ is defined as in the paragraph before 3.2.2.

4.5.1 Theorem. The family of ω- sets $\{M_{[[\tau]]} \mid \tau \in PT_\bot\}$ underlies an $(H_{pTerm})_\bot$ - algebra in $(\omega\text{-}\mathbf{bSET}^\bullet)^{PT_\bot}$.

Proof. The functions

$$(k_\tau)_{Const} : Const_\tau \to M_{[[\tau]]}$$
$$(k_\tau)_{Var} : Var_\tau \to M_{[[\tau]]}$$
$$(k_\tau)_{\sigma\to} : M_{[[\sigma \to \tau]]} \times M_{[[\sigma]]} \to M_{[[\tau]]}$$
$$(k_\tau)_\lambda : Var_\mu \times M_{[[\sigma]]} \to M_{[[\tau]]}$$

defined in 2.2.1 extend to strict functions (which are given the same names)

$$(k_\tau)_{Const} : (Const_\tau)_\bot \to M_{[[\tau]]}$$
$$(k_\tau)_{Var} : (Var_\tau)_\bot \to M_{[[\tau]]}$$
$$(k_\tau)_{\sigma\to} : M_{[[\sigma \to \tau]]} \# M_{[[\sigma]]} \to M_{[[\tau]]}$$
$$(k_\tau)_\lambda : (Var_\mu)_\bot \# M_{[[\sigma]]} \to M_{[[\tau]]}$$

The function $(k_\tau)_{[\tau'/t]\sigma} : M_{[[\forall t.\sigma]]} \to M_{[[\tau]]}$ defined in 3.2.1 extends to a strict function

$$(k_\tau)_{[\tau'/t]\sigma} : M_{[[\forall t.\sigma]]} \to M_{[[\tau]]}$$

and it finally makes sense to define

$$(k_\tau)_{\forall t.\sigma}(\underline{m}) = (\{x_i : \rho_i \mid i \in I\}, \lambda S.<m_{<t\leftarrow C>}(S) \mid C \in \mathbf{C}> : \Pi_{i \in I}[[\rho_i]] \Rightarrow [[\forall t.\sigma]])$$

if $\underline{m} \in {}_t M_{[[\sigma]]}$, and $\perp$ otherwise. Lastly, set $k_{\perp_{PT}} : \{\perp\} \rightarrow M_{K\{\perp\}}$ to be the only such strict function. Note that $(k_\tau)_{\forall t.\sigma}$ is the only component that is not doubly strict. For each $\tau \in PT_\perp$, appropriate combinations of these functions determine

$$k_\tau : (H_{pTerm})_\perp(\{M_{[[\tau]]} \mid \tau \in PT_\perp\})_\tau \rightarrow M_{[[\tau]]}$$

Hence, there is a unique $(H_{pTerm})_\perp$ - algebra homomorphism

$$F: \{(pTerm_\tau)_\perp \mid \tau \in PT_\perp\} \rightarrow \{M_{[[\tau]]} \mid \tau \in PT_\perp\}$$

with components $F_\tau : (pTerm_\tau)_\perp \rightarrow M_{[[\tau]]}$ for each $\tau \in T$.

4.5.2 The free-variable functions

The free variable function in 2.2.2 and 3.3 is clearly extended by the above definitions as follows:

For each type τ, there is a term $\perp_\tau$ of type τ and $FV(\perp_\tau : \tau) = \perp$.

Case $x \in pTerm_{\perp_{PT}}$ $FV[x] = \perp$

Case $g \in pTerm_\sigma$ and $g : \sigma$ has the property that there exists an $x : \tau \in FV(g)$, such that $t \in fv(\tau)$, then $Fv(\lambda t.g) = \perp$

4.5.3 The semantics functions

The semantics functions in 2.2.3 are clearly extended by the above definitions as follows:

For each type τ, there is a term $\perp_\tau$ of type τ and $[[\perp_\tau : \tau]] : K\{\perp\} \Rightarrow [[\tau]]$ is the unique such natural transformation.

Case $x \in pTerm_{\perp_{PT}}$ $[[x]] = id_{K\{\perp\}}$

Case $g \in pTerm_\sigma$ and $g : \sigma$ has the property that there exists an $x : \tau \in FV(g)$, such that $t \in fv(\tau)$, then $[[\lambda t \,.\, f]] = id_{K\{\perp\}}$

Thus the initial algebra of terms is allowed to contain illegitimate terms, but all such terms are interpreted in the semantics by the identity natural transformation on $K\{\perp\}$, which we take as a suitable mathematical translation of "meaningless".

5. References

[1] K.B. Bruce and G. Longo, A modest model of records, inheritance and bounded quantification, in Logic in Computer Science, Proceedings Third Annual Symposium, Edinburgh, Scotland, 1988, Computer Society Press, Washington, D. C., 38 - 50.

[2] H. Ehrig and B. Mahr, *Fundamentals of Algebraic Specification 1*, Springer-Verlag, New York, 1985.

[3] J W. Gray, A categorical treatment of polymorphic operations, in *Mathematical Foundations of Programming Language Semantics*, M. Main, A Melton, M. Mislove and D. Schmidt, Eds., 3rd Workshop, New Orleans 1987, Lecture Notes in Computer Science 298, Springer-Verlag, New York, 1987., 2 - 22.

[4] J. W. Gray, The Integration of Logical and Algebraic Types, Proc. Int. Workshop on Categorical Mathods in Computer Science with Aspects from Topology, Lecture Notes in Computer Science, Springer-Verlag, New York, 1989, to appear.

[5] C. A. Gunter, P. D. Mosses, and D. S. Scott, Semantic domains and denotational semantics, in *Handbook of Theoretical Computer Science*, North Holland, to appear.

[6] J. M. E. Hyland, The effective topos, in *The L.E.J. Brouwer Centenary Symposium*, A. S. Troelstra and D. van Dalen, Eds., North-Holland, Amsterdam, 1982.

[7] J. M. E. Hyland, <completeness properties>

[8] J. Lambek, A fixpoint theorem for complete categories, Math. Zeitschr. 103 (1968), 151 - 161.

[9] J. Lambek and P. Scott, Introduction to Higher Order Categorical Logic, Cambridge Studies in Advanced Mathematics 7, Cambridge University Press, New York, 1986.

[10] G. Longo and E. Moggi, Constructive natural deduction and its 'Modest' interpretation, in *Semantics of Natural and Computer Languages*, J. Meseguer et al, Eds., M. I. T. Press, Cambridge, MA, 1989.

[11] E. Manes and M. Arbib, *Algebraic Approaches to Program Semantics*, Springer-Verlag, New York, 1986.

[12] F. Oles, A category-theoretic approach to the semantics of programming languages, Thesis, Syracuse University, 1982.

[13] F. Oles, Lambda Calculi with implicit type conversions,

[14] J. C. Reynolds, Towards a theory of type structures, in *Colloque sur la Programmation*, Lecture Notes in Computer Science 19, Springer-Verlag 1974, 408 - 425.

[15] J. C. Reynolds, Types, abstraction, and parametric polymorphism, IFIP 83, Paris.

[16] J. C. Reynolds, Polymorphism is not set-theoretic, in [LNCS 173], 145 - 156.

[17] J. C. Reynolds, Three approaches to type structure. In TYPSOFT advanced seminar on the role of semantics in software development, Springer-Verlag, New York, 1985.

Lecture Notes in Computer Science

Vol. 352: J. Díaz, F. Orejas (Eds.), TAPSOFT '89. Volume 2. Proceedings, 1989. X, 389 pages. 1989.

Vol. 353: S. Hölldobler, Foundations of Equational Logic Programming. X, 250 pages. 1989. (Subseries LNAI).

Vol. 354: J.W. de Bakker, W.-P. de Roever, G. Rozenberg (Eds.), Linear Time, Branching Time and Partial Order in Logics and Models for Concurrency. VIII, 713 pages. 1989.

Vol. 355: N. Dershowitz (Ed.), Rewriting Techniques and Applications. Proceedings, 1989. VII, 579 pages. 1989.

Vol. 356: L. Huguet, A. Poli (Eds.), Applied Algebra, Algebraic Algorithms and Error-Correcting Codes. Proceedings, 1987. VI, 417 pages. 1989.

Vol. 357: T. Mora (Ed.), Applied Algebra, Algebraic Algorithms and Error-Correcting Codes. Proceedings, 1988. IX, 481 pages. 1989.

Vol. 358: P. Gianni (Ed.), Symbolic and Algebraic Computation. Proceedings, 1988. XI, 545 pages. 1989.

Vol. 359: D. Gawlick, M. Haynie, A. Reuter (Eds.), High Performance Transaction Systems. Proceedings, 1987. XII, 329 pages. 1989.

Vol. 360: H. Maurer (Ed.), Computer Assisted Learning – ICCAL '89. Proceedings, 1989. VII, 642 pages. 1989.

Vol. 361: S. Abiteboul, P.C. Fischer, H.-J. Schek (Eds.), Nested Relations and Complex Objects in Databases. VI, 323 pages. 1989.

Vol. 362: B. Lisper, Synthesizing Synchronous Systems by Static Scheduling in Space-Time. VI, 263 pages. 1989.

Vol. 363: A.R. Meyer, M.A. Taitslin (Eds.), Logic at Botik '89. Proceedings, 1989. X, 289 pages. 1989.

Vol. 364: J. Demetrovics, B. Thalheim (Eds.), MFDBS 89. Proceedings, 1989. VI, 428 pages. 1989.

Vol. 365: E. Odijk, M. Rem, J.-C. Syre (Eds.), PARLE '89. Parallel Architectures and Languages Europe. Volume I. Proceedings, 1989. XIII, 478 pages. 1989.

Vol. 366: E. Odijk, M. Rem, J.-C. Syre (Eds.), PARLE '89. Parallel Architectures and Languages Europe. Volume II. Proceedings, 1989. XIII, 442 pages. 1989.

Vol. 367: W. Litwin, H.-J. Schek (Eds.), Foundations of Data Organization and Algorithms. Proceedings, 1989. VIII, 531 pages. 1989.

Vol. 368: H. Boral, P. Faudemay (Eds.), IWDM '89, Database Machines. Proceedings, 1989. VI, 387 pages. 1989.

Vol. 369: D. Taubner, Finite Representations of CCS and TCSP Programs by Automata and Petri Nets. X, 168 pages. 1989.

Vol. 370: Ch. Meinel, Modified Branching Programs and Their Computational Power. VI, 132 pages. 1989.

Vol. 371: D. Hammer (Ed.), Compiler Compilers and High Speed Compilation. Proceedings, 1988. VI, 242 pages. 1989.

Vol. 372: G. Ausiello, M. Dezani-Ciancaglini, S. Ronchi Della Rocca (Eds.), Automata, Languages and Programming. Proceedings, 1989. XI, 788 pages. 1989.

Vol. 373: T. Theoharis, Algorithms for Parallel Polygon Rendering. VIII, 147 pages. 1989.

Vol. 374: K.A. Robbins, S. Robbins, The Cray X-MP/Model 24. VI, 165 pages. 1989.

Vol. 375: J.L.A. van de Snepscheut (Ed.), Mathematics of Program Construction. Proceedings, 1989. VI, 421 pages. 1989.

Vol. 376: N.E. Gibbs (Ed.), Software Engineering Education. Proceedings, 1989. VII, 312 pages. 1989.

Vol. 377: M. Gross, D. Perrin (Eds.), Electronic Dictionaries and Automata in Computational Linguistics. Proceedings, 1987. V, 110 pages. 1989.

Vol. 378: J.H. Davenport (Ed.), EUROCAL '87. Proceedings, 1987. VIII, 499 pages. 1989.

Vol. 379: A. Kreczmar, G. Mirkowska (Eds.), Mathematical Foundations of Computer Science 1989. Proceedings, 1989. VIII, 605 pages. 1989.

Vol. 380: J. Csirik, J. Demetrovics, F. Gécseg (Eds.), Fundamentals of Computation Theory. Proceedings, 1989. XI, 493 pages. 1989.

Vol. 381: J. Dassow, J. Kelemen (Eds.), Machines, Languages, and Complexity. Proceedings, 1988. VI, 244 pages. 1989.

Vol. 382: F. Dehne, J.-R. Sack, N. Santoro (Eds.), Algorithms and Data Structures. WADS '89. Proceedings, 1989. IX, 592 pages. 1989.

Vol. 383: K. Furukawa, H. Tanaka, T. Fujisaki (Eds.), Logic Programming '88. Proceedings, 1988. VII, 251 pages. 1989 (Subseries LNAI).

Vol. 384: G.A. van Zee, J.G.G. van de Vorst (Eds.), Parallel Computing 1988. Proceedings, 1988. V, 135 pages. 1989.

Vol. 385: E. Börger, H. Kleine Büning, M.M. Richter (Eds.), CSL '88. Proceedings, 1988. VI, 399 pages. 1989.

Vol. 386: J.E. Pin (Ed.), Formal Properties of Finite Automata and Applications. Proceedings, 1988. VIII, 260 pages. 1989.

Vol. 387: C. Ghezzi, J.A. McDermid (Eds.), ESEC '89. 2nd European Software Engineering Conference. Proceedings, 1989. VI, 496 pages. 1989.

Vol. 388: G. Cohen, J. Wolfmann (Eds.), Coding Theory and Applications. Proceedings, 1988. IX, 329 pages. 1989.

Vol. 389: D.H. Pitt, D.E. Rydeheard, P. Dybjer, A.M. Pitts, A. Poigné (Eds.), Category Theory and Computer Science. Proceedings, 1989. VI, 365 pages. 1989.

Vol. 390: J.P. Martins, E.M. Morgado (Eds.), EPIA 89. Proceedings, 1989. XII, 400 pages. 1989 (Subseries LNAI).

Vol. 391: J.-D. Boissonnat, J.-P. Laumond (Eds.), Geometry and Robotics. Proceedings, 1988. VI, 413 pages. 1989.

Vol. 392: J.-C. Bermond, M. Raynal (Eds.), Distributed Algorithms. Proceedings, 1989. VI, 315 pages. 1989.

Vol. 393: H. Ehrig, H. Herrlich, H.-J. Kreowski, G. Preuß (Eds.), Categorical Methods in Computer Science. VI, 350 pages. 1989.

Vol. 394: M. Wirsing, J.A. Bergstra (Eds.), Algebraic Methods: Theory, Tools and Applications. VI, 558 pages. 1989.

Vol. 395: M. Schmidt-Schauß, Computational Aspects of an Order-Sorted Logic with Term Declarations. VIII, 171 pages. 1989 (Subseries LNAI).

Vol. 396: T.A. Berson, T. Beth (Eds.), Local Area Network Security. Proceedings, 1989. IX, 152 pages. 1989.

Vol. 397: K.P. Jantke (Ed.), Analogical and Inductive Inference. Proceedings, 1989. IX, 338 pages. 1989 (Subseries LNAI).

Vol. 398: B. Banieqbal, H. Barringer, A. Pnueli (Eds.), Temporal Logic in Specification. Proceedings, 1987. VI, 448 pages. 1989.

Vol. 399: V. Cantoni, R. Creutzburg, S. Levialdi, G. Wolf (Eds.), Recent Issues in Pattern Analysis and Recognition. VII, 400 pages. 1989.

Vol. 400: R. Klein, Concrete and Abstract Voronoi Diagrams. IV, 167 pages. 1989.

Vol. 401: H. Djidjev (Ed.), Optimal Algorithms. Proceedings, 1989. VI, 308 pages. 1989.

Vol. 402: T.P. Bagchi, V.K. Chaudhri, Interactive Relational Database Design. XI, 186 pages. 1989.

Vol. 403: S. Goldwasser (Ed.), Advances in Cryptology – CRYPTO '88. Proceedings, 1988. XI, 591 pages. 1990.

Vol. 404: J. Beer, Concepts, Design, and Performance Analysis of a Parallel Prolog Machine. VI, 128 pages. 1989.

Vol. 405: C.E. Veni Madhavan (Ed.), Foundations of Software Technology and Theoretical Computer Science. Proceedings, 1989. VIII, 339 pages. 1989.

Vol. 406: C.J. Barter, M.J. Brooks (Eds.), AI '88. Proceedings, 1988. VIII, 463 pages. 1990 (Subseries LNAI).

Vol. 407: J. Sifakis (Ed.), Automatic Verification Methods for Finite State Systems. Proceedings, 1989. VII, 382 pages. 1990.

Vol. 408: M. Leeser, G. Brown (Eds.), Hardware Specification, Verification and Synthesis: Mathematical Aspects. Proceedings, 1989. VI, 402 pages. 1990.

Vol. 409: A. Buchmann, O. Günther, T. R. Smith, Y.-F. Wang (Eds.), Design and Implementation of Large Spatial Databases. Proceedings, 1989. IX, 364 pages. 1990.

Vol. 410: F. Pichler, R. Moreno-Diaz (Eds.), Computer Aided Systems Theory – EUROCAST '89. Proceedings, 1989. VII, 427 pages. 1990.

Vol. 411: M. Nagl (Ed.), Graph-Theoretic Concepts in Computer Science. Proceedings, 1989. VII, 374 pages. 1990.

Vol. 412: L. B. Almeida, C. J. Wellekens (Eds.), Neural Networks. Proceedings, 1990. IX, 276 pages. 1990.

Vol. 413: R. Lenz, Group Theoretical Methods in Image Processing. VIII, 139 pages. 1990.

Vol. 414: A. Kreczmar, A. Salwicki, M. Warpechowski, LOGLAN '88 – Report on the Programming Language. X, 133 pages. 1990.

Vol. 415: C. Choffrut, T. Lengauer (Eds.), STACS 90. Proceedings, 1990. VI, 312 pages. 1990.

Vol. 416: F. Bancilhon, C. Thanos, D. Tsichritzis (Eds.), Advances in Database Technology – EDBT '90. Proceedings, 1990. IX, 452 pages. 1990.

Vol. 417: P. Martin-Löf, G. Mints (Eds.), COLOG-88. International Conference on Computer Logic. Proceedings, 1988. VI, 338 pages. 1990.

Vol. 418: K. H. Bläsius, U. Hedtstück, C.-R. Rollinger (Eds.), Sorts and Types in Artificial Intelligence. Proceedings, 1989. VIII, 307 pages. 1990. (Subseries LNAI).

Vol. 419: K. Weichselberger, S. Pöhlmann, A Methodology for Uncertainty in Knowledge-Based Systems. VIII, 136 pages. 1990 (Subseries LNAI).

Vol. 420: Z. Michalewicz (Ed.), Statistical and Scientific Database Management, V SSDBM. Proceedings, 1990. V, 256 pages. 1990.

Vol. 421: T. Onodera, S. Kawai, A Formal Model of Visualization in Computer Graphics Systems. X, 100 pages. 1990.

Vol. 422: B. Nebel, Reasoning and Revision in Hybrid Representation Systems. XII, 270 pages. 1990 (Subseries LNAI).

Vol. 423: L. E. Deimel (Ed.), Software Engineering Education. Proceedings, 1990. VI, 164 pages. 1990.

Vol. 424: G. Rozenberg (Ed.), Advances in Petri Nets 1989. VI, 524 pages. 1990.

Vol. 425: C. H. Bergman, R. D. Maddux, D. L. Pigozzi (Eds.), Algebraic Logic and Universal Algebra in Computer Science. Proceedings, 1988. XI, 292 pages. 1990.

Vol. 426: N. Houbak, SIL – a Simulation Language. VII, 192 pages. 1990.

Vol. 427: O. Faugeras (Ed.), Computer Vision – ECCV 90. Proceedings, 1990. XII, 619 pages. 1990.

Vol. 428: D. Bjørner, C. A. R. Hoare, H. Langmaack (Eds.), VDM '90. VDM and Z – Formal Methods in Software Development. Proceedings, 1990. XVII, 580 pages. 1990.

Vol. 429: A. Miola (Ed.), Design and Implementation of Symbolic Computation Systems. Proceedings, 1990. XII, 284 pages. 1990.

Vol. 430: J. W. de Bakker, W.-P. de Roever, G. Rozenberg (Eds.), Stepwise Refinement of Distributed Systems. Models, Formalisms, Correctness. Proceedings, 1989. X, 808 pages. 1990.

Vol. 431: A. Arnold (Ed.), CAAP '90. Proceedings, 1990. VI, 285 pages. 1990.

Vol. 432: N. Jones (Ed.), ESOP '90. Proceedings, 1990. IX, 436 pages. 1990.

Vol. 433: W. Schröder-Preikschat, W. Zimmer (Eds.), Progress in Distributed Operating Systems and Distributed Systems Management. Proceedings, 1989. V, 206 pages. 1990.

Vol. 435: G. Brassard (Ed.), Advances in Cryptology – CRYPTO '89. Proceedings, 1990. XIII, 634 pages. 1990.

Vol. 436: B. Steinholtz, A. Sølvberg, L. Bergman (Eds.), Advanced Information Systems Engineering. Proceedings, 1990. X, 392 pages. 1990.

Vol. 437: D. Kumar (Ed.), Current Trends in SNePS – Semantic Network Processing System. Proceedings, 1989. VII, 162 pages. 1990. (Subseries LNAI).

Vol. 438: D. H. Norrie, H.-W. Six (Eds.), Computer Assisted Learning – ICCAL '90. Proceedings, 1990. VII, 467 pages. 1990.

Vol. 439: P. Gorny, M. Tauber (Eds.), Visualization in Human-Computer Interaction. Proceedings, 1988. VI, 274 pages. 1990.

Vol. 440: E. Börger, H. Kleine Büning, M. M. Richter (Eds.), CSL '89. Proceedings, 1989. VI, 437 pages. 1990.

Vol. 441: T. Ito, R. H. Halstead, Jr. (Eds.), Parallel Lisp: Languages and Systems. Proceedings, 1989. XII, 364 pages. 1990.

Vol. 442: M. Main, A. Melton, M. Mislove, D. Schmidt (Eds.), Mathematical Foundations of Programming Semantics. Proceedings, 1989. VI, 439 pages. 1990.

GPSR Compliance

The European Union's (EU) General Product Safety Regulation (GPSR) is a set of rules that requires consumer products to be safe and our obligations to ensure this.

If you have any concerns about our products, you can contact us on ProductSafety@springernature.com

In case Publisher is established outside the EU, the EU authorized representative is:

Springer Nature Customer Service Center GmbH
Europaplatz 3
69115 Heidelberg, Germany

Batch number: 09624486

Printed by Printforce, the Netherlands